Foundations of Molecular Quantum Electrodynamics

This book presents a comprehensive account of molecular quantum electrodynamics from the perspectives of physics and theoretical chemistry. The first part of the book establishes the essential concepts underlying classical electrodynamics, using the tools of Lagrangian and Hamiltonian mechanics. The second part focuses on the fundamentals of quantum mechanics, particularly how they relate to, and influence, chemical and molecular processes. The special case of the Coulomb Hamiltonian (including the celebrated Born–Oppenheimer approximation) is given a modern treatment. The final part of the book is devoted to non-relativistic quantum electrodynamics and describes in detail its impact upon our understanding of atoms and molecules, and their interaction with light. Particular attention is paid to the Power–Zienau–Woolley (PZW) representations, and both perturbative and non-perturbative approaches to QED calculation are discussed. This book is ideal for graduate students and researchers in chemical and molecular physics, quantum chemistry, and theoretical chemistry.

Guy Woolley is Emeritus Professor of Chemical Physics at Nottingham Trent University. In 1978 he was awarded the Royal Society of Chemistry Marlow Medal, and he was elected Fellow of the American Physical Society in 1995 'for fundamental advances in the proper description of molecules and their interaction with radiation'. He is responsible for the development of the Power–Zienau–Woolley transformation, one of the most widely used methods in molecular quantum electrodynamics.

Foundations of Molecular Quantum Electrodynamics

R. GUY WOOLLEY

Nottingham Trent University

CAMBRIDGE UNIVERSITY PRESS

University Printing House, Cambridge CB2 8BS, United Kingdom

One Liberty Plaza, 20th Floor, New York, NY 10006, USA

477 Williamstown Road, Port Melbourne, VIC 3207, Australia

314–321, 3rd Floor, Plot 3, Splendor Forum, Jasola District Centre,
New Delhi – 110025, India

103 Penang Road, #05–06/07, Visioncrest Commercial, Singapore 238467

Cambridge University Press is part of the University of Cambridge.

It furthers the University's mission by disseminating knowledge in the pursuit of
education, learning, and research at the highest international levels of excellence.

www.cambridge.org
Information on this title: www.cambridge.org/9781009225762
DOI: 10.1017/9781009225786

First published 2022

A catalogue record for this publication is available from the British Library.

ISBN 978-1-009-22576-2 Hardback

To Anne, Claire and Euan

Au terme dernier, vous m'apprenez que cet
univers prestigieux et bariolé se réduit à l'atome
et que l'atome lui-même se réduit à l'électron.
Tout ceci est bon et j'attends que vous continuiez.
Mais vous me parle d'un invisible système
planétaire où des électrons gravitent autour d'un
noyau. Vous m'expliquez ce monde avec une
image. Je reconnais alors que vous en êtes
venus à la poesie: je ne connaîtrai jamais.
Ai-je le temps de m'en indigner ?
Vous avez déjà changé de théorie.
Ainsi cette science qui devait tout m'apprendre
finit dans l'hypothèse, cette lucidité sombre
dans la métaphore, cette incertitude se résout en œuvre d'art.

Albert Camus, *Le mythe de Sisyphe*, © Éditions Gallimard

Contents

Preface

The book begins with an account of how chemistry came about as a science and hit on a structural conception to account for the empirical facts of chemical reactions in terms of the spatial arrangement of atoms. Such an approach is to be contrasted with that of physics, which is founded on a mechanical description of natural phenomena. The basic unit of chemistry is the atom; however, not enough is known about their 'interactions' to formulate a 'mathematical chemistry' beyond stoichiometry, and so recourse has long been taken to subatomic structure in terms of electrons and nuclei. Such particles have the additional physical property of electric charge, and a combination of electrodynamics and quantum theory is unavoidable in a fundamental physical account. These two contrasting methodologies collided in the early twentieth century when physics advanced into what had hitherto been the province of the chemist. In the author's opinion the resulting tension has never been resolved satisfactorily. An exploration of how that tension arises and where it persists seems a good enough reason to try to describe the 'known knowns' and the 'known unknowns' in this story. Although some of the principal features of quantum chemistry will feature, this is not another book about quantum chemistry, a subject for which there is already an extensive textbook, monograph and journal literature.

The book is divided into three parts: 'The Classical World', 'Quantum Theory', and 'Non-relativistic Quantum Electrodynamics'. After the historical account of the development of chemical thought, the next three chapters in Part I summarise some essential facts about classical electrodynamics, starting from Maxwell's explanation of electromagnetic phenomena in terms of the fields $\mathbf{E}$ and $\mathbf{B}$ – the electric and magnetic fields, respectively. Throughout the book extensive use will be made of auxiliary variables to describe the field (the field potentials) and the charge–current density of the charged particles (the polarisation fields), and a recurring theme is the need to keep in mind that the auxiliary variables are arbitrary and unphysical. Chapter 2 describes classical electromagnetism; looking ahead to a hoped-for quantum theory, it is important that the combined system of charged particles and electromagnetic field can be described by the mechanical schemes devised by Lagrange ('Lagrangian mechanics') and Hamilton ('Hamiltonian mechanics'), and these are developed in Chapter 3. Chapter 4 is concerned with the formal description of the electromagnetic field in the absence of charges in terms of a collection of independent harmonic oscillators using the Hamiltonian scheme.

Part II of the book introduces the ideas of the quantum theory. Quantum mechanics is the general dynamical theory of physical systems according to the quantum laws when temperature is ignored; in other words, it is a $T = 0$ theory. Chapter 5 gives an outline of the basic quantum mechanical formalism and concepts, while Chapter 6 reviews methods for approximate calculations that have proved to be useful in the present context, and includes some introductory remarks about the finite temperature theory. Chapter 7 describes the quantisation of the free electromagnetic field both in terms of photons (the 'particle' description) and through reinterpretation of the field variables as quantum mechanical operators (the 'wave' description); they are entirely equivalent. Chapter 8 is concerned with the quantum mechanics of a collection of electrons and nuclei when only electrostatic interactions between the charges are considered – based on the so-called Coulomb Hamiltonian. This has been much studied in physics (the 'many-body problem') and, of course, is of central importance for the physics of atoms and molecules. For the latter a 'clamped-nuclei' formulation is ubiquitous. The assumption that the nuclei, even if identical, can be regarded as fixed, distinguishable classical charges is commonly justified by reference to some version of the celebrated 'Born–Oppenheimer approximation' which is given close scrutiny.

Part III is devoted to non-relativistic quantum electrodynamics (QED) understood as the quantum mechanics of a closed collection of charges interacting with the electromagnetic field, finishing with what it might mean for atoms and molecules given that they are described as collections of electrons and nuclei; this is where the 'known unknowns' become prominent as we cycle back to the first chapter. Chapter 9 examines how quantisation of the combined system of charges and field may be carried out, paying special attention to the gauge transformations associated with the vector potential, and a dual description in terms of polarisation fields. Chapter 10 describes the perturbation theory approach to QED, based on choosing the 'unperturbed' reference system as an idealised 'free field' and an atomic system in the absence of electromagnetic radiation. Chapter 11 gives some introductory remarks on the non-perturbative investigations of non-relativistic QED Hamiltonians that are of active interest in mathematical physics. Finally, in Chapter 12 we confront this mathematically sophisticated account of the building blocks of atoms with what the chemists have found with molecules.

I am indebted to Akbar Salam and Brian Sutcliffe, who gave me much valuable commentary on the early draft of this book; naturally, I am solely responsible for errors that remain. My heartfelt thanks for the love and support of my family and their indulgence over the period of its writing. Many thanks also to my editors, Sarah Armstrong and Nicholas Gibbons of Cambridge University Press, for their encouragement and advice.

PART I

THE CLASSICAL WORLD

1 Introduction

1.1 The Origins of Chemistry

Chemistry is concerned with the composition and properties of matter, and with the transformations of matter that can occur spontaneously or under the action of heat, radiation or other sources of energy. It emerged as a science in recognisably modern form at the end of the eighteenth century. From the results of chemical experiments, the chemist singles out a particular class of materials that have characteristic and invariant properties. This is done through the use of the classical separation procedures – crystallisation, distillation, sublimation and so on – that involve a phase transition. Such materials are called pure substances and may be of two kinds: elements and compounds. A pure substance is an idealisation since perfect purity is never achieved in practice.

Formally, elements may be defined as substances which have not been converted either by the action of heat, radiation or chemical reaction with other substances, or small electrical voltages, into any simpler substance. Compounds are formed from the chemical combination of the elements, and have properties that are invariably different from the properties of the constituent elements; they are also homogeneous. These statements derive from antiquity; thus from Aristotle [1]:

> An element, we take it, is a body into which other bodies may be analysed, present in them potentially or in actuality (which of these, is still disputable), and not itself divisible into bodies different in form.

Similar statements can be found in Boyle and in Lomonosov, for example; they gain significance when the notion of 'simpler' substance is explicated. A substantial account of the history and philosophy of these ideas can be found in a recent Handbook [2].

In the seventeenth century, a scientific attitude emerged that is recognisably 'modern'; it aimed to describe the physical aspects of the natural world through analytical procedures of classification and systematisation in order to find explanations of natural phenomena in purely naturalistic terms [3]. The underlying mechanical philosophy[1] was grounded firmly in a picture of a world of physical objects endowed with

---

[1] The idea that the physical world is a complex machine that could, in principle, be built by a skilled artisan.

well-defined fixed properties that can be described in mathematical terms – shape, size, position, number and so on. It can be seen as a return to the mathematical ideals of the Pythagoreans and of Plato, and a renewal of the ideas of the early Greek atomists, for example Democritus. There was quite explicitly a movement against the still prevailing Aristotelian system of the scholastic philosophers which was closely connected with the religious authorities. The prime movers of this revolution were Galileo and Descartes; both sought a quantitative approach to physics through the use of mathematics applied to mechanical or corpuscular models that would replace a philosophical tradition that had originated in antiquity.

The scientific revolution initiated by Galileo with its quantitative approach to physics represented a fundamental shift from the organism to the machine as the model in terms of which the physical world should be understood. However, the belief that this understanding was founded on 'ultimate explanations' rooted in principles that appear self-evident was demolished by Newton's account in the *Principia* of the physical world based on universal gravitation. Action-at-a-distance (gravity) cannot be reconciled with a strict mechanistic philosophy, instead the goals of science became focused on finding the best theoretical account of experience and experiment, in preference to reliance on common-sense notions of the world. As far as chemistry is concerned, this shift in outlook did not take place until more than a century later.[2] Alchemy was regarded with increasing scepticism throughout the eighteenth century partly due to the accumulation of empirical evidence that spoke against transmutation of metals, but nevertheless aspects of it such as the interpretation of chemical properties in terms of alchemical 'principles' and the phlogiston ideas of Becher and Stahl lingered on.

Eventually the new view of physics prevailed comprehensively; the significance of this for chemistry was the recognition that the alchemical 'principles' and phlogiston could not survive the Newtonian imperative. On the other hand a mathematical formulation of chemical laws on Newtonian lines was never achieved. The phlogiston chemists did have some success in distinguishing between mixtures and pure substances (including some of the elements in the modern sense) by experimental means; this was also true of the later alchemists but to a much more limited extent. Thereby they discovered that there were limits to the amount of separation of a starting material that could be achieved by repeated crystallisation and distillation, and that the end products of these physical separation procedures often had characteristic physical properties such as boiling point, melting point[3] and crystal morphology by which different substances could be recognised. Once it became possible to distinguish reliably between different substances, systematic chemical experimentation could be carried out. Two enduring features emerge from such practice:

---

[2] Lavoisier notably took a steam engine as a model for describing a living body [4]. Newton's lengthy absorption in alchemy, contemporary with his novel physics, had been imbued with the organic ideas of growth and maturation [5], [6].

[3] The German physicist and instrument maker, D. G. Fahrenheit, invented the mercury thermometer in 1714 and devised the temperature scale that bears his name shortly after [7].

1. Chemical transformations generally bring about changes in the properties of substances.
2. Substances vary widely in their chemical reactivity when in contact with other substances.

The change in attitude that accompanied the emergence of chemistry as a science can be seen in Lavoisier's *Traité élémentaire de chemie*; he recognised that differences in masses were experimentally accessible and proposed that the elements should be characterised by their gravimetric properties. He produced a reasonably correct and extensive list of elements with their modern names [8]. Lavoisier was by no means the first chemist to have investigated the weight relationships of reagents and products in chemical transformations. He was, however, the first person to publish an account of how such information could serve as the basis of a systematic analytical approach to chemistry. Even so, light and 'caloric' were still included in his list of elements despite their imponderable nature and the impossibility of isolating 'caloric' experimentally in its free state. After Lavoisier, an element came to be understood as a pure substance that formed products of greater weight than itself in all chemical changes which it underwent. The significance of the implementation of this new meaning for the concept of 'element' was that (a) elements and compounds could be recognised experimentally by a physical property and (b) it then became apparent that the characteristic properties of the elements did not persist in their compounds. This is in stark contrast with the mythical conception it displaced.

The characteristic chemical notion of a pure substance is based on an ideal conception of the chemical and physical properties of matter and their changes under specified experimental conditions (pressure, temperature, in inert containers etc.). Physical properties belong to materials in isolation from other materials and are those properties that can be observed without conversion of the material into other substances, whereas chemical properties refer to the chemical reactions that materials undergo. There we have a fundamental distinction between the goals of chemistry and physics. The core activity of the chemist is the experimental preparation of chemical compounds – *chemical synthesis* – and their characterisation – *chemical analysis*.

This first chapter gives an account of the historical development of the atomic–molecular conception of chemistry that led to the fundamental chemical idea of molecular structure. This is the overarching idea that opens the way to a systematic account of the experimental facts of chemistry; it is a microscopic interpretation in terms of the 'smallest particle' of an element, the 'atom', but does not require any detailed physical description of an 'atom'. There is a parallel history of the search in physics for the characterisation of the 'atom', and a central question for science is how/whether these two histories might be unified. Physics is based on dynamics and requires rules governing the interactions between basic 'particles'. It ascribes a fundamental role to the notions of energy and time, concepts entirely lacking from a structural account. In the absence of a theoretical formulation of the interactions of atoms sufficient for describing chemistry, recourse was taken more than a hundred years ago to subatomic structure – the discovery that atoms were composite entities comprised of positive (nuclei) and negative (electrons) charged particles – as a basis for a physical account of molecular properties. It is thus inevitable than a physical understanding of chemistry must involve some version of quantum theory and the electrodynamics of charged particles. This book is devoted to things we know about non-relativistic electrodynamics that seem relevant to such a unification which remains controversial. There are also 'known unknowns' to be identified along the way, and from this vantage point the 'unknown unknowns' may be uncovered in the future [9].

1.2 Stoichiometry and Atoms

Measurements of changes in weight – stoichiometry[4] – are a characteristic feature of the quantitative study of chemical reactions; such measurements reveal one of the most important facts about the chemical combination of substances, namely that it generally involves fixed and definite proportions by weight of the reacting substances. These changes in weight are found to be subject to two fundamental laws:

Law of conservation of mass: (A. Lavoisier, 1789)

L1 No change in the total weight of all the substances taking part in any chemical process has ever been observed in a closed system.

Law of definite proportions: (J. L. Proust, 1799)

L2 A particular chemical compound always contains the same elements united together in the same proportions by weight.

The *chemical equivalent* (or *equivalent weight*) of an element is the number of parts by weight of it which combines with, or replaces eight parts by weight of oxygen or the chemical equivalent of any other element; the choice of eight parts by weight of oxygen is purely conventional. By direct chemical reaction and the careful weighing of reagents and products, one can determine accurate equivalents directly. Depending on the physical conditions under which reactions are carried out, one may find significantly different equivalent weights for the *same* element corresponding to the formation of several chemically distinct pure substances. These findings are summarised in the laws of chemical combination [10]:

Law of multiple proportions: (J. Dalton, 1803)

L3 If two elements combine to form more than one compound the different weights of one which combine with the same weight of the other are in the ratio of simple whole numbers.

Let $E[A, n]$ be the equivalent weight of element A in compound n [11]; if we consider the different binary compounds formed by elements A and B, the Law of Multiple Proportions implies

$$\frac{E[A, i]}{E[B, i]} = \omega_{ij} \frac{E[A, j]}{E[B, j]}, \tag{1.1}$$

where ω_{ij} is a simple fraction.

---

[4] From Greek $\sigma\tau o\iota\chi\varepsilon\tilde{\iota}o\nu$ – *stoicheion* – an element.

Law of reciprocal proportions: (J. Richter, 1792)

L4 The proportions by weight in which two elements respectively combine with a third element are in a simple ratio to the proportion by weight in which the two elements combine with one another.

In the notation just introduced this means

$$\frac{E[Y,YZ]}{E[Z,YZ]} = \frac{m}{n}\left(\frac{E[Y,XY]/E[X,XY]}{E[Z,XZ]/E[X,XZ]}\right), \qquad (1.2)$$

where m and n are small integers. On the other hand, a knowledge of the proportions by weight of the elements in a given pure substance is not sufficient information to fix the chemical identity of the substance since there may be several, or many, compounds with the same proportions by weight of their elemental constituents; for example, this is true of many hydrocarbon substances which are chemically distinct yet contain one part by weight of hydrogen to twelve parts by weight of carbon, for example, acetylene, benzene, vinylbenzene, cyclooctatetraene and so on. In these cases, there are distinct compounds formed by two elements that exhibit constant chemical equivalents.

At the beginning of the nineteenth century, the chemical elements were given a microscopic interpretation in terms of Dalton's atomic hypothesis that marks the beginning of chemical theory. The impetus for this new insight came from Dalton's investigations of the properties of mixtures of gases and their solubility in water, and his interest in meteorology. The constant composition of the atmosphere was explained by Lavoisier, Berthollet, Davy and other prominent chemists as being due to a loose chemical combination between its elements. Dalton, who had made a detailed study of the *Principia*, combined Newton's atomic picture of fluids with his own ideas about heat to argue correctly that the atmosphere was a physical mixture of gases. His interest in the mechanism of mixing (and solution) of gases prompted him to determine the relative sizes of the atoms of the gases, and for this purpose he had first to determine their relative weights [12], [13]. Only later did he attempt to apply his atomic theory to chemical experiments; his success in correctly deducing the formulae of the oxides of nitrogen (N_2O, NO, NO_2) led him to state the law of multiple proportions.

Henceforth, the elements were to be regarded as being composed of microscopic building blocks, atoms, which were indestructible and had invariable properties, notably weight, characteristic of the individual elements. Similarly, compounds came to be thought of in terms of definite combinations of atoms that we now call molecules. All molecules of the same chemical substance are exactly similar as regards size, mass and so on. If this were not so, it would be possible to separate the molecules of different types by chemical processes of fractionation, whereas Dalton himself found that successively separated fractions of a gaseous substance were exactly similar. Dalton's idea is different from historically earlier interpretations of the atomic concept such as that of early Greeks, like Democritus, or of Boyle and Newton.

Nearly 50 years of confusion followed Dalton until the Sicilian chemist Cannizzaro outlined [14] a method whereby one could reliably determine a consistent set of weights of different kinds of atoms from the stoichiometric data associated with a set of chemical reactions, and he used this method to define the atomic composition of molecules. Cannizzaro's argument was based on Avogadro's hypothesis that equal volumes of gases at the same pressure and temperature contain equal numbers of molecules. From the mathematical point of view, the problem is indeterminate in the sense that one cannot exclude the possibility that the 'true' atomic weights are actually integer submultiples of those proposed. Cannizzaro offered a partial remedy by observing that the probability that one has the 'true' weights is increased by increasing the amount of data about stoichiometric relations. It is the case that a complete account of the mathematical relations that represent stoichiometry does not require any assumption about atoms [15].

Another limitation is that stoichiometry is concerned only with the changes in weight that occur in chemical reactions; it says nothing about the changes in other properties that accompany chemical transformations. Equally, the original atomic theory could say nothing about the chemical affinity of atoms, why some atoms combine and others do not, nor give any explanation of the restriction to simple fractions in the laws of chemical combination of atoms. Affinity had been a major problem for the phlogiston chemists which was not resolved by Dalton's atomism; only much later with the aid of a structural conception of chemical substances would it be amenable to elucidation. That said, this account of stoichiometry was a major theoretical achievement in classical chemistry based on the atomic/molecular conception of matter.

1.3 Molecular Structure and Chemical Bonds

Having sorted out ideas about elements and compounds in terms of atoms and molecules, attention shifted to synthesis – the making of new compounds – and progress thereafter was rapid, especially in the chemistry of compounds containing the element carbon, what we call organic chemistry. It seems pertinent to recognise that the synthesis of new substances has been the principal experimental activity of chemists for more than 200 years. The number of known pure organic and inorganic substances has grown from a few hundred in 1800 to several hundred million today, with a doubling time of about 13 years that had been remarkably constant over the whole span of two centuries [16]. In order to keep track of the growth of experimental results, more and more transformations of compounds into other compounds, some kind of theoretical framework was needed. In the nineteenth century, the only known forces of attraction that might hold atoms together were the electromagnetic and gravitational forces, but these were seen to be absolutely useless for chemistry and so were given up in favour of a basic structural principle. The development of the interpretation of chemical experiments in terms of molecular structure was a highly original step for chemists to take since it had nothing to do with the then known physics based on the Newtonian ideal of the mathematical specification of the forces responsible for the observed motions of matter. It was one of the most far-reaching steps ever taken in science. G. N. Lewis once wrote [17]

No generalization of science, even if we include those capable of exact mathematical statement, has ever achieved a greater success in assembling in a simple way a multitude of heterogeneous observations than this group of ideas which we call structural theory.

In the 1850s the idea of atoms having autonomous valencies had developed, and this led Frankland to his conception of a chemical bond [18], [19]. He wrote [20]

By the term *bond*, I intend merely to give a more concrete expression to what has received various names from different chemists, such as atomicity, an atomic power, and an equivalence. A monad is represented as an element having one bond, a dyad as an element having two bonds, *etc.* It is scarcely necessary to remark by this term I do not intend to convey the idea of a material connection between the elements of a compound, the bonds actually holding the atoms of a chemical compound being, as regards their nature much more like those which connect the members of our solar system.

The idea of representing a bond as a straight line joining atomic symbols is probably due to Crum Brown. Frankland, with due acknowledgement, adopted Crum Brown's representation which put circles round the atom symbols, but by 1867 the circles had been dropped and more or less modern chemical notation became widespread.

There is a long history in chemistry of the view that chemical combination is due to electrical forces. In the early nineteenth century, Berzelius attempted to systematise the chemical knowledge of his time in an electrochemical theory which took Volta's ideas of galvanic action in a battery as its starting point [21]. The rise of organic chemistry, in which the combination of atoms was not obviously of an electrical kind, led to the eclipse of his approach; the theory of types and the theory of radicals both bid to replace it. Later, it was recognised that Berzelius' idea that the quantity of electricity collected in each atom of different elements depended on their mutual electrochemical differences and controlled their chemical affinity was contradicted by the laws of electrolysis discovered by Faraday. The electrical theory of chemical combination was revived and expanded by von Helmholtz in his celebrated 1881 Faraday lecture [22].

In 1875 van 't Hoff published a famous booklet which marks the beginning of stereochemistry [23]. Following a suggestion of Wislicenus, van 't Hoff proposed that molecules were microscopic material objects in the ordinary three-dimensional space of our sensory experience with physicochemical properties that could be accounted for in terms of their three-dimensional structures. For example, if the four valencies of the carbon atom were supposed to be directed towards the corners of a tetrahedron, there was a perfect correspondence between predicted and experimentally prepared isomers, and a beautiful structural explanation for the occurrence of optical activity. It is natural to extend this hypothesis to all molecules and to suppose that optically active molecules are simply distinguished from other species in that they possess structures that are dissymmetric. Here there is a clear implication for the dimensionality of the 'molecular space'. In a two-dimensional world there would be two forms of the molecule CH_2X_2, whereas only one such compound is known. On the other hand, molecules such as C-abde exist in two forms; these facts require a three-dimensional arrangement of the 'bonds'. Evidently, no picture of the atom is required for this construction; indeed molecular structures can be reduced to suitably labeled points (atoms) joined by lines (bonds). Moreover, van 't Hoff's identification of ordinary physical space as the space supporting these structures is optional; any Euclidean 3-space will do. For van 't Hoff, stereochemistry was part of an argument to give a proof of the physical reality of molecules; molecules could not be perceived directly simply because of the limitations inherent in our senses.

Atomic structure seems first to have been related to valency when both Mendeléev and Meyer observed, independently, in 1869 how valency was correlated with position in the periodic table [24]. There was however no agreement about the nature of atoms. In the same year as van 't Hoff inaugurated stereochemistry with his advocacy of the tetrahedral bonding about the carbon atom, Maxwell gave strong support to Lord Kelvin's vortex model of the atom [25] because it offered an atomic model which had permanence in magnitude, the capacity for internal motion or vibration (which Maxwell linked to the spectroscopy of gases), and a sufficient amount of possible characteristics to account for the differences between atoms of different kinds [26].

From the second half of the nineteenth century onwards, the attribution of physical reality to atoms and molecules was highly controversial because of its obvious metaphysical character. While the realist position was advocated strongly by chemists such as van 't Hoff, and physicists such as Maxwell and Boltzmann, it was criticised severely by other noted scientists such as Duhem and Ostwald whose scientific philosophy was related to the positivism of Mach; for them atoms were fictions of the mind, and they preferred to restrict their discussions to the macroscopic domain. Yet again, others preferred to maintain a sharp distinction between what they regarded as objective knowledge and what was only probably known or speculative; for example, Kekulé did not share the strong conviction of his student van 't Hoff about the structural model, but Kekulé was nevertheless an effective user of the model.

On the other hand, chemists had made a change that brought their thinking much more into line with the customary approach in physics; from the 1860s onwards, inductive argument was replaced by a deductive model based on the formulation and testing of hypotheses [27]. Another important point to keep in mind is that chemistry at the start of the nineteenth century was a science of the transformation of substances (Lavoisier), whereas by the end of the century it had become a science of the transformations of molecules (van 't Hoff), so much so that practitioners of the chemical sciences now often do not distinguish between substances and molecules.

Thus over a period of many years, chemists developed a chemical language – a system of signs and conventions for their use – which gave them a representation of their fundamental postulate that atoms are the building blocks of matter; molecules are built up using atoms like the letters of an alphabet. A molecule in chemistry is seen as a structure, as a semi-rigid collection of atoms held together by chemical bonds. So not only can the numbers of different kinds of atoms in a molecule be counted, but their disposition with respect to each other can be imagined, and this leads to pictures of molecules.

The laws that govern the relative dispositions of the atoms in three-dimensional space are the classical valency rules which provide the syntax of chemical structural formulae. In particular, they specify the combinations of atoms that can be realised under 'ordinary' conditions. Valency, the capacity of an atom for stable combination with other atoms, is thus a constitutive property of the atom not requiring further explanation.

To each pure substance there corresponds a structural molecular formula; and conversely, to each molecular formula there corresponds a unique pure substance. It is absolutely fundamental to the way chemists think that there is a direct relationship between specific features of a molecular structure and the chemical properties of the substance to which it corresponds. Of especial importance is the local structure in a molecule involving a few atoms coordinated to a specified centre, for this results in the characteristic notion of a functional group; the presence of such groups in a molecule expresses the specific properties of the corresponding substance (acid, base, oxidant etc.) which, however, is realised only experimentally in an appropriate reaction context.

Furthermore, each pure substance can be referred to one or several categories of chemical reactivity, and can be transformed into other substances which fall successively in other categories. The structural formula of a molecule summarises or represents the connection between the spatial organisation of the atoms and a given set of chemical reactions that the corresponding substance may participate in. This set includes not only the reactions required for its analysis and for its synthesis, but also potential reactions that have not yet been carried out experimentally. This leads to a fundamental distinction between the chemical and physical properties of substances; while the latter can be dealt with by the standard 'isolated object' approach of physics, the chemical properties of a substance make sense only in the context of the network that describes its chemical relationships, actual and potential, with other substances. Since there is no apparent limit in principle to the (exponential) growth in the number of new substances, the chemical network may not be bounded.

1.4 Atomic Structure and Chemistry

The first tentative steps towards a theory of the chemical bond followed Thomson's discovery of the electron in the late 1890s and his claim that the electron was a universal constituent of atoms. There were several independent measurements of the charge/mass ratio of cathode rays contemporary with Thomson's announcement in 1897; crucially, however, he was the first to measure the charge on the electron in an experiment with his student Rutherford using the Wilson cloud chamber device invented in Cambridge [28]. Thomson initially favoured a uniform distribution of positive charge inside an 'atomic sphere' with solely negatively charged electrons – the so-called 'plum pudding model' of an atom. He had found that the mass of the electron was about $1/1700$ of the mass of the hydrogen atom, and since he assumed the positive charge distribution contributed no mass to the atom, this implied that atoms must contain thousands of electrons [29].

In his Romanes Lecture (1902), Lodge suggested that chemical combination must be the result of the pairing of oppositely charged ions, for (quoted in Stranges, [30])

> It becomes a reasonable hypothesis to surmise that the whole of the atom may be built up of positive and negative electrons interleaved together, and of nothing else; an active or charged ion having one negative electron in excess or defect, but the neutral atom having an exact number of pairs.

The notion of positive and negative electrons was an early 'solution' to the evident problem of the electroneutrality of the atom, and also its stability since a positive charge is needed to keep the electrons together [31]. Earnshaw's theorem in classical electrostatics implies that a collection of charges interacting purely through Coulomb's inverse square law cannot have an equilibrium configuration, and so must be moving [32]; on the other hand, classical electrodynamics implies that moving charges must generally lose energy by radiation.[5]

In 1906, Thomson showed that the number of electrons in an atom is of similar magnitude to the relative atomic mass of the corresponding substance, and that the mass of the carriers of positive electricity could not be small compared to the total mass of the atomic electrons. These conclusions came from three independent theoretical results: firstly, a formula he derived for the refractive index of a monatomic gas; secondly, his formula for the absorption of β-particles in matter; and thirdly, the cross section,[6] σ, for the scattering of X-rays by gases [33]:

$$\sigma = \frac{8\pi}{3} \left(\frac{1}{4\pi\varepsilon_0} \frac{e^2}{m_e c^2} \right)^2. \qquad (1.3)$$

Thus, the hydrogen atom could contain only one electron.

The use of a potential energy surface (PES) as key to understanding the dynamics of molecules can be glimpsed in the beginnings of chemical reaction rate theory more than a century ago that go beyond the purely thermodynamic considerations of van 't Hoff and Duhem, and in the first attempts to understand molecular ('band') spectra in dynamical terms in the same period. As early as 1892, Lord Rayleigh had pointed out that the absence of broadening of the spectral lines of molecular gases due to molecular rotation was an outstanding difficulty for spectroscopic theory [34]. The lack of continuous bands in the spectra of gases was taken as clear evidence of a radical failure of either classical mechanics or classical electrodynamics, or both. Later Bjerrum developed Rayleigh's approach to show that the width of infrared absorption bands should be of the order of magnitude to be expected from the superposition of molecular rotations on molecular vibrations [35], [36]. Thus, for a diatomic molecule, Bjerrum found that an absorption band in a molecular gas at thermal equilibrium should be a doublet separated by a frequency interval of

---

[5] See Appendix C where the classical field of a moving charge is investigated.

[6] Known now as the low-energy (Thomson) limit of the Compton scattering cross section calculated according to Quantum Electrodynamics (QED). e and m_e are the charge and mass parameters of the electron, respectively, c is the speed of light and ε_0 the permittivity of vacuum.

$$\Delta v \approx \frac{1}{\pi}\sqrt{\frac{2k_B T}{I}}, \tag{1.4}$$

where I is the molecular moment of inertia, T is the temperature in Kelvin and k_B is Boltzmann's constant.

The idea of basing a theory of chemical reactions (chemical dynamics) on an energy function that varies with the configurations of the participating molecules seems to be due to Marcelin. In his last published work, his thesis, Marcelin showed how the Boltzmann distribution for a system in thermal equilibrium and statistical mechanics can be used to describe the rate, v, of a chemical reaction [37]. The same work was republished in the Annales de Physique shortly after his death [38].[7] His fundamental result can be expressed, in modern terms, as

$$v = M\left(e^{-\Delta G_+^{\#}/RT} - e^{-\Delta G_-^{\#}/RT}\right), \tag{1.5}$$

where R is the molar gas constant (Avogadro's number, N_A times k_B), T is the temperature in Kelvin, the subscripts $+, -$ refer to the forward and reverse reactions and $\Delta G^{\#}$ is the change in the molar Gibbs (free) energy in going from the initial ($+$) or final ($-$) state to the 'activated state'. The pre-exponential factor M is obtained formally from statistical mechanics. Marcelin gave several derivations of this result using both thermodynamic arguments and also the statistical mechanics he had learnt from Gibbs' famous memoir [39].

The most interesting aspect of Marcelin's account is the suggestion that molecules can have more degrees of freedom than those of simple point material particles. In this perspective, a molecule can be assigned a set of coordinates $\mathbf{q} = q_1, q_2, \ldots, q_n$, and their corresponding canonical momenta $\mathbf{p} = p_1, p_2, \ldots p_n$. Then the instantaneous state of the molecule is associated with a 'representative' point in the canonical phase space $\mathcal{P}$ of dimension $2n$, and so as the position, speed or structure of the molecule changes, its representative point traces a trajectory in the $2n$-dimensional phase space [37].

In his phase space representation of a chemical reaction, the transformation of reactant molecules into product molecules was viewed in terms of the passage of a set of trajectories associated with the 'active' molecules through a 'critical surface' $\mathcal{S}$ in $\mathcal{P}$ that divides $\mathcal{P}$ into two parts, one part being associated with the reactants, the other with the products. According to Marcelin, for passage through this surface it is required[8] [37]

> [une molécule] il faudra [....] qu'elle atteigne une certaine région de l'éspace sous une obliquité convenable, que sa vitesse dépasse une certain limite, que sa structure interne corresponde à une configuration instable, etc. ....

Although this discussion looks familiar, it does so only because of the modern interpretation we put upon it. It is important to note that nowhere did Marcelin elaborate on how the canonical variables were to be chosen, nor even how n could be fixed in any given case. The words 'atom', 'electron' and 'nucleus' do not appear anywhere in his

---

[7] René Marcelin was killed in action fighting for France in September 1914.
[8] that a molecule must reach a certain region of space at a suitable angle, that its speed must exceed a certain limit, that its internal structure must correspond to an unstable configuration etc. ....

thesis, in which respect he seems to have followed the scientific philosophy of Gibbs [40] and his countryman Duhem [41]. On other pages in the thesis, Marcelin refered to the 'structure' (also 'architecture') of a molecule and to molecular 'oscillations' but never otherwise invoked the structural conception of a molecule due to van 't Hoff, although he was very well aware of van 't Hoff's physical chemistry.

The activity of physicists in what hitherto had been the province of chemists did not pass unremarked. At the 1909 meeting of the British Association for the Advancement of Science (BAAS), the distinguished organic chemist Armstrong offered them some fairly pointed advice [42]:

> Now that physical inquiry is largely chemical, now that physicists are regular excursionists into our territory, it is essential that our methods and our criteria are understood by them. I make this remark advisedly, as it appears to me that, of late years, while affecting almost to dictate a policy to us, physicists have taken less and less pain to make themselves acquainted with the subject matter of chemistry, especially with our methods of arriving at the root conceptions of structure and the properties as conditioned by structure. It is a serious matter that chemistry should be so neglected by physicists.

Thomson was one of very few physicists with a serious interest in studying the role of electrons in chemistry, and his penchant for qualitative arguments took him steadily away from the mainstream of physics. The 'plum-pudding' model became of purely historical interest when a completely novel conception was introduced by Rutherford in 1911; he successfully explained the back-scattering of α-particles by a thin gold foil in terms of his notion of the atomic nucleus where most of the atomic mass, and all of the positive charge in the atom resided [43]. The nucleus is negligibly small in comparison with the dimensions of an atom (10^{-5}:1). Almost immediately, the astronomer Nicholson proposed a 'planetary' model of the atom,[9] in which electrons orbit the positively charged nucleus [45], [46]. If electrons are placed in a circular orbit of radius a with angular velocity ω, an energetic equilibrium is obtained when the centrifugal force on a specified electron is balanced by the attractive force of the nucleus less the repulsions of the other electrons. Critically, however, with more than one electron the orbit does not have dynamical stability. Nicholson's calculations were directed towards a hypothetical atom he believed was responsible for the spectra of nebulae; this atom, called 'Nebulium', could not be identified with any terrestrial atom. The notion has long been consigned to obscurity, though the calculations proved significant in the following decade.

As an example, consider the helium atom modelled as a pair of electrons in a circular orbit about a nucleus with charge $q = 2e$ which for simplicity is regarded as immobile; then there are six degrees of freedom. The equilibrium arrangement has the electrons on opposite sides of the nucleus in steady motion in the same direction. The modes can be classified as in-plane and perpendicular to the plane of the orbit. One in-plane mode

---

[9] Nicholson was inspired by Maxwell's account of the rings of Saturn [44]; although electrostatics and gravitation are described by the same inverse square law, a crucial difference is that the electrons repel each other while the 'particles' in Saturn's rings experience only attractive forces. This leads to a marked difference in the requirements for the stability of particle orbits.

is unstable; if the electrons are subject to a perturbation in the orbital plane directed at right angles to the line joining the pair, the electrons do not return to the equilibrium configuration, but instead move apart exponentially. This is the typical behaviour of multielectron systems. Notice also that the classical mechanics is quite incomplete since the preceding argument leads to an equation involving the combination $a^2\omega^3$ so that another statement is required to fix either the angular velocity or the radius a. The only parameters available in the Newtonian mechanics of the problem are the charge and mass of the electron, and these are not enough to construct units of length, mass and time required for a complete physical theory. If the velocity of light is admitted, in recognition of electromagnetic phenomena in the atomic regime, then a characteristic length (the classical electron radius) can be constructed,

$$r_e = \frac{1}{4\pi\varepsilon_0} \frac{e^2}{m_e c^2}, \tag{1.6}$$

which, to within a numerical factor, is the square root of the Thomson scattering cross section, (1.3). Its magnitude, however, is $\sim 10^{-15}$ m, far too small to be a characteristic length for atoms and molecules.

Two quite different remedies for this situation were proposed in 1913. Bohr linked atomic structure to Planck's constant, h, the quantum of action. While accepting the correctness of Nicholson's calculations, he proposed the following remarkable hypothesis [47], [48]:

> In any molecular system consisting of positive nuclei and electrons in which the nuclei are at rest relative to each other and the electrons move in circular orbits, the angular momentum of every electron round the centre of its orbit will in the permanent state[10] of the system be equal to $h/2\pi$, where h is Planck's constant.

For each orbit this gives

$$ma^2\omega = \frac{h}{2\pi} \equiv \hbar, \tag{1.7}$$

which is sufficient to fix the radius of the orbit and the frequency ω. The quantisation of the angular momentum is supposed to trump the classical dynamical instability of the orbits in multielectron systems. For the hydrogen atom, Bohr found $2a \approx 1.1 \times 10^{-10}$ m and $\omega \approx 6.2 \times 10^{15}$ s^{-1}. Bohr further took it that higher-energy electron orbits were associated with integer multiples of the quantised angular momentum. This leads to a system of discrete energy levels, and he identified the observed spectral frequencies $\{\nu\}$ with *transitions* between these energy levels according to the quantum law (an expression of the conservation of energy):

$$h\nu_{nm} = E_n - E_m. \tag{1.8}$$

The theory gives quantitative agreement with the observed sequence of spectral lines of atomic hydrogen known as the Balmer and Paschen series, and offers the prediction (verified later) of other series in the IR and UV parts of the spectrum. The same theory applies to He$^+$ with a simple modification of the nuclear charge and the reduced

---

[10] In modern terms, the ground state.

mass of the electron–nucleus pair. In companion papers, Bohr gave a qualitative discussion of multielectron atoms [49] and of the hydrogen molecule [50]. These justly famous papers are commonly referred to now as 'Bohr's trilogy'. In the following years, Bohr developed his approach and gave a comprehensive account of atomic spectra and the periodic table. He was centrally involved in the development of quantum theory and was acutely aware of the paradoxes it entailed prior to the discovery of quantum mechanics; Bohr received the Nobel Prize in Physics in 1922 [51].

Shortly after Bohr's introduction of the idea of stationary state energy levels in atoms determined by Planck's constant h, similar ideas were developed for the vibrations and rotations of more general molecules. Molecular structure was expressed in terms of a mechanical model (for example, the dumbbell model of a diatomic molecule or the symmetric top model of a polyatomic molecule) amenable to classical dynamical calculations supplemented by 'quantum conditions'[11] [52], [54], and this led to rapid progress in the understanding of molecular spectra.

In the same year as the appearance of Bohr's trilogy, Thomson took an entirely different view; he argued that because of the instability problem of classical orbits in an electrostatic field every electron would have to have its own orbit, and hence an extremely complicated picture of an atom would ensue that would be useless for the needs of chemistry. He therefore proposed a modification of electrostatics [54],

> In considering the forces which may exist in the atom, we must remember that we cannot assume that the forces due to the charges of electricity inside the atom are of exactly the same character as those given by the ordinary laws of Electrostatics; these laws may merely represent the average effect of a large number of such charges, and in the process of averaging some of the peculiarities possessed by the individuals may disappear.

His proposal was that the force law between a nucleus of charge Ze and an electron a distance r apart is expressed by the equation

$$F = \frac{Ze^2}{4\pi\varepsilon_0 r^2}\left(1 - \frac{l}{r}\right),\tag{1.9}$$

where the length, l, is a characteristic atomic constant of order 10^{-10} m. With such a force law, a number of electrons can be in stable equilibrium around a nucleus without having to be assigned to orbits of the Bohr type; instead it leads to a model of the atom with a size of the order of l in which electrons are static. It is noteworthy that Thomson made no reference to the hydrogen atom, and no reference to Bohr and Rutherford [55]. The model was of no consequence in physics; however, Thomson made use of it to develop a systematic discussion of the facts of chemistry and ideas about valency which were congenial to chemists in the UK and USA. Thomson thought his formulation provided a unification of chemistry and physics; his mature ideas were recorded in his Franklin lectures delivered in 1923 [56].

One of Thomson's admirers in the USA was Noyes, editor of the *Journal of the American Chemical Society* between 1902 and 1917, and a person like Lewis of considerable

---

[11] This is the approach that we now refer to as the Old Quantum Theory.

influence. In 1917, Noyes reviewed recent developments in ideas about valency in terms of electronic models and wrote [57]

> Physicists in general have directed their attention to rotating or rapidly moving electrons and to the relation between these and spectral lines, the disintegration of atoms and other phenomena involving individual atoms. Chemists, on the other hand, following the suggestion of J. J. Thomson, have considered chiefly the role which the valence electrons probably play in the combination of atoms.

Ideas about the electronic structure of atoms very similar to those of Lodge also formed in the mind of Lewis, but they were not published until 1916,[12] after the Bohr atom ideas had become prominent in physics. It was Lewis' skilful combination of electronic ideas with traditional notions of the bond that proved so persuasive to chemists. In his 1923 book, Lewis wrote the following about the development of his theory [17]:

> In the year 1902 ..., I formed an idea of the inner structure of the atom which, although it contained certain crudities, I have ever since regarded as representing essentially the arrangement of electrons in the atom ...
> The main features of this theory of atomic structure are as follows:
>
> 1. The electrons in an atom are arranged in concentric cubes.
> 2. A neutral atom of each element contains one more electron than a neutral atom of the element next preceding.
> 3. The cube of eight electrons is reached in the atoms of the rare gases, and this cube becomes in some sense the kernel about which the larger cube of electrons of the next period is built.
> 4. The electrons of an outer incomplete cube may be given to another atom, as in Mg^{++}, or enough electrons may be taken from other atoms to complete the cube, as in Cl^-, thus accounting for 'positive and negative valence'.

The model of the atom that is presupposed here is a static one inspired by the work of Thomson. In his 1916 paper, Lewis introduced a new idea and a new means of representation, and these are quite unambiguously Lewis' contributions alone [59]. The new idea was the 'rule of two' in which he asserted that the occurrence of electrons in molecules in even numbers was pretty much universal. The new means of representation was the method of symbolising electrons by dots which is now so familiar to us. The ability to make the correspondence of a pair of dots between two atom symbols and the bond was extremely attractive to working chemists. It is obvious that these ideas owe absolutely nothing to quantum theory, and certainly nothing to Bohr. It would be wrong to believe that this was because those involved in the developments here did not know what was going on in physics; they knew very well and were, on the whole, pretty sceptical about them. At a meeting of the AAAS in New York in December 1916, Lewis devoted his address to the idea of a static atom and commented that [61]:

> Unless we are willing, under the onslaught of quantum theories, to throw over all the basic principles of physical science, we must conclude that the electron in the Bohr

---

[12] The priority in publication of the 'octet rule' is actually by Abegg in 1904 [58]. Lewis did not publish until 1916 at about the same time as Kossel, who had arrived at very similar conclusions [59], [60].

atom not only ceases to obey Coulomb's law, but exerts no influence whatsoever upon another charged particle at any distance.

Lewis considered this absence of effect logically and scientifically objectionable for "that state of motion which produces no physical effect whatsoever may better be called a state of rest" [59]. He was not alone in his scepticism; efforts continued for a decade after Bohr's hydrogen atom paper to get a theory in which the electrons in an atom remained still and were distributed at the corners of a cube. In his Nobel lecture in 1922, however, Bohr delivered a sharp attack on static atom theories, pointing out that in view of Earnshaw's theorem in electrostatics [32], [51],

> Statical positions of equilibrium for the electron are in fact not possible in cases where the forces between the electrons and the nucleus even approximately obey the laws that hold for the attractions and repulsions between electrical charges.

He also suggested that the developing quantum theory offered the possibility of relating the properties of the elements and the experimental results concerning the constituents of atoms, something that was quite beyond the statical atom models. By this Bohr meant physical properties; his examples demonstrate the periodicity (in the sense of the periodic table) of the elements. There is nothing about chemical bonding and valency.

This was clearly, at the very least, an uncomfortable situation, and Sidgwick attempted to avoid the difficulty by shifting the argument away from atomic structure as such, to the idea of molecular structure in which pairs of electrons had common orbits of the Bohr–Sommerfeld type involving the molecular nuclei. He seems to have been the first chemist to point out that it was possible to imagine a dynamical situation in which a pair of electrons could hold a pair of nuclei together; this suggestion was made at a meeting of the Faraday Society in Cambridge in 1923 [62]. At that meeting, Lewis gave an introductory address in which he signaled his accession to a similar point of view which shortly afterwards was elaborated in book form [17]. Lewis was clearly still unhappy with quantum theory, for in the closing pages of his book he could not resist referring to it as "the entering wedge of scientific Bolshevism."

The dynamical model described by Sidgwick had contemporaneously been discussed in terms of the Old Quantum Theory by Pauli and by Nordheim who attempted to classify the various sorts of orbits satisfying the 'quantum conditions' that were possible for electrons shared by two nuclei [63], [64]. Their calculations were unsuccessful as indeed were contemporary calculations on the Bohr stationary states of the helium atom. We noted earlier that the Bohr model of the helium atom is dynamically unstable; a similar behaviour is found with H_2. Nordheim investigated the forces between two hydrogen atoms as they approach each other adiabatically in various orientations consistent with the quantum conditions. Before the atoms get close enough for the attractive and repulsive forces to balance out, a sudden discontinuous change in the electron orbits takes place and the electrons cease to revolve solely round their parent nuclei. Nordheim was unable to find an interatomic distance at which the energy of the combined system was less than that of the separated atoms. At the end of his paper Nordheim wrote [64]

> dass eine rein adiabatische Annäherung nicht zu einer Bindung führen kann, da sich die Atome zwar anfangs anziehen, dann aber infolge der Stosswirkung in genan derselben Weise auseinanderfliegen müssen.

Paraphrased, this means that a purely adiabatic approximation cannot describe the formation of bonds between atoms because if they come together adiabatically, they can just as easily separate.

The parlous situation regarding the application of the Old Quantum Theory to atomic and molecular systems shortly before the discovery of quantum mechanics is dealt with in the books by Sommerfeld [65] and Born [66]; for example, Sommerfeld expressed his hopes as follows:

> To the future falls the task of working out *a complete topology of the interior of the atom* and, beyond this, a system of mathematical chemistry, that is one which will tell us the exact position of the electrons in the atomic envelope and how this qualifies the atom to form molecules and to enter into chemical compounds.
>
> The subject of mathematical physics has been in existence for more than one hundred years; a system of mathematical chemistry that can achieve what we have just mentioned, that can shed light on the still very obscure conception of valency and can, at least in typical cases, predict the reactions that must occur, is only on the point of being created.

He introduced his detailed account of the Bohr models for He, H_2 and H_2^+ with the following words:

> The following calculations concern models that are indeed interesting from the historical aspect but that cannot be maintained empirically and theoretically.

In Sommerfeld's view, chemical bonding and valency remained a mystery that the developing quantum theory had not illuminated [65].

Following the discovery of quantum mechanics by Heisenberg [67], a consistent account of the structure of the atom was rapidly realised, and all previous conceptions of the atom became untenable. By the time Sidgwick published his book (1927), he had decided to face the consequences of the quantum revolution in physics. The preface to his book begins [68]:

> This book aims at giving a general account of the principles of valency and molecular constitution founded on the Rutherford-Bohr atom.... In developing the theory of valency there are two courses open to the chemist. He may use symbols with no definite physical connotation ... or he may adopt the concepts of atomic physics, ... and try to explain chemical facts in terms of these. But if he takes the latter course, as is done in this book, he must accept the physical conclusions in full ....

But he was clearly uneasy when he acknowledged the newly published work of Schrödinger:

> It has yet given no proof that the physical concepts which led (him) to his fundamental differential equation should be taken so literally as to be incompatible with the conceptions of the nature of electrons and nuclei to which the work of the last thirty years has led.

Even at this date, there was still refusal from some distinguished chemists at what had developed in the conception of the nature of matter. Henry Armstrong was still an outspoken critic, writing in 1927 [69]:

> On p. 414 [Nature (1927), vol 120], Prof W. L. Bragg asserts that 'In sodium chloride there appear to be no molecules represented by NaCl. The equality in number of sodium and chlorine atoms is arrived at by a chess-board pattern of these atoms; it is a result of geometry and not of a pairing-off of the atoms.' This statement is more than 'repugnant to common sense'. It is absurd to the n th degree, not chemical cricket. Chemistry is neither chess nor geometry, whatever X-ray physics may be. Such unjustified aspersion of the molecular character of our most necessary condiment must not be allowed any longer to pass unchallenged. A little study of the Apostle Paul may be recommended to Prof. Bragg, as a necessary preliminary even to X-ray work, especially as the doctrine has been insistently advocated at the recent Flat Races at Leeds, that science is the pursuit of truth. It were time that chemists took charge of chemistry once more and protected neophytes against the worship of false gods: at least taught them to ask for something more than chess-board evidence.

Armstrong evidently failed to appreciate that the solid state was qualitatively different from fluids where his organic chemistry flourished, but also that Bragg had accepted completely the chemist's classical notion of structure in the atomic domain; there was nothing in what Bragg had done that had anything to do with the developing quantum theory.

Heitler and London's paper on the quantum chemistry of the H_2 molecule according to Schrödinger's wave mechanics appeared in the same year [70]. In contrast to the earlier investigations of Pauli and Nordheim using the Old Quantum Theory methods, the new quantum mechanics based on adiabatic approach of the atoms yielded a bound molecular ground state; key to London's approach was the fundamental notion that as the nuclei moved they acted as adiabatic parameters in the electronic wave function [71]. Thus quantum chemistry was initiated as an *electronic structure theory*. Shortly before the publication of the Heitler and London paper, Sidgwick had sent London a copy of his new book seeking comment on the consistency of the Lewis-inspired approach and the new developments in wave mechanics; while London praised the book, he preferred his own approach. According to his biographer, London, as a new university teacher, had written recently to Schrödinger suggesting that he did not think quantum mechanics was necessary for chemists' understanding of chemical processes; indeed he thought that a course in quantum mechanics for chemists might frighten them [72].

It fell to Pauling to attempt a reconciliation between the Lewis theory and the approach made by Heitler and London through the development of the Valence Bond model of electronic structure. This he did in a series of papers published between 1928 and 1933 and whose conclusions are brought together in his book, dedicated to Lewis and published in 1939 [73]. In this enormously influential book, Pauling had a clear programme:

> I formed the opinion that, even though much of the recent progress in structural chemistry has been due to quantum mechanics, it should be possible to describe the new developments in a thorough-going and satisfactory manner without the use of advanced mathematics. A small part only of the body of contributions of quantum mechanics to chemistry has been purely quantum mechanical in character; .... The advances which have been made have been in the main the result of essentially chemical arguments, ... The principal contribution of quantum mechanics to chemistry has been the suggestion of new ideas, such as resonance.

Pauling started his exposition from the idea of the electron pair bond as envisaged by Lewis and showed how this can be understood in the context of the Heitler–London calculation as being due to strong orbital overlap. Introducing the idea of orbital hybridisation, he then used the idea of maximum overlap in discussing bonding generally. It should not be thought, however, that all were as convinced as was Pauling in the correspondence between perfect pairing and the bond. Mulliken arrived at very different conclusions from the standpoint of Molecular Orbital theory. He devoted his 1931 review to a description of molecular structure in terms of molecular orbitals, and at the end of the last section, felt constrained to write [74]:

> The fact that valence electrons almost always occur in pairs in saturated molecules appears to have after all no fundamental connection with the existence of chemical binding. ... A clearer understanding of molecular structure ... can often be obtained by dropping all together the idea of atoms or ions held together by valence forces, and adopting the molecular point of view, which regards each molecule as a distinct individual built up of nuclei and electrons.

For Mulliken at least, it was clearly somewhat doubtful even then that the bond was either necessary for, or explicable in terms of the quantum mechanics required to account for chemical binding. At issue here was a question of interpretation rather than technique, for both methods had provided a basis for useful calculations; moreover, from the technical point of view it was soon shown that the two methods could be extended to give ultimately the same description of the electronic structure of the H_2 molecule [75]. Later this proof was extended to the general case of the polyatomic molecule [76], so that the choice of method depended on convenience rather than a point of principle.

1.5 Chemical Physics and Quantum Chemistry

The scope of quantum chemistry in its first two decades can be gauged from two famous books which showed the development of a wide-ranging formalism, although practical calculations were strongly limited by the sheer complexity of the requisite wave-mechanical calculations [77], [78]. What transformed the subject was the development and widening availability of electronic computers in the years after WWII [79]. The successes and contributions of quantum chemistry to modern chemistry are well known and need not be detailed here; a comprehensive overview of the whole subject can be found in a recent Handbook [80]. Quantum chemistry is widely thought to be an explicit justification of Dirac's original claim that quantum mechanics could be used directly and quantitatively to describe the facts of chemistry if only the computations could be done. It is worth reminding ourselves of what Dirac wrote and its context. Dirac started by remarking (in 1929) that quantum mechanics had been nearly completed, the remaining problem being essentially its relationship with relativity ideas [81]. He continued:

> These give rise to difficulties only when high-speed particles are involved, and are therefore of no importance in the consideration of atomic and molecular structure and ordinary chemical reactions, in which it is, indeed, usually accurate if one neglects relativity variation of mass with velocity and assumes only Coulomb forces between the various electrons and atomic nuclei. The underlying physical laws necessary for the mathematical theory of a large part of physics and the whole of chemistry are thus completely known, and the difficulty is only that the exact application of these laws leads to equations much too complicated to be soluble.

The evidence for such a claim was really rather slight, probably amounting to little more than the work of Heitler and London [70] on the electronic structure of the hydrogen molecule,[13] but nevertheless it has been regarded as 'received wisdom' ever since. As we shall see (§5.3.3), an evident irony is that the claim was made in the introduction to a justly famous paper showing the far-reaching implications of permutation symmetry in the new quantum mechanics for systems of identical particles.

Chemistry relies on the atom as its basic unit; however, there is no sufficient account of interactions between atoms that could serve as a fundamental basis for theoretical chemistry. Thus chemical physics and quantum chemistry invoke sub-atomic structure and rely on Schrödinger's equation and an appropriate Hamiltonian for atoms and molecules which are taken to be composed of charged particles, electrons and nuclei. The following quotation comments on the 'derivation' of the wave equation for the hydrogen atom [77]:

> On observing that there is a formal relation between this [Schrödinger] wave equation and the classical energy equation for a system of two particles of different masses and charges, we seize on this as providing a simple, easy, and familiar way of describing the system, and we say that the hydrogen atom consists of two particles, the electron and proton, which attract each other according to Coulomb's inverse-square law. Actually we do not know that the electron and proton attract each other in the same way that two macroscopic electrically charged bodies do, inasmuch as the force between two particles in a hydrogen atom has never been directly measured. All that we do know is that the wave equation for the hydrogen atom bears a certain formal relation to the classical dynamical equations for a system of two particles attracting each other in this way.

[13] Doubtless Dirac was aware of the then recent work of Born and Oppenheimer (1927).

The wave equation for the hydrogen atom with the Hamiltonian based on purely (Coulombic) electrostatic forces between the charged particles yields definite formulae for discrete energy levels expressed in terms of fundamental constants including values for the electron and proton masses and their charges; if we substitute the experimentally determined values (obtained from other experiments), remarkable agreement with spectroscopic data is achieved.

A fundamental theory of atoms and molecules is expected to justify such a result; the restriction to electrostatics is, however, no more than a practical ansatz that can be traced back to the Bohr theory of the atom [47], [48]. Of course, the model did not survive the discovery of quantum mechanics, but it left a seemingly permanent imprint; the quantum theory that developed from it is fundamentally spectroscopic in nature (energy levels, transition matrix elements, the S-matrix, response functions etc.). Bohr's model is mainly remembered for his introduction of Planck's constant, h, and the resulting quantisation of the angular momentum. Much less remarked on today is that Bohr made a decisive break with classical electrodynamics. In modern terms, the idea is this; formally one fixes the gauge of the vector potential, $\mathbf{A}$, by the Coulomb gauge condition,

$$\nabla \cdot \mathbf{A} = 0, \tag{1.10}$$

and it then follows easily that the longitudinal part of the electric field strength due to the electrons and nuclei can be expressed entirely in terms of their coordinates and gives rise to the familiar static Coulomb potential in the Hamiltonian. 'Radiation reaction' due to the transverse part of their electromagnetic field is simply discarded ad hoc, and the role of the radiation field is demoted to the status of an 'external' perturbation inducing transitions between Bohr's stationary states. For a system of charged particles with purely electrostatic interactions, this leads to the so-called 'Coulomb Hamiltonian' and the miracle of quantisation sweeps away the pathologies of its classical ancestor, as demonstrated by the Kato–Rellich theorem [82], [83]. However, when electromagnetic radiation is admitted the situation is much more complicated because of 'self interactions' which lead to formally infinite 'electromagnetic masses' for the charges [84].

Since we are dealing with charged particles, a fundamental theory of atoms and molecules must presumably be based on their electrodynamics, and so we require electrodynamics formulated in terms of Hamiltonian dynamics, since this is the route to Schrödinger's equation. It is conventional to begin with a classical description knowing that the canonical quantisation scheme due to Dirac, based on the correspondence

$$i\hbar \text{ classical Poisson} - \text{bracket} \rightarrow \text{quantum commutator}, \tag{1.11}$$

is a standard procedure for obtaining a quantum theory from a classical analogue that has been cast in Hamiltonian form, and this is the route we shall follow. It has long been recognised, however, that the scheme involves analogy which may not be reliable, since the resulting quantum theory may or may not turn out to be satisfactory. The classical theory is thus no more than a recognisable starting point towards a quantum theory, the required endpoint.

The usual discussion in the literature of classical electrodynamics concentrates on the Lorentz force for the dynamics of the charges, with fields obtained from the relevant (retarded) solutions of the Maxwell equations; much of the discussion is concerned with aligning the theory with special relativity which is an obvious priority in general physics. In classical electrodynamics, the limiting case of point charged particles is pathological, and a major goal of the theory is the treatment of the infinities that arise. For example, the Coulomb energy is divergent for a classical point charge, as will be shown in Chapter 2.

In some sense, this means that the notion of a point particle carrying electric charge is simply inconsistent with classical physics. We now know from quantum mechanics that classical physics cannot be used for lengths shorter than about the reduced Compton wavelength ($\lambdabar_C = \hbar/m_0 c$) for the particle; according to the uncertainty principle, this corresponds to energies greater than the pair production threshold. It is known that maintaining explicit Lorentz invariance and gauge invariance provides the best route to making sense of the divergences that plague the electrodynamics of point charged particles.

Atoms and molecules are characterised minimally by the specification of a definite number of nuclei and electrons. There is no known theory of a system with a fixed finite number of particles interacting through the electromagnetic force that accommodates gauge invariance and is covariant under Lorentz transformations, so that any general account of atoms and molecules will be 'non-relativistic' to some degree. It is usually accepted that the first step in transforming to a Hamiltonian description is to ensure that Newton's law of motion for the charges with the Lorentz force, and the Maxwell equations for the field, are recovered as Lagrangian equations of motion. There is then a standard calculation for the determination of the associated Hamiltonian. This is the subject matter of Chapter 3.

It is important to note that the customary starting point for classical Lagrangian electrodynamics involves symbols for the electric charges $\{e_n\}$ and masses $\{m_n\}$ of the particles which are merely parameters that cannot be assumed to have the experimentally determined values. There is a subtle change of viewpoint here; the original equations of motion, modelled on macroscopic classical electrodynamics, describe the electromagnetic fields associated with prescribed sources through Maxwell's equations, while Newton's laws are used to describe the motion of charged particles in a prescribed electromagnetic field. The Lagrangian formalism, however, describes a closed system for which $\partial L/\partial t = 0$, so that by the usual arguments the Hamiltonian H is the constant energy of the whole system.

For comparison with experimental data, the parameter e is required to be the experimentally observed charge of a particle; a gauge-invariant theory guarantees charge conservation and at non-relativistic energies there are no physical processes that can modify the value of e. This is true in both classical and quantum theories. The situation with the mass parameter m for a particle is quite different since there is a charge–field interaction that leads to an 'electromagnetic mass' additional to the 'mechanical mass' m. It is possible for the 'electromagnetic mass' (due to self-interaction) to become arbitrarily large and this requires m to be negative so that the observed mass = mechanical mass + electromagnetic mass has its observed (positive) value. This pathology certainly occurs in the point charge limit and is the origin of so-called 'runaway' solutions in the classical equations of motion for the charged particles. A feature of the runaway solution is that it has an essential singularity at $e = 0$, so there is no possibility of constructing solutions of the interacting charge and field system that pass smoothly into the solutions of the non-interacting system as $e \to 0$. Some of these problems are inherited by the quantum theory resulting from canonical quantisation of non-relativistic classical electrodynamics.

From the point of view of fundamental theory, it is clear that there is a considerable gulf between Bohr's picture which is the basis for the usual perturbation theory procedures, and the actual characterisation of the non-relativistic QED Hamiltonian H for a collection of charged particles. The conventional perturbation theory of optical physics assumes that the Hilbert space of the full system is the same as that for the 'free' reference system (atoms/molecules and EM field without coupling), as in ordinary quantum mechanics, so that the diagonalisation of the full Hamiltonian, H, expressed in the reference system basis can be expressed as a certain unitary transformation. If one takes the charges to be 'point particles', this is never the case; the usual remedy is to smooth out point charges, which is physically plausible for nuclei, but not so evident for electrons. In the presence of electromagnetic radiation, all the discrete energy levels of the atomic system become thresholds of continuous spectra; they are said to be 'embedded' eigenvalues (resonances). The fate of these discrete states of isolated atoms/molecules thus requires the perturbation theory of *continuous* spectra. A short introduction to these ideas is given in Chapter 11.

As for treating the field as an 'external' perturbation, this is commonly implemented by assuming that the electromagnetic field variables in the Hamiltonian are classical variables. But if the electromagnetic field is regarded as a physical system, it clearly has a specific Hamiltonian which has to be quantum mechanical if all its properties are to be described. That does not contradict the fact that one can realise *states* of the field that have some of the same statistical properties (mean correlation functions of the field) as in Maxwell's classical electrodynamics, and so might be called 'classical states'.

A fundamental shift in chemical perspective occurred during the years either side of World War II. By and large, the historical approach to molecular structure was highly successful for organic chemistry, even though there were puzzles and anomalies that had to be regarded as 'special cases', for example, concerning the structural formulae for polycyclic hydrocarbons such as anthracene; it was much less successful for inorganic compounds. For this reason, the systematic use of physical methods of structure determination, especially X-ray and electron diffraction techniques, in organic chemistry textbooks was much delayed with respect to those of inorganic chemistry and did not become widespread until the late 1950s [85], [86]. This change in methodology seemed to imply a fundamental revision in the notion of molecular structure from being a hypothesis that encoded the actual and potential chemistry of a substance (the set of chemical reactions a substance may participate in) to being an experimental observable to be measured by a physical technique.

Most of the physical techniques of structure determination fall under the general heading of spectroscopy, that is, they involve the monitoring of some kind of radiation that has previously interacted with the chemical substance. With the passage of time, it has become evident that the experimental results derived from these techniques are quite generally either reported directly in terms of classical molecular structure models (e.g. diffraction experiments, microwave spectroscopy, dielectric properties) or in terms of correlations with classical structural features (e.g. infrared, visible-UV, NMR spectroscopies). There is, however, an important distinction to be made. Classical structural formulae deliberately suppress detailed geometric information and instead focus on the configuration (or conformation) of the functional groups so as to convey the relevant information about the position of the substance in the chemical network. The precise structural diagrams derived from physical measurements do not identify functional groups per se and hence do not encode the chemistry of the substance; for that one must refer back to the older conception of a molecule.

While it is widely believed that this change in orientation of the basis for molecular structure is an inevitable outcome of the development of modern physical theory applied to molecular systems, a more reasonable view is that a far-reaching reinterpretation of these experiments has been made for reasons that are largely independent of any requirements of physics (specifically quantum mechanics). Looking back, it is apparent that Armstrong's appeal to physicists [42], quoted in §1.4, was never heard, but equally Sidgwick's claim [68] that the chemist "must accept the physical conclusions in full" does not describe how things have turned out. It is also quite clear that the success of chemistry based on the conception of molecular structure initiated by van 't Hoff is quite independent of the physical nature of the atom which, as we have seen in this chapter, underwent very radical revisions up to the discovery of quantum mechanics. In other words, rather than the seamless integration of chemical theory into physics, all that has happened is that the nineteenth-century rupture between chemistry and physics has been patched over in the framework of quantum chemistry.

References

[1] Aristotle, *De Caelo*, iii.3.302^a, translated by J. L. Stocks (1930) in *The Works of Aristotle*, Vol. 2, edited by W. D. Ross, Clarendon Press.

[2] *Handbook of the Philosophy of Science, Volume 6 – Philosophy of Chemistry*, (2012), edited by A. I. Woody, R. F. Hendry and P. Needham, North-Holland.

[3] Weinberg, S. (2015), *To Explain the World: The Discovery of Modern Science*, Allen Lane, Penguin UK.

[4] Jacob, F. (1974), *The Logic of Living Systems*, translated by Betty E. Spillmann, p. 43, Allen Lane.

[5] Dobbs, B. J. (1975), *The Foundations of Newton's Alchemy: or 'The Hunting of the Green Lion'*, Cambridge University Press.

[6] Dobbs, B. J. (1992), *The Janus Faces of Genius: The Role of Alchemy in Newton's Thought*, Cambridge University Press.

[7] Fahrenheit, D. G. (1724), Phil. Trans. Roy. Soc. (London) **33**, 78.

[8] Lavoisier, A. (1789), *Traité élémentaire de chimie* in *Oeuvres de Lavoisier*, (1862), Paris, Imprimerie Imperiale; English translation republished by Dover Publications Inc. (1965) as *Elements of Chemistry*.

[9] Rumsfeld, D. H. (2002), accessed at https://archive.ph/20180320091111/ http://archive.defense.gov/Transcripts/Transcript.aspx?TranscriptID=2636.

[10] Parkes, G. D. (1961), *Mellor's Modern Inorganic Chemistry*, revised edition by G. D. Parkes, Longman.

[11] Berry, R. S., Rice, S. A. and Ross, J. D. (1980), *Physical Chemistry*, Ch. 1, J. Wiley and Sons Inc.

[12] Coward, H. F. (1927), J. Chem. Education **4**, 23.

[13] Hartley, H. (1967), Proc. Roy. Soc. (London) A**300**, 291.

[14] Cannizzaro, S. (1858), Nuovo Cimento, **7**, 321.

[15] Woolley, R. G. (1995), Mol. Phys. **85**, 539.

[16] Schummer, J. (1999), Educación Quimica **10**, 92.

[17] Lewis, G. N. (1923), *Valence and the Structure of Atoms and Molecules*, Chemical Catalog Co.

[18] Lagowski, J. J. (1966), *The Chemical Bond*, Houghton Mifflin.

[19] Russell, C. A. (1971), *The History of Valency*, Leicester University Press.

[20] Frankland, E. (1866), J. Chem. Soc. **19**, 377.

[21] Berzelius, J. J. (1819), *Essai sur la théorie des proportions chimiques, et sur l'influence chimique de l'électricité*, Méquignon-Merquis.

[22] von Helmholtz, H. L. F. (1881), J. Chem. Soc. **39**, 277.

[23] van 't Hoff, J. H. (1875), *La Chimie dans l'Espace*, Bazendijk.

[24] Mendeléev, D. (1891), *The Principles of Chemistry*, Vol. 2, p. 16, footnote, Longmans Green.

[25] Thomson, W. (1869), Proc. Roy. Soc. Edinburgh, **6**, 94. doi:10.1017/S0370 164600045430.

[26] Maxwell, J. C. (1875), Entry for 'ATOM' in *Encyclopaedia Britannica*, 9th ed., **3**, 36.

[27] Rocke, A. J. (1990), *Beyond History of Science: Essays in Honor of Robert E. Schofield*, Edited by E. Garber, Lehigh University Press.

[28] Thomson, J. J. (1899), Phil. Mag. Series V **48**, 547.

[29] Thomson, J. J. (1904), *Electricity and Matter*, The 1903 Silliman Lectures at Yale University, Yale University Press.

[30] Stranges, A. N. (1982), *Electrons and Valence*, Texas A & M University Press.

[31] Jeans, J. (1901), Phil. Mag. Series VI **2**, 421.

[32] Earnshaw, S. (1842), Trans. Camb. Phil. Soc. **7**, 97.

[33] Thomson, J. J. (1906), Phil. Mag. Series VI **11**, 769.

[34] Lord Rayleigh (1892), Phil. Mag. Series V **34**, 410.

[35] Bjerrum, N. (1912), in Festschrift, W Nernst zu seinem fünfundzwanzigjährigen Doktorjubiläum gewidmet von seinen Schülern, Knapp, Halle, Germany.

[36] Kemble, E. C. (1926), *Molecular Spectra in Gases*, Bull. Nat. Res. Council, **11**, Part 3, No. 57, Ch. 1.

[37] Marcelin, R. (1914), *Contribution à l'étude de la cinétique physico-chimique*, Gauthier-Villars.

[38] Marcelin, R. (1915), Annales de Physique **3**, 120–231.

[39] Gibbs, J. W. (1902), *Elementary Principles in Statistical Mechanics*, C. Scribner.

[40] Navarro, L. (1998), Arch. Hist. Exact Sci. **53**, 147.

[41] Duhem, P. (1911), *Traité d'énergétique*, 2 volumes, Gauthier-Villars.

[42] Armstrong, H. (1909), *Presidential Address to Section B – Chemistry*, British Association for the Advancement of Science, 420.

[43] Rutherford, E. (1911), Phil. Mag. Series VI **21**, 669.

[44] Maxwell, J. C. (1859), *On the Stability of the Motion of Saturn's Rings*, Adams Prize essay (1856), Macmillan & Co.

[45] Nicholson, J. W. (1911), Monthly Notices Roy. Astronom. Soc. **72**, 49.

[46] Nicholson, J. W. (1914), Monthly Notices Roy. Astronom. Soc. **74**, 486.

[47] Bohr, N. (1913), Phil. Mag. Series VI **26**, 1.

[48] Bohr, N. (1913), Nature **92**, 231.

[49] Bohr, N. (1913), Phil. Mag. Series VI **26**, 476.

[50] Bohr, N. (1913), Phil. Mag. Series VI **26**, 857.

[51] Bohr, N. (1922), Nobel Lecture – *The Structure of the Atom*, accessed at `www.nobelprize.org/nobel_prizes/physics/laureates/1922/Bohr-lecture.html`.

[52] Schwarzschild, K. (1916), Sitzungsber. Preuss. Akad. Wiss. **1**, 548.

[53] Heurlinger, T. (1919), Physikalische Z. **20**, 188.

[54] Thomson, J. J. (1913), Phil. Mag. Series VI **26**, 792.

[55] Pais, A. (1986), *Inward Bound. Of Matter and Forces in the Physical World*, Oxford University Press.

[56] Thomson, J. J. (1923), *The Electron in Chemistry*, The Franklin Institute.

[57] Noyes, A. N. (1917), J. Amer. Chem. Soc. **39**, 879.

[58] Abegg, R. (1904), Z. Anorg. Chem. **39**, 330.

[59] Lewis, G. N. (1916), J. Amer. Chem. Soc. **38**, 762.

[60] Kossel, W. (1916), Ann. der Physik **49**, 229.

[61] Lewis, G. N. (1917), Science **46**, 298.

[62] Sidgwick, N. V. (1923), Trans. Farad. Soc. **19**, 469.

[63] Pauli, W. (1922), Ann. der Physik **68**, 177.

[64] Nordheim, L. (1923), Z. Physik **19**, 69.

[65] Sommerfeld, A. (1923), *Atomic Structure and Spectral Lines*, translated from the Third German Edition by H. L. Brose, Methuen & Co. Ltd.

[66] Born, M. (1924), *The Mechanics of the Atom*, translated by J. W. Fisher (1927), revised by D. R. Hartree, G. Bell & Sons, Ltd.

[67] Heisenberg, W. (1925), Z. Physik **33**, 879.

[68] Sidgwick, N. V. (1927), *The Electronic Theory of Valency*, Oxford University Press.

[69] Armstrong, H. E. (1927), Nature, **120**, 478.

[70] Heitler, W. and London, F. (1927), Z. Physik **44**, 455.

[71] London, F. (1928), in *Probleme der modernen Physik*, edited by P. Debye, S. Hirsel.

[72] Gavroglu, K. (1995), *Fritz London: A Scientific Biography*, Cambridge University Press.

[73] Pauling, L. (1939), *The Nature of the Chemical Bond*, Cornell University Press.

[74] Mulliken, R. S. (1931), Chem. Rev. **9**, 347.

[75] Slater, J. C. (1932), Phys. Rev. **41**, 255.

[76] Longuet-Higgins, H. C. (1948), Proc. Phys. Soc. **60**, 270.

[77] Pauling, L. and Bright Wilson, E. (1935), *Introduction to Quantum Mechanics*, McGraw-Hill.

[78] Eyring, H., Walter, J. and Kimball, G. K. (1944), *Quantum Chemistry*, J. Wiley.

[79] Smith, S. J. and Sutcliffe, B. T. (1997), Ch. 5 in *Reviews in Computational Chemistry, Volume 10*, edited by Kenny B. Lipkowitz and Donald B. Boyd, VCH Publishers, Inc.

[80] (2003) *Handbook of Molecular Physics and Quantum Chemistry, 3 Volume Set*, edited by S. Wilson, P. F. Bernath and R. McWeeny, J. Wiley & Sons, Inc.

[81] Dirac, P. A. M. (1929), Proc. Roy. Soc. (London) A**123**, 714.

[82] Kato, T. (1951), Trans. Amer. Math. Soc. **70**, 212–18.

[83] Sutcliffe, B. T. and Woolley, R. G. (2013), *Advances in Quantum Methods and Applications in Chemistry, Physics and Biology*, Progress in Theoretical Chemistry and Physics, **27**, edited by M. Hotokka, E. J. Brändas, J. Maruani and M. Delgado-Barrio, 3.

[84] There is, of course, an infinite 'self-energy' in the Coulomb Hamiltonian which is simply ignored.

[85] Paoloni, L. (1979), *Towards a Culture-based Approach to Chemical Education in Secondary Schools: The Role of Chemical Formulae in the Teaching of Chemistry*, Eur. J. Sci. Education **1**, 365.

[86] Paoloni, L. (1981), *Reflections on Chemical Philosophy*, unpublished memoir.

2 Classical Electromagnetism

2.1 Introduction

Classical physics is the body of knowledge about physics created from the time of Newton in the seventeenth century until the end of the nineteenth century. The core of it is undoubtedly the mechanics of 'particles' created by Newton that is given mathematical form in his famous laws of motion. When the volume of a body is unimportant for the physics, its motion can be specified by giving the position of a point. A particle is characterised purely by its mass, M. As time passes, the point particle follows a path called its orbit under the influence of the forces that act on it. If there are no forces, the particle remains in a state of uniform (unaccelerated) motion and is said to be free. All motion is relative, but before Einstein time was self-evidently absolute. A key notion, due originally to Galileo, is that the laws of motion are the same in all unaccelerated ('inertial') reference frames; the coordinate transformations that relate any two such frames are called Galilean transformations. Collectively, they form an abstract symmetry group usually known as the Galilean group (see §2.3.1).

In the nineteenth century, the property of electric charge was recognised as another attribute of material bodies, and this led to the first researches into electrical and magnetic phenomena such as the characterisation of the force law associated with a system of charges at rest (Coulomb, Gauss), and forces associated with electrical conductors and magnets (Ampère, Biot and Savart, Ørsted). Electromagnetic theory was given a radically new direction by Faraday, who focused attention on the space surrounding electrical conductors and magnets; his lines of force are the first expression of the idea of the electromagnetic field as a primary entity that carries energy, momentum and force. Faraday, however, lacked the mathematical tools required to turn his ideas into a quantitative theory.

This step was taken by Maxwell who showed that the known electrical and magnetic phenomena, and the properties of light, were encompassed by his famous system of equations in which electric and magnetic fields appear as the dependent variables [1]. Maxwell's development of electrodynamics marks a decisive shift away from the Newtonian ideal of a dynamical theory of material points based on ordinary differential equations supplemented by initial conditions. The electromagnetic field is described by continuous functions of space and time that satisfy partial differential equations to which must be adjoined appropriate boundary conditions. These are the celebrated Maxwell equations with which we shall begin.

Since our main interest is to develop a quantum theory of processes involving atoms and molecules in electromagnetic fields, the goal of the next two chapters is to show how classical electrodynamics can be given an Hamiltonian formulation such that the Maxwell equations are just Hamilton's equations of motion. The standard procedure of canonical quantisation as developed by Dirac then gives a corresponding quantum theory for the field and its sources which are taken to form a closed dynamical system.

2.2 Maxwell's Theory

A charge density ρ and a current density $\mathbf{j}$ are sources of the electromagnetic field. The sources and the field in a given inertial frame are related by the Maxwell equations which in the usual vector calculus notation, originally due to Heaviside [2], [3], take the form

$$\mathbf{\nabla} \cdot \mathbf{B} = 0, \tag{2.1}$$

$$\mathbf{\nabla} \wedge \mathbf{E} + \frac{\partial \mathbf{B}}{\partial t} = 0, \tag{2.2}$$

$$\varepsilon_0 \mathbf{\nabla} \cdot \mathbf{E} = \rho, \tag{2.3}$$

$$\mathbf{\nabla} \wedge \mathbf{B} - \frac{1}{c^2} \frac{\partial \mathbf{E}}{\partial t} = \mu_0 \mathbf{j}; \tag{2.4}$$

here $\mathbf{E}$ is the electric field intensity, $\mathbf{B}$ is the magnetic induction and c is the speed of light. These equations hold at every point in space $(\mathbf{x})$ and time (t). With this understanding, $\mathbf{x}$ and t can be suppressed in the field equations to simplify the notation. We use S.I. units throughout so that $\mathbf{E}$ and $c\mathbf{B}$ have the same dimensions. Equation (2.1) is the statement that there are no free magnetic charges, while Eq. (2.2) describes electromagnetic induction (Faraday, Lenz). Coulomb's Law and Gauss's theorem in electrostatics are covered by (2.3) which is generally valid for space- and time-dependent charge density and electric field, while (2.4) gives Ampère's Law. The appearance of the displacement current, $c^{-2}\partial \mathbf{E}/\partial t$, in (2.4) is a crucial modification due to Maxwell himself; it is essential both for the description of light as an electromagnetic phenomenon, and for the expression of local conservation of electric charge.

The Maxwell equations involve two fundamental constants characterising the electrical permittivity (ε_0) and magnetic permeability (μ_0) of free space; they are related by the equation

$$\varepsilon_0 \mu_0 c^2 = 1. \tag{2.5}$$

In the electrodynamics of continuous media, ε_0 and μ_0 must be multiplied by the dielectric constant (ε) and permeability (μ), respectively, of the medium, and it is customary to replace $\mathbf{E}$ and $\mathbf{B}$ by modified fields $\mathbf{D}$ and $\mathbf{H}$. As we are aiming at a microscopic theory of charged particles interacting with the field, we shall continue with the notation in Eqs. (2.1)–(2.4).

Two important symmetry operations are space inversion, or parity, denoted by P and time reversal denoted by T; by considering the experimental arrangements that generate electric and magnetic fields, it is easily seen that the electric field intensity $\mathbf{E}$ changes sign under P and is invariant under T; $\mathbf{E}$ is a time-even polar vector. On the other hand, the magnetic induction $\mathbf{B}$ is unchanged by the parity operation P but changes sign under time reversal T; hence, the magnetic induction $\mathbf{B}$ is a time-odd axial vector. It then follows that the Maxwell equations (2.1)–(2.4) are invariant under the separate operations of P and T [4]. If Maxwell's equations are to be valid in all inertial reference frames, one must give up Galilean relativity for electromagnetism; this fact led Einstein to the remarkable conclusion that there is no absolute notion of simultaneity (as there is in Newton's mechanics), and thence to special relativity. The symmetry group of transformations between different inertial frames is called the Poincaré group; the Galilean group can be thought of as a limiting case of the Poincaré group appropriate for particles with speeds $v \ll c$ (cf. Chapter 3).

It is a classical result that the Maxwell equations for the free field imply conservation laws for 15 mechanical quantities because they admit a larger group of symmetries than those required for invariance under Lorentz transformations [5], [6]. The additional symmetries describe the conformal or angle-preserving transformations in space-time, which are generated by dilations, $x_\mu \rightarrow \lambda x_\mu$, and inversions, $x_\mu \rightarrow x_\mu/x^2$, where λ is a real scalar, and x^2 is given by (2.84). These are symmetries because the Maxwell equations provide no intrinsic length scales which only arise through interactions with charged particles. The Poincaré group essential for special relativity is a subgroup of the infinite dimensional Lie Group SO(4,2) that describes conformal symmetry.

When first derivatives of the fields are admitted, many more conserved currents can be defined [7], [8]. As an example, consider the following scalar quantity:

$$\xi = \frac{1}{2}\varepsilon_0\left(\mathbf{E}\cdot(\mathbf{\nabla}\wedge\mathbf{E}) + c^2\mathbf{B}\cdot(\mathbf{\nabla}\wedge\mathbf{B})\right). \tag{2.6}$$

Calculation of its time derivative yields

$$\frac{\partial\xi}{\partial t} = -\frac{1}{2}\varepsilon_0 c^2\mathbf{\nabla}\cdot\left(\mathbf{E}\wedge(\mathbf{\nabla}\wedge\mathbf{B}) - \mathbf{B}\wedge(\mathbf{\nabla}\wedge\mathbf{E})\right) \equiv -\mathbf{\nabla}\cdot\mathbf{J}^{(Z)}, \tag{2.7}$$

which shows that ξ is the spatial density of a conserved quantity. The flux of the conserved quantity is the vector field $\mathbf{J}^{(Z)}$; it is directed along the propagation direction of the electromagnetic wave. ξ and $\mathbf{J}^{(Z)}$ are components of a third-rank tensor quantity originally called the 'zilch' tensor [9]. A characteristic of these additional conserved quantities is that they are generally restricted to the free field since they do not survive interactions, and it was originally believed that the zilch tensor had no physical significance [10]. More recently, the quantity ξ has been proposed as a candidate [11], [12] for the 'optical chirality' of laser light beams that carry orbital angular momentum ('twisted light'), though ξ does not have the dimensions of a true angular momentum [13]. A connection between the zilch tensor and the Stokes parameter description of the polarisation properties of electromagnetic radiation was, however, recognised and will be described in Chapter 4 [9], [14]; the Stokes parameters can be quantised to give the Stokes operators, so this correspondence is also valid in the quantum theory of radiation (see Chapter 7).

Two important results are easily derived using standard vector identities. Since DivCurl vanishes, we obtain from (2.2),

$$\mathbf{\nabla} \cdot \frac{\partial \mathbf{B}}{\partial t} = \frac{\partial}{\partial t} (\mathbf{\nabla} \cdot \mathbf{B}) = 0. \tag{2.8}$$

Thus (2.1) is independent of time at every point in space. Equations (2.1) and (2.2) are evidently independent of the sources and are kinematical statements about the electromagnetic field. By a similar argument, we infer from (2.3) to (2.5) that

$$\frac{\partial \rho}{\partial t} + \mathbf{\nabla} \cdot \mathbf{j} = 0. \tag{2.9}$$

This important relation is the equation of continuity that expresses charge conservation at every point $(\mathbf{x}, t)$. This may be seen as follows: if we integrate the equation of continuity (2.9) over a volume Ω with fixed boundaries, we obtain

$$\int_{\Omega} \mathbf{\nabla} \cdot \mathbf{j} \, d^3\mathbf{x} = \int_{S} \mathbf{j} \cdot d^2\mathbf{S} = -\int_{\Omega} \frac{\partial \rho}{\partial t} \, d^3\mathbf{x} = -\frac{\partial}{\partial t} \int_{\Omega} \rho \, d^3\mathbf{x} = -\frac{\partial q_\Omega}{\partial t}, \tag{2.10}$$

that is, the flux of charge through the boundary surface S is precisely equal to the rate of change of charge inside Ω; no net charge is created or destroyed. This is the law of conservation of electric charge.

All ordinary matter on a macroscopic scale is found to remain electrically neutral at all temperatures to a very high degree of accuracy.[1] Its coupling to the electromagnetic field is characterised by a quantity we call electric charge. The laws of electrolysis established by Faraday when combined with the atomic theory of matter imply that electricity at the atomic level is discrete [16], [17]; in ordinary matter, charge is found in positive or negative integral multiples of a basic unit of charge ($e = 1.602 \times 10^{-19}$ Coulombs). Apart from movement of charge from one volume to another, creation and destruction of pairs of particles with opposite charges is allowed by the law (2.10). Particles with masses m_k moving slowly compared to the speed of light, however, obey Galilean kinematics and are subject to the law of conservation of mass:

$$M = \sum_k N_k m_k = \text{constant}. \tag{2.11}$$

Since, in general, there are no rational relations between the masses $\{m_k\}$, the constancy of M implies that the particle numbers $\{N_k\}$ are also constant. Hence, particle number (and with it charge) is absolutely conserved in Galilean invariant theories; this holds in both classical and quantum mechanics. Pair creation is a feature specifically of relativistic quantum mechanics.

As shown in Appendix B, a vector field $\mathbf{U}(\mathbf{x})$ can always be decomposed into orthogonal parts that are called the longitudinal and transverse components. Thus, for the

---

[1] This represents a precision of 1 part in 10^{10} for $v/c \sim 10^{-3}$ [15].

electric field intensity we can write

$$\mathbf{E} = \mathbf{E}^{\parallel} + \mathbf{E}^{\perp}, \quad \mathbf{E}^{\perp} \cdot \mathbf{E}^{\parallel} = 0, \tag{2.12}$$

where the orthogonal components satisfy

$$\nabla \cdot \mathbf{E}^{\perp} = 0, \qquad \text{transverse,} \tag{2.13}$$

$$\nabla \wedge \mathbf{E}^{\parallel} = 0 \qquad \text{longitudinal.} \tag{2.14}$$

The magnetic induction is always transverse (2.1). From (2.3), (2.13) and (2.14) we see that the charge density ρ is responsible for the longitudinal component of the electric field intensity.

Maxwell's system of coupled first-order equations may be transformed directly into a pair of uncoupled inhomogeneous second-order equations,

$$\left(\nabla^2 - \frac{1}{c^2} \frac{\partial^2}{\partial t^2} \right) \mathbf{Y} = \mathbf{C}_\mathbf{Y}(\mathbf{j}, \rho), \tag{2.15}$$

with $\mathbf{Y} = \mathbf{E}$ or $\mathbf{B}$. The source terms $\mathbf{C}_\mathbf{Y}$ are

$$\mathbf{C}_\mathbf{E} = \left(\mu_0 \frac{\partial \mathbf{j}}{\partial t} + \frac{1}{\varepsilon_0} \nabla \rho \right), \quad \mathbf{C}_\mathbf{B} = - \mu_0 \nabla \wedge \mathbf{j}. \tag{2.16}$$

The pairs $(\rho, \mathbf{j})$ and $(\mathbf{E}, \mathbf{B})$ are required to be continuous functions of space and time with continuous derivatives such that space and time differentiation is commutative for them.

It follows that in a region of space where there are no charges or currents the electromagnetic fields satisfy a *wave equation*,

$$\left(\nabla^2 - \frac{1}{c^2} \frac{\partial^2}{\partial t^2} \right) \mathbf{Y} = \mathbf{0}, \quad \mathbf{Y} = \mathbf{E}, \mathbf{B}, \tag{2.17}$$

from which Maxwell's constant c can be identified as the speed of the wave motion. Electromagnetic waves are purely transverse since (cf. (2.1), (2.3))

$$\nabla \cdot \mathbf{B} = \nabla \cdot \mathbf{E} = 0. \tag{2.18}$$

These equations define the free field.

A *plane wave* has the special property of depending on the time t and one spatial coordinate, say z, which we call the propagation direction. The spatial derivatives in the plane transverse to the direction of travel vanish, and the wave equation (2.17) simplifies to

$$\frac{\partial^2 Y}{\partial z^2} - \frac{1}{c^2} \frac{\partial^2 Y}{\partial t^2} = 0, \tag{2.19}$$

where Y is any component of $\mathbf{E}$ and $\mathbf{B}$. This equation has solutions of the form

$$Y = Y_1(z - ct) + Y_2(z + ct). \tag{2.20}$$

Elementary calculation shows that the functions Y_1 and Y_2 are trigonometric functions, so that, for example,

$$Y = Y^{(0)} \cos(kz - \omega t), \tag{2.21}$$

where $Y^{(0)}$ is an integration constant and $\omega = kc$ is a solution. The displacement $\mathbf{Y}$ of a travelling-plane electromagnetic wave moving in an arbitrary direction at the point $(\mathbf{x}, t)$ can be represented as

$$\mathbf{Y}(\mathbf{x}, t) = \mathbf{Y}^{(0)} e^{i(\mathbf{k} \cdot \mathbf{x} - \omega t)}, \tag{2.22}$$

where $\mathbf{Y}^0$ is an arbitrary complex-valued constant vector and the wave vector $\mathbf{k}$ has magnitude ω/c.

Putting the solution (2.22) into Maxwell's equations for the free field (2.2), (2.4) yields the important relations for plane waves:

$$\mathbf{E} = -c\,\hat{\mathbf{k}} \wedge \mathbf{B}, \quad \mathbf{B} = \frac{1}{c}\,\hat{\mathbf{k}} \wedge \mathbf{E}. \tag{2.23}$$

Thus, the fields are orthogonal to each other and to $\mathbf{k}$ in agreement with (2.18). For each $\mathbf{k}$ we introduce two unit vectors in the plane at right angles to $\mathbf{k}$, that is, tangent to the sphere of fixed $|\mathbf{k}|$, for example

$$\hat{\boldsymbol{\varepsilon}}(\mathbf{k})_1 = \frac{(k_2, -k_1, 0)}{\sqrt{k_1^2 + k_2^2}}, \tag{2.24}$$

$$\hat{\boldsymbol{\varepsilon}}(\mathbf{k})_1 \wedge \hat{\boldsymbol{\varepsilon}}(\mathbf{k})_2 = \hat{\mathbf{n}}, \tag{2.25}$$

where $\hat{\mathbf{n}} = \mathbf{k}/|\mathbf{k}|$ is directed along the direction labelled 3. We call the $\hat{\boldsymbol{\varepsilon}}(\mathbf{k})_i$, $i = 1, 2$ the *polarisation* vectors for the mode $\mathbf{k}$. It is not possible to choose a continuous basis of polarisation vectors for every possible value of $\mathbf{k}$, essentially because of the topological fact captured by Poincaré's 'can't comb the hairs on a sphere' theorem [18], but this is usually of no consequence since the arbitrary polarisation vectors are unobservable. An important property is their sum rule [19], [20],

$$\sum_{\lambda=1,2} \hat{\varepsilon}(\mathbf{k})_{\lambda i} \hat{\varepsilon}(\mathbf{k})_{\lambda j} = \delta_{ij} - \frac{k_i k_j}{k^2} \equiv \mathcal{B}(\mathbf{k})_{ij}, \tag{2.26}$$

which is a function that is discontinuous only at $k = 0$.

At any instant in time, the electric field vector of a plane wave travelling in the z-direction has components given by

$$
\begin{aligned}
E_x &= E_1^{(0)} \cos(kz - \omega t) \\
E_y &= E_2^{(0)} \cos(kz - \omega t + \alpha) \\
E_z &= 0,
\end{aligned}
\tag{2.27}
$$

and the three parameters $\{E_1^{(0)}, E_2^{(0)}, \alpha\}$ can be used to characterise the polarisation of the wave. For example, at any point along a circularly polarised wave, the electric field $\mathbf{E}$ maintains a fixed magnitude but has a direction that rotates in space with a constant angular frequency. Hence, the tip of the $\mathbf{E}$-vector traces out a circle, and the components (E_x, E_y) of $\mathbf{E}$ oscillate with the same amplitude, $E_1^{(0)} = E_2^{(0)}$, but with a phase difference of $\pm\pi/2$. If the tip of the $\mathbf{E}$-vector rotates in a clockwise sense when viewed by an observer receiving the wave, it is said to be right-circularly polarised; if anticlockwise, it is left-circularly polarised. If $E_1^{(0)} = E_2^{(0)}$ and the phase difference is zero, the resultant is linear polarisation. Otherwise, if $E_1^{(0)} \neq E_2^{(0)}$, the polarisation is elliptical.

2.2.1 The Mechanical Properties of the Electromagnetic Field

We now consider the mechanical properties of the electromagnetic field. The electromagnetic energy in unit volume – the energy density – is defined to be

$$E = \frac{1}{2}\varepsilon_0 \left(\mathbf{E} \cdot \mathbf{E} + c^2 \mathbf{B} \cdot \mathbf{B}\right), \tag{2.28}$$

and is such that its volume integral is the energy content in the specified volume,

$$\mathcal{E} = \int_\Omega E \, \mathrm{d}^3 \mathbf{x}. \tag{2.29}$$

Consider a time-dependent electromagnetic field; the rate of change of the energy density is

$$\frac{\partial E}{\partial t} = \varepsilon_0 \left(\mathbf{E} \cdot \frac{\partial \mathbf{E}}{\partial t} + c^2 \, \mathbf{B} \cdot \frac{\partial \mathbf{B}}{\partial t} \right). \tag{2.30}$$

The time derivatives in this equation may be expressed in terms of the space derivatives of the fields, and the sources, using the Maxwell equations (2.2) and (2.4), and with the aid of an obvious vector identity there results Poynting's theorem,

$$\frac{\partial E}{\partial t} + \boldsymbol{\nabla} \cdot \mathbf{S} = -\mathbf{j} \cdot \mathbf{E}, \tag{2.31}$$

where we have defined the Poynting vector $\mathbf{S}$ for the field by

$$\mathbf{S} = \varepsilon_0 c^2 \, (\mathbf{E} \wedge \mathbf{B}). \tag{2.32}$$

Only the real part of the electromagnetic field is required for classical electromagnetism; the real part can always be trivially separated out after linear operations. Moreover, physical quantities determined from quadratic combinations of the field vectors are evaluated as time averages over a cycle for which the cycle average theorem holds, provided the time dependence is harmonic [20], [21]:

$$\overline{(\Re \, \mathbf{X})(\Re \mathbf{Y})} = \tfrac{1}{2} \, \Re(\mathbf{XY}^*). \tag{2.33}$$

As an example, the intensity of a plane wave is the time average of the magnitude of the Poynting vector, $|\mathbf{S}|$, over one cycle; for the plane wave, since the wave is harmonic, this is simply $\frac{1}{2}|\mathbf{S}|$, where

$$|\mathbf{S}| = \varepsilon_0 c^2 |\mathbf{E} \wedge \mathbf{B}| = \varepsilon_0 c |\mathbf{E} \wedge \hat{\mathbf{k}} \wedge \mathbf{E}| = \varepsilon_0 c |\mathbf{E}^{(0)}|^2. \tag{2.34}$$

The fields $\mathbf{E}$ and $\mathbf{B}$ contribute equally to the average energy of the plane wave which is $\frac{1}{2}c|\mathbf{S}|$, that is, c times the intensity. $\mathbf{S}$ is along the direction of $\mathbf{k}$.

Similarly, the flux $\mathbf{J}^{(Z)}$, (2.7), obtained from the zilch tensor, is also collinear with $\mathbf{k}$; an elementary calculation using (2.27) shows that

$$\mathbf{J}^{(Z)} = \omega E_1^{(0)} E_2^{(0)} \sin(\alpha) \, \hat{\mathbf{k}}. \tag{2.35}$$

Thus, $\mathbf{J}^{(Z)}$ vanishes for a linearly polarised wave ($\alpha = 0$), is directed along $\mathbf{k}$ for a right-circularly polarised wave ($\alpha = \pi/2$), and is of equal magnitude but directed in the opposite direction to $\mathbf{k}$ for a left-circularly polarised wave ($\alpha = -\pi/2$) [22].

Equation (2.31) may be compared with (2.9) and (2.10), and we make a similar interpretation; $\mathbf{S}$ may be taken to be the rate at which electromagnetic energy crosses unit area normal to $\mathbf{S}$. This flux of electromagnetic energy is called electromagnetic radiation.[2] In the absence of a current density in the region of space under consideration, we would have zero on the RHS of (2.31), which would appear as an equation of continuity expressing the conservation of energy for the field. If there is a current density $\mathbf{j}$, the law of conservation of energy is maintained by interpreting the term $-\mathbf{j} \cdot \mathbf{E}$ as the rate at which $\mathbf{j}$ does work on the electromagnetic field, thereby increasing its energy content.

The electromagnetic field momentum density, $\mathbf{u}$, is defined as

$$\mathbf{u} = \frac{\mathbf{S}}{c^2} \equiv \varepsilon_0 \left(\mathbf{E} \wedge \mathbf{B} \right). \tag{2.36}$$

Evidently, it is a vector perpendicular to the plane containing $\mathbf{E}$ and $\mathbf{B}$; its direction defines the propagation direction for electromagnetic radiation. It was known already to Maxwell that the linear momentum of the electromagnetic field can manifest itself physically in the form of radiation pressure on a macroscopic body. If we again make use of the Maxwell equations, we can calculate the rate of change of the field momentum, $\partial \mathbf{u}/\partial t$, using (2.36) and put the result in the symmetrical form,

$$-\frac{\partial \mathbf{u}}{\partial t} = -\varepsilon_0 \left(\mathbf{E}\boldsymbol{\nabla} \cdot \mathbf{E} + c^2 \mathbf{B}\boldsymbol{\nabla} \cdot \mathbf{B} \right)$$
$$+ \left(\mathbf{E} \wedge (\boldsymbol{\nabla} \wedge \mathbf{E}) + c^2 \mathbf{B} \wedge (\boldsymbol{\nabla} \wedge \mathbf{B}) \right) + \mathbf{F}(\rho, \mathbf{j}), \tag{2.37}$$

where $\mathbf{F}$ is the Lorentz force density:

$$\mathbf{F} = \rho \mathbf{E} + \mathbf{j} \wedge \mathbf{B}. \tag{2.38}$$

In the field we can either speak of the flow of momentum, or in terms of stress; however, unlike a material medium, the system of stresses due to the electromagnetic field is that acting through planes fixed with respect to the coordinate axes of the reference frame.

The angular momentum carried by the field can be constructed directly from the linear momentum density $\mathbf{u}$. The angular momentum in a volume element, $d^3\mathbf{x}$, of the field at position $\mathbf{x}$ is

$$d\mathbf{J} = (\mathbf{x} \wedge \mathbf{u}) \, d^3\mathbf{x}, \tag{2.39}$$

so that in a finite volume,

$$\mathbf{J} = \int_\Omega (\mathbf{x} \wedge \mathbf{u}) \, d^3\mathbf{x} = \frac{1}{c^2} \int_\Omega (\mathbf{x} \wedge \mathbf{S}) \, d^3\mathbf{x}. \tag{2.40}$$

It was suggested by Poynting that the circular polarisation of light could be understood in terms of angular momentum, though this is not evident from (2.40) [23]. Nevertheless, when a circularly polarised light beam passes through a thin quartz plate, the plate

---

[2] The symbol $\mathbf{S}$ comes from the German word for radiation, Strahlung.

is observed to rotate slightly due to the transfer to it of angular momentum from the field [24].

2.3 The Scalar and Vector Potential

The components of the fields $(\mathbf{E}, \mathbf{B})$ that arise as solutions of the Maxwell equations (2.1)–(2.4) are not all independent quantities; for example, in the special case of the free field (no charges) only two of the six components are independent. This functional redundancy can be reduced by introducing as auxiliary quantities the scalar and vector potentials, ϕ and $\mathbf{a}$, respectively. If we set

$$\mathbf{B} = \boldsymbol{\nabla} \wedge \mathbf{a}, \tag{2.41}$$

$$\mathbf{E} = -\frac{\partial \mathbf{a}}{\partial t} - \boldsymbol{\nabla}\phi, \tag{2.42}$$

we see that Eqs. (2.1) and (2.2) are satisfied whatever ϕ and $\mathbf{a}$ are, and that the second pair of Maxwell equations (2.3), (2.4) yield

$$\frac{\partial}{\partial t}\left(\boldsymbol{\nabla}\cdot\mathbf{a}\right) + \boldsymbol{\nabla}^2\phi = -\frac{\rho}{\varepsilon_0}, \tag{2.43}$$

$$\boldsymbol{\nabla}^2\mathbf{a} - \boldsymbol{\nabla}\left(\boldsymbol{\nabla}\cdot\mathbf{a}\right) - \frac{1}{c^2}\left(\frac{\partial^2\mathbf{a}}{\partial t^2} + \boldsymbol{\nabla}\frac{\partial\phi}{\partial t}\right) = -\mu_0\,\mathbf{j}. \tag{2.44}$$

A pair $(\phi, \mathbf{a})$ satisfying (2.43), (2.44) is by no means fully determined; we can introduce new potentials $(V, \mathbf{A})$ by setting

$$\mathbf{a} \rightarrow \mathbf{A} = \mathbf{a} - \boldsymbol{\nabla}\chi, \tag{2.45}$$

$$\phi \rightarrow V = \phi + \frac{\partial\chi}{\partial t}, \tag{2.46}$$

where χ is an arbitrary differentiable scalar field, and by (2.41), (2.42) these yield the same electric field intensity $\mathbf{E}$ and magnetic induction $\mathbf{B}$, as the original choice $(\phi, \mathbf{a})$. The transformation from $(\phi, \mathbf{a})$ to $(V, \mathbf{A})$ specified by (2.45), (2.46) is usually called a gauge transformation. The arbitrariness in the potentials $(\phi, \mathbf{a})$ resides in the quantity χ which can be assigned any value; choosing a value for χ is referred to as fixing the gauge of the potentials.

For some applications, it is convenient to express the potentials in terms of the field strengths through integration of Eqs. (2.41) and (2.42). Static, spatially uniform fields $\mathbf{E}$ and $\mathbf{B}$ constitute a practically important special case; for such fields, Eqs. (2.41) and (2.42) can be integrated directly to yield a family of potentials by writing them out in component form and doing partial integrations as an obvious extension of the method given for constructing solenoidal vector fields [25]. The Maxwell equations (2.1) and (2.2) must be imposed as a set of subsidiary conditions. The results of these integrations are the potentials in the static symmetric gauge [26],

$$\mathbf{a} = \frac{1}{2}\,\mathbf{B}\wedge\mathbf{x}, \quad \phi = -\,\mathbf{E}\cdot\mathbf{x}, \tag{2.47}$$

together with a gauge function,

$$\chi = \mathbf{x} \wedge \mathbf{c}(t) \cdot \mathbf{B}, \tag{2.48}$$

where $\mathbf{c}(t)$ arises as an arbitrary integration constant.[3]

Notice that although in this case the physical electromagnetic fields are static, the potentials are in general time-dependent; the possibility of gauge transformation implies that any set of potentials is as valid as any other, and obviously physical arguments, derived from the field strengths, cannot be transferred to the potentials. In classical electrodynamics the force on the charges is the Lorentz force, derived from (2.38), which is determined by the fields $(\mathbf{E}, \mathbf{B})$. The conditions under which the Lorentz force is a conservative force are well known; for example, $\mathbf{E}$ is conservative if its Curl vanishes and, by Maxwell's equations, that requires any magnetic field to be stationary. Thus, the conditions for the energy in a classical electromagnetic field to be conserved are expressed in terms of the physical fields $(\mathbf{E}, \mathbf{B})$, and the space-time behaviour of the potentials is irrelevant. As we will see, quantum mechanics works with energy and momentum rather than force and is based on a Hamiltonian (or possibly Lagrangian) formulation which requires the use of the (arbitrary) potentials as working variables. It may well be convenient to choose the static potentials (2.47) by putting $\mathbf{c}(t) = 0$; it is then important to verify that calculated quantities are independent of this particular choice.

Another approach is to make use of integration by parts to express a particular choice for the potentials as integrals over space that satisfy (2.41) and (2.42) identically [27],

$$\mathbf{A}(\mathbf{x},t) = \int \left(\boldsymbol{\nabla}_{\mathbf{x}'} \wedge \mathbf{B}(\mathbf{x}',t) \right) \frac{1}{4\pi |\mathbf{x}' - \mathbf{x}|} \, d^3\mathbf{x}', \tag{2.49}$$

$$V(\mathbf{x},t) = \int \left(\boldsymbol{\nabla}_{\mathbf{x}'} \cdot \mathbf{E}(\mathbf{x}',t) \right) \frac{1}{4\pi |\mathbf{x}' - \mathbf{x}|} \, d^3\mathbf{x}', \tag{2.50}$$

$$= \int \frac{\rho(\mathbf{x}',t)}{4\pi \varepsilon_0 |\mathbf{x}' - \mathbf{x}|} \, d^3\mathbf{x}', \tag{2.51}$$

together with the gauge function,

$$\chi(\mathbf{x},t) = -\int \sigma(\mathbf{x}') \frac{1}{4\pi |\mathbf{x}' - \mathbf{x}|} \, d^3\mathbf{x}', \tag{2.52}$$

with an arbitrary choice $\sigma = \boldsymbol{\nabla} \cdot \mathbf{a}$ for the divergence of the vector potential. Then, from (2.52), we have that χ is a solution of the Poisson equation with an arbitrary source term,

$$\nabla^2 \chi = \sigma. \tag{2.53}$$

---

[3] $\mathbf{c}(t)$ can be allowed to depend on $\mathbf{x}$ as well. The only restriction is that the scalar field $\chi(\mathbf{x},t)$ must belong to the class of functions for which the order of differentiation with respect to x, y, z, t is immaterial, that is, $\chi(\mathbf{x},t)$ must be integrable.

This arbitrariness can be removed by making a definite choice for σ. A particularly important choice of gauge leads to the Coulomb gauge characterised by the condition,

$$\sigma = \nabla \cdot \mathbf{A} = 0, \tag{2.54}$$

so that $\chi = 0$. In the particular case of the free field ($\rho = 0$) we also have $V = 0$, from (2.50), and so the field strengths in this case can be described by the potentials

$$\phi = 0, \tag{2.55}$$

$$\mathbf{a} = \mathbf{A}. \tag{2.56}$$

Thus, the free field is described completely by the transverse vector field $\mathbf{A}$, that is, with two degrees of freedom that correspond to the two independent polarisation states of the radiation. We shall always use the symbol $\mathbf{A}$ for the Coulomb gauge vector potential.

This freedom in the choice of potentials can be used to simplify Eqs. (2.43), (2.44). For example, using (2.54), Eq. (2.43) is reduced to Poisson's equation for the scalar potential,

$$\nabla^2 V = -\frac{\rho}{\varepsilon_0}. \tag{2.57}$$

If the current is split into its longitudinal and transverse parts,

$$\mathbf{j} = \mathbf{j}^{\parallel} + \mathbf{j}^{\perp}, \tag{2.58}$$

(2.44) separates into two uncoupled equations and

$$\frac{1}{c^2} \frac{\partial^2 \mathbf{A}}{\partial t^2} - \nabla^2 \mathbf{A} = \mu_0 \mathbf{j}^{\perp}, \tag{2.59}$$

$$\varepsilon_0 \frac{\partial}{\partial t} \nabla V = \mathbf{j}^{\parallel}. \tag{2.60}$$

Equations (2.57) and (2.60) are equivalent to the equation of continuity (2.9). In this way, the static Coulomb field of the charges is separated from the radiation field which is described purely in terms of its two physical transverse components.

Another widely used gauge is the Lorentz gauge defined by the condition

$$\nabla \cdot \mathbf{a} + \frac{1}{c^2} \frac{\partial \phi}{\partial t} = 0. \tag{2.61}$$

When (2.61) is combined with (2.43) and (2.44), it leads to wave equations for both ϕ and $\mathbf{a}$,

$$\nabla^2 \phi - \frac{1}{c^2} \frac{\partial^2 \phi}{\partial t^2} = -\frac{\rho}{\varepsilon_0}, \tag{2.62}$$

$$\nabla^2 \mathbf{a} - \frac{1}{c^2} \frac{\partial^2 \mathbf{a}}{\partial t^2} = -\mu_0 \mathbf{j}. \tag{2.63}$$

Just as in the Coulomb gauge theory, the solutions of the inhomogeneous wave equations (2.62)–(2.63) can be written down in terms of known Green's functions, and then (2.41) and (2.42) can be used to compute the physical fields. We do not enter into details, however, as our main interest is a formulation of classical electrodynamics in

Hamiltonian form in preparation for quantisation so that we can discuss the interaction of the electromagnetic field with atoms and molecules using the Schrödinger equation. We shall see that the scalar potential plays no role as a dynamical variable in the Hamiltonian scheme because it has no conjugate momentum, and it is therefore eliminated; the remaining gauge transformations then reside purely with the vector potential, and we now study them in more detail.

It is convenient to make a refinement of (2.52); we consider an instant in time and suppress the time variable t to simplify the notation. The Green's function $\mathbf{g}(\mathbf{x}, \mathbf{x}')$ for the divergence operator satisfies [28]

$$\nabla_\mathbf{x} \cdot \mathbf{g}(\mathbf{x}, \mathbf{x}') = -\delta^3(\mathbf{x} - \mathbf{x}'), \tag{2.64}$$

in terms of Dirac's delta function. The longitudinal component of $\mathbf{g}$ is well defined; it can be written as

$$\mathbf{g}(\mathbf{x}, \mathbf{x}')^\| = \nabla_\mathbf{x} \frac{1}{4\pi|\mathbf{x} - \mathbf{x}'|}. \tag{2.65}$$

Although the differential equation (2.64) is consistent with solutions involving $(\mathbf{x} - \mathbf{x}')$, these are only a subset of the solution set and we may treat $\mathbf{x}$ and $\mathbf{x}'$ as independent variables. In applications we require $\mathbf{g}(\mathbf{x}, \mathbf{x}')$ to have a Fourier transform in the variable $\mathbf{x}$; from (2.65) we have that

$$\mathbf{g}(\mathbf{p}, \mathbf{x}')^\| = -\frac{i\mathbf{p}}{p^2} e^{i\mathbf{p} \cdot \mathbf{x}'}. \tag{2.66}$$

The longitudinal component of $\mathbf{g}(\mathbf{p}, \mathbf{x}')$ is directed along $\mathbf{p}$, so we introduce an orthogonal triad of unit vectors $\{\hat{\boldsymbol{\varepsilon}}_i, i = 1, 2, 3\}$ with $\hat{\boldsymbol{\varepsilon}}_3 = \mathbf{p}/p$; then the transverse component lies in the plane containing $\{\hat{\boldsymbol{\varepsilon}}_1, \hat{\boldsymbol{\varepsilon}}_2\}$,

$$\mathbf{g}(\mathbf{p}, \mathbf{x}')^\perp = \hat{\boldsymbol{\varepsilon}}_1 f(\mathbf{p}, \mathbf{x}') + \hat{\boldsymbol{\varepsilon}}_2 h(\mathbf{p}, \mathbf{x}'), \tag{2.67}$$

in terms of two scalar fields f and h. Since the transverse component satisfies

$$p\, \hat{\boldsymbol{\varepsilon}}_3 \cdot \mathbf{g}(\mathbf{p}, \mathbf{x}')^\perp = 0, \tag{2.68}$$

$f(\mathbf{p}, \mathbf{x}')$ and $h(\mathbf{p}, \mathbf{x}')$ can behave as p^ν, $\nu \geq -1$ near $p = 0$, but otherwise their properties are governed only by the requirement that their Fourier transforms (on $\mathbf{p}$) exist. The transverse component of the Green's function $\mathbf{g}(\mathbf{x}, \mathbf{x}')$ is thus essentially arbitrary.

A choice for $\mathbf{g}$ that turns out to be important in the Hamiltonian formulation of electrodynamics[4] is based on direct integration of (2.64). This is the path-dependent representation

$$\mathbf{g}(\mathbf{x}, \mathbf{x}'; \mathbf{O}, C) = \mathbf{g}(\mathbf{x}, \mathbf{O})^\| + \int_C^{\mathbf{x}'} \delta^3(\mathbf{z} - \mathbf{x}) \, d\mathbf{z}, \tag{2.69}$$

which has a transverse component as well as the longitudinal component (2.65); $\mathbf{g}(\mathbf{x}, \mathbf{O})^\|$ is an integration constant.[5] C is any suitable curve starting from a fixed point $\mathbf{O}$ and ending at the space-point $\mathbf{x}'$ such that the integral exists; the paths C may be

---

[4] This is true in both classical *and* quantum versions.

[5] This is most easily confirmed by evaluating the scalar product of $\hat{\boldsymbol{\varepsilon}}_3$ with the Fourier transform of (2.69).

finite or infinite. To see this, let $s(\mathbf{x})$ be a smooth function vanishing at infinity; then [29]

$$\int \boldsymbol{\nabla}_{\mathbf{x}} \cdot \mathbf{g}(\mathbf{x}, \mathbf{x}') s(\mathbf{x}) \, d^3 \mathbf{x} = \int \left(\nabla^2 \frac{1}{4\pi |\mathbf{x} - \mathbf{O}|} \right) s(\mathbf{x}) \, d^3 \, \mathbf{x}$$

$$- \int \int_{\mathbf{O}}^{\mathbf{x}'} \delta^3(\mathbf{z} - \mathbf{x}) \boldsymbol{\nabla} s(\mathbf{x}) \cdot d\mathbf{z} \, d^3 \mathbf{x} \qquad (2.70)$$

after an integration by parts.[6] The term $s(\mathbf{O})$ coming from the lower limit of the line integral is cancelled out since

$$\nabla^2 \frac{1}{4\pi |\mathbf{x} - \mathbf{O}|} = - \delta^3(\mathbf{x} - \mathbf{O}) \qquad (2.71)$$

and the RHS of (2.70) reduces to

$$- s(\mathbf{x}') = - \int \delta^3(\mathbf{x} - \mathbf{x}') s(\mathbf{x}) \, d^3 \mathbf{x}; \qquad (2.72)$$

thus (2.64) is recovered.

Introducing this Green's function notation, (2.52) after an integration by parts is just

$$\chi(\mathbf{x}) = \int \mathbf{a}(\mathbf{x}') \cdot \mathbf{g}(\mathbf{x}', \, \mathbf{x})^{\|} \, d^3 \mathbf{x}', \qquad (2.73)$$

to which only the longitudinal part of $\mathbf{a}$ contributes. Any linear functional of the vector potential such as (2.73) is called a gauge condition. A general gauge condition can be expressed by

$$\int \mathbf{a}(\mathbf{x}') . \mathbf{g}(\mathbf{x}', \mathbf{x}) \, d^3 \mathbf{x}' = 0, \qquad (2.74)$$

which, after separation into longitudinal and transverse parts, allows us to write the gauge function (2.52) in an equivalent form using the Coulomb gauge vector potential $\mathbf{A}$:

$$\chi(\mathbf{x}) = - \int \mathbf{A}(\mathbf{x}') . \mathbf{g}(\mathbf{x}', \mathbf{x})^{\perp} \, d^3 \, \mathbf{x}'. \qquad (2.75)$$

In this form, $\mathbf{g}^{\perp}$ can be replaced by the complete Green's function $\mathbf{g}$, because $\mathbf{A}$ is transverse. For each of the infinitely many possible choices of $\mathbf{g}(\mathbf{x}, \mathbf{x}')$, there is a vector potential that satisfies (2.74), namely

$$\mathbf{a}(\mathbf{x}') = \mathbf{A}(\mathbf{x}') - \boldsymbol{\nabla}_{\mathbf{x}'} \int \mathbf{A}(\mathbf{x}'') \cdot \mathbf{g}(\mathbf{x}'', \mathbf{x}') \, d^3 \mathbf{x}''. \qquad (2.76)$$

There is no connection between any given physical situation (i.e. experimental setup) and a specific form for the Green's function $\mathbf{g}$.

---

[6] The boundary term vanishes at infinity because $s(\mathbf{x})$ does.

We give two examples[7] of the use of (2.74). Firstly, we can define a gauge by declaring that we will work with $\mathbf{g}(\mathbf{x}, \mathbf{x}')^{\perp} = 0$. From (2.75) we see that this is equivalent to choosing $\chi = 0$, so we are working with the Coulomb gauge vector potential $\mathbf{A}$ that satisfies (2.54), and a scalar potential, V, that is determined purely by the charge density, (2.50). As an example of the line integral form, suppose we specify the straight-line path $\mathbf{z}$ between two spatial points, $\mathbf{x}$ and $\mathbf{x}'$, that is, $\mathbf{z} = \mathbf{x}' - \sigma \mathbf{D}$, $\mathbf{D} = \mathbf{x}' - \mathbf{x}$, $0 \leq \sigma \leq 1$ in $\mathbf{g}(\mathbf{x}' : \mathbf{x})$, (2.69); then evaluation of the gradient in (2.76) yields

$$\mathbf{a}(\mathbf{x}') = -\int_0^1 (1 - \sigma)\mathbf{D} \wedge \mathbf{B}(\mathbf{z}(\sigma))\,\mathrm{d}\sigma. \tag{2.77}$$

Direct computation shows that $\mathbf{a}(\mathbf{x}')$ is indeed a vector potential since it satisfies $\nabla_{\mathbf{x}'} \wedge \mathbf{a}(\mathbf{x}') = \mathbf{B}(\mathbf{x}')$ (see Appendix E). The gauge condition (2.74) can then be put in the simple form

$$\mathbf{D} \cdot \mathbf{a}(\mathbf{x}') = 0. \tag{2.78}$$

Thus, the vector potential at the position $\mathbf{x}'$ is such that its component along the straight line connecting $\mathbf{x}'$ to a fixed point $\mathbf{x}$ vanishes [30], by (2.78). This vector potential is often referred to in the literature as being in the 'Poincaré' gauge [31], [32].

In all calculations involving the Green's function $\mathbf{g}(\mathbf{x}, \mathbf{x}')$, we assume that the spatial paths $\{C\}$ are such that the order of integration is immaterial in expressions of the form

$$\int \mathbf{U}(\mathbf{x}, t)\,\mathrm{d}^3\mathbf{x} \int_C^{\mathbf{x}'} \delta^3(\mathbf{z} - \mathbf{x})\,\mathrm{d}\mathbf{z}, \tag{2.79}$$

as we usually integrate over the $\mathbf{x}$ variable in the Dirac δ before evaluating the line integral. This Green's function when combined with (2.74) leads to a gauge condition involving a line integral over the vector potential,

$$\int_C^{\mathbf{x}} \mathbf{a}(\mathbf{z}) \cdot \mathrm{d}\mathbf{z} = -\int \mathbf{g}(\mathbf{x}', \mathbf{O})^{\|} \cdot \mathbf{a}(\mathbf{x}')\,\mathrm{d}^3\mathbf{x}'. \tag{2.80}$$

Such an integral is independent of C, and so is determined purely by its endpoints, only if $\nabla \wedge \mathbf{a} = 0$, that is, if $\mathbf{a}$ is purely longitudinal (a gradient of an integrable function); conversely, for a non-zero magnetic field, $\mathbf{B} = \nabla \wedge \mathbf{a}$, the LHS of (2.80) contains a path–dependent contribution involving the transverse component of the vector potential.

There are two ideas that we have used in the foregoing discussion which are worth emphasising. Firstly, the choice of a gauge condition is often a very convenient means of simplifying a calculation involving the electromagnetic field. Secondly, for a quantity to be a candidate for a physical observable it must be independent of any such choice of gauge; this is the principle of gauge invariance. We shall return to the general form of gauge condition (2.74) and the vector potential that satisfies it, (2.76), in §3.4 when we discuss the classical Hamiltonian for electrodynamics; its use has the considerable advantage that it is not necessary to specify a particular gauge condition at the outset.

---

[7] There has been endless discussion over decades, which continues today, in non-relativistic quantum electrodynamics as to the relative merits of just these two gauge conditions, even though there are infinitely many gauge conditions not brought to the 'beauty contest'!

2.3.1 Electromagnetism and Relativity

In Newtonian physics, time is absolute and the transformation to a reference frame moving with relative velocity $\mathbf{v}$ is given by the Galileo transformation or 'boost':

$$\mathbf{x}' = \mathbf{x} - \mathbf{v}t, \quad t' = t. \tag{2.81}$$

Galilean boosts together with space and time translations, and space rotations define the group of symmetry transformations of classical mechanics (the 10-dimensional Galilean group, G_∞). The fields $\mathbf{E}$ and $\mathbf{B}$ are not invariant under such a Galileo transformation. Neither are the Maxwell equations (2.1)–(2.4) since they imply that the speed of light, c, is the same in all unaccelerated reference frames, and this is clearly inconsistent with the Galileo transformation.

Poincaré discovered that the symmetry group of transformations which leave the Maxwell equations invariant is another 10-dimensional group, which now bears his name [33]; the Galilean group, G_∞, is the $c \to \infty$ limit of the Poincaré group, G_c. We shall discuss these two groups in more detail in §3.4. Poincaré's transformations consist of space and time translations, space rotations and Lorentz boosts which can be viewed as hyperbolic rotations in the $x-t$, $y-t$ and $z-t$ planes (and so mix up the space and time coordinates). Thus, in contrast to (2.81) we have, for a Lorentz boost along x,

$$x' = \gamma(x - \beta ct), \quad y' = y, \quad z' = z, \quad ct' = \gamma(ct - \beta x) \tag{2.82}$$

with

$$\gamma = \sqrt{\frac{1}{1-\beta^2}}, \quad \beta = \frac{|\mathbf{v}|}{c}. \tag{2.83}$$

This finding was extremely important in the development of special relativity with the formulation of relativistic electromagnetism. Since the hyperbolic functions (cosh, tanh, etc.) correspond to trigonometric functions with imaginary angles, one can also view the six angles of rotation defined in a four-dimensional space (three space rotations, three hyperbolic space-time rotations) as three complex-valued rotation angles in three-dimensional space.

Writing the equations of electromagnetism in covariant form using 4-vectors and tensors is desirable for the development of the Lorentz invariant quantum theory of electrons and photons; however, in this chapter the charges will be treated in a non-relativistic approximation ($\beta \ll 1$), and only a brief indication of this fundamental aspect of electrodynamics will be presented here [34], [15]. The fundamental postulates of special relativity are that (i) the speed of light, c, is a universal constant independent of direction, and (ii) the mechanical and electromagnetic laws of nature are of the same form in all inertial reference frames.

In special relativity, we use a four-dimensional coordinate system in which the usual position vector $\mathbf{x}$ is conjoined with the time t to yield a 'coordinate' 4-vector $x^\mu = (ct, \mathbf{x})$, $\mu = 0, 1, 2, 3$. We adopt the convention that $x^0 = ct$ such that obvious quantities like the 4-scalar product

$$x \cdot x = c^2 t^2 - x_1^2 - x_2^2 - x_3^2 \tag{2.84}$$

and the volume of a specified space-time region

$$\Omega = \int dx_0 dx_1 dx_2 dx_3 \tag{2.85}$$

are invariant under Lorentz transformations, which are the orthogonal transformations that preserve (2.84).

Unlike inertial mass, the electric charge is both a signed quantity and a Lorentz invariant in the sense that

$$\int_\Omega \rho \, dx_1 dx_2 dx_3 = \text{invariant}, \tag{2.86}$$

and so ρ must behave under a Lorentz transformation in the same way as dx_0. Hence, we set

$$j^0 = c\rho \tag{2.87}$$

and regard $(c\rho, \mathbf{j})$ as the components of a charge-current 4-vector j^μ; the scalar combination $j \cdot j = c^2\rho^2 - j^2$ is Lorentz invariant. On the other hand, charged particles are found to have the property of charge invariance, that is, their electric charge is independent of the velocity of the particle with respect to any observer. For example, the velocity variation of the charge/mass ratio e/m is found experimentally to vary precisely according to the relativistic mass formula $m = m_0/\sqrt{1 - \beta^2}$ [15], [35].

The presence of both $+$ and $-$ signs in the scalar product (2.84) marks a fundamental difference from the familiar Euclidean geometry and vector analysis. It is characteristic of *Minkowski space* defined by the metric

$$\eta_{\mu\nu} = \text{diag}(+1, -1, -1, -1). \tag{2.88}$$

A convenient way to take this feature into account automatically is to modify the well-known Einstein summation convention for repeated vector/tensor component labels by introducing raised and lowered indices and using the Minkowski metric; thus to lower indices we write, for example,

$$x_\mu = \eta_{\mu\nu} x^\nu. \tag{2.89}$$

Now the inverse of $\eta_{\mu\nu}$ is also given by the RHS of (2.88),

$$\eta^{\mu\sigma} \eta_{\sigma\nu} = \delta^\mu_\nu. \tag{2.90}$$

Thus, the summation convention in special relativity consists of the following rules: (i) repeated indices always occur in pairs, and (ii) for the indices in any pair, one is raised and the other is lowered. These rules ensure that invariance under Lorentz transformations is explicit in quantities that have all indices paired.

The famous formulation of the Maxwell equations displayed in (2.1)–(2.4), which employs 3-vector calculus, can be cast into the notation of special relativity. First, we need to define the 4-derivative which reads

$$\partial_\mu = \frac{\partial}{\partial x^\mu} \equiv \left(\frac{1}{c} \frac{\partial}{\partial t}, \nabla \right). \tag{2.91}$$

Next, the scalar (ϕ) and vector (**a**) potentials combine into a 4-vector

$$a_\mu = \left(\frac{1}{c}\phi, -\mathbf{a}\right), \quad a^\mu = \left(\frac{1}{c}\phi, \mathbf{a}\right), \tag{2.92}$$

and the gauge transformation (2.45), (2.46) reads

$$a_\mu \to a_\mu - \partial_\mu \chi. \tag{2.93}$$

Elementary computations show that the antisymmetric tensor

$$f_{\mu\nu} = \partial_\mu a_\nu - \partial_\nu a_\mu \tag{2.94}$$

has components that are proportional to the electric and magnetic fields. This so-called *Faraday tensor* can be displayed as a 4×4 matrix:

$$f_{\mu\nu} = \begin{pmatrix} 0 & \frac{1}{c}E_x & \frac{1}{c}E_y & \frac{1}{c}E_z \\ -\frac{1}{c}E_x & 0 & -B_z & B_y \\ -\frac{1}{c}E_y & B_z & 0 & -B_x \\ -\frac{1}{c}E_z & -B_y & B_x & 0 \end{pmatrix}. \tag{2.95}$$

The associated tensor with two raised indices can be calculated using the metric

$$f^{\mu\nu} = \eta^{\mu\sigma}\eta^{\nu\tau}f_{\sigma\tau} \tag{2.96}$$

and has the same form as the matrix (2.95) with the signs of the *electric field components reversed*. The rule (2.96) allows us to move freely between raised and lowered indices, as convenient.

In terms of the field tensor $f_{\mu\nu}$, the homogeneous Maxwell equations (2.1), (2.2) take the form[8]

$$\partial_\alpha \left(\frac{1}{2}\varepsilon^{\alpha\beta\gamma\delta}\right)f_{\gamma\delta} = \partial_\alpha f_{\beta\gamma} + \partial_\beta f_{\gamma\alpha} + \partial_\gamma f_{\alpha\beta} = 0. \tag{2.97}$$

Similarly, when ρ and **j** are combined into the charge-current 4-vector j^μ, Eq. (2.9) becomes the vanishing of a 4-divergence,

$$\partial_\mu j^\mu = 0. \tag{2.98}$$

The other two Maxwell equations, (2.3), (2.4), can be presented as the single equation

$$\partial_\mu f^{\mu\nu} = \mu_0 j^\nu. \tag{2.99}$$

These general results can be illustrated by the simple example of a parallel-plate capacitor. An observer in a reference frame, that is, at rest relative to the capacitor (its rest frame) will describe it as the source of a static electric field ($\mathbf{E}=\mathbf{E}_1; \mathbf{B}=\mathbf{B}_1=\mathbf{0}$) determined by a charge density ρ_1. An observer moving uniformly relative to the capacitor sees a non-zero magnetic field ($\mathbf{B}_2$) in addition to a modified electric field ($\mathbf{E}_2$)

---

[8] The Levi–Civita symbol, conventionally denoted by the lower-case Greek letter ε, is defined on n-dimensional vector spaces and has values given by the rules $\varepsilon_{a_1 a_2 \ldots a_n} = +1$ if ($a_1 a_2 \ldots a_n$) is an *even* permutation of $(1, 2, \ldots n)$, $= -1$ if it is an *odd* permutation, and is zero otherwise. It is most often used in three and four dimensions, ε_{ijk} and $\varepsilon_{\alpha\beta\gamma\delta}$, respectively, in combination with vectors, matrices, tensors using the Einstein index notation and the summation convention.

because this observer sees a current $\mathbf{j}_2$ flowing in opposite directions in the plates as well as a modified charge density ρ_2. Thus, the electric and magnetic fields present, and the charge and current density differ for the different observers; both agree about the Lorentz invariant magnitude of $c^2\rho^2 - j^2$.

The conservation laws for the energy, momentum and angular momentum of the field may be summarised in a compact and elegant form using the covariant notation [21]. The field momentum density $\mathbf{u}$ and the energy density E must be taken together and regarded as the components of the field 4-momentum density vector $P^\mu = (E/c, \mathbf{u})$. This is because there is no Lorentz invariant way to separate energy from momentum. The angular momentum density can be expressed by the space-space components of a second rank tensor. The Lorentz force density $\mathbf{F} = (F_1, F_2, F_3)$ and the scalar $(1/c)\mathbf{j} \cdot \mathbf{E} = F_0$ are also the components of a 4-vector F_ν. A Lorentz invariant, symmetrical traceless second-rank tensor $T_{\mu\nu}$, called the energy-momentum tensor[9] of the field (or the stress tensor), can be constructed from the electromagnetic field tensor $f_{\mu\nu}$, (2.95). In terms of the three-dimensional fields and Eqs. (2.28), (2.32), (2.36), the components of $T_{\mu\nu}$ are $(i, j = 1, 2, 3)$

$$T_{ij} = E\delta_{ij} - \varepsilon_0\left(E_iE_j + c^2B_iB_j\right),$$
$$T_{0j} = -cu_j,$$
$$T_{i0} = -\frac{1}{c}S_i,$$
$$T_{00} = -E, \tag{2.100}$$

and with these correspondences, Eqs. (2.31) and (2.37) combine in the statement that the 4-vector F_ν is the 4-divergence of the tensor $T_{\mu\nu}$,

$$\frac{\partial T_{\mu\nu}}{\partial x_\mu} = F_\nu. \tag{2.101}$$

In the absence of sources, the 4-divergence of the stress tensor vanishes,

$$\frac{\partial T_{\mu\nu}}{\partial x_\mu} = 0, \tag{2.102}$$

and this is the covariant statement of Poynting's theorem and the conservation of linear momentum for the free field. The covariant expression of the conservation of angular momentum is contained in the statement that the third-rank tensor defined by

$$M_{\mu\nu\lambda} = T_{\mu\nu}x_\lambda - T_{\mu\lambda}x_\nu, \tag{2.103}$$

has vanishing divergence, by virtue of (2.102). The three space–space components of $M_{4\nu\lambda}$ (that is, M_{0ij}) are required to describe the angular momentum density, while the three quantities M_{00i} describe the Lorentz boosts, which generate three time-dependent conserved quantities. Of course, the distinction between angular momentum (rotations in spatial planes) and boosts (hyperbolic rotations in space-time planes) is not Lorentz invariant because different inertial observers need not agree about what is purely a

---

[9] $T_{\mu\nu}$ is the stress tensor density; the integrated stress tensor $\Theta_{\mu\nu} = \int_\Omega T_{\mu\nu} \, d^3x$.

plane in space. Thus, all six quantities $M_{4\nu\lambda}$ can be taken together as the relativistic generalisation of angular momentum.

2.4 Charged Particles

2.4.1 The Charge–Current Density

We must now consider the charge and current density which we specialise at once to the form for a collection of point material particles. A particle is characterised by its charge, e, and invariant mass, m which are time-independent; its position varies in time according to some dynamics. If the coordinate vector of the nth particle is $\mathbf{x}_n$, the charge density for N particles ($n = 1,\ldots.N$) is the distribution

$$\rho(\mathbf{x}) = \sum_n e_n \delta^3(\mathbf{x}_n - \mathbf{x}), \tag{2.104}$$

where $\delta^3(\mathbf{x})$ is the three-dimensional Dirac delta function. Its integral over the volume containing the charges is the net charge of the system, q, which is independent of time,

$$\int \rho(\mathbf{x})\,\mathrm{d}^3\mathbf{x} = \sum_n e_n = q. \tag{2.105}$$

Suppose the charge distribution is contained in some volume Ω with a boundary surface S. Gauss's law tells us that the flux of the electric field $\mathbf{E}$ through the surface may be computed using the Maxwell equation (2.3) as

$$\int_S \mathbf{E}\cdot\mathrm{d}\mathbf{S} \equiv \int_\Omega (\boldsymbol{\nabla}\cdot\mathbf{E})\,\mathrm{d}^3\,\mathbf{x} = \frac{1}{\varepsilon_0}q, \tag{2.106}$$

where q is the net charge contained in Ω.

It is impossible for a collection of charged particles to maintain a static equilibrium purely through electrostatic forces; this is the content of Earnshaw's theorem [36]. Bohr alluded to this classical result in his 1922 Nobel lecture (cf §1.3) to rule out an electrostatical explanation for the stability of atoms and molecules. The theorem may be proved by demonstrating a contradiction. Suppose the charges are at rest and consider the motion of the charge e_n in the electric field, $\mathbf{E}$, generated by all of the other particles. Assume that this particular charge has $e_n > 0$. The equilibrium position of this particle is the point $\mathbf{x}_n^0$ where $\mathbf{E}(\mathbf{x}_n^0) = \mathbf{0}$, since the force on the charge is $e_n\mathbf{E}(\mathbf{x}_n)$ (the Lorentz force for this static case). Obviously, $\mathbf{x}_n^0$ cannot be the equilibrium position of any other particle. However, in order for $\mathbf{x}_n^0$ to be a stable equilibrium point, the particle must experience a restoring force when it is displaced from $\mathbf{x}_n^0$ in any direction. For a positively charged particle at $\mathbf{x}_n^0$, this requires that the electric field points radially towards $\mathbf{x}_n^0$ at all neighbouring points. But from Gauss's law applied to a small sphere centred on $\mathbf{x}_n^0$, this corresponds to a negative flux of $\mathbf{E}$ through the surface of the sphere, implying the presence of a negative charge at $\mathbf{x}_n^0$, contrary to our original assumption. Thus, $\mathbf{E}$ cannot point radially towards $\mathbf{x}_n^0$ at all neighbouring points, that is, there must be some neighbouring points at which $\mathbf{E}$ is directed away from $\mathbf{x}_n^0$. Hence, a positively charged particle placed at $\mathbf{x}_n^0$ will always move towards such points. There is, therefore, no static equilibrium configuration.

The time-dependence of the charge density arises solely from the motion of the particles according to

$$\frac{d\rho}{dt} = \sum_n e_n \dot{\mathbf{x}}_n \cdot \nabla_n \delta^3(\mathbf{x}_n - \mathbf{x}), \tag{2.107}$$

since all $\{e_n\}$ are constant in time. If we define a current density $\mathbf{j}$,

$$\mathbf{j}(\mathbf{x}) = \sum_n e_n \dot{\mathbf{x}}_n \delta^3(\mathbf{x}_n - \mathbf{x}), \tag{2.108}$$

then it is easily seen that the pair $(\rho, \mathbf{j})$ satisfy the equation of continuity,

$$\frac{d\rho}{dt} + \nabla \cdot \mathbf{j} = 0. \tag{2.109}$$

2.4.2 The Electromagnetic Fields of a Charged Particle

The unique solution of the Maxwell equations for the electromagnetic field of a point charge e at rest at the origin of the coordinates that vanishes at infinity is

$$\mathbf{E}(\mathbf{x}) = \frac{e\hat{\mathbf{x}}}{4\pi\varepsilon_0 x^2}, \quad \mathbf{B}(\mathbf{x}) = 0. \tag{2.110}$$

These correspond to a charge density

$$\rho(\mathbf{x}) = e\,\delta^3(\mathbf{x}). \tag{2.111}$$

According to (2.28), the energy of this field is therefore

$$\mathcal{E}_0 = \tfrac{1}{2}\varepsilon_0 \int_{\Omega_0} E^2\, d^3\mathbf{x} = \lim_{a \to 0} \mathcal{E}_a, \quad \mathcal{E}_a = \left(\frac{e^2}{8\pi\varepsilon_0 a} \right). \tag{2.112}$$

Here we have imagined the point particle as the limiting case $(a \to 0)$ of a charge density ρ with charge distributed uniformly over the surface of a sphere of radius a, for example,

$$\rho(\mathbf{r}) = \frac{e}{4\pi a^2}\delta(|\mathbf{r} - \mathbf{x}| - a), \quad \int \rho(\mathbf{r})\,d^3\mathbf{r} = e. \tag{2.113}$$

Then $\mathbf{E} = 0$ inside the sphere, and (2.110) applies outside it. The volume Ω_a extends from the surface of the sphere to infinity while the point particle's coordinate, $\mathbf{x}$, locates the centre of the sphere.

Suppose now that the particle is moving at constant speed $\dot{\mathbf{x}}$; then analogously to (2.110), we can calculate the momentum associated with the moving Coulomb field of the charge by integrating the Poynting vector (2.32),

$$\mathbf{S} = \varepsilon_0 c^2 \, (\mathbf{E} \wedge \mathbf{B}), \tag{2.114}$$

over the volume Ω_a. Using $\mathbf{B}(\mathbf{x}) = \dot{\mathbf{x}} \wedge \mathbf{E}(\mathbf{x})/c^2$, we obtain the famous Abraham–Lorentz formula [37],

$$\mathbf{p} = \tfrac{4}{3} m_e \dot{\mathbf{x}}, \tag{2.115}$$

which clearly is not the required non-relativistic equation relating momentum and velocity that involves the observable mass of the charge. The energy $\mathcal{E}_a$ is conventionally interpreted in terms of an 'electromagnetic mass', m_e, via Einstein's relativistic relation between energy and mass,

$$\mathcal{E}_a = m_e c^2, \quad m_e = \frac{1}{2}\left(\frac{e^2}{8\pi\varepsilon_0 a c^2}\right), \tag{2.116}$$

that is an addition to the 'mechanical mass' of the particle because the charge interacts with the field it produces. We define

$$\Delta m = \frac{4}{3} m_e \tag{2.117}$$

and declare that the observable mass of the particle is

$$m_{\text{obs}} = m + \Delta m. \tag{2.118}$$

The self-interaction of a charged particle will be discussed in detail within the classical Hamiltonian formalism in §3.8, and after quantisation in §10.6 and §11.1.

The energy $\mathcal{E}'_a$ and momentum $\mathbf{p}'$ of a charge moving with velocity $\dot{\mathbf{x}}$ can be obtained from the rest-frame values $(\mathcal{E}_a, \mathbf{p})$ by means of a relativity transformation

$$\mathcal{E}'_a = \gamma m_e c^2, \quad \mathbf{p}' = \gamma m_e \dot{\mathbf{x}}, \tag{2.119}$$

with $\gamma = 1/\sqrt{1 - \dot{\mathbf{x}}^2/c^2}$. The discrepancy between (2.115) and (2.119) can be understood by observing that integration over the space outside a sphere is not Lorentz invariant because the sphere does not look spherical to a moving observer (Lorentz contraction), and the Galilean result is a limiting case of the Lorentz invariant result,

$$\gamma \approx 1 + \frac{|\dot{\mathbf{x}}|^2}{2c^2}, \;\; \rightarrow \mathcal{E}'_a = m_e c^2 + \tfrac{1}{2} m_e |\dot{\mathbf{x}}|^2, \quad \mathbf{p}' = m_e \dot{\mathbf{x}}, \tag{2.120}$$

valid in the non-relativistic regime, $|\dot{\mathbf{x}}| \ll c$ [38].

The idea of the charge being distributed over the surface of a sphere, which avoids infinite electromagnetic mass, has the difficulty that the particle is not stable [33]. In the absence of suitable confining forces, it would be expected to fly apart because of Coulomb repulsion, and there are no such forces in a purely electrodynamic theory. Formally one shows the lack of stability by calculating the integrated stress tensor from (2.100) and (2.110) for a particle at rest,

$$\Theta_{ik} = 0, \quad i \neq k, \quad \Theta_{ii} = -\tfrac{1}{3} m_e c^2, \quad i, k = 1, 2, 3. \tag{2.121}$$

An equilibrium system requires $\Theta_{ii} = 0$, so the model requires an additional contribution of non-electromagnetic nature to the stress tensor. This problem can be solved by appeal to an explicitly covariant (Lorentz invariant) formalism and will not be pursued here.

The regularisation of infinities that arise in non-relativistic classical electrodynamics can be interpreted in terms of a smoothing of the delta function in (2.104); the delta function should be understood as a linear functional on a space of suitably smooth functions, $s(x), x \in \Re^n$. It can be represented as the (singular) limit of a class of functions known as 'approximations to the identity'. We write formally

$$\lim_{a \to 0} \delta_a(x - x_0) = \delta(x - x_0), \qquad (2.122)$$

which is to be understood as a shorthand for the relation

$$s(x_0) = \lim_{a \to 0} \int_{\Re^n} s(x) \delta_a(x - x_0) \, dx; \qquad (2.123)$$

the limit is to be taken after the integral is evaluated.[10] There are infinitely many functions that can serve as $\delta_a(x)$; familiar examples are

$$\delta_a(x): \frac{1}{|a|\sqrt{\pi}} e^{-(x/a)^2}, \quad \frac{1}{a\pi} \frac{\sin(x/a)}{(x/a)}, \quad \frac{1}{\pi} \frac{a}{a^2 + x^2}. \qquad (2.124)$$

The delta function is real and even, and $\delta_a(x)$ can likewise be chosen to be real and even.

We imagine the charge density of the particle is centred on the position $\mathbf{x}_n$ and write

$$\rho(\mathbf{x}) = \sum_n e_n \xi_{an}(\mathbf{x} - \mathbf{x}_n), \quad \xi_a(\mathbf{x}) \geq 0, \qquad (2.125)$$

with ξ_a a spherically symmetric function, normalised such that

$$\int \xi_a \, d^3\mathbf{x} = 1. \qquad (2.126)$$

Let $\chi_a(k)$ be the Fourier transform of $\xi_a(\mathbf{x})$; we require the properties ($k = |\mathbf{k}|$)

$$|\chi(k)| \leq 1, \quad \chi_a(0) = \chi_0(k) = 1, \quad \int_0^\infty \chi_a^2(k) \, dk = \frac{\pi}{2a}, \qquad (2.127)$$

to ensure that the usual classical point particle results are recovered.[11] If the function ξ_a is chosen as an approximation to the identity, $\delta_a(x)$, its Fourier transform is real and even.

The proposed regularisation implies a modification to Coulomb's law at short distances; a theorem in electrostatics assures us that, outside a spherically symmetric charge density, the electric field is the same as that generated by placing a point charge of the same magnitude at the origin. So, if we take a pair of extended charges, their mutual potential energy will be Coulombic ($\propto r^{-1}$) for all separation distances r greater

---

[10] The regularisation consists in keeping $a > 0$.

[11] It is worth bearing in mind that the classical result may no longer be relevant after quantisation; for example, quantisation introduces a new length – the Compton wavelength of the particle ($\lambda_C = h/mc$) – and the 'classical limit' $h \to 0$ may not be the same as $a \to 0$.

than the sum of the radii of the two charge densities. For r close to 0, however, there will be a modification to Coulomb's law. Let the charge density of the pair be

$$\rho(\mathbf{x}) = \sum_{n=1,2} e_n \xi_{an}(\mathbf{x} - \mathbf{x}_n). \tag{2.128}$$

The electrostatic (Coulomb) energy of the charge density ρ in (2.128) is

$$\mathcal{E} = \frac{1}{2} \int\int \frac{\rho(\mathbf{x})\,\rho(\mathbf{x}')}{4\pi\varepsilon_0|\mathbf{x} - \mathbf{x}'|} \, d^3\mathbf{x}'\, d^3\mathbf{x}. \tag{2.129}$$

Introducing Fourier transforms, this becomes

$$\mathcal{E} = \frac{e_1 e_2}{\varepsilon_0} \frac{1}{(2\pi)^3} \int \frac{\chi_{a1}(k)\chi_{a2}(k)e^{i\,\mathbf{k}\cdot\mathbf{r}}}{k^2} \, d^3\mathbf{k}, \tag{2.130}$$

where $\mathbf{r} = \mathbf{x}_1 - \mathbf{x}_2$ is the vector separation of the centres of the two particles. For point charges ($\chi_{0n} = 1$), Eq. (2.130) yields Coulomb's law as it should. The self-interaction energy of a single charge with $a > 0$ is given by (2.112).

2.4.3 Polarisation Fields and Classical Multipole Moments

In the traditional formulation of the electrodynamics of bulk matter it is usual to introduce two polarisation fields in place of the charge–current density; this description is accompanied by an explicit division of the charges into 'bound' and 'free' charge densities [37]. A similar procedure is useful here except that no assumptions are made about the localisation of the charges. This point is worth elucidating further; our eventual goal is a quantum theory of atoms and molecules interacting with the electromagnetic field. In such a description, ρ and $\mathbf{j}$ will be reinterpreted as operators, and so the polarisation fields, which are linear functionals of ρ and $\mathbf{j}$, will also become operators. In a quantum theory, the distinction between bound and free particles is expressed by the division of the spectrum of states into discrete and continuum parts. Hence, we do not wish to impose a classical structure on the polarisation fields that will be inappropriate after quantisation.

The idea of the polarisation field is to replace the charge–current density 4-vector j^μ with a second-rank tensor density $p^{\mu\nu}$ through the relation

$$j^\mu = \partial_\nu p^{\mu\nu} \tag{2.131}$$

so as to construct the displacement fields with the combination

$$d^{\mu\nu} = cf^{\mu\nu} + (c\varepsilon_0)^{-1} p^{\mu\nu}. \tag{2.132}$$

The inner product $\varepsilon_0 d_{\mu\nu} d^{\mu\nu}$ is an energy density. Provided $p^{\mu\nu}$ is antisymmetric ($p^{\mu\nu} = -p^{\nu\mu}$) and single-valued, the equation of continuity, (2.9), will be satisfied automatically:

$$0 = \partial_\mu j^\mu = \partial_\mu \partial_\nu p^{\mu\nu}. \tag{2.133}$$

Moreover, Eq. (2.131) is exactly of the same form as the Maxwell equations summarised in (2.99) and so $p^{\mu\nu}$ is an electromagnetic field associated specifically with the current j^μ; however, $p^{\mu\nu}$ is *not* required to satisfy the other Maxwell equations (2.97) so $p^{\mu\nu}$ is not necessarily the Maxwell field. We call $p^{\mu\nu}$ the *polarisation electromagnetic field*. As noted earlier (§2.3.1), an antisymmetric second-rank tensor has six independent components which can be identified as the components of a pair of vectors. As a 4×4 matrix the polarisation tensor is

$$p^{\mu\nu} = \begin{pmatrix} 0 & cP_x & cP_y & cP_z \\ -cP_x & 0 & M_z & -M_y \\ -cP_y & -M_z & 0 & M_x \\ -cP_z & M_y & -M_x & 0 \end{pmatrix}. \tag{2.134}$$

The vector $\mathbf{P}$ is usually called the 'electric polarisation field', while $\mathbf{M}$ is the 'magnetisation' (magnetic polarisation field).

On its own, Eq. (2.131) is not sufficient to determine fully the tensor $p^{\mu\nu}$. Consider an arbitrary field described by a four-vector $(u, \mathbf{U}) = u^\beta, (\beta = 0, 1, 2, 3)$, and form a second-rank tensor by the construction

$$h_{\mu\nu} = \varepsilon_{\mu\nu\alpha\beta} \left(\partial_\alpha u^\beta \right). \tag{2.135}$$

Then a new polarisation tensor defined by

$$p'^{\mu\nu} = p^{\mu\nu} + h^{\mu\nu} \tag{2.136}$$

also satisfies the defining relation (2.131) since

$$(\partial_\nu h^{\mu\nu}) = \varepsilon_{\mu\nu\alpha\beta} \left(\partial_\nu \partial_\alpha u^\beta \right) = 0, \tag{2.137}$$

whatever u^β be chosen as, provided the derivatives are defined.

In three-dimensional vector notation, (2.131) gives the relations

$$\rho = -\boldsymbol{\nabla} \cdot \mathbf{P}, \tag{2.138}$$

$$\mathbf{j} = \frac{d\mathbf{P}}{dt} + \boldsymbol{\nabla} \wedge \mathbf{M}. \tag{2.139}$$

In terms of $\mathbf{P}$ and $\mathbf{M}$, Eq. (2.136) becomes

$$\mathbf{P} \to \mathbf{P}' = \mathbf{P} + \boldsymbol{\nabla} \wedge \mathbf{U}, \tag{2.140}$$

$$\mathbf{M} \to \mathbf{M}' = \mathbf{M} - \frac{d\mathbf{U}}{dt} + \boldsymbol{\nabla} u, \tag{2.141}$$

where u and $\mathbf{U}$ are arbitrary scalar and vector fields, respectively (with appropriate derivatives). Such a polarisation field description has commonly been regarded as being particularly 'physical' or 'natural', for example, through its multipolar representation; however, the interpretation is limited by the arbitrary character of $\mathbf{P}$ and $\mathbf{M}$. The longitudinal component of the electric polarisation field is well defined through (2.138) so the arbitrariness resides in the transverse component $\mathbf{P}^\perp$, and in the magnetisation which is entirely transverse. Any two pairs $\{\mathbf{P}, \mathbf{M}\}$ and $\{\mathbf{P}', \mathbf{M}'\}$ related by (2.140), (2.141) will satisfy (2.138), (2.139), and so cannot be distinguished; thus there is no correspondence with a definite physical (experimental) situation, and their forms cannot be decided by such practical considerations. Instead, there is a requirement to ensure that calculated quantities that claim to be physical observables are *independent* of any particular choice for **P** and **M**.

The close similarity with the discussion of gauge transformations in §2.3 is obvious; as we shall see in §3.3, the fields $(\phi, \mathbf{a})$ and $(\mathbf{P}, \mathbf{M})$ afford alternative ways of expressing a fundamental property of electrodynamics. The arbitrary characters of the field potentials and the polarisation fields for the charges are intimately related; both are consequences of electric charge conservation, and the arbitrary parts of both sets of fields can be parameterised by the Green's function solution, $\mathbf{g}(\mathbf{x}, \mathbf{x}')$, of (2.64).

To see this we note that the formal integral representation of the electric polarisation field,

$$\mathbf{P}(\mathbf{x}) = \int \mathbf{g}(\mathbf{x}, \mathbf{x}') \rho(\mathbf{x}') \, \mathrm{d}^3 \mathbf{x}', \tag{2.142}$$

is a Green's function solution of (2.138). We saw in §2.3 that the Green's function $\mathbf{g}(\mathbf{x}, \mathbf{x}')$ can be expressed in the path-dependent form (2.69), and combination of this equation with (2.142) for a multi-particle system yields for the electric polarisation field

$$\mathbf{P}(\mathbf{x}) = q\mathbf{g}(\mathbf{x}, \mathbf{O})^{\parallel} + \sum_n^N e_n \int_{C_n}^{\mathbf{x}_n} \delta^3 (\mathbf{z} - \mathbf{x}) \, \mathrm{d}\mathbf{z}. \tag{2.143}$$

The first term on the RHS of (2.143) vanishes if either $| \mathbf{x} - \mathbf{O} | \to \infty$ (that is, taking $\mathbf{O} =$ spatial infinity) or $q = 0$ (an overall neutral system).

If now **j** and **P** are separated into longitudinal and transverse parts, a simple calculation shows that

$$\mathbf{N} = \mathbf{j} - \frac{\mathrm{d}\mathbf{P}}{\mathrm{d}t} \tag{2.144}$$

is purely transverse, since the longitudinal parts cancel identically by virtue of the equation of continuity (2.9). We may therefore write $\mathbf{N} = \nabla \wedge \mathbf{M}$, and use of (B.0.14) from Appendix B yields

$$\begin{aligned}
\mathbf{M}(\mathbf{x}) &= \int \frac{\nabla_{\mathbf{x}'} \wedge \nabla_{\mathbf{x}'} \wedge \mathbf{M}(\mathbf{x}')}{4\pi |\mathbf{x} - \mathbf{x}'|} \, \mathrm{d}^3 \mathbf{x}' \\
&= \int \frac{\nabla_{\mathbf{x}'} \wedge \mathbf{j}(\mathbf{x}') - \nabla_{\mathbf{x}'} \wedge \mathbf{K}(\mathbf{x}')}{4\pi |\mathbf{x} - \mathbf{x}'|} \, \mathrm{d}^3 \mathbf{x}',
\end{aligned} \tag{2.145}$$

where

$$\mathbf{K}(\mathbf{x}') = \int \mathbf{g}(\mathbf{x}', \mathbf{x}'') \frac{\mathrm{d}\rho(\mathbf{x}'')}{\mathrm{d}t} \, \mathrm{d}^3 \mathbf{x}''. \tag{2.146}$$

Equations (2.142) and (2.145) are general forms for the arbitrary polarisation fields which display their dependence on the Green's function $\mathbf{g}(\mathbf{x}, \mathbf{x}')$ [28].

The classical definitions of multipole moments [21] that have traditionally been carried over into atomic theory can be derived from a particular choice of finite path in the electric polarisation field (2.143). This choice is the set of straight-line paths of finite length from the arbitrary origin **O** to each of the particle coordinates $\mathbf{x}_n$,

$$\mathbf{z}_n(\sigma) = \mathbf{x}_n - \sigma(\mathbf{x}_n - \mathbf{O}), \quad 0 \le \sigma \le 1. \tag{2.147}$$

Introducing the path (2.147) into the polarisation field and expansion of the integrand in powers of σ, followed by term-by-term integration, yields the multipole series,[12]

$$\sum_n e_n \int_{C_n}^{\mathbf{x}_n} \delta^3(\mathbf{z}_n - \mathbf{x}) \, d\mathbf{z}_n \approx (\mathbf{d} - \mathbf{Q} \cdot \nabla_{\mathbf{x}} + \ldots) \delta^3(\mathbf{x} - \mathbf{O}), \tag{2.148}$$

where

$$\mathbf{d} = \sum_n e_n (\mathbf{x}_n - \mathbf{O}),$$

$$\mathbf{Q} = \tfrac{1}{2} \sum_n e_n (\mathbf{x}_n - \mathbf{O})(\mathbf{x}_n - \mathbf{O}), \tag{2.149}$$

and $\ldots$ in (2.148) represents multipoles involving higher-order powers of $(\mathbf{x}_n - \mathbf{O})$ contracted with $\nabla_{\mathbf{x}}$ vectors [39], [40].

The interaction energy, $\mathcal{E}_{\mathbf{E}}$, of a charge distribution ρ with a static electric field $\mathbf{E}$ can be expressed as

$$\mathcal{E}_{\mathbf{E}} = \int \rho(\mathbf{x}) \phi(\mathbf{x}) \, d^3\mathbf{x}, \tag{2.150}$$

where ϕ is a scalar potential for the field. After integration by parts and use of (2.138), this is also

$$\mathcal{E}_{\mathbf{E}} = - \int \mathbf{P}(\mathbf{x}) \cdot \mathbf{E}(\mathbf{x}) \, d^3\mathbf{x}. \tag{2.151}$$

Evaluation of the integral (2.151) using (2.148) then yields the interaction energy as the familiar multipolar expansion

$$\mathcal{E}_{\mathbf{E}} = q\phi(\mathbf{O}) - \mathbf{d} \cdot \mathbf{E}(\mathbf{O}) - \mathbf{Q} : \mathbf{F}(\mathbf{O}) - \ldots, \tag{2.152}$$

where $\mathbf{F}$ is the electric field gradient.[13] An essentially similar discussion can be given for the coupling of the charges to a static magnetic field and to electromagnetic radiation in terms of the multipole moments and the fields and their gradients [4]. A characteristic feature of these multipole moment definitions is the property that only the first non-vanishing multipole is independent of the coordinate origin. Thus, if the origin is moved from $\mathbf{O}$ to $\mathbf{O}' = \mathbf{O} + \mathbf{r}$, where $\mathbf{r}$ is some constant vector (that is, a uniform translation), the position vector $\mathbf{x}_i$ in the old coordinate system becomes $\mathbf{x}_i' = \mathbf{x}_i - \mathbf{r}$ in the new, and the overall charge q and the moments transform as

$$q' = q \tag{2.153}$$

$$\mathbf{d}' = \mathbf{d} - q\mathbf{r} \tag{2.154}$$

$$\mathbf{Q}' = \mathbf{Q} - \mathbf{dr} + \tfrac{1}{2}\mathbf{rr}. \tag{2.155}$$

---

[12] The calculation is dealt with in Appendix E.

[13] The polarisation field is a distribution so makes sense when integrated over with a suitably smooth function that vanishes at ∞. The derivatives in (2.148) can then be transferred to the smooth function using integration by parts with a change in sign for the *odd* powers.

The energy $\mathcal{E}_E$ has two important properties. First, it is invariant to the change in origin $\mathbf{O} \to \mathbf{O}'$; secondly, since the field gradient has zero trace outside the sources of the field, $\mathcal{E}_E$ is independent of the trace of $\mathbf{Q}$. Thus, it is possible to define a traceless quadrupole tensor [41], [42],

$$\boldsymbol{\Theta} = \tfrac{1}{2} \sum_n e_n (3\mathbf{x}_n \mathbf{x}_n - x_n^2 \mathbf{I}), \tag{2.156}$$

for use in place of $\mathbf{Q}$. It is easy to see that this is simply a convention; since $\mathrm{Tr}\mathbf{F} = 0$, the energy $\mathcal{E}_E$ is invariant against a class of transformations of $\mathbf{Q}$,

$$\mathbf{Q} \to \mathbf{Q}_f = \tfrac{1}{2} \sum_n e_n (\mathbf{x}_n - f(x_n^2)\mathbf{I}) \qquad \mathcal{E}_{E,f} = \mathcal{E}_E, \tag{2.157}$$

where f is an arbitrary scalar function of $x_n^2 = \mathbf{x}_n \cdot \mathbf{x}_n$, and no particular form for f has physical significance since information about the quadrupole moment can only be found through its interaction with a field gradient.

For an overall electrically neutral collection of electrons and nuclei, the arbitrary origin $\mathbf{O}$ can be eliminated as follows. Suppose there are altogether M nuclei with positive charges $\{eZ_a : a = 1, \ldots M\}$; then there must be $K = MZ_a$ electrons with charge $-e$ to give electroneutrality. Let the nuclei have coordinates $\{\mathbf{x}_m^N : m = 1, \ldots M\}$; similarly the electrons have coordinates $\{\mathbf{x}_k^e : k = 1, \ldots K\}$. Now write the charge density as a sum of contributions from individual electrons and nuclei:

$$\begin{aligned}
\rho &= \sum_n e_n \delta^3(\mathbf{x} - \mathbf{x}_n) = \rho(\mathbf{x})^N + \rho(\mathbf{x})^e \\
&= eZ_1 \delta^3(\mathbf{x} - \mathbf{x}_1^N) + eZ_2 \delta^3(\mathbf{x} - \mathbf{x}_2^N) + \ldots + eZ_M \delta^3(\mathbf{x} - \mathbf{x}_M^N) \\
&\quad - e\delta^3(\mathbf{x} - \mathbf{x}_1^e) - e\delta^3(\mathbf{x} - \mathbf{x}_2^e) - \ldots - e\delta^3(\mathbf{x} - \mathbf{x}_K^e).
\end{aligned} \tag{2.158}$$

Next, take the terms for the nuclei and rewrite them as

$$\begin{aligned}
\rho(\mathbf{x})^N &= e\left(\underbrace{\delta^3(\mathbf{x} - \mathbf{x}_1^N) + \ldots \delta^3(\mathbf{x} - \mathbf{x}_1^N)}_{Z_1 \text{ terms}} \right) \\
&\quad + \ldots + e\left(\underbrace{\delta^3(\mathbf{x} - \mathbf{x}_M^N) + \ldots + \delta^3(\mathbf{x} - \mathbf{x}_M^N)}_{Z_M \text{ terms}} \right).
\end{aligned} \tag{2.159}$$

There are now altogether K terms for the nuclei with coefficient $+e$ which can be paired off with the K terms for the electrons with coefficient $-e$, so the charge density can be rearranged to the form

$$\begin{aligned}
\rho(\mathbf{x}) &= e\big(\delta^3(\mathbf{x} - \mathbf{x}_1^N) - \delta^3(\mathbf{x} - \mathbf{x}_1^e)\big) \\
&\quad + e\big(\delta^3(\mathbf{x} - \mathbf{x}_1^N) - \delta^3(\mathbf{x} - \mathbf{x}_2^e)\big) + \cdots \\
&\quad \cdots + e\big(\delta^3(\mathbf{x} - \mathbf{x}_M^N) - \delta^3(\mathbf{x} - \mathbf{x}_K^e)\big) \\
&= e \sum_{m=1}^M \sum_{k=1}^{Z_m} \big(\delta^3(\mathbf{x} - \mathbf{x}_m^N) - \delta^3(\mathbf{x} - \mathbf{x}_{mk}^e)\big).
\end{aligned} \tag{2.160}$$

Combining this paired-particle representation of the charge density with (2.142) yields the polarisation field as

$$\mathbf{P}(\mathbf{x}) = e \sum_{m=1}^{M} \sum_{k=1}^{Z_m} \mathcal{G}(\mathbf{x}; \mathbf{x}_m^N, \mathbf{x}_{mk}^e), \tag{2.161}$$

where $\mathcal{G}$ is defined by

$$\mathcal{G}(\mathbf{x}; \mathbf{x}_m^N, \mathbf{x}_{mk}^e) = \mathbf{g}(\mathbf{x}, \mathbf{x}_m^N) - \mathbf{g}(\mathbf{x}, \mathbf{x}_{mk}^e). \tag{2.162}$$

If we introduce the line integral representation (2.69) for the Green's function, we may write

$$\mathcal{G}(\mathbf{x}; \mathbf{x}_m^N, \mathbf{x}_{mk}^e) = \int_{\mathbf{x}_{mk}^e}^{\mathbf{x}_m^N} \delta^3(\mathbf{z} - \mathbf{x})\, d\mathbf{z}, \tag{2.163}$$

independent of the arbitrary origin $\mathbf{O}$, and where the integral is taken over any path C from the electron at the point $\mathbf{x}_{mk}^e$ to the nuclear position $\mathbf{x}_m^N$. Thus, the general electric polarisation field for an atom or molecule may be written as a sum of 'atomic' contributions,[14]

$$\mathbf{P}(\mathbf{x}) = \sum_{m=1}^{M} \mathbf{P}(\mathbf{x})_m$$

$$\mathbf{P}(\mathbf{x})_m = e \sum_{k=1}^{Z_m} \int_{\mathbf{x}_{mk}^e}^{\mathbf{x}_m^N} \delta^3(\mathbf{z} - \mathbf{x})\, d\mathbf{z}. \tag{2.164}$$

It is obvious from this formulation that the inclusion of the origin about which a multipole expansion is made introduces a redundant feature quite unnecessarily. No physical observable can depend on the arbitrary origin $\mathbf{O}$, and so with the traditional definitions one has to manoeuvre with electric and magnetic multipoles in order to achieve an overall origin-invariant energy – it is, however, still entirely conventional to proceed in this way.

A typical term in (2.164) is

$$\mathbf{P}(\mathbf{x}, C) = e \int_{\mathbf{x}_1}^{\mathbf{x}_2} \delta^3(\mathbf{x} - \mathbf{z})\, d\mathbf{z} \tag{2.165}$$

for some path C, and it is instructive to analyse this simple case; the generalisation to multi-particle systems is immediate. It describes an electric polarisation field of two equal and opposite charges fixed at the positions $\mathbf{x}_1$ and $\mathbf{x}_2$. In applications the most commonly used path is the straight line starting at $\mathbf{x}_1$ and ending at $\mathbf{x}_2$, which may be given the parametric form

$$\mathbf{z}(\sigma) = \mathbf{x}_1 + \sigma(\mathbf{x}_2 - \mathbf{x}_1) \quad 0 \leq \sigma \leq 1. \tag{2.166}$$

Then the polarisation field (2.165) becomes

$$\mathbf{P}(\mathbf{x}, C = \text{st.line}) = e\mathbf{r} \int_0^1 \delta^3(\mathbf{x}_1 + \sigma\mathbf{r} - \mathbf{x})\, d\sigma, \tag{2.167}$$

---

[14] Note that here we have changed the labels for the electronic coordinates to $\{\mathbf{x}_{mk}\}$ for the purposes of the summation notation; this is only a notational convenience. It does not mean that a certain number of electrons must be associated with a specific nucleus. The description here is entirely classical, so there aren't any 'atoms'.

where $\mathbf{r} = \mathbf{x}_2 - \mathbf{x}_1$. Equation (2.167) is easily evaluated using the expansion of $\delta^3(\mathbf{r})$ in spherical polar coordinates to give [43]

$$\mathbf{P}(\mathbf{x}, C = \text{st.line}) = \frac{e\hat{\mathbf{r}}\delta^2(1 - \cos(\omega))}{|\mathbf{x}_1 - \mathbf{x}|^2}, \quad |\mathbf{x}_1 - \mathbf{x}| \leq |\mathbf{r}|$$

$$= 0, \qquad\qquad\qquad |\mathbf{x}_1 - \mathbf{x}| > |\mathbf{r}| \qquad (2.168)$$

where ω is the angle between the vectors $\mathbf{x}_1 - \mathbf{x}$ and $\mathbf{r}$. Note that the angular delta function is dimensionless. The two conditions that $|\mathbf{x}_1 - \mathbf{x}| \leq |\mathbf{r}|$ and that the angle $\omega = 0$ imply that the only points $\mathbf{x}$ for which the electric polarisation field $\mathbf{P}(\mathbf{x})$ is non-zero are those that lie on the path joining $\mathbf{x}_1$ and $\mathbf{x}_2$.

Let C_1 and C_2 be two distinct paths from the charge at $\mathbf{x}_1$ to the charge at $\mathbf{x}_2$ so that $C_1 - C_2$ is a closed loop, with C_2 the straight-line path between the two charges. If we take some other path C_1, the resulting polarisation field is related to that over the straight-line path by

$$\mathbf{P}(\mathbf{x}, C_1) = \mathbf{P}(\mathbf{x}, C_2) - \nabla_\mathbf{x} \wedge \left(e \int_{\Sigma_{12}} \delta^3(\mathbf{y} - \mathbf{x}) \, d\mathbf{S} \right), \qquad (2.169)$$

in agreement with (2.140), where Σ_{12} is the surface enclosed by the closed curve $\mathbf{y}$ formed from C_1 and C_2 [44].

The Fourier transform of the electric polarisation field is often encountered in applications because $\mathbf{P}(\mathbf{x})$ is usually involved in integrals over its scalar product with electromagnetic field variables that have mode expansions labelled by wave vectors $\mathbf{k}$. Transforming (2.165) we have

$$\mathbf{P}(\mathbf{k}, C) = e \int_{\mathbf{x}_1}^{\mathbf{x}_2} e^{i\mathbf{k}\cdot\mathbf{z}} d\mathbf{z} \qquad (2.170)$$

for some path C. Before evaluating (2.170) for the path C_2, consider the limiting case $|\mathbf{k}| \to 0$,

$$\lim_{|\mathbf{k}|\to 0} \mathbf{P}(\mathbf{k}, C) = e \lim_{|\mathbf{k}|\to 0} \int_{\mathbf{x}_1}^{\mathbf{x}_2} e^{i\mathbf{k}\cdot\mathbf{z}} d\mathbf{z}. \qquad (2.171)$$

For *finite-length* paths, appeal to Lebesgue's dominated convergence theorem justifies the interchange of the order of the integration and the limit, and there results

$$\lim_{|\mathbf{k}|\to 0} \mathbf{P}(\mathbf{k}, C) = e \int_{\mathbf{x}_1}^{\mathbf{x}_2} \left(\lim_{|\mathbf{k}|\to 0} e^{i\mathbf{k}\cdot\mathbf{z}} \right) d\mathbf{z}$$

$$= e \int_{\mathbf{x}_1}^{\mathbf{x}_2} d\mathbf{z} = e(\mathbf{x}_2 - \mathbf{x}_1) \equiv \mathbf{d}, \qquad (2.172)$$

where $\mathbf{d}$ is the electric dipole moment. This result is true for *any* finite path; however, even if $|\mathbf{x}_1 - \mathbf{x}_2| < \infty$, it is still possible to have infinite paths for which the above argument fails, and the limit does not exist for such cases.

The explicit evaluation of the Fourier transform (2.170) for the straight-line path between the charges yields

$$\mathbf{P}(\mathbf{k}, C = \text{st.line}) = \int \mathbf{P}(\mathbf{x}, C = \text{st.line}) e^{i\,\mathbf{k}\cdot\mathbf{x}} \, d^3\mathbf{x} = e \, e^{i\mathbf{k}\cdot\mathbf{x}_1} \frac{(e^{i\mathbf{k}\cdot\mathbf{r}} - 1)}{i\mathbf{k}\cdot\mathbf{r}} \mathbf{r} \qquad (2.173)$$

$$\approx e \left[1 + \frac{i}{2} \mathbf{k} \cdot (\mathbf{x}_1 + \mathbf{x}_2) \right] \mathbf{r} + O(k^2) \qquad (2.174)$$

up to the 'electric quadrupole' term whose specific form is dependent on the choice of the straight-line path.

So far nothing has been said about time dependence. If the charges are allowed to move, there is also a current density to be described by (2.139), and one requires the magnetisation, $\mathbf{M}$. The companion to (2.165) is [43, 45]

$$\mathbf{M}(\mathbf{x}, C) = e \int_{\mathbf{x}_1}^{\mathbf{x}_2} \delta^3(\mathbf{z} - \mathbf{x}) \, d\mathbf{z} \wedge \dot{\mathbf{z}}, \qquad (2.175)$$

which generalises directly to the many-charge case. In terms of the path (2.166), one has

$$\mathbf{z}(\sigma) = \mathbf{x}_1 + \sigma\mathbf{r}, \quad d\mathbf{z} = \mathbf{r}\,d\sigma, \qquad (2.176)$$

and

$$\dot{\mathbf{z}} = \dot{\mathbf{x}}_1 + \sigma\dot{\mathbf{r}} \qquad (2.177)$$

if the time variation of the path is assigned purely to the motion of the charges. Then

$$\mathbf{M}(\mathbf{x}, C = \text{st.line}) = e \, (\mathbf{r} \wedge \dot{\mathbf{x}}_1) \int_0^1 \delta^3(\mathbf{x}_1 + \sigma\mathbf{r} - \mathbf{x}) \, d\sigma$$

$$+ e \, (\mathbf{r} \wedge \dot{\mathbf{r}}) \int_0^1 \delta^3(\mathbf{x}_1 + \sigma\mathbf{r} - \mathbf{x}) \sigma \, d\sigma. \qquad (2.178)$$

The first integral is the same as in (2.167), while the second may be evaluated in similar fashion. Their Fourier transforms yield

$$\mathbf{M}(\mathbf{k}, C = \text{st.line}) = ie \, \frac{\mathbf{r} \wedge \dot{\mathbf{x}}_1}{\mathbf{k}\cdot\mathbf{r}} (e^{i\mathbf{k}\cdot\mathbf{x}_1} - e^{i\mathbf{k}\cdot\mathbf{x}_2})$$

$$- e \, \frac{\mathbf{r} \wedge \dot{\mathbf{r}}}{\mathbf{k}\cdot\mathbf{r}} \left(\frac{e^{i\mathbf{k}\cdot\mathbf{x}_1} - e^{i\mathbf{k}\cdot\mathbf{x}_2}}{\mathbf{k}\cdot\mathbf{r}} + ie^{i\mathbf{k}\cdot\mathbf{x}_2} \right). \qquad (2.179)$$

Reversing the direction of the path of integration is the same as interchanging the labels 1,2 for the endpoints, and so $\mathbf{M}$ should be antisymmetric under this exchange; (2.179) may be simplified to the explicitly antisymmetric form,

$$\mathbf{M}(\mathbf{k}, C = \text{st.line}) = ie \frac{\mathbf{r} \wedge \dot{\mathbf{x}}_1 e^{i\mathbf{k}\cdot\mathbf{x}_1} - \mathbf{r} \wedge \dot{\mathbf{x}}_2 e^{i\mathbf{k}\cdot\mathbf{x}_2}}{\mathbf{k}\cdot\mathbf{r}} - e \frac{\mathbf{r} \wedge \dot{\mathbf{r}}}{(\mathbf{k}\cdot\mathbf{r})^2} \left(e^{i\mathbf{k}\cdot\mathbf{x}_1} - e^{i\mathbf{k}\cdot\mathbf{x}_2} \right). \qquad (2.180)$$

After Fourier transformation, Eq. (2.139) reads

$$\mathbf{j}(\mathbf{k}) = \frac{d\mathbf{P}(\mathbf{k})}{dt} + i\mathbf{k} \wedge \mathbf{M}(\mathbf{k}). \qquad (2.181)$$

To the same order in $\mathbf{k}$ as the electric quadrupole term, (2.174), we require simply $\mathbf{M}(\mathbf{0})$, and for the straight-line path this 'magnetic dipole' contribution is

$$\mathbf{M}(\mathbf{0}) = \frac{e}{2} \mathbf{r} \wedge (\dot{\mathbf{x}}_1 + \dot{\mathbf{x}}_2). \qquad (2.182)$$

The forms (2.165) and (2.175) can be obtained directly from a reduction of results in relativistic electrodynamics to the non-relativistic limit without any analogies to the classical dielectric theory. A moving charge traces out a curve in spacetime called the particle's worldline. It is convenient to describe the coordinates of the worldline using a parameter τ, so that $x^\mu(\tau)$, $\mu = 1,2,3 = \mathbf{x}(\tau)$ and $x^0(\tau) = c\tau$. A path between two charges can be described in parameterised form, such that the boundary values of the parameter σ give the positions of the two particles [46]. If the particles move, so does this path, which sweeps out a two-dimensional surface describable by functions $X^\mu(\tau,\sigma), \mu = 0-3$. This is the realm of string theory, much of which is described in entirely classical fashion prior to quantisation [47]. Mansfield [29] has shown that the tensor

$$p^{\mu\nu}(Y) = -e \int d\tau d\sigma \left(\frac{\partial X^\mu}{\partial \sigma} \frac{\partial X^\nu}{\partial \tau} - \frac{\partial X^\nu}{\partial \sigma} \frac{\partial X^\mu}{\partial \tau} \right) \delta^4(X - Y) \tag{2.183}$$

satisfies Eq. (2.131) with the current density for two moving particles of equal and opposite charges

$$j^\mu = e \int_{-\infty}^{+\infty} \dot{X}_1 \delta^4(X - X_1)\, dt - e \int_{-\infty}^{+\infty} \dot{X}_2 \delta^4(X - X_2)\, dt. \tag{2.184}$$

For the reduction to non-relativistic form, we put $X^0(\tau) = c\tau$ and choose the 'static gauge' with $\tau = t$, where t is physical time. Then, since

$$\left(\frac{\partial X^\mu}{\partial \sigma} \right) = \left(\frac{\partial X^0}{\partial \sigma}, \frac{\partial \mathbf{X}}{\partial \sigma} \right) \equiv \left(0, \frac{\partial \mathbf{X}}{\partial \sigma} \right) \tag{2.185}$$

$$\left(\frac{\partial X^\mu}{\partial \tau} \right) = \left(\frac{\partial X^0}{\partial t}, \frac{\partial \mathbf{X}}{\partial t} \right) \equiv \left(c, \frac{\partial \mathbf{X}}{\partial t} \right), \tag{2.186}$$

the corresponding polarisation fields that represent the non-relativistic forms of the current density (2.184) are just (2.165) and (2.175) with $\mathbf{X}$ identified with $\mathbf{z}$. This suggests a novel interpretation of the polarisation fields in terms of the paths themselves being considered as dynamical variables.

2.4.4 Faraday's Lines of Force

The electric polarisation field $\mathbf{P}(\mathbf{x})$ is a contribution to the electric field $\mathbf{E}(\mathbf{x})$ due to the charges other than the transverse electric field associated with free radiation. The flux of $\mathbf{P}(\mathbf{x})$ through a closed surface Σ is given by the integral

$$\int_\Sigma \mathbf{P}(\mathbf{x}) \cdot d\mathbf{S} \equiv \int_V (\mathbf{\nabla} \cdot \mathbf{P}(\mathbf{x}))\, d^3\mathbf{x}, \tag{2.187}$$

by Green's theorem, where V is the volume enclosed by Σ. As an example consider a typical term in (2.164). If V contains both endpoints, the integral over V yields zero, consistent with the assumed electroneutrality. Otherwise the RHS of (2.187) counts the number of charges contained in V, while the LHS counts the number of paths, or flux lines that cut the bounding surface Σ, in this case one. However, $\mathbf{P}$ is not the solution of the Maxwell equation for the electric field of the flux lines.

A possible way of visualising the situation is to regard the flux lines as physical objects; this is essentially the view of Faraday [48] who sought to describe the properties of the electromagnetic field (and other fields) in terms of Faraday lines of force. The lines were endowed with physical characteristics, for example, tension and mass, and could move with both transverse and longitudinal oscillations. In this way, he aimed to account for the effects of the field on charged bodies, and for the propagation of electromagnetic radiation without recourse to an æther. Open lines have equal and opposite charges at their endpoints, while closed paths describe some state of the field. The lines are tangent to the field and so describe its direction; the density of lines measures the magnitude of the electric field.

The vector $\mathbf{g}$, (2.64), occurs in a manifestly gauge-invariant formulation of electrodynamics [49]; Dirac considered the example of a single electron located at a point $\mathbf{X}$ and examined the electric field around it. At a point $\mathbf{x}$ in space, this turns out to exceed the electric field of the vacuum state by an amount $e\varepsilon_0^{-1}\mathbf{g}(\mathbf{x}:\mathbf{X})$. Note that for the case of a single charged (point) particle with charge e located at $\mathbf{X}$, we may write [50]

$$e\mathbf{g}(\mathbf{x},\mathbf{X}) = \int \mathbf{g}(\mathbf{x},\mathbf{x}')\rho(\mathbf{x}')\,d^3\mathbf{x}' \equiv \mathbf{P}(\mathbf{x}), \tag{2.188}$$

since in this case

$$\rho(\mathbf{x}) = e\delta^3(\mathbf{x}-\mathbf{X}). \tag{2.189}$$

The choice of $\mathbf{g}$ specified in (2.65) leads to the result that the excess field is precisely the Coulomb field of the charge. This is the longitudinal component of the polarisation field for a single charge e,

$$\mathbf{P}(\mathbf{x})^{\|} = e\nabla_\mathbf{x}\frac{1}{4\pi|\mathbf{x}-\mathbf{X}|}, \tag{2.190}$$

with Fourier transform

$$\mathbf{P}(\mathbf{k})^{\|} = -ie\mathbf{k}\frac{e^{i\mathbf{k}\cdot\mathbf{X}}}{k^2}. \tag{2.191}$$

A more general choice such as (2.69) leads to the Coulomb field *plus* a field of pure electromagnetic radiation as the excess. Dirac's discussion [49] is consistent with taking the polarisation field for a single charged particle as

$$\mathbf{P}(\mathbf{x}) = e\int_C^\mathbf{X} \delta^3(\mathbf{z}-\mathbf{x})\,d\mathbf{z}, \tag{2.192}$$

where C is a path from ∞ to the particle at $\mathbf{X}$.

The line integral form implies that the electric field is concentrated purely on the path C ending at the charge and is a multivalued quantity because of the path dependence. Dirac interpreted the electric field associated with the path C as a single *Faraday line of force* between the charge and the reference point $\mathbf{O}$, which he took to be spatial infinity. The tensor $p^{\mu\nu}$ with components defined by (2.183) describes the electromagnetic field associated with the line of force connecting pairs of particle positions. He also noted that a closed path would describe a state of the electromagnetic field that is connected with the particles because the elementary charge e occurs in the coefficient of the integral. He further conjectured that a novel quantum electrodynamics might be constructed using the lines of force (the paths $\mathcal{C}$) as the basic dynamical variables from which our conventional notions of charged particles and electromagnetic fields would be derived; however, as here, Dirac still regarded a given path as a fixed classical object.

As just noted the points in space $\mathbf{x}$ for which the electric polarisation field $\mathbf{P}$ is non-zero are restricted to the chosen path $\mathcal{C}$; on the other hand, its longitudinal component $\mathbf{P}^{\parallel}$ extends throughout the whole space. The relationship between the two may be visualised in terms of a *superposition* of straight-line paths to fill out the space. Consider the path parameterisation $\mathbf{z}(\sigma) = \mathbf{X} + \sigma\hat{\mathbf{n}}; -\infty \leq \sigma \leq 0$ which describes a straight path from infinity to the point $\mathbf{X}$ in a direction specified by $\hat{\mathbf{n}}$. Then the polarisation field is

$$\mathbf{P}(\mathbf{x})_{\hat{\mathbf{n}}} = e \int_{-\infty}^{0} \hat{\mathbf{n}} \, \delta^3(\mathbf{z}(\sigma) - \mathbf{x}) \, d\sigma. \tag{2.193}$$

The normalised superposition of the $\mathbf{P}$-fields over all possible directions $\hat{\mathbf{n}}$ in the continuum limit is simply the uniform average

$$\langle \mathbf{P}(\mathbf{x}) \rangle_{\hat{\mathbf{n}}} = \frac{1}{4\pi} \int \mathbf{P}(\mathbf{x})_{\hat{\mathbf{n}}} \, d\Omega. \tag{2.194}$$

The paths within the infinitesimal solid angle $d\Omega$ will fill up a cone at the position $\mathbf{X}$ with volume elements $d^3\mathbf{z} = r^2 \, dr \, d\Omega$, where $r = |\mathbf{z} - \mathbf{X}|$. Equation (2.194) is easily evaluated using an argument given by Belinfante [27] with the result that

$$\langle \mathbf{P}(\mathbf{x}) \rangle_{\hat{\mathbf{n}}} = \mathbf{P}(\mathbf{x})^{\parallel}. \tag{2.195}$$

References

[1] Maxwell, J. C. (1865), Phil. Trans. Roy. Soc. (London), **CLV**, reprinted in *A Dynamical Theory of the Electromagnetic Field*, (1982), ed. T. F. Torrance, Scottish Academic Press.

[2] Heaviside, O. (1892), Phil. Mag. Series V **27**, 324.

[3] Heaviside, O. (1892), Phil. Trans. Roy. Soc. (London) A**183**, 423.

[4] Barron, L. D. (1982), *Molecular Light Scattering and Optical Activity*, Cambridge University Press.

[5] Cunningham, E. (1910), Proc. London Math. Soc. **8**, 77.

[6] Bateman, H. (1910), Proc. London Math. Soc. **8**, 223.

[7] Fushchich, W. I. and Nikitin, A. G. (1992), J. Math. Phys. A: Math. Gen. **25**, L231.

[8] Anco, S. C. and Pohjanpelto, J. (2003), Proc. Roy. Soc. (London) A**459**, 1215.

[9] Lipkin, D. M. (1964), J. Math. Phys. **5**, 696.

[10] Kibble, T. W. B. (1965), J. Math. Phys. **6**, 1022.

[11] Tang, Y. and Cohen, A. E. (2010), Phys. Rev. Letters **104**, 163901.

[12] Yang, N. and Cohen, A. E. (2011), J. Phys. Chem. B**115**, 5304.

[13] Cameron, R. P., Barnett, S. M. and Yao, A. M. (2012), New J. Phys. **14**, 053050 (16 pages).

[14] Woolley, R. G. (1975), Adv. Chem. Phys. **33**, 153.

[15] Jackson, J. D. (1998), *Classical Electrodynamics*, 3rd ed., John Wiley and Sons Inc.

[16] Stoney, G. J. (1894), Phil. Mag. Series V **38**, 418.

[17] von Helmholtz, H. V. F. (1881), J. Chem. Soc. **39**, 277.

[18] Poincaré, H. (1886), J. de Math. **2**, 151.

[19] Power, E. A. (1964), *Introductory Quantum Electrodynamics*, Longmans.

[20] Loudon, R. (1983), *The Quantum Theory of Light*, 2nd ed., Clarendon Press.

[21] Landau, L. and Lifshitz, E. M. (1971), *Course of Theoretical Physics*, vol. 2, *The Classical Theory of Fields*, 3rd ed., Pergamon Press.

[22] Lipkin, D. M. (1964), J. Math. Phys. **5**, 696.

[23] Poynting, J. H. (1909), Proc. Roy. Soc. (London) A**82**, 560.

[24] Beth, R. A. (1936), Phys. Rev. **50**, 115.

[25] Hay, G. E. (1953), *Vector and Tensor Analysis*, Ch. V, §59, Dover Publications Inc.

[26] Johnson, B. R., Hirschfelder, J. O. and Yang, K. H. (1983), Rev. Mod. Phys. **55**, 109.

[27] Belinfante, F. J. (1962), Phys. Rev. **128**, 2832.

[28] Woolley, R. G. (1999), Int. J. Quant. Chem. **74**, 531.

[29] Mansfield, P. R. W. (2012), J. High Energy Phys. 2012:149.

[30] Valatin, J. G. (1954), Proc. Roy. Soc. (London) A**222**, 93.

[31] Brittin, W. E., Rodman Smythe, W. and Wyss, W. (1982), Amer. J. Phys. **50**, 693.

[32] Jackson, J. D. (2002), Amer. J. Phys. **70**, 917.

[33] Poincaré, H. (1906), Rend. Circ. Mat. Palermo **21**, 129.

[34] Lorrain, P. and Corson, D. R. (1970), *Electromagnetic Fields and Waves*, 2nd ed., W. H. Freeman.

[35] Hughes, V. W., Fraser, L. J. and Carlson, E. R. (1988), Z. Phys. D – Atoms, Molecules and Clusters **10**, 145.

[36] Earnshaw, S. (1842), Trans. Camb. Phil. Soc. **7**, 97.

[37] Lorentz, H.A. (1909), *The Theory of Electrons and Its Application to the Phenomena of Light and Radiant Heat*, B. G. Teubner.

[38] Rohrlich, F. (1960), Amer. J. Phys. **28**, 639.

[39] Atkins, P. W. and Woolley, R. G. (1970), Proc. Roy. Soc. (London) A**319**, 549.

[40] Woolley, R. G. (1971), Proc. Roy. Soc. (London) A**321**, 557.

[41] Buckingham, A. D. (1959), J. Chem. Phys. **30**, 1580.

[42] Buckingham, A. D. and Longuet-Higgins, H. C. (1968), Mol. Phys. **14**, 63.

[43] Woolley, R. G. (1971), Mol. Phys. **22**, 1013.

[44] Healy, W. P. (1977), Proc. Roy. Soc. (London) A**358**, 367.

[45] Woolley, R. G. (1975), Ann. Inst. Henri Poincaré **23**, 365.

[46] The path between two charges must not be confused with the worldline of an individual charge.

[47] Zwieback, B. (2009), *A First Course in String Theory*, 2nd ed. Cambridge University Press.

[48] Faraday, M. (1846), Phil. Mag. Series III **28**, 345.

[49] Dirac, P. A. M. (1955), Can. J. Phys. **33**, 650.

[50] The quantity $c_r(x,x')$ in Dirac's equations (18), (19) and (40) is essentially **g**.

Dynamics of Charges and the Electromagnetic Field

3.1 Introduction

The dynamics of the electromagnetic field and charged particles are governed by the Maxwell–Lorentz equations discussed in the previous chapter. For purely classical considerations this is perfectly sufficient, and classical electrodynamics is usually presented as an exploration of the implications of Maxwell's partial differential equations and of the Lorentz Force Law using Newton's equation of motion.

As we shall see in Part II, quantum theory largely dispenses with the notion of force and so the conventional account of classical electrodynamics does not provide a suitable starting point for quantisation. We require an alternative formulation that treats the field and charged particles in a common framework. It is therefore important for the development of the (quantum) dynamical theory of the combined system of charged particles and electromagnetic field that there are such formulations of mechanics due to Lagrange and Hamilton. These methods are developed from energy considerations [1].

In this chapter we summarise the main ideas of the classical Lagrangian and Hamiltonian approaches to mechanics and describe their application to classical electrodynamics. As we shall see, the basic 'coordinates' for the electromagnetic field are the scalar and vector potentials $(\phi, \mathbf{a})$ which should be seen as useful working variables; their gauge transformation properties (Chapter 2) play an important role in the resulting theory which carries with it an intrinsic redundancy because $(\phi, \mathbf{a})$ are not independent quantities. We begin with Lagrange's method.

3.2 Lagrangian Mechanics

The principle of least action (or Hamilton's principle) asserts that a mechanical system can be characterised by a function, L, of coordinates and velocities called the Lagrangian, such that the integral

$$S = \int_{t_1}^{t_2} L(x, \dot{x}, t) \, \mathrm{d}t \tag{3.1}$$

taken over a path from t_1 to t_2 has its minimum value for the actual motion of the system in this time interval. S is a mechanical quantity called the action of the system. By way of illustration, we begin with the simple example of a point particle of mass m moving under a potential energy, $V(x)$, for which the Lagrangian can be taken to be

$$L = \frac{1}{2}m\dot{x}^2 - V(x). \tag{3.2}$$

Let $x(t)$ describe the path that minimises the action S; then the principle of least action asserts that a neighbouring path that shares the same end points $x(t_1), x(t_2)$,

$$x(t)' = x(t) + \delta x(t), \tag{3.3}$$

makes S increase provided that the variations at the end points vanish:

$$\delta x(t_1) = \delta x(t_2) = 0. \tag{3.4}$$

Now on making the replacement $x \to x'$, the action S is altered to

$$\begin{aligned} S \to S' &= \int_{t_1}^{t_2} \left(\frac{1}{2}m(\dot{x} + \delta\dot{x})^2 - V(x + \delta x) \right) dt \\ &= S + \int_{t_1}^{t_2} \left(m\dot{x}\delta\dot{x} - \delta x \frac{dV(x)}{dx} \right) dt + 0(\delta x^2) \\ &\equiv S + \delta S. \end{aligned} \tag{3.5}$$

This calculation is called a variation of S; for S to be the minimum value of the integral we must have

$$\delta S = 0, \tag{3.6}$$

that is,

$$\int_{t_1}^{t_2} \left(m\dot{x}\delta\dot{x} - \delta x \frac{dV(x)}{dx} \right) dt = 0. \tag{3.7}$$

Integration by parts applied to the first term in (3.7) yields

$$\begin{aligned} \int_{t_1}^{t_2} m\dot{x}\delta\dot{x}\, dt &= m\dot{x}\,\delta x\big|_{t_1}^{t_2} - m\int_{t_1}^{t_2} \ddot{x}\delta x\, dt \\ &= -m\int_{t_1}^{t_2} \ddot{x}\delta x\, dt, \end{aligned} \tag{3.8}$$

in view of (3.4). Hence, (3.6) requires

$$\int_{t_1}^{t_2} \left(m\ddot{x}\delta x + \frac{dV(x)}{dx}\delta x \right) dt = 0. \tag{3.9}$$

This must be satisfied for an arbitrary variation δx and so the condition for S to take its minimum value is

$$m\ddot{x} = -\frac{dV(x)}{dx}, \tag{3.10}$$

which is Newton's second law.

Note that for a free particle, integration of (3.10) yields

$$\dot{x} = \frac{p_0}{m}, \quad (x - x_0) = \frac{p_0 t}{m},$$

(3.11)

and so for this simple case the action integral is

$$S = p_0 (x - x_0),$$

(3.12)

that is,

$$\text{Action} = \text{momentum} \times \text{displacement}.$$

(3.13)

Thus, the dimensions of action are those of angular momentum or equivalently energy × time; these are alternative forms for the dimensions of Planck's constant, h, which is often referred to as the 'quantum of action'.

This example is capable of considerable generalisation. It can be extended to systems with many degrees of freedom and to fields and can accommodate the use of generalised (curvilinear) coordinates. Let the coordinates be $x = (x_1, \ldots x_i, \ldots x_N)$; then we write the action integral as

$$S = \int_{t_1}^{t_2} L(x, \dot{x}, t) \, dt.$$

(3.14)

Its extremum may be found by use of the calculus of variations; as before we restrict the variational procedure so that the variations vanish at t_1 and t_2, and obtain

$$\delta S = \sum_{i}^{N} \int_{t_1}^{t_2} \left(\frac{\partial L}{\partial x_i} \delta x_i + \frac{\partial L}{\partial \dot{x}_i} \delta \dot{x}_i \right) dt$$
$$= \sum_{i}^{N} \int_{t_1}^{t_2} \left(\frac{\partial L}{\partial \dot{x}_i} - \frac{d}{dt} \left(\frac{\partial L}{\partial \dot{x}_i} \right) \right) \delta x_i \, dt,$$

(3.15)

after an integration by parts and use of the boundary conditions on the variations. The condition $\delta S = 0$ therefore implies the system of equations

$$0 = \frac{\partial L}{\partial x_i} - \frac{d}{dt} \left(\frac{\partial L}{\partial \dot{x}_i} \right), \quad i = 1, \ldots N,$$

(3.16)

which are known as the Euler–Lagrange equations, or as the Lagrangian equations of motion.

The circumstance that L in Eq. (3.16) is assumed to depend on only coordinates and velocities arises because the target equation of motion involves no more than second-order derivatives with respect to time, for example Newton's law for a mechanical system. It is perfectly possible to work with Lagrangian functions that depend on higher-order time derivatives of the coordinates, say up to the kth order,

$$L = L(x, x^{(1)}, x^{(2)}, \ldots x^{(k)}),$$

(3.17)

where

$$x^{(n)} = \frac{d^n x}{dt^n}.$$

(3.18)

The direct application of the calculus of variations then yields

$$\sum_{i=0}^{k}\left(-\frac{\mathrm{d}}{\mathrm{d}t}\right)^{i}\left(\frac{\partial L}{\partial x^{(i)}}\right)=0, \tag{3.19}$$

provided all boundary variations of the paths up to order $(k-1)$ vanish. It follows that if the highest-order time derivative in L is k, the equations of motion are of order $2k$. It is often the case that the derivatives up to the maximum order for all the coordinates enter into the Lagrangian L; L is then said to be 'normal'. It is not a necessary requirement, however; and if some of the derivatives are missing, the Lagrangian is said to be 'degenerate'. As we will see, degeneracy has important implications for the transition to the Hamiltonian formalism since it implies that there will be equations of constraint among the Hamiltonian variables.

An alternative approach to Lagrangians with higher-order derivatives than velocities is to introduce Lagrange multipliers as additional coordinates to be varied. As a simple example suppose the Lagrangian involves accelerations,

$$L = L(x, \dot{x}, \ddot{x}). \tag{3.20}$$

We regard the velocity, $\dot{x}$, as an independent variable and write

$$q_0 \equiv x, \quad q_1 \equiv \dot{x}, \quad \dot{q}_1 \equiv \ddot{x}, \tag{3.21}$$

so that the Lagrangian becomes

$$L = L(q_0, q_1, \dot{q}_1) + \lambda(q_1 - \dot{q}_0). \tag{3.22}$$

The equations of motion are recovered by deriving the Euler–Lagrange equations for variations of q_0, q_1, λ. This construction can be generalised for Lagrangians containing derivatives up to, say, the kth order by introducing the recursion

$$q_i = \dot{q}_{i-1}, \quad i = 1, 2, \ldots, k-1, \tag{3.23}$$

with the initial value

$$q_0 = x. \tag{3.24}$$

The modified Lagrangian will then contain $(k-1)$ Lagrange multipliers as additional variables; because of the Lagrange multipliers it is degenerate and so will lead to a constrained Hamiltonian system. This will be considered in detail in §3.5, since electrodynamics involves exactly such constrained dynamics.

For a closed system, the Lagrangian function cannot have any explicit dependence on the time t; on the other hand, the characteristic property of a closed system is that its energy is constant. These two statements are closely related; the first implies

$$\frac{\partial L}{\partial t} = 0. \tag{3.25}$$

Now by the chain rule of calculus,

$$\frac{\mathrm{d}L}{\mathrm{d}t} = \sum_{i}^{N}\left(\frac{\partial L}{\partial x_i}\right)\dot{x}_i + \sum_{i}^{N}\left(\frac{\partial L}{\partial \dot{x}_i}\right)\ddot{x}_i. \tag{3.26}$$

With the aid of the Euler–Lagrange equations, (3.16), we rewrite the first term in (3.26) as

$$\sum_i^N \left(\frac{\partial L}{\partial x_i} \right) \dot{x}_i = \sum_i^N \dot{x}_i \frac{\mathrm{d}}{\mathrm{d}t} \left(\frac{\partial L}{\partial \dot{x}_i} \right), \tag{3.27}$$

so that

$$\frac{\mathrm{d}L}{\mathrm{d}t} = \frac{\mathrm{d}}{\mathrm{d}t} \sum_i^N \dot{x}_i \frac{\partial L}{\partial \dot{x}_i}. \tag{3.28}$$

Equation (3.28) can be rearranged to

$$\frac{\mathrm{d}}{\mathrm{d}t} \left(\sum_i^N \dot{x}_i \frac{\partial L}{\partial \dot{x}_i} - L \right) = 0, \tag{3.29}$$

and thus

$$\sum_i^N \dot{x}_i \frac{\partial L}{\partial \dot{x}_i} - L = E, \tag{3.30}$$

where the constant E is identified as the energy of the system.

This calculation is the prototype of the derivation of the mechanical conservation laws in the Lagrangian formalism; Eq. (3.30) is also a central part of Hamilton's scheme for mechanics. It is evident from the Euler–Lagrange equations, (3.16), that the Lagrangian L, and hence also the action S, can be multiplied by an arbitrary constant without any change in the equations of motion; with the identification of (3.30) as energy we see that this is just the freedom to introduce appropriate constants in the definitions of the units used for the action and the energy.

We must now consider how we might construct a Lagrangian function for a given physical system. Firstly, we note a substantial lack of uniqueness in L. Suppose we have two Lagrangian functions for a system that differ by a total time derivative of some function of the coordinates and the time, $f(x,t)$:

$$L_2(x,\dot{x},t) = L_1(x,\dot{x},t) + \frac{\mathrm{d}f(x,t)}{\mathrm{d}t}. \tag{3.31}$$

The corresponding action integrals are related by

$$S_2 = S_1 + f(x(t_2),t_2) - f(x(t_1),t_1), \tag{3.32}$$

and so when the action integrals are varied, $x(t) \rightarrow x(t) + \delta x(t)$ with fixed values at the end points, we obtain

$$\delta S_2 = \delta S_1. \tag{3.33}$$

Hence, the form of the Lagrangian function can be modified by such a total time derivative without affecting the equations of motion. Notice that for dimensional reasons the function f is itself an action. The function f can also involve derivatives of the coordinates and still leave the equations of motion unchanged according to (3.19).

The action S has the important property of behaving as a scalar under certain coordinate transformations. For slowly moving mass point particles, these transformations describe Galilean relativity; on the other hand, if special relativity is appropriate, then S must behave as a true scalar under Lorentz transformations. Such properties allow us to specify suitable Lagrangian functions for ordinary mechanical systems without difficulty. As an example consider first a slowly moving point particle; if the particle is free, the invariance of S under a Galilean transformation that relates two inertial frames requires $L \propto \dot{x}^2$. The proportionality constant is conventionally written as $\frac{1}{2}m$, where m is identified as the mass of the particle; the mass cannot be negative since the action integral S is required to possess a minimum. For a system of N free particles, each particle has its own equation of motion that is independent of all the other particles, and so the total Lagrangian must be a linear superposition of kinetic energy terms,

$$L = \frac{1}{2} \sum_i^N m_i \dot{x}_i^2 \equiv T. \tag{3.34}$$

For a system of interacting particles, the Lagrangian can often be written as the difference between the kinetic energy, T, and a potential energy function, V:

$$L = T - V. \tag{3.35}$$

More generally, the equations of motion for a dynamical system with N degrees of freedom can often be put in the form [2]

$$\mathbf{G}(x, \dot{x}, \ddot{x}, t) = \dot{\mathbf{f}}_2(x, \dot{x}, t) - \mathbf{f}_1(x, \dot{x}, t) = 0. \tag{3.36}$$

This is so, for example, in Newtonian mechanics, where force is rate of change of momentum so that $\mathbf{f}_1$ and $\mathbf{f}_2$ are identified as forces and momenta, respectively. The functions $(\mathbf{f}_1, \mathbf{f}_2)$ are not determined uniquely, however, but only up to equivalence by the transformations

$$\mathbf{f}_1 \to \mathbf{f}_{1'} = \mathbf{f}_1 + \frac{d\mathbf{C}}{dt}, \tag{3.37}$$

$$\mathbf{f}_2 \to \mathbf{f}_{2'} = \mathbf{f}_2 + \mathbf{C}, \tag{3.38}$$

where the components of the N-dimensional vector $\mathbf{C}$ are linearly independent, but otherwise arbitrary. There is thus considerable freedom in the form of the 'forces' and 'momenta' that yield the given equations of motion. Even so it is often possible to write down L directly in terms of some chosen pair $(\mathbf{f}_1, \mathbf{f}_2)$.

The inverse problem in the calculus of variations applied to dynamics poses the following question: given a definite equation of motion expressed in terms of position, velocity and acceleration variables as in (3.36), what are the conditions such that it can be recovered from the usual Euler–Lagrange equations for a corresponding Lagrangian L? For our purposes, the answer is that the equation of motion must satisfy the so-called *Helmholtz conditions* $\{H_i\}$ which are [3]

$$H_1: \quad \frac{\partial G_i}{\partial \ddot{x}_m} \equiv \frac{\partial G_m}{\partial \ddot{x}_i}$$

$$H_2: \quad \frac{\partial G_i}{\partial \dot{x}_m} + \frac{\partial G_m}{\partial \dot{x}_i} \equiv \frac{\mathrm{d}}{\mathrm{d}t}\left(\frac{\partial G_i}{\partial \ddot{x}_m} + \frac{\partial G_m}{\partial \ddot{x}_i}\right)$$

$$H_3: \quad \frac{\partial G_i}{\partial x_m} - \frac{\partial G_m}{\partial x_i} \equiv \frac{1}{2}\frac{\mathrm{d}}{\mathrm{d}t}\left(\frac{\partial G_i}{\partial \dot{x}_m} - \frac{\partial G_m}{\partial \dot{x}_i}\right) \qquad i,m = 1,2,3; \qquad (3.39)$$

they are necessary and sufficient conditions.

We now turn to the Lagrangian description of a field in as preparation for incorporating the dynamics of the electromagnetic field in the framework outlined here. Formally, the transition from a point particle description to a field description consists of the replacements of the orbit $x(t)$ by the field variable ξ, and of time t by the space-time vector $x_\alpha = (\mathbf{x}, t)$. The field variable is a continuous function of x_α. Note that in contrast to particle mechanics, $\mathbf{x}$ and t are now treated on an equal footing as independent variables. We introduce a Lagrangian density,

$$\mathcal{L} = \mathcal{L}(\xi, \xi^{,\alpha}), \quad \xi^{,\alpha} = \frac{\partial \xi}{\partial x_\alpha}; \ x_1, x_2, x_3 = \mathbf{x}, x_4 = t, \qquad (3.40)$$

so that the field Lagrangian is its integral over space:

$$L = \int \mathcal{L}(\xi, \xi^{,\alpha}) \mathrm{d}^3\mathbf{x}. \qquad (3.41)$$

The action integral is then defined in the same way as in particle mechanics,

$$S = \int \mathcal{L}(\xi, \xi^{,\alpha}) \mathrm{d}\Omega, \quad \mathrm{d}\Omega = \mathrm{d}^3\mathbf{x}\,\mathrm{d}t, \qquad (3.42)$$

and the associated Euler–Lagrange equations are

$$\frac{\partial \mathcal{L}}{\partial \xi} - \left(\frac{\partial \mathcal{L}}{\partial \xi^{,\alpha}}\right)^{,\alpha} = 0, \qquad (3.43)$$

or in vector notation

$$\frac{\partial \mathcal{L}}{\partial \xi} - \frac{\partial}{\partial t}\left(\frac{\partial \mathcal{L}}{\partial(\partial \xi/\partial t)}\right) - \boldsymbol{\nabla}\cdot\left(\frac{\partial \mathcal{L}}{\partial(\boldsymbol{\nabla}\xi)}\right) = 0. \qquad (3.44)$$

The Euler–Lagrange equations, (3.44), can also usefully be written in terms of the Lagrangian L, in which case functional derivatives are required. A brief summary of some relevant relations is given in Appendix A which shows that we may simply replace $\mathcal{L}$ by L, and ∂ by δ in (3.43), (3.44). A calculation corresponding to (3.25)–(3.30) shows that there exists a set of conserved quantities if the Lagrangian for the field has no explicit dependence on the space-time coordinates x_α. Although we do not go into details here, it should be mentioned that this conservation condition leads directly to the energy-momentum tensor for the field. This is called Noether's theorem and is of fundamental importance for identifying symmetries and conserved quantities in Lagrangian field theory [4].

3.3 Lagrangian Electrodynamics

As an example that brings together dynamics and the discussion in §2.4.1, consider a point particle with mass M, electric charge e and position $\mathbf{x}$, moving with velocity $\dot{\mathbf{x}}$ in a specified electromagnetic field $(\mathbf{E}, \mathbf{B})$; the equations of motion, (3.36), can be formulated by choosing $\mathbf{f}_1$ as the Lorentz force and $\mathbf{f}_2$ as the particle momentum:

$$\mathbf{f}(\mathbf{x}, \dot{\mathbf{x}}, t)_1 = e\mathbf{E}(\mathbf{x}, t) + e\dot{\mathbf{x}} \wedge \mathbf{B}(\mathbf{x}, t) \tag{3.45}$$

$$\mathbf{f}(\dot{\mathbf{x}}, t)_2 = \mathbf{p}(\dot{\mathbf{x}}). \tag{3.46}$$

Equations (3.45), (3.46) are subject to the transformations, Eqs. (3.37), (3.38).

In non-relativistic mechanics, $\mathbf{p} = M\dot{\mathbf{x}}$; the equation of motion in the form of (3.36), required for the Helmholtz conditions, is therefore

$$G_p = M\ddot{x}_p - eE_p(x, t) - e\varepsilon_{pqr}\dot{x}_q B_r(x, t) = 0. \tag{3.47}$$

The derivatives required for the Helmholtz conditions are

$$\frac{\partial G_i}{\partial x_m} = -e\frac{\partial E_i}{\partial x_m} - e\varepsilon_{iqr}\dot{x}_q\frac{\partial B_r}{\partial x_m}$$

$$\frac{\partial G_i}{\partial \dot{x}_m} = -e\varepsilon_{imr}B_r$$

$$\frac{\partial G_i}{\partial \ddot{x}_m} = M\delta_{im}. \tag{3.48}$$

It is easy to see that H_1 is satisfied since

$$\frac{\partial G_i}{\partial \ddot{x}_m} = \frac{\partial G_m}{\partial \ddot{x}_i} = M\delta_{im}, \tag{3.49}$$

which simply gives the constant M for $i = m$; otherwise it gives 0. Using this result, the terms in H_2 are

$$\frac{\partial G_i}{\partial \dot{x}_m} + \frac{\partial G_m}{\partial \dot{x}_i} = -e\varepsilon_{imr}B_r - e\varepsilon_{mir}B_r = 0$$

$$\frac{\mathrm{d}}{\mathrm{d}t}\left(\frac{\partial G_i}{\partial \ddot{x}_m} + \frac{\partial G_m}{\partial \ddot{x}_i}\right) = 2\frac{\mathrm{d}}{\mathrm{d}t}(2M\delta_{im}) = 0. \tag{3.50}$$

The third Helmholtz condition, H_3, is valid, provided that the Maxwell equations (2.1) and (2.2), that is,

$$\boldsymbol{\nabla} \wedge \mathbf{E} + \frac{\partial \mathbf{B}}{\partial t} = 0; \quad \boldsymbol{\nabla} \cdot \mathbf{B} = 0, \tag{3.51}$$

are satisfied by the electromagnetic fields $\mathbf{E}, \mathbf{B}$. This can be seen in the following way. First we write the condition in an equivalent form,

$$(\boldsymbol{\nabla} \wedge \mathbf{G})_p = \frac{1}{2}\frac{\mathrm{d}}{\mathrm{d}t}\varepsilon_{pim}\frac{\partial G_i}{\partial \dot{x}_m}, \tag{3.52}$$

which is convenient for calculation. Now the LHS of (3.52) is,[1] from (3.47),

$$(\nabla \wedge \mathbf{G})_p = -e(\nabla \wedge \mathbf{E})_p - e\varepsilon_{pim}\varepsilon_{iqr}\dot{x}_q \frac{\partial B_r}{\partial x_m}$$

$$= -e(\nabla \wedge \mathbf{E})_p + e\dot{x}_p \frac{\partial B_r}{\partial x_r} - e\dot{x}_m \frac{\partial B_p}{\partial x_m}. \qquad (3.53)$$

Returning to vector notation, the second term in (3.53) is $\dot{x}_p(\nabla \cdot \mathbf{B})$, and the third term is the p component of $(\frac{d\mathbf{B}}{dt} - \frac{\partial \mathbf{B}}{\partial t})$; so finally

$$(\nabla \wedge \mathbf{G})_p = e\left[(\nabla \wedge \mathbf{E}) + e\frac{\partial \mathbf{B}}{\partial t}\right]_p + \dot{x}_p(\nabla \cdot \mathbf{B}) - e\left(\frac{d\mathbf{B}}{dt}\right)_p. \qquad (3.54)$$

The RHS of (3.52) is

$$\frac{1}{2}\frac{d}{dt}\varepsilon_{pim}\frac{\partial G_i}{\partial \dot{x}_m} = -e\frac{1}{2}\frac{d}{dt}\varepsilon_{pim}(\varepsilon_{imr}B_r)$$

$$= -e\left(\frac{d\mathbf{B}}{dt}\right)_p. \qquad (3.55)$$

Equations (3.54) and (3.55) are equal, provided that (3.51) holds.

This calculation confirms that, though the Lorentz force in general is non-conservative, it is still possible to have a Lagrangian function from which Newton's law with the Lorentz force can be recovered in the usual way using the Euler–Lagrange equations. From the previous discussion, we know that the Lagrangian for a collection of charged particles in an electromagnetic field will not be unique; nevertheless a valid form can be written directly using a line integral over the Lorentz force,

$$L = \frac{1}{2}\sum_n m_n \dot{x}_n^2 + \sum_n \int_{C_n}^{\mathbf{x}_n} \mathbf{f}_{1n}(\mathbf{z},\dot{\mathbf{z}},t) \cdot d\mathbf{z}, \qquad (3.56)$$

where the paths $\{C_n\}$ are from some arbitrary origin to the particle coordinates $\{\mathbf{x}_n\}$. In general, $\mathbf{f}_{1n}(\mathbf{z},\dot{\mathbf{z}},t) \cdot d\mathbf{z}$ is not an exact differential, and so the Lagrangian is path-dependent.[2] Nevertheless the Euler–Lagrange equations give directly

$$\frac{d}{dt}\left(\frac{\partial L}{\partial \dot{\mathbf{x}}_n}\right) = m\ddot{\mathbf{x}}_n$$

$$\frac{\partial L}{\partial \mathbf{x}_n} = \mathbf{f}_{1n}(\mathbf{x}_n,\dot{\mathbf{x}}_n,t), \qquad (3.57)$$

and so the Lorentz force acting on particle n is recovered. On the other hand, if a particular path is specified, extra terms appear on the RHS of (3.57); but these vanish because they contain the Maxwell equations (2.1) and (2.2) as factors, and so the equation of motion is preserved if these Maxwell equations are included in the Lagrangian theory as necessary subsidiary conditions [2], [5]. This is very satisfactory since (2.1) and (2.2) imply that the electromagnetic fields can be described in terms of the field potentials $\{\phi, \mathbf{a}\}$ which are suitable Lagrangian coordinates for use in the principle of least action.

---

[1] This follows from the identity $\varepsilon_{ijk}\varepsilon_{imn} = \delta_{jm}\delta_{kn} - \delta_{jn}\delta_{km}$.
[2] That is, L does not have a definite value, but it does have definite first derivatives.

The form of the Lorentz force, (3.45), suggests that (3.56) can be written in terms of the polarisation fields reviewed in §2.4.3. Indeed if we take $q = 0$, we have at once

$$
\begin{aligned}
L_{\text{int}} &= \sum_n \int_{C_n}^{\mathbf{x}_n} \mathbf{f}_{1n}(\mathbf{z}, \dot{\mathbf{z}}, t) \cdot \mathrm{d}\mathbf{z} \\
&= \int \mathbf{P}(\mathbf{x}) \cdot \mathbf{E}(\mathbf{x}, t) \, \mathrm{d}^3 \mathbf{x} + \int \mathbf{M}(\mathbf{x}) \cdot \mathbf{B}(\mathbf{x}, t) \, \mathrm{d}^3 \mathbf{x} \\
&= L_{\mathbf{E}} + L_{\mathbf{B}}.
\end{aligned}
\tag{3.58}
$$

In terms of the Faraday tensor for the field and the polarisation tensor described in Chapter 2, (3.58) is just

$$
L_{\text{int}} = -\frac{\varepsilon_0}{2} \int f_{\mu\nu} p^{\mu\nu} \, \mathrm{d}^3 \mathbf{x}.
\tag{3.59}
$$

For $q \neq 0$, a small obvious modification involving $\mathbf{E}^{\parallel}$ is required (see (2.143)). The path dependence of (3.56) obviously carries over to (3.58), (3.59); it can be recognised as a contribution to L in the following way.

Suppose we adopt (2.142) as the general form for the electric polarisation field; then the part of the Lagrangian potential in (3.58) depending on $\mathbf{E}$ in full is

$$
L[\mathbf{E}, \rho] = \iint \mathbf{E}(\mathbf{x}, t) \cdot \mathbf{g}(\mathbf{x}, \mathbf{x}') \rho(\mathbf{x}') \, \mathrm{d}^3 \mathbf{x}' \, \mathrm{d}^3 \mathbf{x}.
\tag{3.60}
$$

Turning the argument round, we could regard (3.60) as the fundamental form for this part of the Lagrangian potential and take (2.142) as a convenient abbreviation for one of the integrations that leaves the electric field explicit in the Lagrangian. We could equally well simplify (3.60) by defining a scalar functional of the electric field,

$$
\phi(\mathbf{x}', t) = \int \mathbf{E}(\mathbf{x}, t) \cdot \mathbf{g}(\mathbf{x}, \mathbf{x}') \, \mathrm{d}^3 \mathbf{x},
\tag{3.61}
$$

and present (3.60) in a form with the charge density explicit,

$$
L[\rho, \phi] = \int \rho(\mathbf{x}') \phi(\mathbf{x}', t) \, \mathrm{d}^3 \mathbf{x}'.
\tag{3.62}
$$

For the moment we make the choice $\mathbf{g}(\mathbf{x}, \mathbf{x}')^{\perp} = 0$ in (3.61), that is, we use the Coulomb gauge condition, and calling the result $V(\mathbf{x}', t)$, we have at once

$$
\mathbf{E}(\mathbf{x}, t)^{\parallel} = -\nabla V(\mathbf{x}, t),
\tag{3.63}
$$

where $V(\mathbf{x}, t)$ is recognised as a conventional scalar potential for the electromagnetic field. We may write the electric field in terms of the Coulomb gauge potentials,

$$
\mathbf{E} = -\frac{\partial \mathbf{A}}{\partial t} - \nabla V,
\tag{3.64}
$$

so that (3.60) takes the form

$$L[\mathbf{A}, V, \rho] = -\int \rho(\mathbf{x}) V(\mathbf{x}, t) \, d^3\mathbf{x}$$

$$- \iint \frac{\partial \mathbf{A}(\mathbf{x}, t)}{\partial t} \cdot \mathbf{g}(\mathbf{x}, \mathbf{x}') \rho(\mathbf{x}') \, d^3\mathbf{x}' \, d^3\mathbf{x}. \tag{3.65}$$

In similar fashion (2.145) with (3.58) yields

$$L[\mathbf{B}, \mathbf{j}, \rho] = \iint \mathcal{Z}(\mathbf{x} - \mathbf{x}') \mathbf{B}(\mathbf{x}, t) \cdot \left(\boldsymbol{\nabla}_{\mathbf{x}'} \wedge \mathbf{j}(\mathbf{x}') \right) d^3\mathbf{x}' \, d^3\mathbf{x}$$

$$- \iiint \mathcal{Z}(\mathbf{x} - \mathbf{x}') \, \mathbf{B}(\mathbf{x}, t) \cdot (\boldsymbol{\nabla}_{\mathbf{x}'} \wedge \mathbf{g}(\mathbf{x}', \mathbf{x}'')) \frac{d\rho(\mathbf{x}'')}{dt} \, d^3\mathbf{x}'' \, d^3 \, \mathbf{x}' \, d^3\mathbf{x}, \tag{3.66}$$

and in complete analogy with (3.60) we can introduce a vector-valued functional of the magnetic field and express (3.66) in simplified form with the charge and current densities explicit. It is convenient to make an integration by parts in (3.66) and use the definition of the Coulomb gauge vector potential, (2.49), so that (3.66) becomes

$$L[\mathbf{A}, \mathbf{j}, \rho] = \int \mathbf{j}(\mathbf{x}, t) \cdot \mathbf{A}(\mathbf{x}, t) \, d^3 \, \mathbf{x}$$

$$- \iint \mathbf{A}(\mathbf{x}, t) \cdot \mathbf{g}(\mathbf{x}, \mathbf{x}') \frac{d\rho(\mathbf{x}')}{dt} \, d^3\mathbf{x}' \, d^3\mathbf{x}. \tag{3.67}$$

Combining (3.65) and (3.67), the full Lagrangian potential (3.58) is transformed to

$$L_{\mathrm{int}} = -\int \rho(\mathbf{x}) V(\mathbf{x}, t) \, d^3\mathbf{x} + \int \mathbf{j}(\mathbf{x}) \cdot \mathbf{A}(\mathbf{x}, t) \, d^3\mathbf{x}$$

$$- \iint \frac{\partial \mathbf{A}(\mathbf{x}, t)}{\partial t} \cdot \mathbf{g}(\mathbf{x}, \mathbf{x}') \rho(\mathbf{x}') \, d^3\mathbf{x}' \, d^3\mathbf{x}$$

$$- \iint \mathbf{A}(\mathbf{x}, t) \cdot \mathbf{g}(\mathbf{x}, \mathbf{x}') \frac{d\rho(\mathbf{x}')}{dt} \, d^3\mathbf{x}' \, d^3\mathbf{x}. \tag{3.68}$$

If (2.142) is reintroduced into the last term, this is

$$L[\mathbf{A}, V]_{\mathrm{int}} = -\int \rho(\mathbf{x}) V(\mathbf{x}, t) \, d^3\mathbf{x} + \int \mathbf{j}(\mathbf{x}) \cdot \mathbf{A}(\mathbf{x}, t) \, d^3\mathbf{x}$$

$$- \frac{d}{dt} \int \mathbf{P}(\mathbf{x}) \cdot \mathbf{A}(\mathbf{x}, t) \, d^3 \, \mathbf{x}. \tag{3.69}$$

We have seen that the addition of a total time derivative to a Lagrangian does not contribute to the variation of the action integral associated with the Lagrangian, and so the Euler–Lagrange equations yield the same equations of motion. Thus, the last term in (3.69) does not affect the classical equations of motion; for that reason it is usually dropped, and the customary Lagrangian for charged particles in a given field is based on the first two terms in (3.69) [6]–[9]. The time derivative is required if the

Lagrangian is to display its full form invariance under arbitrary gauge transformations of the potentials; moreover, by retaining it we may expose the physical significance of this form invariance.

Suppose we modify the potentials $(\mathbf{A}, V)$ by making a gauge transformation (§2.3) according to

$$\mathbf{A} \to \mathbf{a} - \boldsymbol{\nabla}\chi, \quad V \to \phi + \frac{\partial \chi}{\partial t}. \tag{3.70}$$

Then the Lagrangian potential $L[\mathbf{A}, V]_{\text{int}}$, (3.69), transforms as

$$L[\mathbf{A}, V]_{\text{int}} \to L[\mathbf{a}, \phi]_{\text{int}} + L^{\chi}, \tag{3.71}$$

where

$$L[\mathbf{a}, \phi]_{\text{int}} = \int \rho(\mathbf{x})\phi(\mathbf{x}, t)\, \mathrm{d}^3\mathbf{x} + \int \mathbf{j}(\mathbf{x}) \cdot \mathbf{a}(\mathbf{x}, t)\, \mathrm{d}^3\mathbf{x}$$

$$- \frac{\mathrm{d}}{\mathrm{d}t} \int \mathbf{P}(\mathbf{x}) \cdot \mathbf{a}(\mathbf{x}, t)\, \mathrm{d}^3\mathbf{x}, \tag{3.72}$$

which is of exactly the same form as (3.69) but written in terms of the new potentials, and

$$L^{\chi} = -\int \frac{\partial \chi}{\partial t}\rho\, \mathrm{d}^3\mathbf{x} - \int \mathbf{j} \cdot \boldsymbol{\nabla}\chi\, \mathrm{d}^3\mathbf{x} + \frac{\mathrm{d}}{\mathrm{d}t} \int \mathbf{P} \cdot \boldsymbol{\nabla}\chi\, \mathrm{d}^3\mathbf{x}. \tag{3.73}$$

Using integration by parts on the second and third terms leads to two integrals taken over surfaces at infinity which we suppose to vanish,[3] and there results

$$L^{\chi} = -\int \frac{\partial \chi}{\partial t}\rho\, \mathrm{d}^3\mathbf{x} + \int \chi \boldsymbol{\nabla} \cdot \mathbf{j}\, \mathrm{d}^3\mathbf{x} + \frac{\mathrm{d}}{\mathrm{d}t} \int \rho\chi\, \mathrm{d}^3\mathbf{x}$$

$$= \int \chi\left(\boldsymbol{\nabla} \cdot \mathbf{j} + \frac{\mathrm{d}\rho}{\mathrm{d}t}\right) \mathrm{d}^3\mathbf{x} = 0, \tag{3.74}$$

by virtue of the equation of continuity, (2.9). We have thus shown that

$$L_{\text{int}} = \int \mathbf{P} \cdot \mathbf{E}\, \mathrm{d}^3\mathbf{x} + \int \mathbf{M} \cdot \mathbf{B}\, \mathrm{d}^3\mathbf{x},$$

$$= -\int \rho\phi\, \mathrm{d}^3\mathbf{x} + \int \mathbf{j} \cdot \mathbf{a}\, \mathrm{d}^3\mathbf{x} - \frac{\mathrm{d}}{\mathrm{d}t} \int \mathbf{P} \cdot \mathbf{a}\, \mathrm{d}^3\mathbf{x}, \tag{3.75}$$

irrespective of the choice of the field potentials $(\phi, \mathbf{a})$. The gauge invariance of the Lagrangian potential, (3.69), describing the interaction of a charge density with the electromagnetic field is seen to be intimately connected with the conservation of electric charge. In §2.4.3 we showed that the polarisation fields also admit a class of transformations derived from a 'polarisation potential' $(u, \mathbf{U})$. If we define new polarisation fields $(\mathbf{P}', \mathbf{M}')$ according to (2.140), (2.141), a straightforward calculation shows that

---

[3] In any case, the surface integrals make no contribution to the variation of the action integral S, and so may be dropped.

the relationship in (3.75) is still valid if we simply replace $(\mathbf{P}, \mathbf{M})$ by $(\mathbf{P}', \mathbf{M}')$. Hence, L_{int} is also form-invariant against changes in $(u, \mathbf{U})$.

For the electromagnetic field, the action S must transform as a scalar under Lorentz transformations; such a scalar involving purely the field variables is the inner product formed from the Faraday tensor (Chapter 2).

$$S_{\text{rad}} = -\frac{\varepsilon_0 c^2}{4} \int f_{\mu\nu} f^{\mu\nu} \, \mathrm{d}^3 \mathbf{x} \, \mathrm{d}t. \tag{3.76}$$

The least action formulation of mechanics requires the equations of motion to be (at least) second-order differential equations in the Lagrangian variables; this is true of the Maxwell equations expressed in terms of $(\phi, \mathbf{a})$ but not in terms of $(\mathbf{E}, \mathbf{B})$, so we use the potentials $(\phi, \mathbf{a})$ as working variables related to $(\mathbf{E}, \mathbf{B})$ by (2.41), (2.42). This also means that (2.1) and (2.2) are automatically satisfied. In three-dimensional notation, (3.76) corresponds to choosing

$$L_{\text{rad}} = \tfrac{1}{2}\varepsilon_0 \int \left(\mathbf{E} \cdot \mathbf{E} - c^2 \mathbf{B} \cdot \mathbf{B} \right) \mathrm{d}^3 \mathbf{x} \tag{3.77}$$

as the Lagrangian, where $|\mathbf{E}|^2$ must enter with a positive sign so that the action S may have a minimum.[4]

With these results, we can now treat a system of charged particles and the electromagnetic field as a closed system using a Lagrangian L with the following structure:

$$L = L_{\text{charges}} + L_{\text{rad}} + L_{\text{int}}. \tag{3.78}$$

In the Galilean relativistic scheme, L_{charges} is given by (3.34); the interaction Lagrangian, L_{int}, is (3.72), and the field Lagrangian, L_{rad}, is (3.77). Then the Lagrangian written out explicitly in full in terms of all the Lagrangian coordinates is

$$L = \tfrac{1}{2} \sum_n^N m_n \dot{x}_n^2 - \int \rho \phi \, \mathrm{d}^3 \mathbf{x} + \int \mathbf{j} \cdot \mathbf{a} \, \mathrm{d}^3 \mathbf{x}$$
$$+ \tfrac{1}{2}\varepsilon_0 \int \left[\left(\frac{\partial \mathbf{a}}{\partial t} + \boldsymbol{\nabla}\phi \right) \cdot \left(\frac{\partial \mathbf{a}}{\partial t} + \boldsymbol{\nabla}\phi \right) - c^2 \boldsymbol{\nabla} \wedge \mathbf{a} \cdot \boldsymbol{\nabla} \wedge \mathbf{a} \right] \mathrm{d}^3 \mathbf{x}, \tag{3.79}$$

where for the moment we have dropped the total time derivative in (3.72). The use of the field potentials ensures that (2.1) and (2.2) are automatically valid; the other pair of Maxwell equations, (2.3) and (2.4), follow from application of the Euler–Lagrange equations (3.44) to (3.79). As an example, consider the Euler–Lagrange equation for the scalar potential ϕ. The required functional derivatives of L, (3.79), are

$$\frac{\delta L}{\delta \phi} = -\rho; \quad \frac{\delta L}{\delta(\partial \phi/\partial t)} = 0; \quad \frac{\delta L}{\delta(\boldsymbol{\nabla}\phi)} = \varepsilon_0 \left(\frac{\partial \mathbf{a}}{\partial t} + \boldsymbol{\nabla}\phi \right) \equiv -\varepsilon_0 \mathbf{E}, \tag{3.80}$$

and these together with (3.44) yield the Maxwell equation (2.3); the other Maxwell equation follows in similar fashion. Similarly the particle equations of motion in terms

---

[4] $|\mathbf{E}|^2$ contains $(\partial \mathbf{a}/\partial t)^2$, and if this term were negative, one could make the action S take arbitrarily large negative values by sufficiently rapid variation of $\mathbf{a}$ with time in (3.42), and so S could not have a minimum [10].

of the Lorentz force follow from the application of (3.16) to (3.79), taking the $\{\mathbf{x}_n\}$ as the Lagrangian coordinates for the particles. In these calculations we assume that the particle and field coordinates are independent quantities when applying the variational calculus.

In writing the full Lagrangian in the form (3.79) no particular gauge condition, is required since the action S it leads to is gauge-invariant. Suppose, however, that we use the interaction Lagrangian in the form of (3.71) and include variations of the arbitrary function χ in the action integral. The extra terms so generated may be written as

$$\delta S' = \int \left(\frac{d\rho}{dt} + \boldsymbol{\nabla} \cdot \mathbf{j} \right) d^4x \, \delta\chi, \tag{3.81}$$

which must vanish for the actual motion. This requires that the integrand vanishes, and so this calculation leads directly to the equation of continuity, (2.9). Thus, not only does the Lagrangian (3.79) reproduce the expected equations of motion, but it also encodes a fundamental conservation law.

The integral in the total time derivative term in (3.69),

$$F = \int \mathbf{P}(\mathbf{x}) \cdot \mathbf{a}(\mathbf{x},t) \, d^3\mathbf{x}, \tag{3.82}$$

has the dimensions of action; using (2.143) for the case of a single point charge e with position $\mathbf{q}$ we have

$$F(\mathbf{q}) = e \int_{\mathbf{O}}^{\mathbf{q}} \mathbf{a}(\mathbf{z}) \cdot d\mathbf{z} + e \int \mathbf{a} \cdot \boldsymbol{\nabla}_{\mathbf{x}} \frac{1}{4\pi|\mathbf{x} - \mathbf{O}|} \, d^3\mathbf{x}. \tag{3.83}$$

The general vector potential may be written in terms of the unique Coulomb gauge vector potential and an arbitrary longitudinal vector field,

$$\mathbf{a} = \mathbf{A} + \boldsymbol{\nabla}\chi, \tag{3.84}$$

according to (2.45), and so the action F is

$$F(\mathbf{q}) = e \int_{\mathbf{O}}^{\mathbf{q}} d\omega + e\chi(\mathbf{q}). \tag{3.85}$$

The differential 1-form,

$$d\upsilon = \mathbf{A}(\mathbf{z}) \cdot d\mathbf{z}, \tag{3.86}$$

is defined in terms of an infinitesimal path element $d\mathbf{z}$ along a path $\mathcal{C}$ from the arbitrary reference point $\mathbf{O}$ to the particle's position $\mathbf{q}$.

Its most important property is demonstrated by evaluating $F(\mathbf{q})$ for two different paths $\mathcal{C}_1$ and $\mathcal{C}_2$ connecting $\mathbf{O}$ to the position of the charge $\mathbf{q}$. We have

$$\Delta F_{1,2} = F(\mathbf{q})_1 - F(\mathbf{q})_2$$

$$= e \int_{\mathbf{O},\mathcal{C}_1}^{\mathbf{q}} d\upsilon - e \int_{\mathbf{O},\mathcal{C}_2}^{\mathbf{q}} d\upsilon$$

$$= e \int_{\mathbf{O}, \mathcal{C}_1}^{\mathbf{q}} \mathrm{d}\upsilon + e \int_{\mathbf{q}}^{\mathbf{O}, \mathcal{C}_2} \mathrm{d}\upsilon \equiv e \oint \mathrm{d}\upsilon$$

$$= e \int_{\Sigma_{12}} \mathbf{B} \cdot \mathrm{d}\mathbf{S}, \tag{3.87}$$

where Σ_{12} is the area bounded by the curves $\mathcal{C}_1$ and $\mathcal{C}_2$. Thus, if $\mathbf{q}$ lies in a region where the magnetic field is non-zero, the action $F(\mathbf{q})$ *does not have a definite value*. This is not important for the classical Lagrangian since the equations of motion are unaffected. As we shall see, however, the differential 1-form $\mathrm{d}\upsilon$ turns out to be of fundamental significance in the Hamiltonian formalism applied to both classical and quantum charges interacting with either classical or quantised electromagnetic fields.

We noted in Chapter 2 that when considering radiation and charged particles together one can define a displacement tensor $d^{\mu\nu}$ as a modification of the Faraday tensor for the field alone that has an inner product interpreted as an energy density. One can therefore define an action by setting

$$S = -\frac{\varepsilon_0}{4} \int d_{\mu\nu} d^{\mu\nu} \, \mathrm{d}^3\mathbf{x} \, \mathrm{d}t. \tag{3.88}$$

Using (2.132) this can be multiplied out into three terms, of which two are exactly the action integrals obtained from (3.59) and (3.76); using the polarisation tensor (2.134) the third, new, term becomes

$$S_{\mathrm{pol}} = -\frac{1}{4\varepsilon_0 c^2} \int p_{\mu\nu} p^{\mu\nu} \, \mathrm{d}^3\mathbf{x} \, \mathrm{d}t = \frac{1}{2\varepsilon_0} \int \mathbf{P} \cdot \mathbf{P} \, \mathrm{d}^3\mathbf{x} \, \mathrm{d}t - \frac{1}{2\varepsilon_0 c^2} \int \mathbf{M} \cdot \mathbf{M} \, \mathrm{d}^3\mathbf{x} \, \mathrm{d}t. \tag{3.89}$$

The magnetisation $\mathbf{M}$ is linear in the particle velocity, and so the second term is formally of order $(v/c)^2$ compared to the first term; the polarisation fields $\mathbf{P}$ and $\mathbf{M}$ for point particles are distributions, so this action can be expected to be a highly singular object. If we employ the lines-of-force interpretation of the fields, (3.89) is the action integral for a collection of lines of force. Since it does not appear possible to write it as a total time derivative, the action (3.88) describes physics different from that of the conventional Lagrangian we developed previously.

The action integral for the line of force between two equal and opposite charges at the end points of the line is the simplest case of (3.89). This has recently been studied using the relativistic form, (2.183), for $p^{\mu\nu}$ and shown to contain divergent factors; it is not necessary to give the details of the calculations, so we simply quote the result that (3.89) can be transformed to the form [11], [12]

$$\mathcal{S}_{\mathrm{pol}} = -\frac{e^2}{2\varepsilon_0} \delta^2(0) \mathcal{S}_{\mathrm{NG}} + \text{a contact interaction term} + \ldots, \tag{3.90}$$

where $\mathcal{S}_{\mathrm{NG}}$ is the standard Nambu–Goto action for a classical relativistic string [13], and $\delta^2(0)$ is the singular spatial delta function in two dimensions evaluated at the origin; note that δ^2 has the dimensions of an inverse area. The leading term of the non-relativistic limit of (3.90) is just

$$\mathcal{S}_{\mathrm{NG}}^{\mathrm{non-rel}} \sim \frac{1}{2\varepsilon_0} \int \mathbf{P} \cdot \mathbf{P} \, \mathrm{d}^3\mathbf{x} \, \mathrm{d}t = \frac{e^2}{2\varepsilon_0} \delta^2(0) \int \mathrm{d}L \, \mathrm{d}t, \tag{3.91}$$

where L is the arc length along the line of force; for the straight-line path between the two charges, this is simply r, and so the leading term of (3.89) grows with increasing r. This is investigated in the next section.

3.3.1 The Energy of the Electric Polarisation Field

Leaving aside the magnetisation contribution, the combination of (3.89) and (2.164) may be decomposed into a sum of 'atomic' terms, associated with a single nucleus and a sum over the pairwise cross terms, involving two distinct nuclei:

$$\frac{1}{2\varepsilon_0}\int \mathbf{P}(\mathbf{x})\cdot\mathbf{P}(\mathbf{x})\,d^3\mathbf{x} = \frac{1}{2\varepsilon_0}\int \sum_m^M |\mathbf{P}(\mathbf{x})_m|^2\,d^3\mathbf{x} + \frac{1}{2\varepsilon_0}\int \sum_{m\neq m'}^M \mathbf{P}(\mathbf{x})_m\cdot\mathbf{P}(\mathbf{x})_{m'}\,d^3\mathbf{x}$$

$$= \mathcal{E}_{\mathbf{P}}. \tag{3.92}$$

These sums must be further differentiated as there are three kinds of terms to evaluate as follows: Type 1 – terms that refer to one nucleus and one electron ('self-terms'); Type 2 – terms that refer to one nucleus and two electrons; and Type 3 – terms that refer to two different nuclei and two electrons. Elementary dimensional considerations show that the energy $\mathcal{E}_{\mathbf{P}}$ must be of the form $e^2/\varepsilon_0 L$ where L is some length determined by the choice of $\mathbf{P}$. Using the Fourier transform of $\mathbf{P}$, we may write

$$\mathcal{E}_{\mathbf{P}} = \frac{1}{2\varepsilon_0}\frac{1}{(2\pi)^3}\int \mathbf{P}(\mathbf{k})\cdot\mathbf{P}(-\mathbf{k})\,d^3\mathbf{k}, \tag{3.93}$$

which is convenient for calculation.

It is easy to see that (3.93) for the single charge is divergent simply by counting powers of $|\mathbf{k}|$ in (2.191); there are many possible ways by which it may be regulated so that it becomes possible to isolate finite contributions (if any) that survive when the regulator is removed. As described in §2.4.2, the regularisation of infinities that arise in non-relativistic classical electrodynamics can be interpreted in terms of a smoothing of the delta function in (2.165). Introducing $\chi_a(k)$ as before, we then have

$$\mathcal{E}_{\mathbf{P},a}^{\parallel} = \frac{e^2}{2\varepsilon_0}\frac{1}{(2\pi)^3}\int \frac{\chi_a^2(k)}{k^2}\,d^3\mathbf{k}$$

$$= \frac{e^2}{8\pi\varepsilon_0 a}. \tag{3.94}$$

In the limiting case of the point particle ($a \to 0$), this is the familiar classical infinite Coulombic self-energy.

For the neutral two-particle system discussed in Chapter 2, there is only the Type 1 contribution,

$$\mathcal{E}_{\mathbf{P}} = \frac{e}{2\varepsilon_0}\int_{\mathbf{x}_1}^{\mathbf{x}_2} \mathbf{P}(\mathbf{z})\cdot d\mathbf{z}, \tag{3.95}$$

with $\mathbf{P}$ given by (2.165). Since $\nabla\wedge\mathbf{P}$ only vanishes if $\mathbf{P}$ is the gradient of a single-valued field (purely longitudinal), we see that $\mathcal{E}_{\mathbf{P}}$ in general is path-dependent; thus, we cannot assign it a definite value in general. According to the earlier discussion of the electric polarisation field (Chapter 2), $\mathcal{E}_\mathbf{P}$ is the energy associated with a line of force. It is of interest to calculate it for the popular straight-line path between $\mathbf{x}_1$ and $\mathbf{x}_2$ using the Fourier transform (2.173). The longitudinal and transverse contributions to $\mathcal{E}_\mathbf{P}$ can be identified by introducing a resolution of the identity into (3.93) to separate the scalar product with

$$\mathbf{I} = \hat{\mathbf{k}}\hat{\mathbf{k}} + (\mathbf{I} - \hat{\mathbf{k}}\hat{\mathbf{k}}). \tag{3.96}$$

The regulated longitudinal contribution from (3.95) is then

$$\mathcal{E}_{\mathbf{P},a}^{\parallel} = \left(\frac{e^2}{4\pi\varepsilon_0}\right)\left(\frac{1}{2\pi^2}\right)\int \chi_a^2(k)\frac{1 - \cos(\mathbf{k}\cdot\mathbf{r})}{k^2}\,\mathrm{d}^3k. \tag{3.97}$$

The angular integrations are elementary, and there results

$$\begin{aligned}
\mathcal{E}_\mathbf{P}^{\parallel} &= \frac{e^2}{2\pi^2\varepsilon_0}\lim_{a\to 0}\int_0^\infty \chi_a^2(k)\left(1 - \frac{\sin(kr)}{kr}\right)\mathrm{d}k \\
&= \frac{e^2}{4\pi\varepsilon_0}\left(\frac{1}{a}\bigg|_{a\to 0} - \frac{1}{r}\right),
\end{aligned} \tag{3.98}$$

where the limit $a \to 0$ has already been taken in the second term as it does not need the regulator. Here the first term is the infinite self-energy of two charges in the limit $a \to 0$ (cf (3.94)), and the second term is the Coulomb energy of the two charges a distance r apart.

Similarly, the regulated transverse part is

$$\mathcal{E}_{\mathbf{P},a}^{\perp} = \left(\frac{e^2}{4\pi\varepsilon_0}\right)\left(\frac{1}{2\pi^2}\right)r^t r^s\int \chi_a^2(k)(\mathbf{1} - \hat{\mathbf{k}}\hat{\mathbf{k}})_{ts}\frac{1 - \cos(\mathbf{k}\cdot\mathbf{r})}{(\mathbf{k}\cdot\mathbf{r})^2}\,\mathrm{d}^3k. \tag{3.99}$$

$\mathcal{E}_{\mathbf{P},a}^{\perp}$ is a path-dependent quantity, and the limit $a \to 0$ corresponds to taking an infinitely thin line. For $a > 0$, the line may be thought of as a tube of cross-section area $\propto a^2$. The angular integrations are again straightforward, with the result that

$$\mathcal{E}_\mathbf{P}^{\perp} = \frac{e^2}{2\pi^2\varepsilon_0}\lim_{a\to 0}\int_0^\infty \chi_a^2(k)\left((kr)\,\mathrm{Si}(kr) + \frac{\sin(kr)}{kr} + \cos(kr) - 2\right)\mathrm{d}k, \tag{3.100}$$

where

$$\mathrm{Si}(t) = \int_0^t \frac{\sin(s)}{s}\,\mathrm{d}s \tag{3.101}$$

is the usual sine integral. The remaining integrals are as follows:

$$\begin{aligned}
\int_0^\infty \chi_a^2(k)(kr)\,\mathrm{Si}(kr)\,\mathrm{d}k &= \mathsf{Y}\!\left(\frac{a}{r}\right), \\
\lim_{a\to 0}\int_0^\infty \chi_a^2(k)\frac{\sin(kr)}{kr}\,\mathrm{d}k &= \frac{\pi}{2r}, \\
\lim_{a\to 0}\int_0^\infty \chi_a^2(k)\cos(kr)\,\mathrm{d}k &= \pi\delta(r), \\
-2\lim_{a\to 0}\int_0^\infty \chi_a^2(k)\,\mathrm{d}k &= -\frac{\pi}{a}\bigg|_{a\to 0},
\end{aligned} \tag{3.102}$$

and so

$$\mathcal{E}_{\mathbf{P}}^{\perp} = \frac{e^2}{4\pi\varepsilon_0} \left(2\pi \, \mathsf{Y}(\frac{a}{r}) \Big|_{a\to0} + \frac{1}{r} + 2\pi\delta(r) - \frac{2}{a}\Big|_{a\to0} \right). \tag{3.103}$$

The second term in (3.103) is a Coulomb energy; notice it is precisely equal and of opposite sign to the second term in (3.98). The third term is a contact interaction while the last term is a self-interaction energy, again infinite in the limit. Adding (3.98) and (3.103), we finally obtain the full energy for the primitive Type 1 case, (2.165), as

$$\mathcal{E}_{\mathbf{P}} = \frac{e^2}{4\pi\varepsilon_0} \left(2\pi\mathsf{Y}(\frac{a}{r}) \Big|_{a\to0} + 2\pi\delta(r) - \frac{1}{a}\Big|_{a\to0} \right), \tag{3.104}$$

which is notable because the static Coulomb interaction energy of the particles usually assumed to originate from $\mathcal{E}_{\mathbf{P}}^{\parallel}$ has cancelled completely. If we take the limiting value of (3.104) for $r \to \infty$, the first two terms vanish, and we are left with the infinite self-energies of the two charges infinitely far apart, in agreement with (3.94). For the infinitely thin straight-line integration path ($a = 0$), the function Y has the form

$$\mathsf{Y} \sim \frac{Z_{\mathcal{C}}}{r}, \quad Z_{\mathcal{C}} = \int_0^\infty x \, \mathsf{Si}(x) \, \mathrm{d}x, \tag{3.105}$$

where $Z_{\mathcal{C}}$ is a divergent coefficient. Comparison of (3.98) and (3.100) shows that the cancellation of the Coulomb terms is *independent* of the choice of the regulator function, $\chi_a(k)$. We remark that these calculations are based on a straight-line path $\mathcal{C}$ in the integrations; the longitudinal contribution is obviously path-independent. There is no reason to suppose that this cancellation of the static Coulombic energy between the longitudinal and transverse contributions to the energy could be avoided by a choice of some other path.

We gain a deeper understanding of the formalism based on polarisation fields by analysing Y with the aid of a regulator, for example

$$\chi_a(k) = e^{-ak/\pi}, \tag{3.106}$$

which satisfies the conditions (2.127). With an obvious change of variable, we then have

$$Y(\frac{a}{r}) = \frac{1}{r} \int_0^\infty x \, \mathsf{Si}(x) e^{-sx}, \quad s = \frac{2a}{\pi r}$$

$$= \frac{1}{r} \left[\frac{\arctan(\frac{1}{s})}{s^2} + \frac{1}{s(1+s^2)} \right]. \tag{3.107}$$

For s near 0, this is

$$Y(\frac{a}{r}) \approx \frac{1}{r} \left[\frac{\pi}{s^2} + \mathrm{O}\left(\frac{1}{s}\right) + \dots \right] = \frac{\pi^3 r}{4a^2} + \mathrm{O}\left(\frac{\pi}{2a}\right) + \dots, \tag{3.108}$$

which shows that for fixed $a > 0$ this is an energy that *increases with increasing r*. This is not the result implied by the literature of the last sixty years since the work of Power and Zienau which first introduced the use of the polarisation fields (in a multipolar approximation) in the non-relativistic quantum electrodynamics of atoms and molecules [14], [15]. The appearance of the polarisation fields in classical electrodynamics is not accidental, and as we will see in §3.6, $\mathcal{E}_\mathbf{P}$ occurs quite naturally in the classical Hamiltonian; a fortiori, after canonical quantisation it appears in non-relativistic quantum electrodynamics.

Type 2 and Type 3 contributions to the energy $\mathcal{E}_\mathbf{P}$ arise from multi-particle polarisation fields; the simplest examples are the cross-terms required for the evaluation of $\mathcal{E}_\mathbf{P}$ with

$$\mathbf{P}(\mathbf{x}) = e \int_{\mathbf{x}_1^e}^{\mathbf{x}^N} \delta^3(\mathbf{z} - \mathbf{x}) \, d\mathbf{z} + e \int_{\mathbf{x}_2^e}^{\mathbf{x}^N} \delta^3(\mathbf{z} - \mathbf{x}) \, d\mathbf{z} \quad \text{Type 2}, \tag{3.109}$$

$$\mathbf{P}(\mathbf{x}) = e \int_{\mathbf{x}_1^e}^{\mathbf{x}^N} \delta^3(\mathbf{z} - \mathbf{x}) \, d\mathbf{z} + e \int_{\mathbf{x}_2^e}^{\mathbf{x}^M} \delta^3(\mathbf{z} - \mathbf{x}) \, d\mathbf{z} \quad N \neq M \quad \text{Type 3}. \tag{3.110}$$

Physical examples of (3.109) and (3.110) are the helium atom and the hydrogen molecule, respectively, since, repeating the argument starting at (2.70), one recovers their charge densities. For the general multi-charge system, all three types of term will arise. These are contact interaction contributions to the energy since each term in the polarisation field has support on its own line of force, and thus the integrand in $\mathcal{E}_\mathbf{P}$ is non-zero only at the points where the lines touch (Type 2, at $\mathbf{x}^N$) or intersect (Type 3). So, for example, using (2.168) the cross-terms from (3.109) yield

$$\mathcal{E}_\mathbf{P} \sim \left(\frac{e^2}{4\pi\varepsilon_0} \right) \frac{\hat{\mathbf{r}}_1 \cdot \hat{\mathbf{r}}_2}{|\mathbf{x}^N - \mathbf{x}_1|^2 |\mathbf{x}^N - \mathbf{x}_2|^2} \delta^3(0), \tag{3.111}$$

where the vectors $\mathbf{r}_i, i = 1, 2$ are $\mathbf{x}^N - \mathbf{x}_i, i = 1, 2$.

Confirmation that this evaluation of $\mathcal{E}_\mathbf{P}$ consists of an energy that increases with increasing separation of two particles with equal and opposite electric charges, and a contact interaction when they meet (+ their classical infinite self-energies) can be found in a literature quite remote from classical electrodynamics, namely that concerned with the Yang–Mills non-Abelian gauge theory of strong interactions involving quarks and gluons, and in string theory. The increasing energy feature is the basis of the understanding of 'quark confinement'. A construction used in the Yang–Mills theory can equally well be applied to (Abelian) quantum electrodynamics; thinking applied originally to a quark–antiquark pair suggests a state describing two particles with equal and opposite charges fixed at $\mathbf{x}_1$ and $\mathbf{x}_2$ that contains a Wilson line,

$$\exp\left(\frac{ie}{\hbar} \int_{\mathbf{x}_1}^{\mathbf{x}_2} d\mathbf{z} \cdot \mathbf{a} \right), \tag{3.112}$$

as a phase factor, with the vector potential $\mathbf{a}$ in an arbitrary gauge, will have a confining contribution to the electric field energy that is exactly of the form shown in (3.95) which leads to (3.108) [16]. Obviously $\mathcal{E}_\mathbf{P}$ is only *part* of the total energy, and we have seen that the full Lagrangian L is completely equivalent to Maxwell's equations, so there must be cancellations arising from terms involving both the transverse electromagnetic fields and the polarisation fields that maintain this equivalence.

3.4 Hamiltonian Mechanics

Quantisation of the Lagrangian using Feynman's path integral method is a very important part of modern quantum theory; a discussion of the path integral method applied to non-relativistic quantum electrodynamics can be found in Hoyle and Narlikar [17]. Such an approach, however, is still rather unusual in atomic and molecular physics which is based largely on the canonical formalism with a Hamiltonian and the Schrödinger equation. Accordingly, we now describe how classical electrodynamics may be put into Hamiltonian form in preparation for quantisation.

The canonical formalism requires a Legendre transformation from position variables and their time derivatives (velocities) to a new description in which the velocities are replaced by the momenta (p) conjugate to the position variables (x) defined by the differential relation

$$p = \frac{\partial L}{\partial \dot{x}}, \tag{3.113}$$

in terms of which the Euler–Lagrange equations become

$$\dot{p} = \frac{\partial L}{\partial x}. \tag{3.114}$$

Accompanying this change of variables is a completely new viewpoint of dynamics. For a system with N degrees of freedom, the Lagrangian scheme describes the motion as an orbit in the configuration space formed by the N position variables. In the Hamiltonian scheme, however, both positions and momenta are regarded as independent 'coordinates' in a $2N$-dimensional phase space. Equation (3.113) must be solved for the velocities in terms of the conjugate position and momentum variables

$$\dot{x} = \dot{x}(x, p), \tag{3.115}$$

so that the energy function E, (3.30), can be written in terms of the new variables. In this section we will suppose that L is purely a function of position and velocity variables, and that the velocity associated with every coordinate occurs in the Lagrangian in such a way that the inversion of (3.113) can indeed be achieved.[5] The resulting quantity is called the Hamiltonian for the system

$$\sum_i^N p_i \dot{x}_i - L = H \equiv H(x, p). \tag{3.116}$$

For a closed system, H has no explicit time dependence and is the (conserved) total energy of the system; this is the only case we discuss. If the Hessian matrix is singular, the Lagrangian is said to be degenerate; in such a case equations of constraint involving

---

[5] The formal requirement is that the Hessian matrix $||\mathbf{W}||_{ik} = \partial^2 L/\partial \dot{x}_i \partial \dot{x}_k$ must be *non-singular*.

the canonical variables, but not the velocities, have to be considered as well. Likewise a more general theory, cast in terms of constraint equations, is required if accelerations or higher derivatives of x are involved in the Lagrangian. Hamiltonian mechanics with constraints will be discussed in the next section, §3.5.

The total differential of the Lagrangian function in terms of positions and velocities is

$$dL = \sum_i \frac{\partial L}{\partial x_i} dx_i + \sum_i \frac{\partial L}{\partial \dot{x}_i} d\dot{x}_i$$
$$= \sum_i \dot{p}_i dx_i + \sum_i p_i d\dot{x}_i. \tag{3.117}$$

Now from (3.116) we can obtain the total differential of the Hamiltonian in the two forms

$$dH = \sum_i \dot{x}_i dp_i + \sum_i p_i d\dot{x}_i - dL, \tag{3.118}$$

and

$$dH = \sum_i \left(\frac{\partial H}{\partial x_i}\right) dx_i + \sum_i \left(\frac{\partial H}{\partial p_i}\right) dp_i. \tag{3.119}$$

Use of (3.117) to replace dL in (3.118) and comparison with (3.119) yield the Hamiltonian equations of motion,

$$\dot{x}_i = \frac{\partial H}{\partial p_i}, \quad \dot{p}_i = -\frac{\partial H}{\partial x_i}. \tag{3.120}$$

All dynamical variables are now regarded as functions of the position and momentum variables; the equation of motion for an observable $f = f(x, p)$ is therefore

$$\frac{df}{dt} = \sum_i \frac{\partial f}{\partial x_i} \dot{x}_i + \sum_i \frac{\partial f}{\partial p_i} \dot{p}_i. \tag{3.121}$$

Introducing the Hamiltonian equations of motion (3.120) leads to

$$\frac{df}{dt} = \sum_i \frac{\partial f}{\partial x_i} \frac{\partial H}{\partial p_i} - \sum_i \frac{\partial f}{\partial p_i} \frac{\partial H}{\partial x_i} \equiv \{f, H\}, \tag{3.122}$$

where on the RHS we have introduced the Poisson bracket (P.B.) of f and H. The P.B. of any two dynamical variables f and g is defined as

$$\{f, g\} = \sum_i \frac{\partial f}{\partial x_i} \frac{\partial g}{\partial p_i} - \sum_i \frac{\partial f}{\partial p_i} \frac{\partial g}{\partial x_i}. \tag{3.123}$$

From this definition the following properties may be easily established:

$$\{f, g\} = -\{g, f\}, \tag{3.124}$$
$$\{f, f\} = 0, \tag{3.125}$$
$$\{f, g+h\} = \{f, g\} + \{f, h\}, \tag{3.126}$$
$$\{f, gh\} = g\{f, h\} + \{f, g\}h, \tag{3.127}$$

together with Jacobi's identity,

$$\{f, \{g,h\}\} + \{g, \{f,h\}\} + \{h, \{f,g\}\} = 0. \tag{3.128}$$

The fundamental canonical[6] variables, which have the properties

$$\{x_i, x_j\} = \{p_i, p_j\} = 0$$
$$\{x_i, p_j\} = \delta_{ij}, \tag{3.129}$$

are the basis for Dirac's canonical quantisation scheme [18], to be discussed in §5.2.

There is considerable freedom in the choice of the position and momentum variables. A canonical transformation is a change of variables

$$x_i \to X(\mathbf{x}, \mathbf{p})_i, \quad p_i \to P(\mathbf{x}, \mathbf{p})_i, \quad i = 1, \dots N, \tag{3.130}$$

with the following properties:

1. The equations of motion remain in Hamiltonian form,

$$\dot{X}_i = \frac{\partial H(\mathbf{X}, \mathbf{P})}{\partial P_i}, \quad \dot{P}_i = -\frac{\partial H(\mathbf{X}, \mathbf{P})}{\partial X_i}, \quad i = 1, \dots N. \tag{3.131}$$

2. All P.B. relations are left unchanged, so that for the new canonical variables

$$\{X_i, X_j\} = \{P_i, P_j\} = 0 \tag{3.132}$$
$$\{X_i, P_j\} = \delta_{ij}. \tag{3.133}$$

In general, the new position variables are mixtures of the old position and momentum variables (and similarly for the new momentum variables) such that the terms 'position' and 'momentum' lose their elementary physical meaning.

As a simple example consider a one-dimensional harmonic oscillator; in units with $m = k = 1$, and choosing ordinary Cartesian variables, the Hamiltonian is

$$H = \tfrac{1}{2}\left(p^2 + x^2\right). \tag{3.134}$$

The equations of motion are coupled,

$$\dot{x} = \frac{\partial H}{\partial p} = p; \quad \dot{p} = -\frac{\partial H}{\partial x} = -x, \tag{3.135}$$

but are easily reduced to the usual equation for simple harmonic motion (SHM):

$$\ddot{x} + x = 0. \tag{3.136}$$

Now consider the following change of variables:

$$x \to X = \tfrac{i}{\sqrt{2}}(x + ip), \quad p \to P = \tfrac{1}{\sqrt{2}}(x - ip). \tag{3.137}$$

It may be verified by direct computation that the P.B.s of the new variables are the same as the old, that is, (3.129), and that the equations of motion for the new variables are in Hamiltonian form, so the transformation is canonical. Under this transformation,

$$H \to H' = -iXP. \tag{3.138}$$

---

[6] They are also referred to as 'conjugate variables'.

The Hamiltonian equations of motion for the new variables are uncoupled,

$$\dot{X} = \frac{\partial H'}{\partial P} = -iX, \tag{3.139}$$

$$\dot{P} = -\frac{\partial H'}{\partial X} = iP, \tag{3.140}$$

and thus the dynamical problem is solved by elementary integration:

$$X(t) = ae^{-it}, \quad P(t) = be^{it}. \tag{3.141}$$

Inverting (3.137), it is apparent that these equations give the familiar harmonic solutions of (3.136).

The canonical variables X and P are classical analogues of the annihilation and creation operators for the quantised harmonic oscillator. One can make a further canonical transformation to so-called action-angle variables (J, θ) by setting $X = \sqrt{J}e^{i\theta}, P = i\sqrt{J}e^{-i\theta}$ such that the Hamiltonian is reduced to cyclic form,

$$H'' = J, \tag{3.142}$$

with even simpler equations of motion,

$$\dot{J} = -\frac{\partial H''}{\partial \theta} = 0, \quad J = J_0, \tag{3.143}$$

$$\dot{\theta} = \frac{\partial H''}{\partial J} = 1, \quad \theta = t - t_0. \tag{3.144}$$

The action variable J is the classical analogue of the quantum number operator for the oscillator.

A canonical transformation can be brought about by a function, F, called the generator of the transformation. Making use of suitable Legendre transformations, we may write F in one of four standard forms involving the old and new canonical variables; these forms may be summarised in the following way. We write the generator $F = F(s_1, S_1, s_2, S_2, \ldots, t)$, where s_n stands for one of the variables x_n and p_n, and S_n is one of the variables X_n, P_n; then the equations

$$t_k = \pm \frac{\partial F}{\partial s_k}, \tag{3.145}$$

$$T_k = \mp \frac{\partial F}{\partial S_k}, \tag{3.146}$$

$$H' = H + \frac{\partial F}{\partial t} \tag{3.147}$$

define a canonical transformation. Here t_k is conjugated to s_k, and T_k to S_k; the upper sign applies to the case where the differentiation is taken with respect to a coordinate, and the lower one to the case of differentiation with respect to a momentum variable [19]. It is easily checked that the overall transformation from the Cartesian variables (p, x) to the action-angle variables (J, θ) is produced by a generator involving the old (x) and the new (θ) position variables,

$$F(x, \theta) = \frac{1}{2}x^2 \cot(\theta), \tag{3.148}$$

with

$$p \equiv \frac{\partial F}{\partial x} = x \cot(\theta), \tag{3.149}$$

$$J \equiv -\frac{\partial F}{\partial \theta} = \tfrac{1}{2} x^2 \csc^2(\theta). \tag{3.150}$$

As we shall see in Chapter 4, the free electromagnetic field can be described as an (infinite) collection of harmonic oscillators, and transformations closely related to the preceding ones will prove very useful.

The general theory of transformation to action-angle variables is afforded by the Hamilton–Jacobi equation, developed originally in the nineteenth century as a powerful tool in celestial mechanics. Suppose we make a transformation in which the new Hamiltonian vanishes,

$$H + \frac{\partial S}{\partial t} = 0, \tag{3.151}$$

with

$$H = H(\mathbf{x}, \nabla S; t). \tag{3.152}$$

The quantity $S = S(\mathbf{x}, t)$ is called Hamilton's principal function. The Hamilton–Jacobi equation is a single partial differential equation that provides a route to the solutions to the dynamical equations. It is to be contrasted with Lagrange's and Hamilton's equations which are systems of ordinary differential equations of order n and $2n$, respectively. Evidently, we have

$$\frac{\partial S}{\partial t} = -H, \quad \frac{\partial S}{\partial \mathbf{x}} = \mathbf{p}. \tag{3.153}$$

The integrability condition for the existence of a solution S is simply

$$\frac{\partial}{\partial \mathbf{x}} \frac{\partial S}{\partial t} = \frac{\partial}{\partial t} \frac{\partial S}{\partial \mathbf{x}} \tag{3.154}$$

or

$$\dot{\mathbf{p}} = -\nabla_{\mathbf{x}} H, \tag{3.155}$$

which is the Hamiltonian equation (3.120).

Now consider the total differential of S; we have

$$dS(\mathbf{x}, t) = \sum_i \frac{\partial S}{\partial x_i} dx_i + \frac{\partial S}{\partial t} dt, \tag{3.156}$$

so that

$$\frac{dS(\mathbf{x}, t)}{dt} = \sum_i \frac{\partial S}{\partial x_i} \dot{x}_i + \frac{\partial S}{\partial t} \tag{3.157}$$

$$= \sum_i p_i \dot{x}_i - H \equiv L(\mathbf{x}, t). \tag{3.158}$$

Thus

$$S(\mathbf{x}, t) = \int^t L(\mathbf{x}, t') dt', \tag{3.159}$$

and we recognise the solution of the HJ equation, S, as the classical action (apart from an undetermined integration constant). If the Hamiltonian is conservative, we may write (cf. (3.116))

$$S + Et \equiv S + Ht = \int^t (L + H) \, dt', \tag{3.160}$$

$$= \int \mathbf{p} \cdot d\mathbf{x} \equiv S_0, \tag{3.161}$$

so that

$$S(\mathbf{x}, t) = S_0(\mathbf{x}, t) - Et. \tag{3.162}$$

In (3.161) the integral is over the path followed in the motion without regard to the time variable; that is, how fast the path is traversed is ignored. By separation of variables one then has the time-independent form of the HJ equation,

$$H(\mathbf{x}, \boldsymbol{\nabla} S_0) = E, \tag{3.163}$$

as a partial differential equation that determines the dynamics.

The central importance of the P.B. in Hamiltonian mechanics follows from the following circumstance. Let the generator, F_ε, of a canonical transformation depend linearly on an infinitesimal parameter ε such that

$$x_i \rightarrow X_i^\varepsilon, \quad p_i \rightarrow P_i^\varepsilon. \tag{3.164}$$

Then the infinitesimal change for any dynamical variable Λ under this transformation is given by the P.B. of Λ with F_ε,

$$\delta\Lambda \equiv \Lambda(\mathbf{X}^\varepsilon, \mathbf{P}^\varepsilon) - \Lambda(\mathbf{x}, \mathbf{p}) = \{\Lambda, F_\varepsilon\}. \tag{3.165}$$

Since the P.B. has the same value for any set of canonical variables, the generator F_ε can involve any combination of the old and new canonical variables. The equation of motion (3.120) for a dynamical variable β can be put in this form if we take ε to be an infinitesimal time increment dt,

$$d\beta = \{\beta, H\} \, dt. \tag{3.166}$$

Continuous composition of infinitesimal transformations (by integration) leads to finite continuous canonical transformations. Thus, we can define an evolution operator $T(H)$ by setting

$$T(H, t)\beta = \exp(\{\beta, H\}t), \tag{3.167}$$

where the exponential operator is defined by its expansion as a series of nested P.B.s,

$$\exp(\{\beta, H\}t) = \beta + \{\beta, H\}t + \frac{1}{2!}\{\{\beta, H\}, H\}t^2 + \dots. \tag{3.168}$$

Thus, the time evolution of the system can be seen as a succession of canonical transformations generated by the Hamiltonian itself. If β has vanishing P.B. with the Hamiltonian, it remains constant in time, that is, β is a constant of the motion.

Among all the possible canonical transformations we particularly single out those that leave the system unchanged, that is, the symmetry transformations. All evidence confirms the equivalence of coordinate systems that differ in any or all of the following ways: a translation of the time origin, a translation of the space origin, a rotation of the space axes, and a constant relative velocity between the two coordinate systems [20]. Such point transformations describe the kinematical symmetries that are required by the relativity principle. Their infinitesimal generators are a set of ten quantities comprising the Hamiltonian, H, and three vectors we denote by $\{P_i, J_i, K_i; i = 1, 2, 3\}$ which together form a Lie algebra; in classical mechanics the P.B. acts as the Lie bracket. This algebra describes the symmetry group of transformations that preserve relativistic invariance. As discussed in §2.2, this is the Poincaré group denoted by G_c; when particles move at speeds that are low compared to the speed of light we may use the $c \to \infty$ limit, G_∞, of G_c. This is the Galilean group.

For both groups a canonical representation is provided by the following P.B. relations,

$$\{P_i, H\} = \{J_i, H\} = 0, \tag{3.169}$$

$$\{P_i, P_j\} = 0, \tag{3.170}$$

$$\{J_i, J_j\} = \varepsilon_{ijk} J_k, \quad \{J_i, P_j\} = \varepsilon_{ijk} P_k, \tag{3.171}$$

$$\{J_i, K_j\} = \varepsilon_{ijk} K_k, \quad \{K_j, H\} = P_j, \tag{3.172}$$

for $i, j, k = 1, 2, 3$. In the case of the Poincaré group we must also have

$$\{K_i, K_j\} = -\varepsilon_{ijk} J_k / c^2, \quad \{K_i, P_j\} = \delta_{ij} H, \tag{3.173}$$

whereas for the Galilean group, these P.B.s reduce to

$$\{K_i, K_j\} = 0, \quad \{K_i, P_j\} = \delta_{ij} M, \tag{3.174}$$

where M is a quantity that has vanishing P.B.s with all ten generators (hence M is a neutral element of the group). Equation (3.169) shows that $\mathbf{P}$ and $\mathbf{J}$ are constants of the motion, and (3.170), (3.171) are consistent with their interpretation as linear momentum and angular momentum, respectively.

We may introduce a position vector $\mathbf{R}$ that is canonically conjugate to $\mathbf{P}$, with

$$\{R_i, R_j\} = 0, \quad \{R_i, P_j\} = \delta_{ij}. \tag{3.175}$$

Reference to the group algebra then suggests that we represent $\mathbf{J}$ in the form

$$\mathbf{J} = \mathbf{R} \wedge \mathbf{P} + \mathbf{S}, \tag{3.176}$$

where $\mathbf{S}$ must be a translationally invariant vector that obeys the vector product rule for angular momentum vectors,

$$\mathbf{S} \wedge \mathbf{S} = \mathbf{S}. \tag{3.177}$$

Obviously with this construction, we are looking forward to the quantum theory, where $\mathbf{S}$ can be interpreted as a spin angular momentum vector. The generator of Galilean boosts may be put in the form [20]

$$\mathbf{K} = \mathbf{P}t - M\mathbf{R}. \tag{3.178}$$

Its Hamiltonian equation of motion reads

$$\frac{d\mathbf{K}}{dt} = \frac{\partial \mathbf{K}}{\partial t} + \{\mathbf{K}, H\} \tag{3.179}$$

$$= \mathbf{P} - M\frac{d\mathbf{R}}{dt} \equiv 0. \tag{3.180}$$

Thus, M is the mass of the system, and $\mathbf{R}$ is a coordinate that moves with the constant velocity

$$\dot{\mathbf{R}} = \frac{\mathbf{P}}{M}. \tag{3.181}$$

In a many-body system, $\mathbf{R}$ and $\mathbf{P}$ become variables for the centre of mass located by

$$M\mathbf{R} = \sum_n m_n \mathbf{x}_n, \quad M = \sum_n m_n. \tag{3.182}$$

The canonical transformations may also give rise to so-called dynamical symmetries in the following way [21]. Suppose we have found a set of constants of the motion for a dynamical system with Hamiltonian H,

$$I(x_1, \ldots x_n; p_1, \ldots p_n)_i = \text{constants}, \quad i = 1, \ldots r. \tag{3.183}$$

Then by definition

$$\dot{I}_i = \{I_i, H\} = 0, \quad i = 1, \ldots r. \tag{3.184}$$

For a given integral I_i, we may define an infinitesimal canonical transformation,

$$X_n = x_n + \delta x_n; \quad P_n = p_n + \delta p_n, \tag{3.185}$$

where, as before,

$$\delta x_n = \varepsilon\{I_i, x_n\}; \quad \delta p_n = \varepsilon\{I_i, p_n\}, \tag{3.186}$$

and ε is an infinitesimal parameter. Under such a transformation, an orbit in phase space is taken into an infinitely near orbit lying on the same energy surface. Hence, the system is invariant under this transformation, and I_i is the generator of a symmetry which may be distinct from the kinematical symmetries required for relativistic invariance.

It may happen that the constants of the motion (3.183) satisfy the Lie relations

$$\{I_i, I_j\} = \sum_k C_{ijk} I_k, \tag{3.187}$$

where the $\{C_{ijk}\}$ are known as the structure-constants; then the $\{I_i\}$ are the generators of a Lie group associated with the dynamical symmetry. In Chapter 4, we shall see that the property of polarisation of light can be described as a manifestation of an intrinsic SU(2) symmetry that arises as a dynamical symmetry of the free electromagnetic field. As regards other symmetries, it is evident that the generators of the symmetry Lie groups are invariant under permutations of identical particles, and so each set of $\{n_i\}$ identical particles is associated with the symmetric group S_{n_i}. The full permutation group of the system is the direct product of the symmetric group for each set of identical particles, a finite group of order $\Pi_i n_i!$ In classical mechanics, the permutation symmetry is rather trivial; in the quantum context, it is of fundamental importance (cf §5.3.3).

The Hamiltonian formalism may be developed for a field theory with some obvious modifications of the foregoing. Starting from the field Lagrangian, we choose the field variable σ to be the 'position' variable, and define its conjugate 'momentum' π as the functional derivative

$$\pi = \frac{\delta L}{\delta\left(\partial\sigma/\partial t\right)}, \tag{3.188}$$

and then the Hamiltonian is

$$H = \int \pi\dot\sigma\, \mathrm{d}^3\mathbf{x} - L. \tag{3.189}$$

The terms 'position' and 'momentum' are purely formal, and σ and π are best thought of as 'conjugate variables'. The canonical equations of motion can be expressed using functional derivatives as

$$\frac{\partial\sigma}{\partial t} = \frac{\delta H}{\delta\pi}, \quad \frac{\partial\pi}{\partial t} = -\frac{\delta H}{\delta\sigma}, \tag{3.190}$$

which are equations that hold at every space-time point $(\mathbf{x},t)$. The generalisation of (3.129) to a field for a specified instant in time is straightforward,

$$\{\sigma(\mathbf{x}),\sigma(\mathbf{x}')\} = \{\pi(\mathbf{x}),\pi(\mathbf{x}')\} = 0, \tag{3.191}$$

$$\{\sigma(\mathbf{x}),\pi(\mathbf{x}')\} = \delta^3(\mathbf{x}-\mathbf{x}'), \tag{3.192}$$

but more complicated P.B. relations apply to dynamical variables at different times because the dynamical evolution of the system is involved in an essential way.

3.5 Constrained Hamiltonian Dynamics

The passage to the Hamiltonian formalism described in §3.4 is the familiar and usual case. In the first instance, the introduction of the Hamiltonian through (3.116) leads to a function defined in a $3N$-dimensional space, regarding the coordinates, velocities and momenta as independent variables

$$H(x,\dot x,p) = \sum_{i=1}^{N} p_i\dot x_i - L(\{x,\dot x\}). \tag{3.193}$$

In the usual case, the equations in (3.113) can be fully inverted because the $N \times N$ Hessian matrix

$$||\mathbf{W}||_{ik} = \frac{\partial^2 L}{\partial\dot x_i\partial\dot x_k} \tag{3.194}$$

is of rank N, and so there are N equations of the form

$$\dot x_i = \dot x_i(x,p). \tag{3.195}$$

These may be substituted into (3.193), and H is transformed to a function defined on a $2N$-dimensional space. However, there are physical theories, including electrodynamics, where this transformation breaks down because one or more of the relations defining the momenta, (3.113),

$$p_i = \frac{\partial L}{\partial \dot{x}_i}, \tag{3.196}$$

are equations that are independent of the velocities $\{\dot{x}_i\}$; the Lagrangian L is said to be 'degenerate'. This breakdown occurs when the Hessian matrix of the Lagrangian is singular. Equation (3.193) is still valid; it is supplemented by a certain number[7] of relations derived from (3.113) of the form

$$\phi_m(x, p) = 0, \quad m = 1, 2, \ldots, k \leq N, \tag{3.197}$$

that is, relations involving only the canonical variables $\{x_i\}, \{p_i\}$.

The theory of constrained Hamiltonian mechanics is originally due to Dirac; the terminology introduced by Dirac is still widely used and we will adhere to it here [22]–[24]. For a modern account, see [25]. The relations (3.197) are called primary constraints; their form and number are determined by the chosen form for the Lagrangian L since they arise from performing the calculation (3.113). The primary constraints define a $(2N - k)$-dimensional subspace, Γ_P, of the $2N$-dimensional phase space. If the rank of the Hessian matrix is R, it is possible to solve (3.113) for R of the velocities. Let this group of velocities be labelled with a Roman subscript and be denoted as $\{\dot{x}_a : a = 1, \ldots, R\}$ while the remainder are identified by a Greek subscript as $\{\dot{x}_\alpha : \alpha = R + 1, \ldots, N\}$. Then we may write

$$\dot{x}_a = \dot{x}_a(x, \{p_b\}, \{\dot{x}_\beta\}), \quad a, b = 1, \ldots, R; \quad \beta = R + 1, \ldots N, \tag{3.198}$$

and on the subspace Γ_P where the constraint equations hold, the Hamiltonian (3.193) may be put in the form

$$H_0 = \sum_{a=1}^{R} p_a \dot{x}_a + \sum_{\alpha=R+1}^{N} (p_\alpha - \phi_\alpha) \dot{x}_\alpha - L(x, \{\dot{x}_b\}, \{\dot{x}_\beta\}), \tag{3.199}$$

since $\phi_\alpha = 0$ on Γ_P. Dirac proved that, although at an arbitrary representative point in the whole space H_0 involves all the variables shown in (3.199), on Γ_P it is actually *independent* of the $\{\dot{x}_\beta\}$, and so is a function of only the canonical variables,

$$H_0 = H_0(x, \{p_a\}). \tag{3.200}$$

The details are given in [23], [25].

It is very convenient to introduce a notation that distinguishes equations valid only on the subspace Γ_P, where the constraint equations hold, from those that are generally valid. Dirac called the former 'weak' equalities and used the symbol $\approx$ for them,[8]

---

[7] If the singular Hessian matrix is of rank R, there will be $k = N - R$ constraints.

[8] This is a special use of the symbol $\approx$ not to be confused with its customary usage of 'approximately equal'.

whereas the latter are 'strong' or ordinary equalities. Thus, two functions u, v of the canonical variables are said to be weakly equal if, and only if,

$$u(x, p) = v(x, p), \quad \phi_\alpha(x, p) = 0, \tag{3.201}$$

and their relationship is denoted

$$u(x, p) \approx v(x, p). \tag{3.202}$$

A useful relation in the algebra of weak equations is the strong equality

$$u \approx 0, \quad v \approx 0 \quad \Rightarrow \quad u\,v = 0. \tag{3.203}$$

With this notation, we may write the Hamiltonian equations of motion in the standard form with ordinary equality replaced by weak equality, thus

$$\dot{q}_i \approx \frac{\partial H_T}{\partial p_i}, \quad \dot{p}_i \approx -\frac{\partial H_T}{\partial q_i}, \tag{3.204}$$

where

$$H_T = H + \sum_m v_m \phi_m, \quad H \approx H_0. \tag{3.205}$$

$H(q, p)$ is any function of the canonical variables such that on the surface Γ_P it coincides with H_0. In the case where $H = H_0$, the coefficients v_m are just the velocities $\dot{q}_\beta$ previously defined. In P.B. notation the equations of motion, (3.204), are

$$\dot{q}_i \approx \{q_i, H_T\}, \quad \dot{p}_i \approx \{p_i, H_T\}, \tag{3.206}$$

and for a general dynamical variable Ω

$$\dot{\Omega}(q, p) \approx \{\Omega(q, p), H_T\}. \tag{3.207}$$

It is to be noted that the partial differentiations with respect to the variables q_i, p_i do not apply to the coefficients v_m; furthermore, the constraint equations, (3.197), can only be used after the partial differentiations have been carried out.

The consistency of the scheme requires that the equations of constraint should hold for all times; this requires that the conditions

$$\dot{\phi}_m \approx \{\phi_m, H_T\} \approx 0, \quad m = 1, \ldots k, \tag{3.208}$$

must hold at least as weak equalities. If a constraint has a vanishing P.B. with all the other constraints, it is called first class; in such a case (3.208) will lead to a further equation involving the canonical variables (but not the $\{v_m\}$) which must vanish weakly for all times. Such a constraint is called a *secondary constraint*. For first-class constraints, the associated coefficients v_m remain undetermined; they are associated with symmetries of the dynamical system. The secondary constraints too must be tested with (3.208) until all the consistency conditions have been established.[9] If a constraint

---

[9] If consistency cannot be achieved, the original Lagrangian must have been inconsistent.

has a non-vanishing P.B. with at least one other constraint, it is called second class; the consistency conditions for second-class constraints lead to equations determining the unknown coefficients $\{v_m\}$. The first class/second class terminology can be extended to general dynamical variables Ω; if the P.B.s of Ω with all the constraints vanish weakly, it is first class; otherwise it is second class.

Suppose there is a system of functions $\{\theta(q,p)_j, \ j = 1,\ldots,J\}$, and form a matrix $\mathbf{C}$ with elements constructed from their P.B.s,

$$C_{ij} = \{\theta_i, \theta_j\}. \tag{3.209}$$

Assume that the matrix is non-singular, that is, the determinant

$$\Delta = \mathrm{Det}||\mathbf{C}|| \neq 0, \tag{3.210}$$

so that $\mathbf{C}$ has an inverse

$$C_{kj}^{-1} \, C_{ik} = \delta_{ij}. \tag{3.211}$$

Since the matrix $\mathbf{C}$ is antisymmetric, J must be an even number. Then we may form a new type of P.B. according to the rule

$$\{A,B\}^* = \{A,B\} - \sum_{rs} \{A, \theta_r\} C_{rs}^{-1} \{\theta_s, B\}. \tag{3.212}$$

Now choose B to be one of the defining functions, $B = \theta_k$, and consider its $*$-bracket with any dynamical quantity A,

$$
\begin{aligned}
\{A, \theta_k\}^* &= \{A, \theta_k\} - \sum_{rs} \{A, \theta_r\} C_{rs}^{-1} \{\theta_s, \theta_k\} \\
&= \{A, \theta_k\} - \sum_{rs} \{A, \theta_r\} C_{rs}^{-1} \, C_{sk} \\
&= \{A, \theta_k\} - \sum_{r} \{A, \theta_r\} \delta_{rk} = 0.
\end{aligned} \tag{3.213}
$$

The $*$-bracket shares all the fundamental algebraic properties of the P.B., whatever the θ_s might be.

Taking all the constraints together, we may divide them into two groups; those that are first class, $\{\phi_a^1\}$, and the remainder which are second class, $\{\chi_n\}$. It may be possible to find one or more linear combinations of the second-class constraints that is first class. We suppose this to have been done, so that we have the minimum number of second-class constraints $\{\chi_n\}$. The condition for this is that the matrix of mutual P.B.s of the $\{\chi_n\}$ has an inverse which requires a non-vanishing determinant;[10] if this is so, we may take the surviving second-class constraints to be the functions $\{\theta_i\}$. This leads to the definition of the Dirac bracket,

---

[10] Suppose that the set is not a minimal set. Then there exists a linear combination $a_g \chi_g$ that is first class and so by definition

$$a_g \{\chi_g, \chi_n\} \approx 0 \tag{3.214}$$

has a non-trivial solution a_g. From the theory of linear equations this is only possible if $\mathrm{Det}||\mathbf{D}||$, where $\mathbf{D}$ is the matrix of mutual P.B.s of the $\{\chi_n\}$, vanishes weakly.

$$[A,B]^* = \{A,B\} - \sum_{rs}\{A,\chi_r\}C_{rs}^{-1}\{\chi_s,B\}, \tag{3.215}$$

where the matrix $\mathbf{C}$

$$\mathbf{C}_{rs} = \{\chi_r,\chi_s\}, \tag{3.216}$$

is constructed from the second-class constraints. From (3.213) we infer that the Dirac bracket of any dynamical variable with a second-class constraint is zero, and so the second-class constraints can be implemented as strong, that is, ordinary equalities:

$$\chi_n = 0. \tag{3.217}$$

Accordingly, they may now be dropped from the Hamiltonian which consists of H and a sum of first-class constraints with arbitrary coefficients, and the equation of motion becomes

$$\dot{\Omega} = [\Omega,H]^* + \sum_a v_a\{\Omega,\phi_a^1\}. \tag{3.218}$$

For a system with only second-class constraints, the second member on the RHS vanishes; for a system with only first-class constraints, the Dirac bracket coincides with the P.B. It is worth noting that Dirac showed that the introduction of Dirac brackets is simply a device for removing redundant dynamical variables; constrained dynamics arises when the Lagrangian formulation involves additional variables, among which there are linear dependencies [23]. The Dirac brackets are like the original P.B.s (3.124)–(3.127), being antisymmetric, associative and satisfying the Jacobi identity and the product rule,

$$[fg,h]^* = [f,h]^*g + f[g,h]^*, \tag{3.219}$$

which is a non-commutative version of the familiar Leibniz product rule in calculus. They are to be used in exactly the same way as the standard P.B.s.

3.6 Hamiltonian Electrodynamics

In electrodynamics we use the particle coordinates $\{\mathbf{x}_n\}$ and the field potentials $\{\phi,\mathbf{a}\}$ as the 'position' variables. When the Lagrangian (3.79) is written out in full in terms of the working variables, we have

$$L = L\left(\{\mathbf{x}_n,\dot{\mathbf{x}}_n\};\phi,\mathbf{a},\frac{\partial\mathbf{a}}{\partial t}\right) \tag{3.220}$$

and see at once that no momentum conjugate to the scalar potential ϕ can be defined via (3.188) because the time derivative of ϕ does not appear in L. The general technique for dealing with a degenerate Lagrangian was summarised in the previous section, §3.5, and will be used here.[11] A toy example is worked through in Appendix H. Dirac's

[11] Cf. the presentations of Dirac's method in [27], [28].

scheme for dealing with this situation is very appropriate for the case of electrodynamics because it will permit us to develop a canonical scheme for an arbitrary gauge condition, such as (2.74). The alternative approach, which to be sure is the traditional one in atomic and molecular physics and quantum chemistry, is to impose a gauge condition at the outset; this simplifies the treatment considerably but does not give a completely satisfactory basis for subsequent investigations into the gauge invariance of calculated quantities.

We start with the field potentials $(\phi, \mathbf{a})$ as the Lagrangian coordinates, and formally introduce their conjugate momenta in the usual way with the Lagrangian (3.79) and the definition (3.188). So we have

$$\boldsymbol{\pi} = \frac{\delta L}{\delta \dot{\mathbf{a}}} \approx \varepsilon_0 \left(\boldsymbol{\nabla}\phi + \dot{\mathbf{a}} \right), \tag{3.221}$$

$$\pi_0 = \frac{\delta L}{\delta \dot{\phi}} \approx 0, \tag{3.222}$$

while the particle momenta follow from (3.79) and (3.113),

$$\mathbf{p}_n = \frac{\partial L}{\partial \dot{\mathbf{x}}_n} \approx m\dot{\mathbf{x}}_n + e_n \mathbf{a}(\mathbf{x}_n), \quad n = 1, \ldots N, \tag{3.223}$$

where (2.108) has been used for the current density in L. These equations have been written with Dirac's weak equality sign $\approx$ because (3.222) implies a restriction on the validity of all the other equations that we have to take account of.

The combination of (3.116) and (3.189) leads to the Hamiltonian,

$$H \approx \sum_n^N \mathbf{p}_n \cdot \dot{\mathbf{x}}_n + \int \boldsymbol{\pi} \cdot \dot{\mathbf{a}} \, d^3\mathbf{x} + \int \pi_0 \dot{\phi} \, d^3\mathbf{x} - L, \tag{3.224}$$

and, on eliminating the 'velocities' in favour of the canonical momenta, we obtain

$$H \approx \frac{1}{2} \sum_n^N \frac{1}{m_n} \left(\mathbf{p}_n - e_n \mathbf{a}(\mathbf{x}_n) \right)^2 + \frac{1}{2}\varepsilon_0 \int \left(\varepsilon_0^{-2} \boldsymbol{\pi} \cdot \boldsymbol{\pi} + c^2 \mathbf{B} \cdot \mathbf{B} \right) d^3\mathbf{x}$$

$$+ \int \phi \left(\boldsymbol{\nabla} \cdot \boldsymbol{\pi} + \rho \right) d^3\mathbf{x}, \tag{3.225}$$

where $\mathbf{B} = \boldsymbol{\nabla} \wedge \mathbf{a}$, and an integration by parts has been used to obtain the last term. We have also used (3.222) to eliminate the term in π_0 provisionally (see what follows).

The Lagrangian assumption that the particle and field variables can be varied independently becomes in the Hamiltonian scheme the statement that all P.B.s involving both field and particle variables are zero. The non-zero fundamental P.B.s are taken to be canonical,

$$\{x_n^r, p_m^s\} = \delta_{nm} \, \delta_{rs}, \tag{3.226}$$

$$\{a(\mathbf{x},t)^r, \pi(\mathbf{x}',t)^s\} = \delta_{rs}\delta^3(\mathbf{x} - \mathbf{x}'), \tag{3.227}$$

$$\{\phi(\mathbf{x},t), \pi_0(\mathbf{x}',t)\} = \delta^3(\mathbf{x} - \mathbf{x}'), \tag{3.228}$$

while all other P.B. relations involving the field and particle variables are assumed to vanish. The canonical momenta in these relations may be expressed in terms of their conjugate positions as derivatives (functional derivatives for the field variable):

$$\mathbf{p}_n \rightarrow -\frac{\partial}{\partial \mathbf{x}_n}$$

$$\boldsymbol{\pi} \rightarrow -\frac{\delta}{\delta \mathbf{a}}. \tag{3.229}$$

At this stage both $\mathbf{a}$ and $\boldsymbol{\pi}$ have three components with no gauge specified for the vector potential.

So far this is all as in the normal passage to a Hamiltonian version of dynamics; however, we now have to confront the occurrence of a primary constraint, (3.222); for consistency it must be true for all time. This requires

$$\dot{\pi}_0 = -\frac{\delta H}{\delta \phi} = 0, \tag{3.230}$$

and hence, from (3.225), we have a secondary constraint

$$\Omega_2 = \boldsymbol{\nabla}.\boldsymbol{\pi} + \rho \approx 0. \tag{3.231}$$

This equation is valid at an arbitrary instant in time; for consistency it too must be valid for all time, so that we require

$$\dot{\Omega}_2 = \{\Omega_2, H\} \approx 0. \tag{3.232}$$

Using the canonical P.B. relations (3.226), (3.227) and the Hamiltonian H, (3.225), however, we find that (3.232) reduces to $0 = 0$ so that it can be written with an ordinary $=$ sign in place of $\approx$. Thus, (3.222) and (3.231) are the only equations of constraint in Hamiltonian electrodynamics; they have the important property that their P.B. relations with the Hamiltonian, and among themselves when taken at different space points, vanish, for example

$$\{\pi(\mathbf{x})_0, \Omega(\mathbf{x}')_2\} \approx 0. \tag{3.233}$$

Such constraints are referred to as first-class constraints. According to Dirac's general theory, a physical observable must have a vanishing P.B. with each first-class constraint. This being the case, the constraints may be added to the Hamiltonian with arbitrary coefficients and will not affect the equations of motion of the physical observables. Thus the general Hamiltonian deduced from the Lagrangian for electrodynamics, (3.79), contains two arbitrary functions, $u(\mathbf{x})$ and $w(\mathbf{x})$ say,

$$\tilde{H} = H + \int w\pi_0 \, d^3\mathbf{x} + \int u\Omega_2 \, d^3\mathbf{x}. \tag{3.234}$$

The two constraints are, however, quite different in nature.

The pair (ϕ, π_0) tell us nothing about the dynamics of the system, for using (3.122) with (3.228), (3.225) and (3.234), we find

$$\dot{\phi} = \{\phi, \tilde{H}\} = w, \tag{3.235}$$

where w is arbitrary, and

$$\dot{\pi}_0 = \{\pi_0, \tilde{H}\} = \Omega_2 \approx 0, \qquad (3.236)$$

which must be taken with (3.232); hence we can discard them. The other equation of constraint, (3.231), occurs explicitly in the Hamiltonian (3.234) with coefficient u. According to (3.232), Ω_2 is a non-trivial constant of the motion, and so it must describe some invariance of the system. This invariance can be exposed as follows. Let us take the general linear superposition involving Ω_2 with a suitably smooth function σ, that is,

$$G = \int \sigma \, \Omega_2 \, \mathrm{d}^3 \mathbf{x}, \qquad (3.237)$$

as the generator of an infinitesimal canonical transformation according to (3.185), (3.186). Then we find

$$\tilde{H} \to \tilde{H}' = \tilde{H}, \qquad (3.238)$$

$$\boldsymbol{\pi} \to \boldsymbol{\pi}' = \boldsymbol{\pi}, \qquad (3.239)$$

$$\mathbf{a} \to \mathbf{a}' = \mathbf{a} + \varepsilon \boldsymbol{\nabla} \sigma, \qquad (3.240)$$

$$\mathbf{x}_n \to \mathbf{x}'_n = \mathbf{x}_n, \qquad (3.241)$$

$$\mathbf{p}_n \to \mathbf{p}'_n = \mathbf{p}_n - \varepsilon e_n \boldsymbol{\nabla} \sigma(\mathbf{x}_n). \qquad (3.242)$$

Equation (3.240) is exactly of the same form as (2.45) that defines a gauge transformation of the vector potential. Thus, Ω_2 in the Hamiltonian scheme is essentially the generator of gauge transformations which arise as a manifestation of a dynamical symmetry; simultaneously, it leads to a compensating transformation of the momentum variables for the charged particles, (3.242). This relationship is an example of the transformation rule (3.38), discussed previously in connection with the equations of motion. Notice that the velocity variables for the charges are invariant under this canonical transformation,

$$\dot{\mathbf{x}}_n \equiv \frac{1}{m_n} (\mathbf{p}_n - e_n \mathbf{a}(\mathbf{x}_n)) \to \dot{\mathbf{x}}'_n = \dot{\mathbf{x}}_n, \qquad (3.243)$$

and it is the gauge-invariant combination of the variables,

$$\mathbf{p}_n - e_n a(\mathbf{x}_n) \equiv \overline{\mathbf{p}}_n, \qquad (3.244)$$

that occurs in the Hamiltonian.

The particle coordinates $\{\mathbf{x}_n\}$ and the field canonical momentum, $\boldsymbol{\pi}$, are left unchanged since their P.B.s with G vanish. If we put $\rho = 0$ in (3.231), these equations are applicable to the electromagnetic field in a volume where there are no charges, that is, the 'free field'; the corresponding quantities for the free field will be denoted by adding a subscript 0 to Ω and G, so that, for example, the canonical transformation with G_0 again gives (3.240), and obviously there is nothing to be said about particle variables. The relationship between G_0 and G is actually another canonical transformation, as will be described later.

If we were only interested in a classical version of electrodynamics in canonical form, we could stop at this point with (3.225), (3.234) as the Hamiltonian, having discarded the term in π_0. This canonical formalism is an essentially complete replacement of the Maxwell–Lorentz account of charged particles and the electromagnetic field. The classical Hamiltonian incorporates the possibility of an 'external free field' since the field variables can have contributions from an electromagnetic field due to sources that are far from the volume of physical space that the 'system' (the collection of N charged particles) is supposed to reside in. For a quantum theory, however, this leaves us with an awkward question: is the arbitrary quantity $u(\mathbf{x})$ in (3.234) to be treated as an operator? This is problematic, as we have no rule with which to construct it. One possible approach is to define the physical states as those that are annihilated by the quantum mechanical operator corresponding to Ω_2, that is,

$$\Omega(\mathbf{x})_2 \Psi = 0, \tag{3.245}$$

which renders u, whatever it is, harmless.

It is useful to develop quantum electrodynamics along these lines for an investigation of gauge invariance (see §9.3.1); for practical calculations, an alternative approach is to deal with the constraint (3.231) directly, as discussed in §3.5. We take (3.231) to be a function θ_k and bring in another function θ_j which has the property that its P.B. with Ω_2 does not vanish; it can be regarded as another equation of constraint. In Dirac's terminology, this yields a canonical scheme with two second-class constraints. The constraints can be allowed for by a redefinition of the P.B.s such that with the new definition all the brackets of the dynamical variables with the two constraints vanish; the constraints can then be taken consistently as ordinary equations $= 0$, and the arbitrary quantities u and w drop out of the Hamiltonian. Since Ω_2 depends on the field canonical momentum, $\boldsymbol{\pi}$, and the particle coordinates $\{\mathbf{x}_n\}$, the new constraint must involve the vector potential $\mathbf{a}$ and/or the particle momenta $\{\mathbf{p}_n\}$.

In electrodynamics, the arbitrary feature of the description of the dynamics is the freedom to make gauge transformations, so the extra constraint is usually taken[12] to be

---

[12] The other possibility, which will only be speculated about here, would lead to an entirely different form for electrodynamics; to see this, note that the charge density is proportional to the electric charge, e, and the natural quantity to involve in an additional constraint, θ_k, would then involve the current density $\mathbf{j}(\mathbf{x})$. For simplicity assume that the constraint is linear in $\mathbf{j}$; its non-vanishing P.B. with the charge density ρ would then be proportional to e^2, so that having formed the matrix of the P.B.s of the two constraints, we see that its inverse is proportional to e^{-2}. Thus, this is a 'strong-coupling' formulation since there is no perturbation expansion in positive powers of e; such a formulation was sketched by Dirac in [23] but was not taken further. Looking forward to a quantum mechanical version, one sees that the coupling can be expressed in terms of the *inverse* of the dimensionless fine structure constant $\alpha = e^2/4\pi\varepsilon_0 hc$ and so is *linear* in Planck's constant h. This would seem to be a much better starting place for describing the 'classical limit' of quantum electrodynamics than the conventional approach which views $\alpha(\propto h^{-1})$ as the important parameter. The whole approach is reminiscent of the quantum mechanical non-relativistic current algebra [29]–[31].

a gauge condition; as discussed previously, a general gauge condition can be expressed by (2.74). The two second-class constraints will therefore be taken to be,

$$\Omega(\mathbf{x})_1 = \int \mathbf{a}(\mathbf{x}') \cdot \mathbf{g}(\mathbf{x}', \mathbf{x}) \, d^3 \, \mathbf{x}', \qquad (3.246)$$

$$\Omega(\mathbf{x})_2 = \rho(\mathbf{x}) + \nabla \cdot \boldsymbol{\pi}(\mathbf{x}). \qquad (3.247)$$

Their mutual P.B.s $K(\mathbf{x}, \mathbf{x}')_{ij} = \{\Omega(\mathbf{x})_i, \Omega(\mathbf{x}')_j\}$ calculated using (3.227) yield a non-singular matrix with elements

$$\mathbf{K} = \begin{bmatrix} 0 & -\delta^3(\mathbf{x} - \mathbf{x}') \\ \delta^3(\mathbf{x} - \mathbf{x}') & 0 \end{bmatrix}. \qquad (3.248)$$

The inverse of this matrix satisfies

$$\int K(\mathbf{x}'', \mathbf{x})_{ij} \, K(\mathbf{x}, \mathbf{x}')_{jk}^{-1} \, d^3 \mathbf{x} = \delta_{ik} \delta^3(\mathbf{x}'' - \mathbf{x}'), \qquad (3.249)$$

and is given by

$$\mathbf{K}^{-1} = \begin{bmatrix} 0 & \delta^3(\mathbf{x} - \mathbf{x}') \\ -\delta^3(\mathbf{x} - \mathbf{x}') & 0 \end{bmatrix}. \qquad (3.250)$$

The Dirac bracket of two dynamical variables j and k which we write as $[j, k]^*$ is defined in (3.215); here it takes the form

$$[j, k]^* = \{j, k\} - \int\int \{j, \Omega(\mathbf{x})_r\} \, K(\mathbf{x}, \mathbf{x}')_{rs}^{-1} \, \{\Omega(\mathbf{x}')_s, k\} d^3 \mathbf{x}' d^3 \, \mathbf{x}, \qquad (3.251)$$

the P.B.s on the RHS being calculated using (3.226), (3.227).

Using (3.249) with (3.251) shows that the Dirac bracket of any dynamical variable with a constraint $(\Omega(\mathbf{x})_i, i = 1, 2)$ vanishes identically. This means that the equations of constraint can be taken as ordinary equations. Thus, we can now set

$$\Omega(\mathbf{x})_1 = \int \mathbf{a}(\mathbf{x}') \cdot \mathbf{g}(\mathbf{x}', \mathbf{x}) \, d^3 \, \mathbf{x}' = 0, \qquad (3.252)$$

as the condition that fixes the gauge of the vector potential in the Hamiltonian, according to the choice of the Green's function $\mathbf{g}(\mathbf{x}', \mathbf{x})$. Similarly, the original constraint Ω_2 can now also be set equal to zero as an ordinary equation,

$$- \nabla . \boldsymbol{\pi} = \rho, \qquad (3.253)$$

and so the Hamiltonian reduces to the form

$$H = \frac{1}{2} \sum_n^N \frac{1}{m_n} (\mathbf{p}_n - e_n \mathbf{a}(\mathbf{x}_n))^2 + \frac{1}{2} \varepsilon_0 \int \left(\varepsilon_0^{-2} \boldsymbol{\pi} \cdot \boldsymbol{\pi} + c^2 \mathbf{B} \cdot \mathbf{B} \right) d^3 \mathbf{x}. \qquad (3.254)$$

The non-zero Dirac brackets of the working variables in the Hamiltonian (3.254) are easily found to be

$$[x_n^r, p_m^s]^* = \delta_{nm} \delta_{rs}, \qquad (3.255)$$

$$[a(\mathbf{x})^r, \pi(\mathbf{x}')^s]^* = \delta_{rs}\delta^3(\mathbf{x} - \mathbf{x}') - \nabla_{\mathbf{x}}^r g(\mathbf{x}', \mathbf{x})^s, \tag{3.256}$$

$$[p_n^r, \pi(\mathbf{x})^s]^* = -e_n \nabla_n^r g(\mathbf{x}, \mathbf{x}_n)^s. \tag{3.257}$$

It is clear that with a simple change of variables the Hamiltonian scheme (3.254)–(3.257) can be written purely in terms of the gauge-invariant variables $\{\mathbf{x}_n, \bar{\mathbf{p}}_n, \mathbf{E}, \mathbf{B}\}$,

$$H = \frac{1}{2}\sum_n |\bar{\mathbf{p}}_n|^2 + \frac{1}{2}\varepsilon_0 \int (\mathbf{E}\cdot\mathbf{E} + c^2\mathbf{B}\cdot\mathbf{B})\, \mathrm{d}^3\,\mathbf{x}. \tag{3.258}$$

Superficially, the Hamiltonian (3.258) appears to describe 'free' charges and the electromagnetic field; however, the interaction is now carried through the Dirac brackets of the modified momentum components which are proportional to the charges $\{e_n\}$

$$[\bar{p}_n^t, \bar{p}_m^r]^* = e_n \delta_{nm}\varepsilon_{rts}B(\mathbf{x}_n)^s, \tag{3.259}$$

$$[\bar{p}_n^r, E(\mathbf{x})^s]^* = \varepsilon_0^{-1}e_n\delta_{rs}\delta^3(\mathbf{x}_n - \mathbf{x}), \tag{3.260}$$

$$[x_n^s, \bar{p}_m^r]^* = \delta_{nm}\delta_{sr}. \tag{3.261}$$

As expected, the fundamental Dirac bracket for the field strengths is independent of the Green's function $\mathbf{g}(\mathbf{x} : \mathbf{x}')$, that is, gauge invariant:

$$[E(\mathbf{x})^r, B(\mathbf{x}')^s]^* = \varepsilon_0^{-1}\varepsilon_{rst}\nabla_{\mathbf{x}'}^t\delta^3(\mathbf{x} - \mathbf{x}'). \tag{3.262}$$

If one changes to this gauge-invariant form, one must remember to carry over the two Maxwell equations that serve as initial conditions:

$$\nabla\cdot\mathbf{B} = 0, \tag{3.263}$$

$$\varepsilon_0\nabla\cdot\mathbf{E} = \rho. \tag{3.264}$$

Let us once more recover the equations of motion. A straightforward calculation using (3.190) and (3.254) shows that

$$\frac{\partial\pi}{\partial t} = -\varepsilon_0 c^2\nabla\wedge\mathbf{B} + \mathbf{j}, \tag{3.265}$$

with $\mathbf{j}$ given by (2.108) in terms of the gauge-invariant particle velocity, (3.243). Comparing (3.253) and (3.265) with the Maxwell equations (2.3) and (2.4), respectively, the conjugate momentum $\pi(\mathbf{x})$ can be recognised as the total electric field, to within a constant

$$\pi = -\varepsilon_0\mathbf{E}. \tag{3.266}$$

Thus, (3.253) is essentially a statement of Gauss's law. However, it is important to keep in mind in the following that $\mathbf{E}$, as essentially the conjugate momentum to the vector potential $\mathbf{a}$, is related to it by the Dirac bracket (3.256) and that the scalar potential has been eliminated. Thereby nothing has been lost. The Maxwell equation (2.1) follows from the use of a vector potential; (2.2) follows from the Hamiltonian equation of motion for $\mathbf{B}$ using the Dirac bracket (3.256), again independently of the gauge.

The usual Newtonian equation of motion involves the force on the particle which is given by

$$m_n\ddot{\mathbf{x}}_n = [m_n\dot{\mathbf{x}}_n, H]^*. \tag{3.267}$$

This is easily evaluated since the velocity variable, $\dot{\mathbf{x}}_n$, has a vanishing P.B. with the constraint $\Omega(\mathbf{x})_1$. Both parts of the Hamiltonian (3.254) contribute to this Dirac bracket. The first term in (3.254) may be written in terms of the velocity variables and gives a contribution,

$$[m_n \dot{x}_n^r, \sum_k^N \frac{1}{2m_k} \dot{x}_k^2]^* = e_n \left(\frac{\partial a(\mathbf{x}_n)^t}{\partial x_n^r} - \frac{\partial a(\mathbf{x}_n)^r}{\partial x_n^t} \right) \dot{x}_n^t, \qquad (3.268)$$

where we recognise the magnetic induction contribution to the Faraday field tensor on the RHS. For the evaluation of the contribution from the field energy term in H, we use

$$[\dot{x}_n^t, \pi(\mathbf{x}')^s]^* = - e\,\delta_{ts}\delta^3(\mathbf{x}_n - \mathbf{x}'), \qquad (3.269)$$

which is valid independently of the gauge. Collecting the two pieces together, we finally get

$$m_n \ddot{\mathbf{x}}_n = e_n\,(\dot{\mathbf{x}}_n \wedge \mathbf{B}(\mathbf{x}_n)) + e_n \mathbf{E}(\mathbf{x}_n), \qquad (3.270)$$

where we have used (3.266) to identify the electric field $\mathbf{E}$; this is the expected Lorentz force law. Essentially similar calculations with (3.258) lead to the same results. The classical equations of motion are gauge invariant, and the issue of gauge invariance arises only if the vector potential is introduced into them in order to simplify the calculation of their solutions.

The Hamiltonian systems (3.254)–(3.257) and (3.258)–(3.262) constitute completely general statements of classical electrodynamics in Hamiltonian form. We emphasise that they are purely classical structures; as we have seen, they lead to the expected Lorentz force law and Maxwell's equations as the equations of motion for the charges and the electromagnetic field, respectively, with their well-known solutions. There are no stable atoms and no photons – both are new concepts that only emerge after the canonical quantisation of the classical Hamiltonian scheme.

An arbitrary gauge for the vector potential is fixed by making a choice for the Green's function $\mathbf{g}(\mathbf{x}, \mathbf{x}')$ in (3.252) so every gauge has its own set of Dirac brackets. We can try to simplify the scheme by making a definite choice of gauge; for example, we noted previously that the condition $\mathbf{g}^\perp = 0$ yields the Coulomb gauge vector potential $\mathbf{A}$. Inserting just $\mathbf{g}(\mathbf{x}, \mathbf{x}')^\parallel$ in (3.256), (3.257), we obtain the well-known P.B. relation for the Coulomb gauge vector potential and its conjugate,

$$[A(\mathbf{x})^r, \pi(\mathbf{x}')^s]^* = \delta_{rs}\delta^3(\mathbf{x} - \mathbf{x}') - \nabla_\mathbf{x}^r \nabla_{\mathbf{x}'}^s \frac{1}{4\pi|\mathbf{x} - \mathbf{x}'|}$$

$$\equiv \delta_{rs}^\perp(\mathbf{x} - \mathbf{x}'), \qquad (3.271)$$

and

$$[p_n^r, \pi(\mathbf{x})^s]^* = - e_n \nabla_n^r \nabla_\mathbf{x}^s \frac{1}{4\pi|\mathbf{x} - \mathbf{x}_n|}, \qquad (3.272)$$

which evidently originates from the purely Coulombic (longitudinal) electric field of the charge located at $\mathbf{x}_n$. On the other hand, the Dirac bracket (3.256) is consistent with the general vector potential (2.76) expressed in terms of the Coulomb gauge vector potential. It is therefore possible to develop Hamiltonian electrodynamics using the

Coulomb gauge vector potential and resorting to (2.76) if a general vector potential is required for an investigation of the gauge invariance of some calculated quantity.

In the Coulomb gauge, the Hamiltonian (3.254) takes a particularly simplified form; for future reference we denote it by $H[0]$ since this gauge requires $\mathbf{g}^\perp = 0$,

$$H[0] = \sum_n^N \frac{|\mathbf{p}_n|^2}{2m_n} + \frac{1}{2}\varepsilon_0 \int (\mathbf{E} \cdot \mathbf{E} + c^2 \mathbf{B} \cdot \mathbf{B}) \, d^3\mathbf{x} + V[0], \qquad (3.273)$$

$$V[0] = -\sum_n \frac{e_n}{m_n} \mathbf{p}_n \cdot \mathbf{A}(\mathbf{x}_n) + \sum_n \frac{e_n^2}{2m_n} \mathbf{A}(\mathbf{x}_n) \cdot \mathbf{A}(\mathbf{x}_n). \qquad (3.274)$$

Writing $\boldsymbol{\pi} = \boldsymbol{\pi}^\parallel + \boldsymbol{\pi}^\perp$, and using (3.257) with $\mathbf{g} = \mathbf{g}^\parallel$ and (3.266), the second term in (3.273) yields

$$\frac{1}{2}\varepsilon_0 \int \mathbf{E}^\parallel \cdot \mathbf{E}^\parallel \, d^3\mathbf{x} = \sum_{n,m} \frac{e_n e_m}{4\pi\varepsilon_0 |\mathbf{x}_n - \mathbf{x}_m|} = V_{\text{Coul}}, \qquad (3.275)$$

$$\frac{1}{2}\varepsilon_0 \int (\mathbf{E}^\perp \cdot \mathbf{E}^\perp + c^2 \mathbf{B} \cdot \mathbf{B}) \, d^3\mathbf{x} = H_{\text{rad}}. \qquad (3.276)$$

3.7 The Geometry of Classical Gauge Transformations

3.7.1 Weyl's Geometric Interpretation of Electromagnetism

The dictionary tells us that the word gauge comes from old French; in modern French it appears as *jauge*. The meaning of gauge is given as a measuring apparatus or standard of measure for the diameter of wire, the width of a railtrack, the contents of a wine cask, the pressure of a gas cylinder and so on. The corresponding term in German is *Eichung*, commonly translated as 'calibration'. The notion of a gauge transformation as a kind of calibration is due to Weyl [32]. Weyl sought to give a geometrical basis to electromagnetism in the spirit of the geometrisation of gravity that Einstein achieved in his theory of general relativity [33], [34]; the result was the first fully worked-through unified theory of gravity and electromagnetism.

In general relativity, the space-time coordinates $x^\mu = (ct, \mathbf{x})$, $(\mu = 0, 1, 2, 3)$ are given the structure of a Riemann geometry.[13] Its characteristic feature is the notion of a curvature of space-time which is such that when a vector is displaced by parallel transport from one point to an arbitrary distant point, the direction of the vector changes by an amount that depends directly on the Riemann–Christoffel curvature tensor.[14] Only in a flat space-time is the curvature tensor zero and the vector unchanged; in this case

---

[13] In this geometry, the four-dimensional space-time is described by a metric tensor $g_{\alpha\beta}$ which has ten independent components that depend on the space-time coordinates. The connection with physics is made by requiring the $g_{\alpha\beta}$ to act as potentials for the gravitational field, so that we can say that gravity is derived from the geometry of space-time [35].

[14] An instructive personal demonstration of parallel transport is this. Hold out an arm to the side parallel with the ground, with the fist closed and the thumb pointing up. Now move the arm round rigidly to the front, and then raise it vertically so that the arm is next to the ear; finally, lower the arm down to its

parallel transport is said to be integrable [36]. This result is interpreted physically as corresponding to the absence of a gravitational field. On the other hand, the magnitude of a vector is not changed by parallel displacement, so that Riemann geometry allows the magnitudes of two vectors to be compared at arbitrarily separated points. Weyl wished to eliminate this 'at-a-distance' feature to give a 'true' infinitesimal geometry; to do so he proposed to modify the geometry of space-time in such a way that gravity and electromagnetism would be unified in a common geometric framework [37].

The new feature in the description that does not arise in standard general relativity is that lengths can no longer be compared at arbitrarily distant points. At each space-time point, observers can choose their standard of lengths in an arbitrary fashion and express all lengths in their (infinitesimal) neighbourhood in terms of this standard *gauge or calibration*. A change in the calibration changes the squared magnitude of a 4-vector by a scale factor ξ, which may be a function of the coordinates x^α,

$$l \to l' = \xi\, l, \tag{3.277}$$

but it does not change the coordinates themselves, and it preserves angles. ξ is a positive number. The transformation law (3.277), 'multiplication by a positive number', defines a 1-parameter Abelian Lie group – the group of dilations.

Under the action of this group, distances between neighbouring events are scaled by ξ:

$$ds' = \xi\, ds. \tag{3.278}$$

In Weyl's new geometry, the squared length, l, of a vector located at x is supposed to change under an infinitesimal parallel displacement by an amount proportional to l and to a differential 1-form,

$$dl = -\, l\, a_\alpha\, dx^\alpha, \tag{3.279}$$

where $a_\alpha = a(x)_\alpha$ is some vector field to be identified by the physical interpretation of the formalism. Essentially, (3.279) is a rule that specifies how a vector is to be transported to an infinitesimally nearby point; in the technical language of differential geometry it is a *connection*.

If the vector is transported by parallel displacement from a point x to a distant point x' via two different paths C and C', we find that not only are the directions of the two resulting vectors different, but their magnitudes have also changed. Integration of (3.279) shows their lengths to be related by [36]

$$l_{(2)} = l_{(1)}\, \exp\left(-\oint a_\alpha\, dx^\alpha\right), \tag{3.280}$$

where the integral is taken round the closed path from x to x' via C, and then back to x via C'. The magnitude of vectors will be unchanged in some region Ω only if

$$\oint a_\alpha\, dx^\alpha = 0, \tag{3.281}$$

original position. After these operations the thumb will be found to be rotated through 90° and pointing to the back; it has been parallel transported round a closed path over a curved surface.

for closed paths lying in Ω. This will be true if the vector field a_α is a pure gradient, (df/dx^α), for then the 1-form $a_\alpha\,dx^\alpha$ is a perfect derivative; the connection is said to be flat.

Now, with the aid of Stokes' theorem we also have

$$\oint a_\alpha\,dx^\alpha = \iint \left(\frac{\partial a_\alpha}{\partial x^\beta} - \frac{\partial a_\beta}{\partial x^\alpha} \right) dx^\alpha\,dx^\beta, \tag{3.282}$$

where the integration is taken over the two-dimensional surface bounded by $\mathcal{C}$ and $\mathcal{C}'$. We define a field curvature tensor by

$$f_{\alpha\beta} = \frac{\partial a_\alpha}{\partial x^\beta} - \frac{\partial a_\beta}{\partial x^\alpha} \tag{3.283}$$

and conclude that length is integrable only if $f_{\alpha\beta}$ vanishes everywhere in the region Ω. As anticipated by the notation, Weyl interpreted the tensor $f_{\alpha\beta}$ as the Faraday tensor for the electromagnetic field,[15]

$$f_{0i} = -\frac{1}{c}E_i, \quad f_{ij} = \varepsilon_{ijk}B_k, \tag{3.284}$$

and the vector field a_α as the field potential. In a region of space where the electromagnetic field vanishes, length is integrable, and Einstein's general relativity is valid as usual.

From (3.279) we have that

$$\frac{dl'}{l'} = -a'_\alpha\,dx^\alpha, \tag{3.285}$$

which with the aid of (3.277) may be transformed successively to give

$$\begin{aligned}
\frac{dl'}{l'} &\equiv d\log_e l' = d\log_e l + d\log_e \xi \\
&= -a_\alpha\,dx^\alpha + \frac{\partial}{\partial x^\alpha}\log_e \xi\,dx^\alpha.
\end{aligned} \tag{3.286}$$

Thus, the dilation (3.277), or equivalently the calibration of distances (3.278), leads to a gauge transformation of the potentials [36]

$$a_\alpha \to a'_\alpha = a_\alpha - \frac{\partial f}{\partial x^\alpha}, \tag{3.287}$$

where we have put $f = \log_e \xi$. The laws of physics must therefore have a double invariance: they must be invariant with respect to arbitrary smooth coordinate transformations (relativistic invariance), and they must also be invariant under gauge transformations. Weyl completed his theory by providing an action function S satisfying this double requirement from which the Lagrangian equations of motion follow in the usual way.

The SI unit of the magnetic field **B** is the tesla. Consequently, if the vector field a_α is to be identified with the 4-potential of the electromagnetic field, the circuit integral

---

[15] To within a proportionality constant that is to be determined.

in (3.280) must have the dimensions[16] of magnetic flux, and the Weyl scale factor in (3.280) can be written in the form

$$W_\gamma = \exp\left(-\frac{e}{\gamma} \oint (\mathbf{a} \cdot \mathbf{dx} - \phi \, dt) \right), \tag{3.288}$$

where e is the fundamental unit of charge; it then follows that both γ and

$$e \oint (\mathbf{a} \cdot \mathbf{dx} - \phi \, dt) \tag{3.289}$$

have the dimensions of action which are energy × time, or angular momentum. The Weyl theory is pure infinitesimal geometry and gives no clue as to the value of the constant γ.

Given the occurrence of e in the scale factor, it is natural to suppose that this is a theory about charged particles, in addition to electromagnetism and gravity. Moreover, in Maxwell's theory of the electromagnetic field there are only two fundamental constants, namely the electrical permittivity of vacuum, ε_0, and the speed of light, c, and it is not possible to form an action from these two quantities alone. The inclusion of charged particles offers the fundamental unit of charge, e, the particle's invariant mass, m_0, and the gravitational constant, G, as additional quantities from which various actions can be defined, and one might hope that γ would be closely related to one of them [38], for example,

$$\gamma = \frac{e^2}{4\pi\varepsilon_0 c}. \tag{3.290}$$

Obviously one must go beyond dimensional considerations and propose a definite physical model that can then be evaluated.

However, despite the formal beauty of the geometry, Weyl's theory has a severe difficulty first pointed out by Einstein [32], [37]. The problem is that if the idea of non-integrable length is correct, the behaviour of atomic clocks would depend on their history according to the paths in space-time they had followed, because the line element ds, (3.278), is connected directly to the physical measurement of distance and time. This is in contradiction with the empirical evidence, particularly the existence of stable atomic species with characteristic spectra. More precisely, an atomic clock measures time in an absolute way, and if one takes the speed of light to be unity, one gets an absolute standard of length; there is then no place for the arbitrary metric standards introduced by Weyl [39]. For such reasons, Weyl's geometrical theory of electromagnetism was deemed untenable.

Shortly after the discovery of wave mechanics, London showed that if the relationship with general relativity was given up, and instead the Weyl scale factor (3.288) was attached to either the de Broglie wave function or the Schrödinger wave function by choosing the constant $\gamma = i\hbar$, then one recovered the wave mechanical account of a particle in an electromagnetic field [40]. In a classic paper, Weyl showed [41] how electromagnetism in quantum mechanics should be understood as a gauge theory

---

[16] Magnetic flux has dimensions of $|\mathbf{B}| \times$ area.

and enunciated the principle that gauge invariance ties together electricity and matter, rather than electricity and gravity as he had earlier [32] proposed. This will be developed in Chapter 9.

3.7.2 Gauge Transformations in the Hamiltonian Formalism

The significance of the Poisson bracket is that it provides the rule for differentiation of a function of the phase space variables. We must compare the value of the function at one point, $\mathbf{z} = (\mathbf{x}, \mathbf{p})$, with its value at an infinitesimally displaced point, $\mathbf{z} + d\mathbf{z}$; infinitesimal displacements are evaluated with the differential operators $\partial/\partial\mathbf{x}, \partial/\partial\mathbf{p}$. According to (3.259), we may identify the gauge-invariant 'momentum' $\bar{\mathbf{p}}$ as the generator of an infinitesimal translation of the particle

$$\mathbf{x} \rightarrow \mathbf{x} + d\mathbf{x}, \tag{3.291}$$

through an infinitesimal canonical transformation with the P.B. relation

$$d\mathbf{x} = \{\mathbf{x}, \bar{\mathbf{p}} \cdot d\mathbf{x}\}. \tag{3.292}$$

An infinitesimal translation $d\mathbf{x}$ of a general phase space function Γ is given by

$$\Gamma(\mathbf{x} + d\mathbf{x}) = \Gamma(\mathbf{x}) + \{\Gamma, \bar{\mathbf{p}} \cdot d\mathbf{x}\}. \tag{3.293}$$

If one transports Γ around an infinitesimal rectangle with sides $d\mathbf{x}, d\mathbf{x}'$, the result after one complete circuit is a change in Γ of

$$\delta\Gamma = \{\Gamma, \{\bar{p}^r, \bar{p}^s\}\} dx^r dx'^s. \tag{3.294}$$

With the aid of (3.259) this becomes

$$\delta\Gamma = e\{\Gamma, \mathbf{B}(\mathbf{x}) \cdot d\boldsymbol{\sigma}\}, \tag{3.295}$$

where the area $d\boldsymbol{\sigma}$ is

$$d\boldsymbol{\sigma} = d\mathbf{x} \wedge d\mathbf{x}'. \tag{3.296}$$

A non-zero value for (3.295) implies that translation of Γ by $d\mathbf{x}$ followed by a translation of $d\mathbf{x}'$ is not the same as translation first by $d\mathbf{x}'$ followed by $d\mathbf{x}$; it is a basic geometrical fact that successive translations on curved surfaces do not commute, so we conclude that classical electrodynamics in Hamiltonian form involves a curved phase space characterised in some way by the magnetic field.

Corresponding to the infinitesimal version (3.295), there is a finite integrated form involving the integral

$$e \int_{\Sigma} \mathbf{B} \cdot d\mathbf{S}, \tag{3.297}$$

where the integral is taken over a surface Σ bounded by a closed curve $\mathcal{P}$. By Stokes theorem this is also

$$e \oint_{\mathcal{P}} \mathbf{a}(\mathbf{x}) \cdot d\mathbf{x} = e \oint_{\mathcal{P}} dv, \tag{3.298}$$

where

$$\mathbf{B}(\mathbf{x}) = \boldsymbol{\nabla} \wedge \mathbf{a}(\mathbf{x}) \qquad (3.299)$$

expresses the usual relationship between the magnetic field and a vector potential. The close connection with (3.84)–(3.86) is evident. In the terms of differential geometry, the 1-form $\mathrm{d}\upsilon$ is the connection that specifies how to make infinitesimal displacements in the phase space, and the magnetic field $\mathbf{B}$ is the associated curvature of the space.

We saw earlier that when two Lagrangian functions (L, L') are related by

$$L \rightarrow L' = L - \frac{\mathrm{d}F}{\mathrm{d}t}, \qquad (3.300)$$

they yield the same equations of motion; their corresponding Hamiltonians are related by a canonical transformation $H \rightarrow H'$ with F as the generator. The relationship between the general Hamiltonian, $H[\mathbf{g}]$, and the Coulomb gauge Hamiltonian, $H[0]$, Eq. (3.274), can be put in this form. The required transformation function, F, is the action integral we have already met in several different contexts in this chapter,

$$F = \int \mathbf{P}(\mathbf{x}) \cdot \mathbf{A}(\mathbf{x}) \, \mathrm{d}^3\mathbf{x}, \qquad (3.301)$$

where $\mathbf{A}$ is the Coulomb gauge vector potential. According to (2.142), the electric polarisation field, $\mathbf{P}(\mathbf{x})$, is a linear functional of the Green's function $\mathbf{g}(\mathbf{x}, \mathbf{x}')$, and so too is F, $F \equiv F[\mathbf{g}]$.

The relationship between $H[0]$ and $H[\mathbf{g}]$ is not a gauge transformation since F has a vanishing P.B. with the vector potential $\mathbf{A}$. This is so even though we define the general vector potential $\mathbf{a}$ using (2.76); the new feature in the canonical scheme is that we only meet the combination $e_n \mathbf{a}(\mathbf{x}_n)$ with the vector potential evaluated at the position of the charge e_n, and the transformation is properly associated with a modification of the particle momentum. Using (2.76), we have

$$e_n \mathbf{A}(\mathbf{x}_n) - e_n \mathbf{a}(\mathbf{x}_n) = \boldsymbol{\nabla}_{\mathbf{x}_n} e_n \int \mathbf{A}(\mathbf{x}'') \cdot \mathbf{g}(\mathbf{x}'', \mathbf{x}_n) \, \mathrm{d}^3 \, \mathbf{x}'' \qquad (3.302)$$

$$= \boldsymbol{\nabla}_{\mathbf{x}_n} e_n \iint \mathbf{A}(\mathbf{x}'') \cdot \mathbf{g}(\mathbf{x}'', \mathbf{x}') \, \delta^3(\mathbf{x}' - \mathbf{x}_n) \, \mathrm{d}^3\mathbf{x}'' \mathrm{d}^3\mathbf{x}'. \qquad (3.303)$$

This can be written in terms of the charge density ρ, (2.104) since

$$\boldsymbol{\nabla}_{\mathbf{x}_n} \rho(\mathbf{x}) = \boldsymbol{\nabla}_{\mathbf{x}_n} \sum_k e_k \delta^3(\mathbf{x} - \mathbf{x}_k) \equiv \boldsymbol{\nabla}_{\mathbf{x}_n} e_n \delta^3(\mathbf{x} - \mathbf{x}_n), \qquad (3.304)$$

and using (2.142), we finally get

$$e_n \mathbf{a}(\mathbf{x}_n) = e_n \mathbf{A}(\mathbf{x}_n) - \boldsymbol{\nabla}_{\mathbf{x}_n} \int \mathbf{P}(\mathbf{x}) \cdot \mathbf{A}(\mathbf{x}) \, \mathrm{d}^3\mathbf{x}, \qquad (3.305)$$

$$= e_n \mathbf{A}(\mathbf{x}_n) + \{\mathbf{p}_n, F[0]\}, \qquad (3.306)$$

in P.B. notation. We now view this change as a classical canonical transformation leading to modified particle and canonical momenta, and a transformed Hamiltonian which is expressed in terms of the electric polarisation field and the bracket in (3.306). The resulting Hamiltonian is exactly what is obtained by straightforwardly using L_{int} in the form (3.58) [5]. In this way, $\mathbf{g}^\perp$ is moved from the Dirac brackets of the variables, (3.255)–(3.257) into the Hamiltonian while the Dirac brackets for the Coulomb gauge are left unchanged.

According to (3.85), $F[0]$ may be expressed as a line integral over the 1-form $\mathrm{d}v$ with the vector potential in the Coulomb gauge ($f = 0$ here). Now the differential 1-form $\mathrm{d}v$ may be taken as the generator of an infinitesimal canonical transformation of a phase space variable Ω, according to the usual rule,

$$\Omega \to \Omega' = \Omega + \mathrm{d}\Omega, \tag{3.307}$$

with $\mathrm{d}\Omega$ determined by the P.B. (or Dirac bracket as required)

$$\mathrm{d}\Omega = e\{\Omega, \mathrm{d}v\}. \tag{3.308}$$

Composition of this continuous transformation along some path $\mathcal{C}$ ending at the particle with charge e, leads to a finite canonical transformation which may be expressed using $F[0]$; indeed if we define the Lie derivative operator $L_{F[0]}$ by

$$L_{F[0]}\Omega = \{\Omega, F[0]\}, \tag{3.309}$$

then $e^{L_{F[0]}}\Omega$ is the new phase space function obtained by transforming Ω using the power series expansion of the exponential according to

$$\tilde{\Omega} = \Omega + \{\Omega, F[0]\} + \frac{1}{2!}\{\{\Omega, F[0]\}, F[0]\} + \dots . \tag{3.310}$$

It is readily verified that the transformation is canonical.

A simple illustration of the relationship (3.310) is afforded by the Gauss's law constraints, G_0 and G, defined by (3.237). Recall that as weak equations the vector potential is not constrained by a gauge condition, and its conjugate acts as the functional derivative operator (3.229). Then the relationship

$$G = e^{L_{F[0]}} G_0 \tag{3.311}$$

is easily established, using (2.138) and (3.229) with (3.310). Similar calculations show that the Hamiltonian $H[\mathbf{g}^\perp]$ for an arbitrary $\mathbf{g}^\perp$ displayed in (3.254) is obtained by choosing Ω as the Coulomb gauge Hamiltonian, (3.274),

$$\Omega = H[0], \tag{3.312}$$

in (3.310). In this case the ... in the series (3.310) are zero since with this choice of Ω the second-order term is purely a function of the 'position' variables for the particles and field, and so has vanishing P.B. with the generator $F[0]$.

The transformed Hamiltonian is independent of the field potential,

$$H[\mathbf{g}] = \sum_n \frac{1}{2m_n}\left(\mathbf{p}_n + \mathbf{Q}_n(\mathbf{B})\right)^2 - \int \mathbf{P}\cdot\mathbf{E}^\perp \mathrm{d}^3\mathbf{x}$$
$$+ \frac{1}{2\varepsilon_0}\int \mathbf{P}\cdot\mathbf{P}\,\mathrm{d}^3\mathbf{x} + \frac{1}{2}\varepsilon_0\int\left(\mathbf{E}^\perp\cdot\mathbf{E}^\perp + c^2\mathbf{B}\cdot\mathbf{B}\right)\mathrm{d}^3\mathbf{x}, \tag{3.313}$$

where $\mathbf{Q}_n(\mathbf{B})$ is obtained from the bracket in (3.306). In Appendix E, this derivative is shown to yield a term causing the cancellation of the Coulomb gauge vector potential and a linear functional of the magnetic field $\mathbf{B}$,

$$\frac{\partial}{\partial \mathbf{x}_n} \int \mathbf{P} \cdot \mathbf{A} \, d^3 x = e_n \mathbf{A}(\mathbf{x}_n) + \mathbf{Q}_n(\mathbf{B}), \qquad (3.314)$$

and so one can write

$$\sum_n \frac{1}{2m_n} (\mathbf{p}_n + \mathbf{Q}_n(\mathbf{B}))^2 = \sum_n \frac{|\mathbf{p}_n|^2}{2m_n} - \int \mathbf{M} \cdot \mathbf{B} \, d^3 x + \int\!\!\int \boldsymbol{\chi} : \mathbf{B}\mathbf{B} \, d^3 x \, d^3 x',$$

where $\mathbf{M}$ is indeed the magnetic polarisation field introduced earlier, but here expressed in terms of the canonical variables $(\{\mathbf{x}_n\}, \{\mathbf{p}_n\})$. Explicit forms for $\mathbf{M}$ and $\boldsymbol{\chi}$ can be found in [42].

3.8 The Hamiltonian Equations of Motion for Electrodynamics

Equation (3.273) is the Hamiltonian that is familiar from numerous textbook discussions of the interactions of atoms and molecules with electromagnetic radiation. What is less usual in the chemical physics literature is any consideration of what might be the consequences of the Hamiltonian equations of motion derived from it. In fact, as has long been known, the classical theory has pathological solutions, so it is plausible that some of the problems will be inherited by a quantum theory that is based on it following the canonical quantisation algorithm. This is not inevitable; the quantum theory of the atom based on purely Coulombic interactions between electrons and nuclei is one of the most striking successes of the quantum theory; it will be described in detail in Chapter 8. This is so despite the fact that the classical mechanics of the many-body problem with pairwise potential energies that vary as $1/r$ is pathological. Those difficulties are swept away by quantisation, as shown by the famous Kato–Rellich theorem which might be taken as a justification for the bold approach of Bohr referred to in Chapter 1. What underlies success here is the implicit assumption that the mass and charge parameters to be used in the Coulomb Hamiltonian are those determined from experiment; however, in the Lagrangian (or Hamiltonian) formulations they enter purely as free parameters, and it remains to be seen how they should be related to experimental values when interactions with electromagnetic radiation are included. This, of course, requires a quantum theory, but nevertheless it is useful to examine first how the precursor classical theory behaves. To conclude this chapter, we examine the classical Hamiltonian equations derived from the Coulomb gauge Hamiltonian for a closed, coupled system of charged particles and the electromagnetic field.

In the Coulomb gauge, Hamilton's equations of motion for a collection of point charges $\{e_n, m_n, n = 1 \ldots N\}$ with positions $\{\mathbf{x}_n, n = 1 \ldots N\}$, at time t, are

$$\dot{\mathbf{p}}_n = \{\mathbf{p}_n, H[0]\} = \boldsymbol{\nabla}_{\mathbf{x}_n} \left(e_n \dot{\mathbf{x}}_n \cdot \mathbf{A}(\mathbf{x}_n) + V_{\text{Coul}} \right), \quad n = 1, \ldots N, \qquad (3.315)$$

$$\dot{\mathbf{x}}_n = \{\mathbf{x}_n, H[0]\} = \frac{1}{m_n}\left(\mathbf{p}_n - e_n \mathbf{A}(\mathbf{x}_n)\right), \quad n = 1, \dots N. \tag{3.316}$$

For each charged particle, we thus require the vector potential $\mathbf{A}(\mathbf{x}_n)$ evaluated at the charge's position, $\mathbf{x}_n$. The solution of the equation of motion for the vector potential (Appendix C) when substituted into (3.315) and (3.316) for particle n implies three contributions: a solution of the homogeneous equation describing the field in the absence of charges, a contribution from the charge n which must be evaluated at $\mathbf{x}_n$ – the 'self-field', and a contribution from all the other charges $\{k \neq n\}$. In the following we examine the case of a single charge e with position $\mathbf{q}$ in order to characterise the 'self-field'. From the results in Appendix C, the vector potential due to the charge with retarded boundary conditions evaluated at the position of the charge is

$$\mathbf{A}(\mathbf{q}(t)) = \left(\frac{e}{\varepsilon_0 c^2}\right)\int_{-\infty}^{t} \mathbf{G}^{\perp}(\mathbf{Q}_{t't}; t, t') \cdot \dot{\mathbf{q}}(t')\, dt', \tag{3.317}$$

where the dyadic $\mathbf{G}^{\perp}$ may be written as ($\chi_0(k) = 1$ for the point particle)

$$\mathbf{G}^{\perp}(\mathbf{Q}_{t't}; t, t') = \frac{c}{(2\pi)^3}\int\left(\frac{1 - \hat{\mathbf{k}}\hat{\mathbf{k}}}{k}\right)e^{i\mathbf{k}\cdot\mathbf{Q}_{t't}}\sin[kc(t-t'))]\, d^3\mathbf{k}, \tag{3.318}$$

with

$$\mathbf{Q}_{t't} = \mathbf{q}(t') - \mathbf{q}(t). \tag{3.319}$$

The non-linear dependence on $\mathbf{q}(t)$ in (3.318) greatly complicates a discussion of the equations of motion; in the following it will be useful to write

$$e^{i\mathbf{k}\cdot\mathbf{Q}_{t't}} = 1 + \left(e^{i\mathbf{k}\cdot\mathbf{Q}_{t't}} - 1\right) \equiv 1 + K_{\mathbf{q}}(\mathbf{k}, t') \tag{3.320}$$

and decompose $\mathbf{G}^{\perp}$ and the vector potential in an obvious way into their linear and non-linear parts

$$\mathbf{A}(\mathbf{q}, t) = \mathbf{A}^{L}(\mathbf{q}, t) + \mathbf{A}^{NL}(\mathbf{q}, t), \tag{3.321}$$

where, for example,

$$\mathbf{A}^{L}(\mathbf{q}, t) = C \iint_{-\infty}^{t}\frac{(1 - \hat{\mathbf{k}}\hat{\mathbf{k}})}{k} \cdot \dot{\mathbf{q}}(t')\sin[kc(t-t')]\, dt'\, d^3\mathbf{k}, \tag{3.322}$$

with

$$C = \left(\frac{e}{8\pi^3 \varepsilon_0 c}\right). \tag{3.323}$$

With the inclusion of an arbitrary solution of the homogeneous (free) field equation, $\mathbf{A}_h$, the complete specification of the equations of motion for a charged particle is

$$\dot{\mathbf{p}} = e\nabla_{\mathbf{q}}\left(\dot{\mathbf{q}}\cdot(\mathbf{A}(\mathbf{q}, t)_h + \mathbf{A}(\mathbf{q}, t))\right) \tag{3.324}$$

$$m\dot{\mathbf{q}} = \mathbf{p} - e\left(\mathbf{A}(\mathbf{q}, t)_h + \mathbf{A}(\mathbf{q}, t)\right). \tag{3.325}$$

The solutions of the equations of motion based on the linear part of the vector potential, $\mathbf{A}^{L}$, are of interest themselves and provide crucial information about the properties of the full problem, for example, the asymptotic ($t \to \infty$) stability of its solutions [43].

3.8.1 A Particular Integral

Before attempting any kind of general discussion of the equations of motion, it may be helpful to work through a special case defined by the condition

$$\dot{\mathbf{q}}(t) = \mathbf{v}, \tag{3.326}$$

where $\mathbf{v}$ is a constant vector. Integration of the particle velocity gives

$$\mathbf{q}(t) = \mathbf{v}t + \mathbf{q}_0, \tag{3.327}$$

where $\mathbf{q}_0$ is a constant vector specifying an initial position. We then have

$$\mathbf{Q}_{t't} = \mathbf{v}(t' - t), \tag{3.328}$$

and so

$$\int_{-\infty}^{t} e^{-i\mathbf{k}\cdot\mathbf{v}(t-t')} \sin[kc(t-t')]\, \mathrm{d}t' = \frac{kc}{(kc)^2 - (\mathbf{k}\cdot\mathbf{v})^2}, \tag{3.329}$$

where the device of 'adiabatic switching' has been used to evaluate the integral.

The integration over $\mathbf{k}$ in (C.0.25) is divergent for a point particle and must be regulated; we do so in the manner described earlier (§2.4.2) which simply means the integrand is modified by multiplication with $\chi_a^2(k)$. We then have

$$c \int \left(\frac{\chi_a^2(k)}{k^2c^2 - (\mathbf{k}\cdot\mathbf{v})^2} \right) (1 - \hat{\mathbf{k}}\hat{\mathbf{k}}) \cdot \mathbf{v}\, \mathrm{d}^3 k = \left(\frac{\pi^2}{ac} \right) I(\beta)\mathbf{v}, \tag{3.330}$$

where we used (2.127) for the k-integration, and

$$I(\beta) = \int_{-1}^{+1} \left(\frac{1-x^2}{1-\beta^2 x^2} \right) \mathrm{d}x \approx \frac{4}{3} + \frac{4\beta^2}{15} + O(\beta^4), \tag{3.331}$$

where $\beta = |\mathbf{v}|/c$. Collecting these results together (and neglecting β for the non-relativistic regime) we get

$$e\mathbf{A} = \left(\frac{e^2}{6\pi\varepsilon_0 ac^2} \right) \mathbf{v}, \tag{3.332}$$

which is a constant vector, independent of the particle's position and the time.

The equations of motion (3.324) and (3.325) now read (with the homogeneous field dropped)

$$\dot{\mathbf{p}} = 0, \tag{3.333}$$

$$m\dot{\mathbf{q}} = \mathbf{p} - e\,\mathbf{A} \tag{3.334}$$

with the vector potential we have just calculated. In terms of a parameter Δm with dimensions of mass (cf §2.4.2)

$$\Delta m = \frac{e^2}{6\pi\varepsilon_0 ac^2}, \tag{3.335}$$

and a constant momentum vector $\mathbf{p}_0$, Eq. (3.334) becomes

$$m\mathbf{v} = \mathbf{p}_0 - \Delta m\, \mathbf{v}, \tag{3.336}$$

which may be rearranged to give

$$\mathbf{p}_0 = m^{\text{obs}}\mathbf{v}, \tag{3.337}$$

where we claim that the observed mass of the particle m^{obs} is precisely the sum of the 'mechanical' mass, m, and the ' electromagnetic' mass, Δm, $m^{\text{obs}} = m + \Delta m$. This is a self-consistent particular integral of the equations of motion. Note that the calculation fails in the point particle limit, $a \to 0$. The electric and magnetic fields due to a charge with constant velocity are evidently zero since all space-time derivatives of $\mathbf{A}$ vanish; the entire electromagnetic interaction results in a change in the mass of the particle.

3.8.2 Mass Renormalisation

The relationship between the mechanical mass m and the observed mass m^{obs} is the basis for *mass renormalisation*. We take account explicitly of the contribution to the mass of the charged particle due to the electromagnetic self interaction, so that the 'structure' parameter does not appear in the equations of motion. For the theory based on an extended charge distribution, this is achieved by extracting from the equation of motion a term simply proportional to $\dot{\mathbf{q}}(t)$ and identifying its coefficient as the mass correction due to the self-interaction. Clearly this is not possible if the point charge limit is taken first; historically, mass renormalisation was devised within the point charge model and had to proceed by quite different means [44].

We use integration by parts on the t' integration in (3.317), choosing the 'dv' factor as $\sin[kc(t-t')]$. The boundary term is easily evaluated since it vanishes in the far past and the exponential and cosine terms simply give 1 at $t' = t$. Hence, after the remaining elementary integration over $\mathbf{k}$, this contribution to the vector potential reduces to

$$uv\bigg| = \left(\frac{\Delta m}{e}\right)\dot{\mathbf{q}}(t). \tag{3.338}$$

The integrated part does not simplify and can probably only be usefully evaluated in some approximation. The renormalised equation of motion for the coordinate $\mathbf{q}$ is therefore

$$m^{\text{obs}}\dot{\mathbf{q}} = \mathbf{p} \tag{3.339}$$

$$-\left(\frac{eC}{c}\right)\int\int\int_{-\infty}^{t}\left(\frac{\chi_a^2(k)}{k^2}\right)\cos[kc(t-t')]\frac{d\boldsymbol{\varepsilon}(\mathbf{k},t')}{dt'}\,dt'\,d^3\mathbf{k}, \tag{3.340}$$

where we have put

$$m^{\text{obs}} = m + \Delta m, \tag{3.341}$$

and

$$\boldsymbol{\varepsilon}(\mathbf{k},t') = \left((1 + K_{\mathbf{q}}(\mathbf{k},t'))\,(\mathbf{1} - \hat{\mathbf{k}}\hat{\mathbf{k}})\cdot\dot{\mathbf{q}}(t')\right). \tag{3.342}$$

The classical radius, r_e, for a charged particle is related to its observed mass, m_{obs}, by equating the particle's invariant mass energy to its electrostatic energy:

$$m^{obs} c^2 = \left(\frac{e^2}{4\pi\varepsilon_0 r_e} \right). \tag{3.343}$$

We use r_e to define a dimensionless parameter $\mu = r_e/a$, in terms of which we may write

$$\frac{\Delta m}{m^{obs}} = \frac{2\mu}{3} \equiv \zeta, \tag{3.344}$$

so that the 'mechanical mass' m is

$$m = (1 - \zeta) \, m^{obs}. \tag{3.345}$$

So far no approximation beyond the use of non-relativistic mechanics has been made. An obvious simplification is to set

$$K_{\mathbf{q}}(\mathbf{k},t) = 0, \tag{3.346}$$

in which case

$$\frac{d\varepsilon(\mathbf{k},t')}{dt'} \rightarrow (\mathbf{1} - \hat{\mathbf{k}}\hat{\mathbf{k}}) \cdot \ddot{\mathbf{q}}(t'). \tag{3.347}$$

Physically, the approximation amounts to the neglect of retardation over the extension of the charge distribution; it corresponds to the neglect of terms in β in §3.8.1. It is called the linear approximation because the derivatives of $\mathbf{q}$ only occur linearly in the resulting integral expression for the vector potential which has no functional dependence on $\mathbf{q}$ itself. Accordingly, there are no magnetic field effects in the self-interaction in this approximation; in other contexts it is known as the 'electric dipole approximation'. Note also that for the same reason the RHS of (3.315) vanishes so that the particle momentum is a constant of the motion. The important thing to recognise is that (3.340) is not like a conventional Newtonian equation of motion; rather it is an integro-differential equation for the velocity vector $\dot{\mathbf{q}}$.

3.8.3 The Point Charge Model

An important limiting case of the calculation just described is the point charge limit with $\chi_0(k) = 1$. In this limit we have $Q_{t't} = 0$, and the coefficient of Δm is simply proportional to $\ddot{\mathbf{q}}$ [42]. Strictly speaking, we can no longer take the particle momentum to be constant in time, since the homogeneous field $\mathbf{A}(\mathbf{q},t)_h$ contributes[17] also to (3.315), so we write the equation of motion for a point charge as

$$\ddot{\mathbf{q}}(t) - \omega_0 \, \dot{\mathbf{q}}(t) = - \left(\frac{\omega_0}{m} \right) \, \mathbf{p}(t) + \left(\frac{e}{m} \right) \, \mathbf{A}(\mathbf{q},t)_h, \tag{3.348}$$

where

$$\omega_0 = \left(\frac{c}{2a} \right) \left(\frac{m}{\Delta m} \right). \tag{3.349}$$

---

[17] It leads to the magnetic field contribution to the Lorentz force law; if the magnetic field effects are deemed to be negligible, $\mathbf{A}(\mathbf{q},t)_h = \mathbf{A}(t)_h$.

Let

$$\dot{\mathbf{q}}(t) = \mathbf{z}(t), \tag{3.350}$$

so that

$$\mathbf{q}(t) = \int^t \mathbf{z}(t') \, dt' + \mathbf{q}_0, \tag{3.351}$$

where $\mathbf{q}_0$ is an integration constant. The solution for the velocity is

$$\mathbf{z}(t) = e^{\omega_0 t} \left[e^{-\omega_0 t_0} \mathbf{z}(t_0) + \int_{t_0}^t e^{-\nu \omega_0} \left(\left(\frac{\omega_0}{m} \right) \mathbf{p}(\nu) + \left(\frac{e}{m} \right) \mathbf{A}(\mathbf{q}, \nu)_h \right) d\nu \right], \tag{3.352}$$

which in general shows runaway behaviour, $\mathbf{z}(+\infty) = \infty$; the omission of the free-field vector potential does not alter this conclusion. Since ω_0 contains e^{-2}, the coordinate has an essential singularity at $e = 0$, so this is a non-perturbative solution.

The situation can be 'saved' if we allow the specification of a particular value for the velocity $\mathbf{z}$ at the instant t_0 as an extra initial condition. This is contrary to the spirit of Hamilton's equations which are a pair of coupled first-order differential equations to be solved with initial data $\mathbf{q}(t_0), \mathbf{p}(t_0)$. We chose $\mathbf{z}(t_0)$ so that[18]

$$e^{-\omega_0 t_0} \mathbf{z}(t_0) = \int_{t_0}^\infty e^{-\omega_0 \nu} \left(\frac{\omega_0}{m} \ \mathbf{p}(\nu) - \left(\frac{e}{m} \right) \mathbf{A}(\mathbf{q}, \nu)_h \right) d\nu. \tag{3.353}$$

Substitution of this choice for $\mathbf{z}(t_0)$ in (3.352) yields the velocity as

$$\begin{aligned}
\mathbf{z}(t) &= \int_t^\infty e^{\omega_0(t-\nu)} \left(\frac{\omega_0}{m} \ \mathbf{p}(\nu) - \left(\frac{e}{m} \right) \mathbf{A}(\mathbf{q}, \nu)_h \right) d\nu \\
&= \int_0^\infty e^{-\omega_0 \tau} \left(\frac{\omega_0}{m} \mathbf{p}(\tau + t) - \left(\frac{e}{m} \right) \mathbf{A}(\mathbf{q}, \tau + t)_h \right) d\tau.
\end{aligned} \tag{3.354}$$

This is well-behaved asymptotically; it is, however, acausal since the velocity at time t depends on the dynamical variables on the RHS at later times; this is called 'pre-acceleration'. Note that in general the coordinate $\mathbf{q}$ in $\mathbf{A}_h$ cannot be ignored, and since it depends on the integration variable τ, integration of the velocity vector actually leads to a nonlinear integral equation for the coordinate variable. The same problems arise in the standard Lorentz force law discussion of a classical point charge; there the force turns out to involve the third time derivative of the coordinate, so additional information again has to be supplied in order to fix the solution. In general, the Lorentz force law permits runaway solutions which can be suppressed as here, but only at the expense of the problem of pre-acceleration [46], [47].

3.8.4 The Extended Charge Model

In the linear approximation the momentum is a constant of the motion, $\mathbf{p}_0$, so that we need only consider the equation of motion for the coordinate. We make the following substitutions in the linear part of the vector potential, $\mathbf{A}^L$

$$x = \left(\frac{2a}{\pi} \right) k, \quad y = \left(\frac{\pi c}{2a} \right) (t - t'). \tag{3.355}$$

---

[18] This idea is due to Dirac in a slightly different context [45].

The equation of motion derived from (3.317) in this approximation with Δm as before is then

$$m\dot{\mathbf{q}}(t) = \mathbf{p}_0 - \Delta m \int_0^\infty \int_0^\infty x \chi_a^2\left(\tfrac{\pi x}{2a}\right) \sin(xy)\dot{\mathbf{q}}\left(t - \tfrac{2ay}{\pi c}\right) dy\,dx, \qquad (3.356)$$

which is a linear integro-differential equation with a delay. We define a linear operator L by the relation

$$\mathsf{L}(\phi(t)) = m\phi(t) + \Delta m\, I_a(\phi(t)), \qquad (3.357)$$

where

$$I_a(\phi(t)) = \int_0^\infty \int_0^\infty x \chi_a^2\left(\tfrac{\pi x}{2a}\right) \sin(xy)\phi\left(t - \tfrac{2ay}{\pi c}\right) dy\,dx, \qquad (3.358)$$

so that (3.356) is concisely expressed as

$$\mathsf{L}(\dot{\mathbf{q}}(t)) = \mathbf{p}_0. \qquad (3.359)$$

If we can solve this equation, the orbit of the particle will again be (3.351).

The linear equation (3.356) can be solved by the method of characteristic functions [43]. The characteristic equation of L is found directly by studying its action on the exponential function e^{st}, where s is a parameter that will determine the solutions, if any exist; in general s will be a complex number. Consider then

$$\mathsf{L}(e^{st}) = m\, e^{st} + \Delta m\, I_a(e^{st}). \qquad (3.360)$$

The y integration is elementary and there results

$$\mathsf{L}(e^{st}) = e^{st}\left[m + \Delta m \int_0^\infty \frac{x^2\, \chi_a^2\left(\tfrac{\pi x}{2a}\right)}{x^2 + \left(\tfrac{2as}{\pi c}\right)^2}\, dx \right]. \qquad (3.361)$$

Thus, the exponential function, e^{st}, provides a family of solutions of the homogeneous equation

$$\mathsf{L}(e^{st}) = 0, \qquad (3.362)$$

for all s values $\{s_n\}$, that are zeroes of the characteristic function

$$h(s_n) = m + \Delta m \int_0^\infty \frac{x^2\, \chi_a^2\left(\tfrac{\pi x}{2a}\right)}{x^2 + \left(\tfrac{2as_n}{\pi c}\right)^2}\, dx = 0. \qquad (3.363)$$

Since both m and Δm can be expressed in terms of the dimensionless parameter ζ as in §3.8.2, (3.363) can be reduced to, for $\zeta \neq 1$,

$$1 = \left(\frac{\zeta}{1-\zeta}\right) \int_0^\infty \frac{x^2\, \chi_a^2\left(\tfrac{\pi x}{2a}\right)}{x^2 + \left(\tfrac{2as_n}{\pi c}\right)^2}\, dx. \qquad (3.364)$$

In general, (3.364) is a transcendental equation with an infinite set of solutions.

As an illustration consider the finite spherical shell model for which

$$\chi_a\left(\tfrac{\pi x}{2a}\right) = \frac{\sin\left(\tfrac{\pi x}{2}\right)}{\tfrac{\pi x}{2}}. \qquad (3.365)$$

Carrying out the x-integration, (3.364) becomes

$$\eta = \left(\frac{\zeta}{1-\zeta}\right)\left(e^{-\eta} - 1\right), \tag{3.366}$$

where

$$\eta = \frac{2as}{c}. \tag{3.367}$$

Writing $\eta = \eta_R + i\eta_I$, we have

$$\eta_R = \left(\frac{\zeta}{(1-\zeta)}\right)\left(e^{-\eta_R}\cos(\eta_I) - 1\right), \tag{3.368}$$

$$\eta_I = -\left(\frac{\zeta}{(1-\zeta)}\right)e^{-\eta_R}\sin(\eta_I). \tag{3.369}$$

These coupled equations are the same as those obtained from an analysis of the Lorentz force law formulation of the same problem [48], [49]. They lead to the conclusion that there are well-behaved, stable solutions for the motion of a finite charged particle under the influence of its radiation reaction, provided a certain condition is satisfied. A runaway solution corresponds to the frequency s having a positive real part. A little consideration of (3.368) shows that for $\eta_R > 0$ there are no solutions provided that $\zeta < 1$; runaway solutions are only possible for $\zeta > 1$ which corresponds to a negative mechanical mass. There are always solutions for $\eta_R < 0$; if $\eta_I = 0$, they decay exponentially in time, whereas for $\eta_I \neq 0$ there are infinitely many oscillatory damped solutions. From the definition (3.344) of the parameter ζ, the condition for stable solutions in the spherical shell model, within the linear approximation, can also be expressed as the requirement that the radius of the shell, a, be at least 1.5 times the classical radius of the particle, r_e. If this condition is met, all continuous solutions of the homogeneous equation of motion are stable, that is, they $\rightarrow 0$ as $t \rightarrow +\infty$. Whereas a specified solution of a first-order differential equation (the expected result for each variable in the Hamiltonian scheme) can be fixed by giving the value of the solution at a point as an initial condition, Eq. (3.356) requires an initial value function to be specified over a definite interval, here $0 < t \leq \frac{2a}{c}$. Once $\dot{q}(t)$ is specified over this time interval, a continuous solution for all $t \geq 0$ is determined [43].

3.8.5 Classical Hamiltonian Electrodynamics Revisited

A fundamental result in the Hamiltonian formulation of mechanics is that the time evolution of the system can be regarded as the unfolding of a sequence of infinitesimal canonical transformations for which the Hamiltonian itself is the generator. Recall that if G is the generator of such a transformation, the change in any dynamical variable Ω, a function of the canonical variables, is given by the P.B. relation

$$\delta\Omega = \{\Omega, G\}. \tag{3.370}$$

If we choose $H\,\mathrm{d}t$ as the generator, we get the following relations for the result of transformation of the basic phase space variables $(q(t)_n, p(t)_n)$,

$$Q(t)_n = q(t)_n + \frac{\partial H}{\partial p_n}\,\mathrm{d}t = q(t)_n + \dot{q}_n\,\mathrm{d}t = q(t + \mathrm{d}t)_n, \tag{3.371}$$

$$P(t)_n = p(t)_n - \frac{\partial H}{\partial q_n}\,\mathrm{d}t = p(t)_n + \dot{p}_n\,\mathrm{d}t = p(t + \mathrm{d}t)_n, \tag{3.372}$$

corresponding to the 'passive' interpretation (LHS) in terms of transformation to new variables, and an 'active' interpretation (RHS) in terms of the time evolution of $q(t)_n, p(t)_n$. A transformation from old (q_n, p_n) to new (Q_n, P_n) variables is canonical if the P.B. relations are preserved by the transformation,

$$\{q_n, q_m\} = \{p_n, p_m\} = 0, \quad \{q_n, p_m\} = \delta_{nm},$$
$$\rightarrow \{Q_n, Q_m\} = \{P_n, P_m\} = 0, \quad \{Q_n, P_m\} = \delta_{nm}. \tag{3.373}$$

Now it is easily seen that if we take the Hamiltonian for a charge interacting with its own electromagnetic field, the above relations are not satisfied. The velocity in (3.316) is the gauge-invariant quantity defined by (3.244) which by (3.259) has components which no longer have vanishing P.B.s with each other. $\dot{q}$ also occurs in the infinitesimally time-translated field variables, and so the field and particle variables will have some non-zero P.B.s, contrary to the original assumptions. There is therefore a fundamental problem with the conventional classical Hamiltonian formulation which amounts to an incomplete specification of the set of dynamical variables; in other words, we need to identify additional variables such that we can make independent variations in the action integral. In the point particle limit the vector potential for the interacting system is proportional to the particle acceleration [42]. If such a Hamiltonian is to be derived from a Lagrangian, it too must involve the particle acceleration.

Recall that a Lagrangian is only determined to within a total time derivative; two Lagrangian functions related by

$$L' = L + \frac{\mathrm{d}f}{\mathrm{d}t}, \tag{3.374}$$

where f has the dimensions of action, yield the same dynamics according to the principle of least action. Thus, if we allow f to be a function of position, $\mathbf{q}$, *and* velocity, $\dot{\mathbf{q}}$, so that the total time derivative is interpreted as

$$\frac{\mathrm{d}}{\mathrm{d}t} = \frac{\partial}{\partial t} + \dot{\mathbf{q}} \cdot \frac{\partial}{\partial \mathbf{q}} + \ddot{\mathbf{q}} \cdot \frac{\partial}{\partial \dot{\mathbf{q}}}, \tag{3.375}$$

the transformed Lagrangian L' will involve the *acceleration* of the particle, in addition to the position and velocity. This implies, however, that the usual Legendre transformation equations for passing to the Hamiltonian are insufficient.

The classical generalisation to Lagrangian functions containing higher derivatives of the coordinate $\mathbf{q}$ beyond the velocity is due to Ostrogradsky [50]. However, Ostrogradsky's method is only valid with non-degenerate Lagrangian functions, and here we know we already have a constrained system because of the gauge transformations of the field potentials. Essentially, the higher-order derivatives must be interpreted as independent variables and appropriate Lagrange multipliers introduced to validate the identifications made; as a result further equations of constraint arise (cf. the example in §H). A discussion of mass renormalisation for a point charged particle in the electric dipole approximation that incorporates constraints in the canonical formalism yields equations of motion involving only observable mass and charge parameters; the bare mechanical mass reappears in the P.B.s of the redefined dynamical variables [51]. This is a refinement of the original mass renormalisation procedure outlined by Kramers [44].

3.8.6 Classical Lagrangian Electrodynamics Revisited

The choice of F required to implement (3.374) is dictated by the requirement that the self-interaction effects mediated through the coupling between the current and the vector potential due to the charge

$$\int \mathbf{j} \cdot \mathbf{a}_s \, d^3 \mathbf{x}, \tag{3.376}$$

should be eliminated from L, and that additional terms must include the mass renormalisation $\Delta m \, |\dot{\mathbf{q}}(t)|^2$ for a regulated charge distribution,

$$\rho = e\xi(\mathbf{x} - \mathbf{q(t)}). \tag{3.377}$$

Explicit calculation shows that this can be achieved through a consideration of the partial time derivative contribution from the generating function

$$F = \frac{1}{c^2} \int \mathbf{j}(\mathbf{x}) \cdot \mathbf{Z}(\mathbf{x}, t) \, d^3 \mathbf{x}, \tag{3.378}$$

where $\mathbf{Z}(\mathbf{x}, t)$ is

$$\mathbf{Z}(\mathbf{x}, t) = \frac{1}{(2\pi)^3} \int \mathbf{Z}(\mathbf{k}, t) e^{-i\mathbf{k} \cdot \mathbf{x}} d^3 \mathbf{k}, \quad \mathbf{Z}(\mathbf{k}, t) = \frac{\boldsymbol{\pi}(\mathbf{k}, t)}{k^2}, \tag{3.379}$$

and $\mathbf{Z}(\mathbf{k}, t)$ is constructed from the solution of Hamilton's equations for the field variables (Appendix C), explicitly

$$\mathbf{Z}(\mathbf{k}, t) = \left(\frac{e}{\varepsilon_0}\right) \chi_a(k) \frac{(1 - \hat{\mathbf{k}}\hat{\mathbf{k}})}{k^2} : \int_{-\infty}^{t} \dot{\mathbf{q}}(t') e^{i\mathbf{k} \cdot \mathbf{q}(t')} \cos[kc(t - t')] \, dt'. \tag{3.380}$$

Comparison with (C.0.17) shows that the field $\mathbf{Z}$ is a Hertz vector for the electric field since it satisfies

$$\mathbf{Z}(\mathbf{x}, t) = - \int \frac{\mathbf{E}(\mathbf{x}', t)}{4\pi\varepsilon_0 |\mathbf{x} - \mathbf{x}'|}. \tag{3.381}$$

In k-space, the current contains the regulator $\chi_a(k)$ as a multiplier

$$\mathbf{j}(\mathbf{k}, t) = e\chi_a(k)\dot{\mathbf{q}}(t) e^{i\mathbf{k} \cdot \mathbf{q}(t)}, \tag{3.382}$$

since it is the Fourier transform of

$$\mathbf{j}(\mathbf{x}, t) = e\dot{\mathbf{q}}(t)\xi(\mathbf{x} - \mathbf{q}(t)). \tag{3.383}$$

dF/dt has contributions from the explicit time dependence of $\mathbf{Z}$ and from the time dependence of the particle's orbit, $\mathbf{q}(t)$. First of all, we have

$$c^2 \frac{\partial F}{\partial t} = \frac{\partial}{\partial t} \int \mathbf{j}(\mathbf{x}) \cdot \mathbf{Z}(\mathbf{x},t) \, d^3\mathbf{x} = \int \mathbf{j}(\mathbf{x}) \cdot \frac{\partial \mathbf{Z}(\mathbf{x},t)}{\partial t} \, d^3\mathbf{x}, \tag{3.384}$$

and using (3.379) this gives

$$\frac{\partial F}{\partial t} = -\int \mathbf{j}(\mathbf{x}) \cdot \mathbf{A}(\mathbf{x},t)_s \, d^3\mathbf{x} + \frac{1}{8\pi^3 \varepsilon_0 c^2} \int \mathbf{j}(\mathbf{x}) \cdot \int \frac{e^{-i\mathbf{k}\cdot\mathbf{x}}}{k^2} \mathbf{j}(\mathbf{k})^\perp \, d^3\mathbf{k} \, d^3\mathbf{x}, \tag{3.385}$$

where

$$\mathbf{j}^\perp(\mathbf{k},t) = (\mathbf{1} - \hat{\mathbf{k}}\hat{\mathbf{k}}) : \mathbf{j}(\mathbf{k},t), \tag{3.386}$$

and we have written subscript s explicitly to denote the vector potential due to the charge, $\mathbf{A}_s$.

The second term in (3.385), which may also be written as

$$\frac{1}{\varepsilon_0 c^2} \iint \frac{\mathbf{j}(\mathbf{x}) \cdot \mathbf{j}(\mathbf{x}')^\perp}{4\pi|\mathbf{x}-\mathbf{x}'|} \, d^3\mathbf{x}' \, d^3\mathbf{x}, \tag{3.387}$$

is singular for point particles. However, for an extended charge model it provides the required mass renormalisation counterterm since, using (3.386),

$$\frac{e^2}{8\pi^3 \varepsilon_0 c^2} \dot{\mathbf{q}}(t)\dot{\mathbf{q}}(t) : \int \chi_a^2(k) \frac{(\mathbf{1} - \hat{\mathbf{k}}\hat{\mathbf{k}})}{k^2} \, d^3\mathbf{k} = \frac{e^2}{8\pi^3 \varepsilon_0 c^2} \frac{8\pi}{3} |\dot{\mathbf{q}}(t)|^2 \int_0^\infty \chi_a^2(k) \, dk$$
$$= \Delta m |\dot{\mathbf{q}}(t)|^2, \tag{3.388}$$

where Δm is given by (3.335), as before, so that finally

$$\frac{\partial F}{\partial t} = -\int \mathbf{j}(\mathbf{x}) \cdot \mathbf{A}(\mathbf{x},t)_s \, d^3\mathbf{x} + \Delta m |\dot{\mathbf{q}}(t)|^2. \tag{3.389}$$

Notice that the vector potential here, $\mathbf{A}_s$, is in the Coulomb gauge, whereas the original Lagrangian was written with a vector potential $\mathbf{a}(\mathbf{x},t)$ in an arbitrary gauge; a gauge transformation $\{\phi,\mathbf{a}\} \rightarrow \{V,\mathbf{A}\}$ alters the Lagrangian by a total time derivative, by virtue of the equation of continuity, so is of no importance for the dynamics.

Using the definition of the current again, we may write

$$\int \mathbf{j}(\mathbf{x}) \cdot \mathbf{Z}(\mathbf{x},t) \, d^3\mathbf{x} = e\dot{\mathbf{q}}(t) \cdot \mathbf{Z}_\xi(\mathbf{q},t), \tag{3.390}$$

where

$$\mathbf{Z}_\xi(\mathbf{q}(t),t) = \frac{1}{(2\pi)^3} \int \chi_a(k) \mathbf{Z}(\mathbf{k},t) e^{-i\mathbf{k}\cdot\mathbf{q}(t)} \, d^3\mathbf{k}. \tag{3.391}$$

Hence, the contributions to dF/dt from the time dependence of the coordinate $\mathbf{q}(t)$ are

$$\dot{\mathbf{q}} \cdot \frac{\partial \mathbf{Z}_\xi(\mathbf{q}(t),t)}{\partial \mathbf{q}} = -\frac{i}{(2\pi)^3} \int \mathbf{j}(\mathbf{k},t)^* \cdot \mathbf{Z}(\mathbf{k},t) \mathbf{k} \cdot \dot{\mathbf{q}}(t) \, d^3\mathbf{k} \tag{3.392}$$

$$\ddot{\mathbf{q}} \cdot \frac{\partial \mathbf{Z}_\xi(\mathbf{q}(t),t)}{\partial \dot{\mathbf{q}}} = e\ddot{\mathbf{q}}(t) \cdot \mathbf{Z}_\xi(\mathbf{x},t). \tag{3.393}$$

Collecting all the terms together, we have

$$\frac{\mathrm{d}F(\mathbf{q},\dot{\mathbf{q}},t)}{\mathrm{d}t} = -\int \mathbf{j}(\mathbf{x}) \cdot \mathbf{A}(\mathbf{x},t)_s \,\mathrm{d}^3\mathbf{x} + \Delta m |\dot{\mathbf{q}}(t)|^2 + e\ddot{\mathbf{q}}(t) \cdot \mathbf{Z}_\xi(\mathbf{q},t)$$
$$+ e\dot{\mathbf{q}}(t) \cdot \nabla_{\mathbf{q}}\dot{\mathbf{q}}(t) \cdot \mathbf{Z}_\xi(\mathbf{q},t). \tag{3.394}$$

In the linear (electric dipole) approximation, the last term in (3.394) is zero.

The new Lagrangian is defined by setting

$$L_1 = L_0 + \frac{\mathrm{d}F}{\mathrm{d}t}, \tag{3.395}$$

with $\mathrm{d}F/\mathrm{d}t$ given by (3.394) and L_0 by (3.79) for a single charge; if we put

$$M = m + 2\Delta m, \tag{3.396}$$

this becomes

$$L_1 = \tfrac{1}{2}M\dot{\mathbf{q}}^2 - \int \rho\phi\,\mathrm{d}^3\mathbf{x} + e\ddot{\mathbf{q}} \cdot \mathbf{Z}_\xi(\mathbf{q},t) + e\dot{\mathbf{q}} \cdot \nabla_{\mathbf{q}}\dot{\mathbf{q}} \cdot \mathbf{Z}_\xi(\mathbf{q},t)$$
$$+ \tfrac{1}{2}\varepsilon_0 \int \left[(\dot{\mathbf{a}} + \nabla\phi) \cdot (\dot{\mathbf{a}} + \nabla\phi) - c^2 \nabla \wedge \mathbf{a} \cdot \nabla \wedge \mathbf{a} \right] \mathrm{d}^3\mathbf{x}. \tag{3.397}$$

The Euler–Lagrange equations yield the following equations of motion:

$$\left(\frac{\partial L_1}{\partial \mathbf{q}}\right) - \frac{\mathrm{d}}{\mathrm{d}t}\left(\frac{\partial L_1}{\partial \dot{\mathbf{q}}}\right) + \frac{\mathrm{d}^2}{\mathrm{d}t^2}\left(\frac{\partial L_1}{\partial \ddot{\mathbf{q}}}\right) = 0, \tag{3.398}$$

$$\left(\frac{\delta L_1}{\delta \phi}\right) - \frac{\partial}{\partial t}\left(\frac{\delta L_1}{\delta(\partial\phi/\partial t)}\right) - \nabla \cdot \left(\frac{\delta L_1}{\delta(\nabla\phi)}\right) = 0, \tag{3.399}$$

$$\left(\frac{\delta L_1}{\delta \mathbf{a}}\right) - \frac{\partial}{\partial t}\left(\frac{\delta L_1}{\delta(\partial\mathbf{a}/\partial t)}\right) - \nabla \cdot \left(\frac{\delta L_1}{\delta(\nabla\mathbf{a})}\right) = 0. \tag{3.400}$$

Taking these in turn, we get the following:

(a) The equation of motion for the particle's orbit

The calculations are elementary if rather tedious; the only point to notice is that $\mathbf{Z}_\xi$ has an explicit dependence on t, and so its previously evaluated partial time derivative will be required. There results

$$m^{\mathrm{obs}}\,\ddot{\mathbf{q}} = e\,\mathbf{E}_\xi(\mathbf{q}) + e\,\dot{\mathbf{q}} \wedge \mathbf{B}_\xi(\mathbf{q}), \tag{3.401}$$

where as before

$$m^{\mathrm{obs}} = M - \Delta m = m + \Delta m, \tag{3.402}$$

and

$$\mathbf{E}_\xi(\mathbf{q}) = -\nabla_{\mathbf{q}}\phi_\xi(\mathbf{q}) - \frac{\partial \mathbf{a}_\xi(\mathbf{q})}{\partial t}, \quad \mathbf{B}_\xi = \nabla_{\mathbf{q}} \wedge \mathbf{a}_\xi(\mathbf{q}). \tag{3.403}$$

(b) Gauss's law from the variation of the scalar potential, ϕ

Functional differentiation yields

$$\frac{\delta L_1}{\delta \phi} = -\rho, \quad \frac{\delta L_1}{\delta(\nabla\phi)} = -\varepsilon_0 \mathbf{E} \tag{3.404}$$

which combine in the Euler–Lagrange equation to give

$$\varepsilon_0 \, \mathbf{\nabla} \cdot \mathbf{E} = \rho. \tag{3.405}$$

(c) The equation of motion for the vector potential, $\mathbf{a}$

The functional derivatives involve only terms in the field strengths:

$$\frac{\delta L_1}{\delta \mathbf{a}} = -\varepsilon_0 c^2 \mathbf{\nabla} \wedge \mathbf{B}$$

$$\frac{\delta L_1}{\delta(\partial \mathbf{a}/\partial t)} = \varepsilon_0 \left(\frac{\partial \mathbf{a}}{\partial t} + \mathbf{\nabla}\phi \right) = -\varepsilon_0 \, \mathbf{E} \tag{3.406}$$

and these with (3.400) yield the dynamical Maxwell equation for the *free electromagnetic field*,

$$\mathbf{\nabla} \wedge \mathbf{B} = \frac{1}{c^2} \frac{\partial \mathbf{E}}{\partial t}. \tag{3.407}$$

The other two Maxwell equations appear as identities through the use of the potentials. Thus, the equation of motion for the particle contains the renormalised (observed) mass which accounts for the interaction between the charge and its own field, so the remaining field equations are free.

Finally, we return to the Lagrangian for a collection of N charged particles; as we have seen, the self-interaction arises from the coupling between the current and the vector potential, specifically the third term in (3.79). For particles that have extended charge distributions modelled by the functions $\{\chi(k)_{an}, n = 1, \ldots N\}$ and using Fourier transforms, we may write

$$\int \mathbf{j} \cdot \mathbf{A} \, d^3\mathbf{x} = \frac{1}{(2\pi)^3} \int \sum_n^N e_n \chi(k)_{an} \dot{\mathbf{x}}_n \cdot \mathbf{A}(\mathbf{k}) e^{-i\mathbf{k}\cdot\mathbf{x}_n} \, d^3\mathbf{k}. \tag{3.408}$$

In this expression, the vector potential is the sum of the contributions from the charged particles, calculated in Appendix C, and the free field,

$$\mathbf{A}(\mathbf{k}) = \mathbf{A}_h + \sum_m^N \mathbf{A}(\mathbf{k},t)_m, \tag{3.409}$$

where $\mathbf{A}_m$ is the immediate generalisation of (C.0.6),

$$\mathbf{A}(\mathbf{k},t)_m = \left(\frac{e_m \chi(k)_{am}}{\varepsilon_0 kc} \right) (1 - \hat{\mathbf{k}}\hat{\mathbf{k}}) : \int_{-\infty}^t \dot{\mathbf{x}}_m(t') e^{i\mathbf{k}\cdot\mathbf{x}_m(t')} \sin[kc(t-t')] \, dt'. \tag{3.410}$$

In order to isolate the self-interaction, it is convenient to rewrite (3.409) in the form

$$\mathbf{A}(\mathbf{k}) = \mathbf{A}_n + \mathbf{A}_{\text{ext}}, \tag{3.411}$$

where the 'external' vector potential is

$$\mathbf{A}_{\text{ext}} = \mathbf{A}_h + \sum_{m \neq n}^N \mathbf{A}(\mathbf{k},t)_m. \tag{3.412}$$

Mass renormalisation now requires a generating function F which is such that its total time derivative cancels the contribution of $\mathbf{A}_n$ in (3.408) and replaces it with the mass counterterms for each particle,

$$\Delta m_n |\dot{\mathbf{x}}_n(t)|^2, \tag{3.413}$$

where

$$\Delta m_n = \frac{e_n^2}{6\pi\varepsilon_0 c^2 a_n^2}. \tag{3.414}$$

In order to generalise (3.378) to the n-particle case, we set for each particle

$$\mathbf{j}(\mathbf{k},t)_n = e_n \chi(k)_{an} \dot{\mathbf{x}}_n(t) e^{i\mathbf{k}\cdot\mathbf{x}_n(t)}, \tag{3.415}$$

and define

$$\mathbf{Z}(\mathbf{x}_n(t),t)_n = \frac{1}{(2\pi)^3} \int \mathbf{Z}_n(\mathbf{k},t) e^{-i\,\mathbf{k}\cdot\mathbf{x}_n(t)} \, \mathrm{d}^3\mathbf{k}, \tag{3.416}$$

where

$$\mathbf{Z}(\mathbf{k},t)_n = \frac{e_n \chi(k)_{an}}{\varepsilon_0 k^2} (\mathbf{1} - \hat{\mathbf{k}}\hat{\mathbf{k}}) : \int_{-\infty}^{t} \dot{\mathbf{x}}_n(t') e^{i\mathbf{k}\cdot\mathbf{x}_n(t')} \cos[kc(t-t')] \, \mathrm{d}t'. \tag{3.417}$$

Then the many-particle transformation function F is

$$F = \frac{1}{c^2} \int \sum_n \mathbf{j}(-\mathbf{k},t)_n \cdot \mathbf{Z}(\mathbf{k},t)_n \, \mathrm{d}^3\mathbf{k}, \tag{3.418}$$

since its partial time derivative is

$$-\frac{1}{(2\pi)^3} \int \sum_n \mathbf{j}(-\mathbf{k},t)_n \cdot \mathbf{A}(\mathbf{k},t)_n \, \mathrm{d}^3\,\mathbf{k} + \sum_n \Delta m_n |\dot{\mathbf{x}}_n(t)|^2, \tag{3.419}$$

as required. There are additionally contributions like the last two terms in (3.394) which are summed over all N particles. These bring the accelerations of the particles into the modified N-particle Lagrangian which can then be discussed in the same way as the single-particle Lagrangian which we discussed previously.

References

[1] Landau, L. and Lifshitz, E. M. (1969), *Course of Theoretical Physics*, vol. 1, *Mechanics*, 2nd ed., Pergamon Press.

[2] Cawley, R. (1979), Phys. Rev. A**20**, 2370.

[3] Whittaker, E. T. (1970), *A Treatise on the Analytical Dynamics of Particles and Rigid Bodies*, p. 45, 4th ed., Cambridge University Press.

[4] Weinberg, S. (1995), *Quantum Theory of Fields*, vol. 1, *Foundations*, Cambridge University Press.

[5] Woolley, R. G. (1975), Ann. Inst. Henri Poincaré A**23**, 365.

[6] Schwarzschild, K. (1903), Nachrichten von der Gesellschaft der Wissenschaften zu Göttingen, Mathematische-Physikalische Klasse, 125.

[7] Heitler, W. (1936), *The Quantum Theory of Radiation*, Clarendon Press (and numerous later printings/editions).

[8] Power, E. A. (1964), *Introductory Quantum Electrodynamics*, Longmans.

[9] Roman, P. and Leveille, J. P. (1974), J. Math. Phys. **15**, 1760.

[10] Landau, L. and Lifshitz, E. M. (1971), *Course of Theoretical Physics*, vol. 2, *The Classical Theory of Fields*, 3rd ed., Pergamon Press.

[11] Edwards, J. P. and Mansfield, P. R. W. (2015), Phys. Lett. **B746**, 335.

[12] Edwards, J. P. and Mansfield, P. R. W. (2015), J. High Energy Phys. **2015**, 127.

[13] Zwieback, B., (2009), *A First Course in String Theory*, 2nd ed. Cambridge University Press.

[14] Power, E. A. and Zienau, S. (1957), Il Nuovo Cimento **6**, 7.

[15] Power, E. A. and Zienau, S. (1959), Phil. Trans. Roy. Soc. (London) **A251**, 427.

[16] Haagensen, P. E. and Johnson, K. (1997), arXiv:hep-th/9702204.

[17] Hoyle, F. and Narlikar, J. V. (1995), Rev. Mod. Phys. **67**, 113.

[18] Dirac, P. A. M. (1967), *The Principles of Quantum Mechanics*, 4th ed., revised, Clarendon Press.

[19] Born, M. (1924), *The Mechanics of the Atom*, translated by J. W. Fisher (1927), revised by D. R. Hartree, G. Bell & Sons, Ltd.

[20] Schwinger, J. (1970), *Particles, Sources and Fields*, Addison-Wesley.

[21] Jauch, J. M. and Hill, E. L. (1940), Phys. Rev. **57**, 641.

[22] Dirac, P. A. M. (1950), Can. J. Math. **2**, 129.

[23] Dirac, P. A. M. (1952), Ann. Inst. Henri Poincaré **13**, 1.

[24] Dirac, P. A. M. (1964), *Lectures on Quantum Mechanics*, Academic Press Inc.

[25] Rothe, H. J. and Rothe, K. D. (2010), *Classical and Quantum Dynamics of Constrained Hamiltonian Systems*, Lecture Notes in Physics **81**, World Scientific Ltd.

[26] Govaerts, J. and Rashid, M. S. (1994), arXiv:hep:th/9403009v2.

[27] Atkins, P. W. and Woolley, R. G. (1970), Proc. Roy. Soc. (London) **A319**, 549.

[28] Woolley, R. G. (1971), Proc. Roy. Soc. (London) **A321**, 557.

[29] Dashen, R. and Sharp, D. H. (1968), Phys. Rev. **165**, 1857.

[30] Sharp, D. H. (1980), in *Foundations of Radiation Theory and Quantum Electrodynamics*, edited by A. O. Barut, p. 183, Springer.

[31] Goldin, G. A. (2006), Bulg. J. Phys. **33**, 81.

[32] Weyl, H. (1918), Sitzungsber. Preuss. Akad. Wiss. **26**, 465.

[33] Einstein, A. (1916), Ann. der Physik **49**, 769.

[34] Einstein, A. (2015), *Relativity: The Special and the General Theory*, translated by R. W. Lawson, Princeton University Press, 100th Anniversary Edition.

[35] Killingbeck, J. and Cole, G. H. A. (1971), *Mathematical Techniques and Physical Applications*, Ch. 3, Academic Press.

[36] Harris, E. G. (1975), *Introduction to Modern Theoretical Physics*, vol. 1, ch. 13, John Wiley and Sons, Inc.

[37] O'Raifeartaigh, L. and Straumann, N. (2000), Rev. Mod. Phys. **72**, 1; see also Goenner, H. F. M. (2004), Living Rev. Relativity **7**, 2 accessed at `www.livingreviews.org/lrr-2004-2`.

[38] Schrödinger, E. (1922), Z. Physik **12**, 13.

[39] Dirac, P. A. M. (1973), Proc. Roy. Soc. (London) A**333**, 403.

[40] London, F. (1927), Z. Physik **12**, 375.

[41] Weyl, H. (1929), Z. Physik **56**, 330.

[42] Woolley, R. G. (1975), Adv. Chem. Phys. **33**, 153.

[43] Bellman, R. and Cooke, K. L. (1963), *Differential-Difference Equations*, R-374-PR, report for US Air Force Project RAND, The RAND Corporation.

[44] Kramers, H. A. (1950), Report to the Solvay Congress 1948, reprinted in *Collected Scientific Papers of H. A. Kramers*, p. 845, North-Holland Publishing Co.

[45] Dirac, P. A. M. (1938), Proc. Roy. Soc. (London) A**167**, 148.

[46] Rohrlich, F. (1990), *Classical Charged Particles*, 2nd ed., Addison-Wesley.

[47] Jackson, J. D. (1998), *Classical Electrodynamics*, 3rd ed., J. Wiley.

[48] Moniz, E. J. and Sharp, D. H. (1974), Phys. Rev. D**10**, 1133.

[49] Rohrlich, F. (1997), Amer. J. Phys. **65**, 1051.

[50] Ostrogradsky, M. (1850), Mem. Acad. Sci. St. Petersb. **4**, 385.

[51] Schiller, R. and Schwartz, M. (1961), Phys. Rev. **126**, 1582.

4 The Classical Free Electromagnetic Field

4.1 Introduction

This chapter is devoted to a description of the energy, polarisation and angular momentum of electromagnetic radiation in a region of space where there are no charges or currents (the free field). In general, the wave equation (2.17) must be solved subject to boundary conditions that specify the particular physical system of interest; these solutions can be constructed by superposition of the plane-wave solutions described in Chapter 2. The entire discussion is classical and is intended to serve as preparation for our later account of the quantum theory of radiation. In particular we introduce a pair of variables that, when quantised, become the familiar photon annihilation and creation operators. There are no photons in the classical theory.

The starting point is the gauge-invariant Hamiltonian for the free field that follows from (3.258) when all the $\{\overline{p}_n\}$ variables for the charges are deleted; then all that remains is the energy of the free field

$$H = \tfrac{1}{2}\varepsilon_0 \int \left(\mathbf{E}\cdot\mathbf{E} + c^2\mathbf{B}\cdot\mathbf{B}\right)\, \mathrm{d}^3\,\mathbf{x} \tag{4.1}$$

with Dirac bracket

$$[E(\mathbf{x})^r, B(\mathbf{x}')^s]^* = \varepsilon_0^{-1}\varepsilon_{rst}\nabla_{\mathbf{x}'}^t\delta^3(\mathbf{x}-\mathbf{x}') \tag{4.2}$$

and the subsidiary conditions,

$$\nabla\cdot\mathbf{E} = \nabla\cdot\mathbf{B} = 0. \tag{4.3}$$

We saw in Chapter 2 that fundamental solutions of the wave equations satisfied by the fields $(\mathbf{E},\mathbf{B})$ are trigonometric functions. In an enclosure or 'cavity' there are standing-wave solutions that follow from imposing the condition that the fields are completely confined to the cavity; then only certain wave vectors are allowed in the solutions. We rely on a classical theorem due to Weyl that claims the main results are independent of the shape of the enclosure, provided its dimensions are much greater than the wavelengths of the radiation of interest [1]. It then suffices to specify a cavity shape of simple symmetrical geometry, for example a cube of side L, or a sphere of radius R. We begin with a cube and choose coordinates (x,y,z) so that the interior of the enclosure is defined by the relations

$$0 < x, y, z < L \tag{4.4}$$

and its walls are parallel to the coordinate planes.

4.2 The Oscillator Variables

In order to obtain standing waves, we require the normal components of the fields to vanish at the walls, so that, for example, we must have for the electric field **E**

$$E_y = E_z = 0 \text{ for } x = 0 \text{ and } x = L, \tag{4.5}$$

$$E_z = E_x = 0 \text{ for } y = 0 \text{ and } y = L, \tag{4.6}$$

$$E_x = E_y = 0 \text{ for } z = 0 \text{ and } z = L, \tag{4.7}$$

and similarly for the components of the magnetic induction **B**. The solutions of the wave equation with these boundary conditions we call modes. A general solution of the wave equation can then be expressed as a superposition of triple products of trigonometric functions with time-dependent vector coefficients $\mathbf{u}(\mathbf{k},t)$, thus for the electric field

$$E_x = \sum_{\mathbf{k}} u(\mathbf{k},t)_x \, \cos(k_x x) \, \sin(k_y y) \, \sin(k_z z), \tag{4.8}$$

$$E_y = \sum_{\mathbf{k}} u(\mathbf{k},t)_y \, \sin(k_x x) \, \cos(k_y y) \, \sin(k_z z), \tag{4.9}$$

$$E_z = \sum_{\mathbf{k}} u(\mathbf{k},t)_z \, \sin(k_x x) \, \sin(k_y y) \, \cos(k_z z). \tag{4.10}$$

There are corresponding expressions for the components of **B** with **u** replaced by another vector function **v**. The wave vector $\mathbf{k} = (k_x, k_y, k_z)$ has components

$$k_x = \frac{N_x \pi}{L}, \quad k_y = \frac{N_y \pi}{L}, \quad k_z = \frac{N_z \pi}{L}, \quad N_x, N_y, N_z = 0, 1, 2, \dots, \tag{4.11}$$

from which we deduce that the wave number k is

$$k = \frac{\pi}{L} \sqrt{(N_x)^2 + (N_y)^2 + (N_z)^2}. \tag{4.12}$$

The sums over **k** are triple sums over the integers in (4.11).

Since electromagnetic waves are transverse, (4.3), we must have

$$\mathbf{k} \cdot \mathbf{u} = \mathbf{k} \cdot \mathbf{v} = 0. \tag{4.13}$$

Furthermore, **u** and **v** are not independent quantities since **E** and **B** must satisfy the Maxwell equations (2.1)–(2.4). This requirement can be satisfied if we introduce a new vector $\mathbf{c}(\mathbf{k},t)$ such that

$$\mathbf{u}(\mathbf{k},t) = -\frac{\partial \mathbf{c}(\mathbf{k},t)}{\partial t}, \quad \mathbf{v}(\mathbf{k},t) = \mathbf{k} \wedge \mathbf{c}(\mathbf{k},t) \tag{4.14}$$

with

$$\mathbf{k} \cdot \mathbf{c}(\mathbf{k}, t) = 0, \tag{4.15}$$

to maintain transversality.

It is now a straightforward computation to express the Hamiltonian (4.1) as a sum over the modes labelled by $\mathbf{k}$; the integration domain is restricted to the volume of the cavity, Ω. Using the expansions (4.8)–(4.10) and (4.14), we find

$$\frac{1}{2} \varepsilon_0 \int_\Omega \mathbf{E} \cdot \mathbf{E} d^3 \mathbf{x} = \frac{\varepsilon_0 \Omega}{16} \sum_{\mathbf{k}} |\dot{\mathbf{c}}(\mathbf{k})|^2, \tag{4.16}$$

$$\frac{1}{2} \varepsilon_0 c^2 \int_\Omega \mathbf{B} \cdot \mathbf{B} d^3 \mathbf{x} = \frac{\varepsilon_0 \Omega}{16} \sum_{\mathbf{k}} \omega^2 |\mathbf{c}(\mathbf{k})|^2, \tag{4.17}$$

where $\omega = kc$ as usual.[1] The plane perpendicular to the wave vector $\mathbf{k}$ is spanned by the polarisation vectors $\{\hat{\boldsymbol{\varepsilon}}(\mathbf{k})_i, \ i = 1, 2\}$ defined in Chapter 2, and in this basis we have

$$\mathbf{c}(\mathbf{k}, t) = q(\mathbf{k})_1 \hat{\boldsymbol{\varepsilon}}(\mathbf{k})_1 + q(\mathbf{k})_2 \hat{\boldsymbol{\varepsilon}}(\mathbf{k})_2, \tag{4.18}$$

so that finally the Hamiltonian for the free field can be written as a sum over contributions from the individual modes,

$$
\begin{aligned}
H &= \frac{\varepsilon_0 \Omega}{8} \sum_{\mathbf{k}} \frac{1}{2} \left(|\dot{\mathbf{c}}(\mathbf{k})|^2 + \omega^2 |\mathbf{c}(\mathbf{k})|^2 \right) \\
&= \frac{\varepsilon_0 \Omega}{8} \sum_{\mathbf{k}, i} \frac{1}{2} \left(\dot{q}(\mathbf{k})_i^2 + \omega^2 q(\mathbf{k})_i^2 \right),
\end{aligned}
\tag{4.19}
$$

where we have used (4.18).

If we define new canonical variables by setting

$$X(\mathbf{k})_i = \sqrt{\frac{\varepsilon_0 \Omega}{8}} q(\mathbf{k})_i$$

$$P(\mathbf{k})_i = \sqrt{\frac{\varepsilon_0 \Omega}{8}} \dot{q}(\mathbf{k})_i \tag{4.20}$$

with canonical Poisson bracket

$$\{X(\mathbf{k})_i, P(\mathbf{k}')_j\} = \delta_{ij} \delta_{\mathbf{k}, \mathbf{k}'}, \quad i, j = 1, 2, \tag{4.21}$$

the Hamiltonian for the field, (4.19), takes the form

$$H = \sum_{\mathbf{k}, i = 1, 2} H(\mathbf{k})_i, \quad H(\mathbf{k})_i = \frac{1}{2} \left(P(\mathbf{k})_i^2 + \omega^2 X(\mathbf{k})_i^2 \right). \tag{4.22}$$

Evaluation of the Hamiltonian equations of motion,

$$\dot{P}(\mathbf{k})_i = -\frac{\partial H}{\partial X(\mathbf{k})_i}, \quad \dot{X}(\mathbf{k})_i = \frac{\partial H}{\partial P(\mathbf{k})_i}, \tag{4.23}$$

[1] The special point value $\mathbf{k} = 0$, corresponding to all three integers in (4.11) being zero, describes a *constant* electromagnetic field; since the boundary condition here requires the field to vanish at the walls, such a field must vanish everywhere. Hence the case $k = 0$ is excluded. The speed of light cannot vanish (it is c in all reference frames) so that the circular frequency ω is strictly positive.

confirms that (4.22) is the Hamiltonian of an infinite set of independent harmonic oscillators associated with the modes.

We noted in Chapter 3 that annihilation and creation operators are important quantities in the quantum theory of the harmonic oscillator, and that classical analogues can be introduced by a simple transformation of the oscillator variables. A similar step is useful here in preparation for quantisation of the field.[2] We set

$$\mathbf{a_k} = \sqrt{\frac{\omega}{2}}\mathbf{X(k)} + i\sqrt{\frac{1}{2\omega}}\mathbf{P(k)} \tag{4.24}$$

and use $\mathbf{a_k}$ and its complex conjugate, $\mathbf{a_k^*}$, as new variables. In terms of their components relative to the polarisation vectors,

$$\mathbf{a_k} = \sum_{\lambda=1,2} \hat{\boldsymbol{\varepsilon}}_\lambda a_{\mathbf{k},\lambda}, \tag{4.25}$$

they yield the non-zero P.B.,

$$\{a_{\mathbf{k},\lambda}, a_{\mathbf{k}',\lambda'}^*\} = -i\,\delta_{\lambda\lambda'}\,\delta_{\mathbf{k},\mathbf{k}'}, \tag{4.26}$$

while the field Hamiltonian (4.22) becomes

$$H = \sum_{\mathbf{k},\lambda} \omega\, a_{\mathbf{k},\lambda}^* a_{\mathbf{k},\lambda}. \tag{4.27}$$

The variables $a_{\mathbf{k},\lambda}$, $a_{\mathbf{k},\lambda}^*$ continue to be useful when the field is coupled to charges.

If instead of the above calculation one starts with the free-field Hamiltonian expressed in terms of the vector potential $\mathbf{a}$ and its conjugate $\boldsymbol{\pi}$ that satisfy a gauge-dependent Dirac bracket, one comes to the same final result [2]. It is simplest to work in the Coulomb gauge with a purely transverse vector potential, $\mathbf{A}$,

$$\boldsymbol{\nabla} \cdot \mathbf{A} = 0. \tag{4.28}$$

This vector potential has two non-zero components and satisfies the same wave equation as the free fields to which it is related by the usual equations:

$$\mathbf{E} = -\frac{\partial \mathbf{A}}{\partial t}, \quad \mathbf{B} = \boldsymbol{\nabla} \wedge \mathbf{A}. \tag{4.29}$$

It is then straightforward to show that the conjugate field variables $(\mathbf{A}, \boldsymbol{\pi})$ are related to the oscillator variables $(\mathbf{X}, \mathbf{P})$ by the Fourier expansions [3]:

$$\mathbf{A(x)} = \frac{1}{\sqrt{\varepsilon_0\Omega}} \sum_{\mathbf{k}} \left(\mathbf{X(k)}\,\cos(\mathbf{k}\cdot\mathbf{x}) - \frac{1}{\omega}\mathbf{P(k)}\,\sin(\mathbf{k}\cdot\mathbf{x})\right), \tag{4.30}$$

$$\boldsymbol{\pi}(\mathbf{x}) = \sqrt{\frac{\varepsilon_0}{\Omega}} \sum_{\mathbf{k}} \omega\left(\mathbf{X(k)}\,\sin(\mathbf{k}\cdot\mathbf{x}) + \frac{1}{\omega}\mathbf{P(k)}\,\cos(\mathbf{k}\cdot\mathbf{x})\right). \tag{4.31}$$

The change of variables (4.24) also yields Fourier expansions of the field variables in terms of running waves,

$$\mathbf{A(x)} = \sqrt{\frac{1}{2\varepsilon_0\Omega c}} \sum_{\mathbf{k},\lambda} \frac{\hat{\boldsymbol{\varepsilon}}(\mathbf{k})_\lambda}{\sqrt{k}} \left(a_{\mathbf{k},\lambda}e^{i\mathbf{k}\cdot\mathbf{x}} + a_{\mathbf{k},\lambda}^* e^{-i\mathbf{k}\cdot\mathbf{x}}\right), \tag{4.32}$$

---

[2] The variables $\mathbf{a_k}$ and $\mathbf{a_k^*}$ introduced here are not a canonical pair, however.

$$\boldsymbol{\pi}(\mathbf{x}) = -i \sqrt{\frac{\varepsilon_0 c}{2\Omega}} \sum_{\mathbf{k},\lambda} \sqrt{k} \hat{\boldsymbol{\varepsilon}}(\mathbf{k})_\lambda \left(a_{\mathbf{k},\lambda} e^{i\mathbf{k}\cdot\mathbf{x}} - a_{\mathbf{k},\lambda}^* e^{-i\mathbf{k}\cdot\mathbf{x}} \right). \tag{4.33}$$

Apart from the introduction of Planck's constant, h, these forms are the ones commonly used in the quantum theory of the field with $a_{\mathbf{k},\lambda}$ and $a_{\mathbf{k},\lambda}^*$ reinterpreted as an adjoint pair of operators.

All the other mechanical quantities for the field can be transformed in this way using the mode expansion; for example, the momentum for the field in the volume Ω is an integral over the Poynting vector (2.32):

$$\mathbf{G} = \frac{1}{c^2} \int_\Omega \mathbf{S} d^3\mathbf{x} = \varepsilon_0 \int_\Omega \mathbf{E} \wedge \mathbf{B} d^3\mathbf{x}. \tag{4.34}$$

Using the mode expansion, this may be evaluated directly, and there results

$$\mathbf{G} = \frac{1}{c} \sum_{\mathbf{k},\lambda} \hat{\mathbf{k}} \, H(\mathbf{k})_\lambda. \tag{4.35}$$

Thus, the field momentum associated with a mode is directed along the propagation direction and its magnitude satisfies the formula $E = pc$ from special relativity. The form of (4.35) makes it obvious that the momentum for the free field is a constant of the motion.

Starting from (4.12), it follows that the number N_K of field modes with wave number less than some maximum value K is found by counting the number of positive integer solutions of the inequality

$$K > \frac{\pi}{L} \sqrt{(N_x)^2 + (N_y)^2 + (N_z)^2}. \tag{4.36}$$

This number is found as follows: square and rearrange the inequality to give

$$(N_x)^2 + (N_y)^2 + (N_z)^2 < \left(\frac{KL}{\pi} \right)^2. \tag{4.37}$$

Treating the triples of integers as coordinates in a space, we see that we need to consider the sets of values of $\{N_x, N_y, N_z\}$ that are enclosed by the surface of a sphere of radius KL/π. Provided that the spacing between the points is $<< KL/\pi$, this number is essentially the volume of the part of the sphere corresponding to the positive values of the coordinates; this is the positive octant, so

$$N_K = \frac{1}{8} \frac{4\pi}{3} \left(\frac{KL}{\pi} \right)^3 = \frac{1}{6} \frac{K^3}{\pi^2} \Omega, \tag{4.38}$$

where Ω is the volume of the enclosure. On the other hand, the number of modes that have wave numbers lying between k and $k + dk$ is

$$dN_k = \frac{1}{6} \frac{(k+dk)^3}{\pi^2} \Omega - \frac{1}{6} \frac{k^3}{\pi^2} \Omega = \frac{1}{2} \frac{k^2}{\pi^2} \Omega dk. \tag{4.39}$$

Taking account of both possible polarisations for the mode, the number of modes per unit volume (the density of states) with wave numbers between k and $k + dk$ is therefore

$$\rho_k dk = \frac{k^2}{\pi^2} dk; \tag{4.40}$$

in terms of the circular frequency $\omega = kc$ this is also

$$\rho_\omega \, d\omega = \frac{\omega^2}{\pi^2 c^3} \, d\omega. \tag{4.41}$$

As shown by Lord Rayleigh, the representation of the energy of the free electromagnetic field in an enclosure in the form of a sum of harmonic oscillator contributions leads directly to a conclusion that is disastrous for classical physics [4]. Supposing the walls of the enclosure are maintained at a constant temperature T, what is the spectral distribution of the radiation within the enclosure ? This question can be answered by the methods of statistical mechanics, provided the radiation is coupled to a thermostat. Since the modes are independent (have no mutual interaction), thermal equilibrium must be realised indirectly. Physically the coupling can be attributed to the weak coupling between the field and the charged particles in the walls of the enclosure and in atoms or molecules of the gas in the enclosure (the pressure can never be strictly zero).

We saw in Chapter 3 that there is indeed a canonical description of the combined system of charges and field with a Hamiltonian that represents the constant energy of the combined system. In this situation according to classical physics, one simply applies the classical equipartition of energy theorem to the radiation field, and so conclude that each radiation oscillator coupled to a thermostat at temperature T must have average energy $k_B T$, where k_B is Boltzmann's constant. Thus, the average energy density of the modes with circular frequency in the interval ω to $\omega + d\omega$, must be just the simple product (density of modes in interval $[\omega, \omega + d\omega]) \times$ (average energy for mode of frequency ω); that is,

$$\overline{E}(\omega) \, d\omega = \rho_\omega k_B T \, d\omega \tag{4.42}$$

$$= \frac{\omega^2}{\pi^2 c^3} \, k_B T \, d\omega. \tag{4.43}$$

This is the Rayleigh–Jeans law for blackbody radiation which shows an energy density that increases indefinitely with frequency leading to the ultraviolet catastrophe [4]. The density of modes calculation leading to (4.41) is not the problem; indeed the factor $\rho_\omega \, d\omega$ appears in the (correct) Planck distribution, so the difficulty lies entirely with the classical equipartition theorem. Rayleigh's result is an inevitable consequence of the combination of statistical mechanics with classical electrodynamics, and signals a profound failure in classical physics that was only resolved with the development of quantum theory.

4.3 Polarisation

Let us now specialise the Hamiltonian (4.22) to a single mode with wave vector $\mathbf{k}$ and frequency $\omega = kc$. We suppress the mode index $\mathbf{k}$ for simplicity and write

$$H = \frac{1}{2}\left(P_1^2 + \omega^2 X_1^2\right) + \frac{1}{2}\left(P_2^2 + \omega^2 X_2^2\right) \equiv H_1 + H_2 \tag{4.44}$$

in an obvious notation. This is the Hamiltonian for an isotropic two-dimensional harmonic oscillator confined to move in the plane perpendicular to **k**, which is a mechanical problem that is exactly soluble. The Hamiltonian is invariant under a set of transformations isomorphic to the Lie group SU(2); the representation involves three functions of the phase space variables which, as constants of the motion, generate symmetry transformations of the Hamiltonian. It is readily verified by direct calculation that these functions are the following bilinear combinations of the canonical variables:

$$T_1 = \frac{1}{\omega}(P_1 P_2 + \omega^2 X_1 X_2), \tag{4.45}$$

$$T_2 = X_1 P_2 - X_2 P_1 \equiv (\mathbf{X} \wedge \mathbf{P})_3, \tag{4.46}$$

$$T_3 = \frac{1}{\omega}(H_1 - H_2), \tag{4.47}$$

where time derivatives are calculated according to (3.122),

$$\dot{T}_i = \{T_i, H\} = 0, \quad i = 1, 2, 3. \tag{4.48}$$

It will be convenient to append the Hamiltonian to this group with the notation

$$T_0 = \frac{H}{\omega} \equiv \frac{1}{\omega}(H_1 + H_2). \tag{4.49}$$

where we have divided by the frequency ω as necessary to make all the $\{T_\mu\}$ dimensionally consistent as actions or angular momenta.

The oscillator energies H_1 and H_2 are separately conserved, while T_0 and T_3 are their sum and difference combinations, respectively. One of the constants of the motion, T_2 is an angular momentum directed along **k** (the 3-axis) associated with the kinematical symmetry of rotations in the plane transverse to the direction $\hat{\mathbf{k}}$. It has nothing to do with the orbital angular momentum of the field (see §4.4); the motion in a plane wave is everywhere parallel to **k** and so there must be zero component[3] of orbital angular momentum along **k**. T_1 and T_3 have no obvious physical interpretation in the 'physical' configuration space (the two-dimensional plane) because they mix coordinates and momenta. They generate dynamical symmetry transformations on the phase space (cf §3.4).

The constants of the motion $\{T_i, \ i = 1, 2, 3\}$ provide a representation of the Lie algebra su(2) characterised by the Poisson-bracket relation

$$\{T_i, T_j\} = 2T_k, \tag{4.50}$$

the indices forming cyclic permutations on $\{1, 2, 3\}$; they may be regarded as the generators of infinitesimal canonical transformations. For the isotropic oscillator, these transformations can be extended to the whole phase space as globally well-defined

---

[3] The plane wave has the familiar Rayleigh expansion in spherical polar coordinates with the polar axis directed along **k** as a series of partial waves corresponding to orbital angular momentum l and zero component of angular momentum along **k**, $e^{ikz} = \sum_l i^l (2l + 1) j(kr)_l Y_{l0}(\cos \theta)$. A wave that is of finite extent in the plane transverse to **k** does have a component of angular momentum along **k**, however [5].

symmetry transformations that constitute the Lie group SU(2) [6]. This group has the single Casimir invariant involving the square of the Hamiltonian

$$T_0^2 = T_1^2 + T_2^2 + T_3^2. \tag{4.51}$$

A physical interpretation of these results can be developed in the following way. The Hamiltonian equations of motion are solved in terms of coordinate functions describing two uncoupled oscillators

$$X_1 = X_1^{(0)} \cos(\omega t + \phi), \quad X_2 = X_2^{(0)} \cos(\omega t + \phi - \delta), \tag{4.52}$$

where the phase difference between the two oscillations δ lies in the interval $[-\pi/2, +\pi/2]$. The values of the $\{T_\mu\}$ are fixed by the initial conditions. In terms of the integration constants $(X_1^{(0)}, X_2^{(0)}, \delta)$, we have

$$T_0 = \frac{\omega}{2}\left((X_1^{(0)})^2 + (X_2^{(0)})^2\right), \tag{4.53}$$

$$T_1 = \omega X_1^{(0)} X_2^{(0)} \cos(\delta), \tag{4.54}$$

$$T_2 = \omega X_1^{(0)} X_2^{(0)} \sin(\delta), \tag{4.55}$$

$$T_3 = \frac{\omega}{2}\left((X_1^{(0)})^2 - (X_2^{(0)})^2\right), \tag{4.56}$$

of which only three are independent because of (4.51). The energies H_1 and H_2 are independent of the phase δ, but T_2 changes sign when $\delta \to -\delta$; rotations in the plane which differ only in their sense (clockwise versus anticlockwise) occur as degenerate pairs.

The elimination of the time t between the equations of (4.52) yields the orbit equation

$$\left(\frac{X_1}{X_1^{(0)}}\right)^2 - 2\left(\frac{X_1 X_2}{X_1^{(0)} X_2^{(0)}}\right)\cos(\delta) + \left(\frac{X_2}{X_2^{(0)}}\right)^2 = \sin^2(\delta). \tag{4.57}$$

In the particular case $\sin(\delta) = 0$ ($\delta = 0$ or π), the orbit degenerates to a pair of straight lines,

$$X_2 = \left(\frac{X_2^{(0)}}{X_1^{(0)}}\right) X_1 e^{i\delta}, \tag{4.58}$$

that are at right angles to each other. More generally, for $\sin(\delta) \neq 0$ (4.57) is a quadratic form in (X_1, X_2) with discriminant

$$\Delta = -\left(\frac{2\sin(\delta)}{X_1^{(0)} X_2^{(0)}}\right)^2 < 0, \tag{4.59}$$

and hence it describes an ellipse in the X_1, X_2 plane.

The major axis of the ellipse is inclined to the 1-axis by an angle θ given by

$$\tan(2\theta) = \frac{2 X_1^{(0)} X_2^{(0)} \cos(\delta)}{X_1^{(0)^2} - X_2^{(0)^2}} = \frac{T_1}{T_3}. \tag{4.60}$$

The angle θ is conventionally called the azimuth of the ellipse. In terms of coordinates (x_1, x_2) aligned with the axes of the ellipse given by the rotation

$$(x_1, x_2) = R(\theta)\, (X_1, X_2), \tag{4.61}$$

the ellipse is

$$\frac{x_1^2}{b_1^2} + \frac{x_2^2}{b_2^2} = 1, \tag{4.62}$$

where

$$\sqrt{\omega}\, b_{1,2} = \sqrt{T_0 + T_2} \pm \sqrt{T_0 - T_2} \tag{4.63}$$

relates the major and minor axes of the ellipse to the constants of the motion $\{T_\mu\}$. The ellipticity of the ellipse, η, is defined by the relation

$$\tan(2\eta) = \frac{b_2}{b_1}. \tag{4.64}$$

Since

$$\tan(2\eta) = \frac{\tan(\eta)}{1 - \tan(\eta)^2} \equiv \frac{b_1 b_2}{b_1^2 - b_2^2}, \tag{4.65}$$

the ellipticity may also be related to the constants of the motion with the aid of (4.51) and (4.63),

$$\tan(2\eta) = \frac{T_2}{\sqrt{T_1^2 + T_3^2}}. \tag{4.66}$$

The physical interpretation of the foregoing is that it provides a description of the polarisation of light. As we have seen, the oscillator coordinate $\mathbf{X}$ is related to the field variables by Fourier transformation, and the orbit described by (4.57) can be identified with the path swept out by the electric field intensity vector associated with the plane wave mode, (2.22); this path lies in the plane perpendicular to $\mathbf{k}$. There are various different schemes[4] for describing polarised light. The parameters $\{T_\mu\}$ used here are essentially the same as the Stokes parameters $\{S_\mu\}$ [7] since the $\{T_\mu\}$ and the $\{S_\mu\}$ both satisfy (4.60) and (4.66) for the angles that characterise the polarisation ellipse. They correspond directly to the set of four intensity measurements required to determine completely the polarisation state of a light beam in terms of its intensity, azimuth (θ) and ellipticity (η) (see Figure 4.1). In terms of the components of the electric field vector (2.27), the Stokes parameters can be represented as

$$S_0 = \frac{1}{4}\left((E_1^{(0)})^2 + (E_2^{(0)})^2\right), \tag{4.67}$$

$$S_1 = \frac{1}{2}E_1^{(0)} E_2^{(0)}\, \cos(\alpha), \tag{4.68}$$

$$S_2 = \frac{1}{2}E_1^{(0)} E_2^{(0)}\, \sin(\alpha), \tag{4.69}$$

---

[4] The four Stokes parameters are closely related to the elements of a 2×2 polarisation matrix. There are also the descriptions based on the Jones vector, and the Mueller matrices [8].

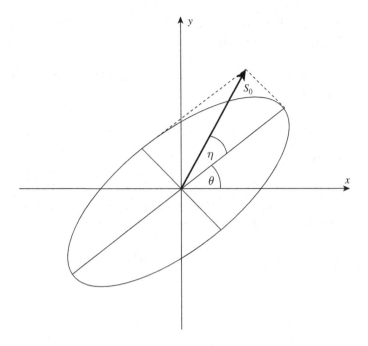

Figure 4.1 Specification of the azimuth and ellipticity of polarised light. [2]

$$S_3 = \frac{1}{4}((E_1^{(0)})^2 - (E_2^{(0)})^2), \tag{4.70}$$

such that

$$\tan(2\theta) = \frac{S_1}{S_3}, \tag{4.71}$$

$$\tan(2\eta) = \frac{S_2}{\sqrt{S_1^2 + S_3^2}}. \tag{4.72}$$

The optical constants of an anisotropic medium measure its dichroism (ΔD) and its birefringence (Δn) which are directly related to the Stokes parameters,

$$\Delta D = D_\perp - D_\parallel = \log_{10}\frac{S_0 - S_3}{S_0 + S_3}, \tag{4.73}$$

$$\Delta n = \frac{\lambda}{2d}\arctan\frac{S_2}{S_1}, \tag{4.74}$$

where λ is the wavelength of light of the probe beam, d is the sample thickness, ΔD is the difference in optical densities for light polarised parallel and perpendicular to the induced optical axis and Δn is the corresponding difference in refractive index.

In the traditional approach to the measurement of the polarisation of light, two optical elements are involved: an analyzer (Nicol prism, polaroid sheet) for which the emergent beam is linearly polarised along the transmission axis of the analyzer, and a compensator (quarter-wave plate) which alters the phase relationship between coherent orthogonal polarisation components of the beam. Let $I(a,b)$ denote the intensity of the light transmitted through a compensator which produces a retardation of b in the 2-component relative to the 1-component, followed by an analyzer with transmission axis oriented at an angle a to the 1-axis. The Stokes parameters are related to the following measurements:

$$S_0 \propto I(0,0) + I(\pi/2,0), \tag{4.75}$$

$$S_1 \propto I(\pi/4,0) - I(3\pi/4,0), \tag{4.76}$$

$$S_2 \propto I(3\pi/4,\pi/2) - I(\pi/4,\pi/2), \tag{4.77}$$

$$S_3 \propto I(0,0) - I(\pi/2,0). \tag{4.78}$$

S_0 measures the total intensity. S_3 gives the excess intensity transmitted by an analyzer with $\theta = 0$ compared with one with $\theta = \pi/2$, while S_1 has a similar interpretation with respect to azimuths of $\theta = \pi/4$ and $3\pi/4$. S_2 is the excess in intensity transmitted by a device that accepts right-circularly polarised light over that transmitted by a device that accepts left-circularly polarised light [8]. The connection of S_2 (T_2) with intrinsic angular momentum is evident; after quantisation it can be associated with the spin of the photon.

The use of beam splitters permits the simultaneous measurement of the four Stokes parameters so that the time evolution of polarisation changes in dynamic photo-processes can be monitored in real time. The physical basis of these methods is the dependence of the reflection and transmission coefficients of beam splitters on beam polarisation. The light beam whose state of polarisation is to be determined is divided into four beams by three beam splitters. The intensities of these four beams are measured and the corresponding electrical signals can be transformed directly into the four Stokes parameters [9].

A single mode of the field is strictly monochromatic and the polarisation of the light is perfect. The Stokes parameters for it satisfy the equality (cf. (4.51))

$$S_0^2 = S_1^2 + S_2^2 + S_3^2. \tag{4.79}$$

In practice, we usually deal with light beams that are quasi-monochromatic at best; that is, they are superpositions of modes with various frequencies contributing. The overall polarisation can then range from perfect polarisation, as previously described, to unpolarised radiation in which the tip of the net **E**-vector shows no preferred directional properties. Any state in between is said to be partially polarised. It is still possible to use the Stokes parameters, but for partial polarisation (4.79) must be replaced by

$$S_0^2 > S_1^2 + S_2^2 + S_3^2 \tag{4.80}$$

because of the unpolarised contribution to the overall intensity.

Equations (4.79) and (4.80) suggest a geometric representation of the state of polarisation. We view the parameters (S_1, S_2, S_3) as the Cartesian coordinates of the surface of a sphere of radius S_0; this sphere is called the Poincaré sphere and is shown in Figure 4.2. Completely polarised light is represented by points on the surface of the Poincaré sphere; the origin of the sphere stands for completely unpolarised light, and

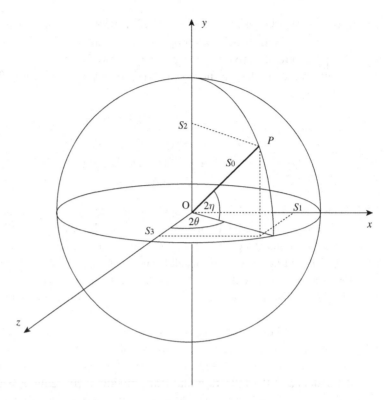

Figure 4.2   The Poincaré sphere. [2]

all other interior points represent partially polarised light. The characteristics of the polarised light are specified by the ellipticity angle ($0 \leq \eta \leq \pi$) and the azimuth angle ($-\pi/4 \leq \theta \leq +\pi/4$) just introduced. Linear polarisation is represented by points in the equatorial plane of the sphere ($\tan(2\eta) = 0$), circular polarisation is represented by the poles ($\tan(2\eta) = \infty$), while for all other values of η the light has elliptical polarisation. The degree of polarisation, P, is defined by

$$P = \frac{1}{S_0}\sqrt{S_1^2 + S_2^2 + S_3^2}, \tag{4.81}$$

with $0 \leq P \leq 1$. The extreme limits correspond to completely unpolarised light when $P = 0$, and to completely polarised light with $P = 1$, respectively.

The Lie group generators $\{T_i, i = 1, 2, 3\}$ and the group invariant, T_0, can be defined for every mode of the field. From (4.22) and (4.49) we have that the conserved energy of the field is

$$\mathcal{E} \equiv H_{\mathrm{rad}} = \sum_{\mathbf{k},i} H(\mathbf{k})_i \equiv \sum_{\mathbf{k}} \omega T(\mathbf{k})_0, \tag{4.82}$$

and this can also be written as an integral involving the fields $\{\mathbf{E}, \mathbf{B}\}$, according to (2.29). Likewise (4.35) shows that the conserved field momentum can be expressed as

$$\mathbf{G} = \frac{1}{c} \sum_{\mathbf{k}} \hat{\mathbf{k}} T(\mathbf{k})_0, \tag{4.83}$$

and this too has an alternative expression involving an integral over the field strengths. We can define similar sums over the generators $\{T_i\}$, and since the modes are independent these too will be constants of the motion.

Notice from the examples of the energy, $\mathcal{E}$, and the momentum, $\mathbf{G}$, that the mode sums over the generators can involve both scalar and vector functions of $\mathbf{k}$; even more generally, the Lie group generators can be multiplied by arbitrary tensor combinations of $\mathbf{k}$ before being summed over the modes. By judicious choices of such multipliers it is possible to transform the mode sums over the $\{T_i\}$ into combinations of the field variables and their derivatives. As an example we consider the intrinsic angular momentum, $T(\mathbf{k})_2$, for the mode $\mathbf{k}$. A straightforward calculation using (4.30) and (4.31) shows that the sum

$$\Phi = -\sum_{\mathbf{k}} k^2 \, T(\mathbf{k})_2 \equiv -\sum_{\mathbf{k}} k^2 \, \hat{\mathbf{k}} \cdot \left(\mathbf{X}(\mathbf{k}) \wedge \mathbf{P}(\mathbf{k}) \right) \tag{4.84}$$

can be written in the equivalent form

$$\Phi = \int_{\Omega} \xi(\mathbf{x}) \, \mathrm{d}^3 \mathbf{x}, \tag{4.85}$$

where ξ is defined in (2.6); similar calculations can be carried through with the other generators. This is a possible way of understanding the occurrence of the 'non-classical' constants of the motion referred to in §2.2. It should be noted, however, that although the intrinsic angular momentum associated with the modes occurs in (4.84) the resulting quantity is *not* a physical angular momentum because of the factor of k^2.

4.4 Angular Momentum

The angular momentum properties of the free electromagnetic field can be described by examining the properties of the vector potential $\mathbf{A}$ under spatial rotations;[5] they provide an essential tool for understanding measurements of the angular distribution and polarisation characteristics of emitted and scattered electromagnetic radiation. Under a rotation of the coordinate axes the coordinate vector $\mathbf{x}$ is rotated to a new position, and the components of the vector field $\mathbf{A}(\mathbf{x})$ are mixed; these two effects can be represented by [10]

$$\mathbf{A}(\mathbf{x})' = R\mathbf{A}(R^{-1}\mathbf{x}). \tag{4.86}$$

---

[5] Here the vector potential $\mathbf{A}$ is the transverse Coulomb gauge vector potential, which for the free field is a gauge-invariant quantity.

Thus, for an infinitesimal rotation about the z-axis

$$\mathbf{A}(x,y,z)' = \{\mathbf{I} + (N^z + \mathbf{M}^z)\delta\theta\}\mathbf{A}(x,y,z), \tag{4.87}$$

where the vector $\mathbf{N}$ is

$$\mathbf{N} = \mathbf{x} \wedge \boldsymbol{\nabla}, \tag{4.88}$$

and $\mathbf{M}$ can be represented in terms of 3×3 matrices, for example

$$\mathbf{M}^x = \begin{bmatrix} 0 & 0 & 0 \\ 0 & 0 & 1 \\ 0 & -1 & 0 \end{bmatrix}, \quad \mathbf{M}^y = \begin{bmatrix} 0 & 0 & -1 \\ 0 & 0 & 0 \\ 1 & 0 & 0 \end{bmatrix}, \quad \mathbf{M}^z = \begin{bmatrix} 0 & 1 & 0 \\ -1 & 0 & 0 \\ 0 & 0 & 0 \end{bmatrix}. \tag{4.89}$$

The matrices $\mathbf{M}^r$ are simply representations of the antisymmetric (Levi–Civita) tensor, such that $\mathbf{M}^r_{st}$ is ε_{rst}.

It is convenient to introduce a modified total rotation operator $\mathbf{J}$ by setting

$$\mathbf{J} = \mathbf{L} + \mathbf{S}, \tag{4.90}$$

where[6]

$$\mathbf{L} = -i\mathbf{N}, \quad \mathbf{S} = -i\mathbf{M}, \tag{4.91}$$

and $i = \sqrt{-1}$ as usual. Since the components of the matrices $\mathbf{M}^r$ are constants, independent of $\mathbf{x}$, it follows that $\mathbf{L}$ and $\mathbf{S}$ commute. It is readily verified that all three vectors $(\mathbf{J}, \mathbf{L}, \mathbf{S})$ satisfy the commutation relation,

$$\mathbf{X} \wedge \mathbf{X} = i\,\mathbf{X}. \tag{4.92}$$

The components of $\mathbf{J}$ have the important property of commuting with the Curl operator,

$$J_n(\boldsymbol{\nabla}\wedge) = (\boldsymbol{\nabla}\wedge)J_n, \quad n = x, y, z. \tag{4.93}$$

Consequently, the rotational properties of the electromagnetic fields $(\mathbf{E}, \mathbf{B})$ can be obtained directly from that of $\mathbf{A}$ using (4.29).

We define complex spherical unit vectors with the properties

$$\hat{e}_q^* = (-1)^q \hat{e}_{-q}, \quad q = 0, \pm 1, \tag{4.94}$$

$$\hat{e}_q^* \cdot \hat{e}_{q'} = \delta_{qq'}. \tag{4.95}$$

These unit vectors may be realised in component form as

$$\hat{e}_{+1} = -\frac{1}{\sqrt{2}}\begin{bmatrix} 1 & i & 0 \end{bmatrix}, \quad \hat{e}_{-1} = \frac{1}{\sqrt{2}}\begin{bmatrix} 1 & -i & 0 \end{bmatrix}, \quad \hat{e}_0 = \begin{bmatrix} 0 & 0 & 1 \end{bmatrix}, \tag{4.96}$$

---

[6] We use the usual notations for total, orbital and spin angular momentum vectors although $\mathbf{N}$ and $\mathbf{M}$ are dimensionless, and (4.90) requires a proportionality constant of suitable dimensions to yield an angular momentum. Quantisation fixes the proportionality constant as $\hbar = h/2\pi$, where Planck's constant, h, has dimensions of angular momentum; there is no natural classical unit. After multiplication by i, as in (4.91), the matrices (4.89) will be recognised as a form of the $j = 1$ (vector) representation of a quantum mechanical angular momentum operator.

from which it follows that $\hat{\mathbf{e}}_{\pm 1}$ are eigenvectors of S_z with eigenvalues of ± 1, respectively,

$$S_z\,\hat{\mathbf{e}}_{+1} = +\hat{\mathbf{e}}_{+1}, \tag{4.97}$$

$$S_z\,\hat{\mathbf{e}}_{-1} = -\hat{\mathbf{e}}_{-1}, \tag{4.98}$$

and

$$S_z\,\hat{\mathbf{e}}_0 = 0. \tag{4.99}$$

The complex spherical unit vectors can be related by a unitary transformation to a set of real orthogonal unit vectors; for example, if we set

$$\hat{\boldsymbol{\varepsilon}}_1 = \begin{bmatrix} 1 & 0 & 0 \end{bmatrix}, \quad \hat{\boldsymbol{\varepsilon}}_2 = \begin{bmatrix} 0 & 1 & 0 \end{bmatrix}, \quad \hat{\mathbf{n}} = \begin{bmatrix} 0 & 0 & 1 \end{bmatrix}, \tag{4.100}$$

the explicit relationship[7] is

$$\hat{\boldsymbol{e}}_{+1} = -\frac{1}{\sqrt{2}}(\hat{\boldsymbol{\varepsilon}}_1 + i\hat{\boldsymbol{\varepsilon}}_2), \quad \hat{\boldsymbol{e}}_0 = \hat{\mathbf{n}}, \quad \hat{\boldsymbol{e}}_{-1} = \frac{1}{\sqrt{2}}(\hat{\boldsymbol{\varepsilon}}_1 - i\hat{\boldsymbol{\varepsilon}}_2). \tag{4.101}$$

Let us now represent the vector potential in the form of the plane wave (2.22); it has two linearly polarised components along $\hat{\boldsymbol{\varepsilon}}_1$ and $\hat{\boldsymbol{\varepsilon}}_2$,

$$\mathbf{A}(\mathbf{x})_i = A\hat{\boldsymbol{\varepsilon}}_i e^{i(\mathbf{k}\cdot\mathbf{x}-\omega t)}, \quad i = 1,2, \tag{4.102}$$

and zero component along $\mathbf{k}$. As described in §2.3, circular polarisation occurs when the two linearly polarised components are superposed $\pi/2$ out of phase. Hence, we may take

$$\mathbf{A}(\mathbf{x})_{R,L} = \frac{1}{\sqrt{2}}A\left(\hat{\boldsymbol{\varepsilon}}_1 + e^{\pm i\pi/2}\hat{\boldsymbol{\varepsilon}}_2\right)e^{i(\mathbf{k}\cdot\mathbf{x}-\omega t)}, \tag{4.103}$$

for right- and left-circularly polarised waves. In view of (4.96) and (4.103), the vector potentials for right and left circularly polarised waves are also eigenfunctions of S_z.

The vector operator $\mathbf{L}$ defined by (4.88) and (4.91) has well-known properties since, apart from a factor of $\hbar$, it is the usual quantum mechanical expression for the orbital angular momentum of a point particle with position coordinate $\mathbf{x}$. The spherical harmonics, Y_{lm}, defined by

$$Y_{lm}(\theta,\phi) = (-1)^m \left[\frac{(2l+1)(l-m)!}{4\pi(l+m)!}\right]^{\frac{1}{2}} P_l^m(\cos\theta)e^{im\phi}$$

$$Y_{lm}(\theta,\phi)^* = (-1)^m Y_{l-m}(\theta,\phi), \tag{4.104}$$

are eigenfunctions of L_z and L^2 satisfying[8]

$$L_z Y_{lm} = m\,Y_{lm}, \tag{4.105}$$

$$L^2 Y_{lm} = l(l+1)\,Y_{lm}. \tag{4.106}$$

---

[7] The (unitary) transformation matrix for this case is $\mathbf{U} = \frac{1}{\sqrt{2}}\begin{bmatrix} -1 & i & 0 \\ 0 & 0 & \sqrt{2} \\ 1 & i & 0 \end{bmatrix}$.

[8] The phase convention is that of Condon and Shortley [11]; Y_{1m} behaves in exactly the same way as the spherical unit vectors (4.96). Note the overall $-$ sign in $\hat{\mathbf{e}}_{+1}$ that ensures this.

Eigenvectors of J^2 and J_z defined in (4.90) can be found immediately by application of the usual angular momentum coupling rules to the complex spherical unit vectors (4.96) and the spherical harmonics (4.104). The resultant is a vector spherical harmonic,

$$\mathbf{Y}_{JlM} = \sum_{mq} Y(\theta, \phi)_{lm}\, \hat{e}_q\, (l\, m\, 1\, q|\, l1\, J\, M), \tag{4.107}$$

where $(l\, m\, 1\, q|l\, 1\, J\, M)$ is a Clebsch–Gordan coefficient [12]. The vector spherical harmonics have the properties

$$J^2\, \mathbf{Y}_{JlM} = J(J+1)\, \mathbf{Y}_{JlM}, \tag{4.108}$$

$$J_z\, \mathbf{Y}_{JlM} = M\, \mathbf{Y}_{JlM}, \tag{4.109}$$

and form a complete orthonormal set of functions,

$$\int \mathbf{Y}^*_{JlM}(\theta,\phi) . \mathbf{Y}_{J'l'M'}(\theta,\phi)\, \mathrm{d}\Omega = \delta_{JJ'}\delta_{ll'}\delta_{MM'}, \tag{4.110}$$

where the integration is over the unit sphere. J_z is the component along $\hat{\mathbf{e}}_0$. Since the spherical harmonics are solutions of the Laplace equation or the wave equation in spherical polar coordinates, we also have

$$\nabla^2\, \mathbf{Y}_{JlM} = 0. \tag{4.111}$$

The Clebsch Gordan coefficients vanish unless $|l-1| \le J \le |l+1|$ and $m+q = M$. From the point of view of parity (the inversion symmetry operation), the vector spherical harmonics divide into two types:

$$\mathbf{Y}_{JJM}: \quad \text{parity } (-1)^J, \tag{4.112}$$

$$\mathbf{Y}_{JJ\pm 1M}: \quad \text{parity } (-1)^{J+1}. \tag{4.113}$$

As an example consider $\mathbf{Y}_{010}$; the properties of the Clebsch–Gordan coefficients require that we must then have $q = -m$ and a sum over $m = 0, \pm 1$ in (4.107). The non-zero Clebsch–Gordan coefficients are [12]

$$(111-1|1100) = (11-11|1100) = \frac{1}{\sqrt{3}}, \quad (1100|1100) = -\frac{1}{\sqrt{3}}. \tag{4.114}$$

Using the well-known forms for $Y_{1m}(\theta,\phi)$ and (4.101) to return to the Cartesian unit vectors, we obtain

$$\mathbf{Y}_{010} = -\frac{1}{\sqrt{4\pi}}(\cos\theta\, \hat{\mathbf{e}}_0 + \sin\theta\cos\phi\, \hat{\mathbf{e}}_1 + \sin\theta\sin\phi\, \hat{\mathbf{e}}_2), \tag{4.115}$$

or

$$\mathbf{Y}_{010} = -\frac{\hat{\mathbf{x}}}{\sqrt{4\pi}}, \tag{4.116}$$

where $\hat{\mathbf{x}}$ is the unit radial vector; by similar calculation we find, for example,

$$\mathbf{Y}_{212} = \sqrt{\frac{3}{16\pi}}\sin(\theta)\, e^{i\phi}\, \hat{\mathbf{e}}_{+1}. \tag{4.117}$$

The complete set of vector spherical harmonics can be used as an expansion set for solutions of the wave equation. In place of the plane-wave solutions (4.102) and (4.103) we search for an harmonic vector potential of the form

$$\mathbf{A}(\mathbf{x}) = \mathbf{U}(\mathbf{x})e^{-i\omega t}, \tag{4.118}$$

where

$$\mathbf{U}(\mathbf{x}) = \sum_{J,l,M} f(r)_{JlM}\mathbf{Y}_{JlM}(\theta,\phi) \tag{4.119}$$

is to be a solution of

$$\left(\nabla^2 + k^2\right)\mathbf{U}(\mathbf{x}) = 0. \tag{4.120}$$

In view of (4.111) the radial functions f are independent of J and M and must satisfy Bessel's |equation,

$$\frac{1}{r^2}\frac{\mathrm{d}}{\mathrm{d}r}\left(r^2\frac{\mathrm{d}f(r)_l}{\mathrm{d}r}\right) + \left(k^2 - \frac{l(l+1)}{r}^2\right)f(r)_l = 0, \tag{4.121}$$

with solutions

$$f(r)_l = a\,j(kr)_l + b\,n(kr)_l, \tag{4.122}$$

in terms of spherical Bessel (j_l) and Neumann (n_l) functions. The constants a and b are to be chosen according to the physical problem being discussed; in scattering problems, for example, the vector potential must behave at large r as an outgoing wave:

$$|\mathbf{A}| \sim \frac{e^{i(kr-\omega t)}}{r}. \tag{4.123}$$

For such a case we set $a = i$, $b = -1$; that is,

$$f(r)_l = i\,j(kr)_l - n(kr)_l \equiv \frac{u(r)_l^+}{kr}, \tag{4.124}$$

since

$$u(r)_l^+ \approx e^{i(kr-\pi l/2)}, r \to +\infty. \tag{4.125}$$

If the vector potential is chosen to satisfy the Coulomb gauge condition (4.28), the angular momentum components of the free fields $\mathbf{E}$ and $\mathbf{B}$ will be simply proportional to those of $\mathbf{A}$. The appropriate forms for the vector potential components $\mathbf{A}_{JM}$ are [10]

$$\mathbf{A}(M)_{JM} = f(r)_J\,\mathbf{Y}_{JJM}(\theta,\phi), \tag{4.126}$$

$$\mathbf{A}(E)_{JM} = \sqrt{\frac{J+1}{2J+1}}f(r)_{J-1}\mathbf{Y}_{JJ-1M}(\theta,\phi), \tag{4.127}$$

$$-\sqrt{\frac{J}{2J+1}}f(r)_{J+1}\mathbf{Y}_{JJ+1M}(\theta,\phi), \tag{4.128}$$

with parities given by (4.112) and (4.113) which distinguish between the electric (E) and magnetic (M) multipole components of the field. These functions can now be used directly to construct the corresponding fields **E** and **B**. Using (4.29) with (4.124), (4.125), we have at once

$$\mathbf{E} = i\omega\mathbf{A}, \tag{4.129}$$

so that

$$\mathbf{E}(M)_{JM} = i\omega\mathbf{A}(M)_{JM}, \tag{4.130}$$

$$\mathbf{E}(E)_{JM} = i\omega\mathbf{A}(E)_{JM}. \tag{4.131}$$

Note that an arbitrary integration constant in f must be chosen so that **E** has appropriate dimensions. The evaluation of the magnetic induction **B** relies on (4.103) and the fact that the Curl operation changes the parity by a factor of (-1); there results

$$-\mathbf{B}(M)_{JM} = ik\mathbf{A}(E)_{JM}, \tag{4.132}$$

$$\mathbf{B}(E)_{JM} = ik\mathbf{A}(M)_{JM}. \tag{4.133}$$

The vector spherical harmonic $\mathbf{Y}_{JJM}$ satisfies the relation

$$\mathbf{Y}_{JJM} \propto \mathbf{L}\, Y_{JM}. \tag{4.134}$$

Since $\mathbf{x} \cdot \mathbf{L} = 0$, it follows that $\mathbf{A}(M)_{JM}$ has no radial component (along the direction of the radius vector **x**); hence $\mathbf{E}(M)_{JM}$ and $\mathbf{B}(E)_{JM}$ are transverse to **x** for all values of $r = |\mathbf{x}|$. The other pair of field strengths in (4.130), (4.131) do have radial components which may be shown to behave as $1/r^2$ for large r; their transverse components vary as $1/r$ for large r. These vector spherical harmonic expansions can be used to calculate the basic mechanical quantities of the field which we imagine is confined in a large sphere centred on the origin of the coordinates. The energy of the field is always (4.1), and this yields for a given J [10]

$$E_J = \varepsilon_0\omega^2 \int \left(|\mathbf{A}(M)_{JM}|^2 + |\mathbf{A}(E)_{JM}|^2 \right) \mathrm{d}^3\mathbf{x}$$

$$= \varepsilon_0\omega^2 \int r^2 \, |f(r)_l|^2 \, \mathrm{d}r, \tag{4.135}$$

independent of M. The time average $<\mathbf{m}>$ of the corresponding angular momentum of the field in such a sphere,

$$\mathbf{m} = \frac{1}{c^2} \int_\Omega \mathbf{x} \wedge \mathbf{S} \, \mathrm{d}^3\mathbf{x}, \tag{4.136}$$

where **S** is the Poynting vector (2.32), has zero components perpendicular to $\hat{\mathbf{e}}_0$, and a component ME_J/ω along $\hat{\mathbf{e}}_0$.

Finally, if we choose the fields in the form of outgoing waves, the outward flux of energy can be obtained from the time average $<\mathbf{S}>$ of the Poynting vector **S**; the rate of flow of energy in the solid angle $\mathrm{d}\Omega$ is then

$$I(\theta, \phi) \, \mathrm{d}\Omega = r^2 | <\mathbf{S}> |. \tag{4.137}$$

We write the vector potential in the form (4.124) with

$$\mathbf{U}(\mathbf{x}) = \sum_{JM} c(E)_{JM}\, \mathbf{A}(E)_{JM} + c(M)_{JM}\, \mathbf{A}(M)_{JM}, \tag{4.138}$$

and the radial function $f(r)_l$ satisfying (4.124). Then for large r the intensity distribution is

$$I(\theta,\phi) = 2\varepsilon_0 c^2 \left| \sum_{JM} i^{-J}\, c(M)_{JM}\mathbf{Y}_{JJM} \right. \tag{4.139}$$

$$\left. + i^{-J+1}\, c(E)_{JM}\left(\sqrt{\frac{J+1}{2J+1}}\mathbf{Y}_{JJ-1M} + \sqrt{\frac{J}{2J+1}}\mathbf{Y}_{JJ+1M} \right) \right|^2, \tag{4.140}$$

which when integrated over a full solid angle yields the total intensity as

$$\int I(\theta,\phi)\,\mathrm{d}\Omega = 2\varepsilon_0 c^2 \sum_{JM} \left(|c(M)_{JM}|^2 + |c(E)_{JM}|^2 \right). \tag{4.141}$$

After quantisation these expansions allow us to write the free-field vector potential directly in an angular momentum representation (§7.5).

References

[1] Weyl, H. (1912), Math. Annalen **71**, 441.

[2] Woolley, R. G. (2003), in *Handbook of Molecular Physics and Quantum Chemistry, 3 Volume Set*, edited by S. Wilson, P. F. Bernath and R. McWeeny, J. Wiley & Sons, Inc.

[3] Landau, L. and Lifshitz, E. M. (1971), *Course of Theoretical Physics*, vol. 2, *The Classical Theory of Fields*, 3rd ed., Pergamon Press.

[4] Lord Rayleigh (1900), Phil. Mag. Series V **49**, 539.

[5] Heitler, W. (1954), *The Quantum Theory of Radiation*, 3rd ed., Clarendon Press, reprinted (1984) by Dover Publications, Inc.

[6] Amiet, J.-P. and Weigert, S. (2002), J. Math. Phys. **43**, 4110.

[7] Stokes, G. G. (1852), Trans. Camb. Phil. Soc. **9**, 399.

[8] Barron, L. D. (1982), *Molecular Light Scattering and Optical Activity*, Cambridge University Press.

[9] Nikolova, L., Todorov, P., Sharlandjiev, P. and Stoyanov, S. (1992), Appl. Opt. **31**, 6698.

[10] Hamilton, J. (1959), *The Theory of Elementary Particles*, Ch. 1, Clarendon Press.

[11] Condon, E. U. and Shortley, G. H. (1935), *Theory of Atomic Spectra*, Cambridge University Press.

[12] Edmonds, A. R. (1960), *Angular Momentum in Quantum Mechanics*, Princeton University Press.

PART II

QUANTUM THEORY

5 The Quantum Mechanical Formalism

5.1 Introduction

The transition from classical physics to quantum theory took place over the first quarter of the twentieth century. Quantum mechanics was completed in 1926, but it was some years before the full extent of the conceptual revolution that had occurred was widely recognised [1], [2]. Its signature, Planck's constant, h, has the dimensions of the mechanical quantity action, joule-seconds (J·s) in the SI system, or equivalently, angular momentum. It first entered physics through Planck's resolution of the ultraviolet catastrophe in the classical theory of blackbody radiation (Chapter 4). That there is something genuinely novel, even revolutionary, about its introduction into the description of electromagnetic radiation may be appreciated from the observation that no such quantity can be constructed from the fundamental constants (c, ε_0) in Maxwell's theory of the electromagnetic field (Chapter 2).

Let us first revisit briefly the discussion of atomic mechanics presented in Chapter 1. The Rutherford model of the nuclear atom taken with the stability of atoms as demonstrated by chemistry, and the observed forms of the spectra of atoms and molecules represent critical problems for classical electrodynamics. The atom was seen as a massive atomic nucleus carrying a positive charge, $+Ze$, together with Z electrons that execute orbits around the nucleus under the influence of their mutual electrostatic forces. The emission spectrum of such a system is derived from the component of the electromagnetic field due to the charges that is observable at great distances from the atom; in this region (the wave zone) the field is purely transverse and can be identified as the emitted radiation. A spectrum is a harmonic analysis of this radiation and its features can be understood by looking at the Fourier analysis of the particle motion. A periodic orbit can be represented by a Fourier series with a characteristic frequency ν, and this leads to the result that in classical electrodynamics the spectrum expected consists of the fundamental frequency ν together with its overtones $n\nu$ ($n = 2, 3, \ldots$); for multiply periodic motions, there are additional combination frequencies. So far we have ignored the reaction of this radiation field on the motion of the charges; the emission of radiation and the requirement of overall energy balance means that the kinetic energy of the particles must decrease steadily with time.[1] This is a small effect on the

---

[1] Radiation reaction is also the cause of an instability – collapse of the electrons into the nucleus – of a classical Rutherford atom.

timescale of the period of the motion, so an instantaneous snapshot of the spectrum would yield very nearly equally spaced lines with frequency separation v; however, continuous observation of the spectrum would yield only a transient continuous band of frequencies [3]. As we know, the actual spectra of atoms are quite different; they are stable line spectra with unequal spacings that accumulate at series limits.

Recognising the limitations of classical theory, Bohr proposed that atoms must be dealt with as quantum phenomena through the introduction of Planck's constant into their mechanical description [4]. The development of this idea led to the Old Quantum Theory; it consists of a classical mechanical description involving charged particles (electrons and nuclei) supplemented by two new conditions. The mechanical system is specified by coordinates and momenta, and a Hamiltonian $H(x, p)$ in the usual way (Chapter 3). Energy is a continuous variable and integration of Hamilton's equations of motion yields a continuous infinity of trajectories. Bohr's first idea was to select a subset of closed trajectories by a quantisation condition involving integer multiples of h. This condition is generally only satisfied for a set of discrete energies $\{E_n\}$. These privileged trajectories are declared to be the stable, physically realisable stationary states of the atom (molecule). The second idea was that transitions may take place between pairs of stationary states; Bohr appealed to the law of conservation of energy to identify the energy difference of a pair with the energy of an absorbed/emitted quantum of radiation,

$$E_n - E_m = h v_{nm}, \tag{5.1}$$

in accordance with Planck's description of electromagnetic radiation.

The two supplementary conditions are clearly ad hoc, and are only justified by the extraordinary agreement between the prediction of this model for the spectrum of the hydrogen atom and the experimental observations. As noted in Chapter 1, there were further successes with molecular spectra. Whatever the selection of 'stationary states' through a condition involving Planck's constant might signify, it is evident that in other respects Bohr's model was quite consistent with classical ideas. In a stationary state, the laws of classical mechanics hold good, as does classical electrodynamics except that the radiation reaction is dispensed with. The motions of electrons in atoms, and the molecular models, can be visualised in terms of perfectly conventional classical particle trajectories – hence the 'solar system' representation of the atomic stationary states, and the picture of a molecule as a translating, vibrating, rotating collection of atoms joined by bonds. It was probably this framework that Sidgwick had in mind (Chapter 1) when he suggested that chemists would have to "accept the physical consequences in full."

The Old Quantum Theory of atomic mechanics was based on the classical Hamilton–Jacobi (HJ) equation,

$$H(x, \nabla S) = E. \tag{5.2}$$

Its solutions determine the dynamics of the classical system described by H; S is the classical action for the motion (cf §3.4). The theory claimed that for a partially (or completely) integrable system one can obtain the energy levels of the stationary states by imposing Bohr–Sommerfeld quantisation conditions involving Planck's constant, h, on the action integrals

$$J_k = \oint p_k \, dx_k = nh, \tag{5.3}$$

where the integral is taken over a closed orbit and the n are integers. In general, the $\{p_k\}$ are functions of the coordinates $\{x_k\}$ and may be found from the solutions of the HJ equation since

$$p_k = \frac{\partial S}{\partial x_k}. \tag{5.4}$$

In classical mechanics, the state of a system of particles with Hamiltonian variables $x_n, p_n : n = 1, \ldots N$ is represented by a point in a $2N$-dimensional phase space. Dynamics is represented by the motion of this representative point along some curve or 'ray' in phase space. The transformation of classical mechanics into the Old Quantum Theory was largely based on the use of quantum rules for the action-angle variables that come from the solution of the HJ equation as just shown. It has long been known that there are striking formal similarities between the phase space description of particle dynamics and the motion of waves [5], [6]; thus, Hamilton's variational principle $\delta \int L \, dt = 0$ is formally identical to Fermat's principle for a fictitious wave motion in the phase space. From the perspective of wave theory, however, the HJ equation is nothing other than Huyghen's principle for wave motion in phase space; this is the regime of geometrical optics which is valid when the wavelength is small compared to the physical dimensions of the 'apparatus', so that wave motion can be calculated by ray tracing.

This line of reasoning inspired a different line of attack that came a decade after Bohr's discovery; de Broglie conjectured that the characteristic quantum formula due to Planck, $E = h\nu$, should be equated to Einstein's energy–mass equivalence formula, $E = m_0 c^2$, for the case of a free particle with invariant mass m_0. This cannot be done in a Lorentz invariant fashion if we insist on the classical notion of a particle as a strictly localised entity with coordinate x. de Broglie's inspired proposal was to associate a particle with a delocalised phase wave, $\psi(x)$ [7]; insisting that this wave has the correct transformation properties under a Lorentz transformation leads to the conclusion that the wavelength, λ, of ψ is related to the particle momentum p according to the de Broglie equation,

$$p = \frac{h}{\lambda}. \tag{5.5}$$

Under a Lorentz boost K from rest, the de Broglie wave, ψ_0, undergoes the transformation

$$\psi_0 = e^{iEt/\hbar} \xrightarrow{K} e^{i(Et-px)/\hbar} \equiv e^{iS/\hbar}, \tag{5.6}$$

where we recognise the classical action for a free particle in the exponent. Dropping the inessential time factor, the de Broglie wave for a system described by a conservative Hamiltonian H is given by

$$\psi = e^{iS_0/\hbar}, \tag{5.7}$$

where S_0 is the solution of the time-independent form of the Hamilton–Jacobi equation.[2] For a bound particle S_0 is generally expressible as a radical; ψ is made single-valued by choosing the integration constants in the solution so as to make the period equal to integral multiples of Planck's constant h, that is,

$$S_0 \equiv \oint \mathbf{p} \cdot \mathbf{dx} = nh. \tag{5.8}$$

For an electron moving with speed v in a circular Bohr orbit of radius r, the angular momentum condition reads

$$L = mvr = n\hbar. \tag{5.9}$$

If this is combined with de Broglie's equation, (5.5), there results a standing-wave condition for the electron

$$n\lambda = 2\pi r, \tag{5.10}$$

since $2\pi r$ is the distance travelled in one complete orbit. So de Broglie's phase waves might account for the stability of the atom and offered an alternative wave explanation of the Bohr–Sommerfeld quantisation conditions that came originally from the action-angle formalism derived from classical particle dynamics.

Both de Broglie and Schrödinger thought it was essential that the theories they developed were consistent with special relativity. Thus, the HJ equation they employed was that for a relativistic particle; they also considered the involvement of external electromagnetic fields. Prior to de Broglie's proposal of a phase wave for a particle, Schrödinger had considered Weyl's geometric theory of electromagnetism (§3.7.1) in the context of the Old Quantum Theory and pointed out what he called "a remarkable property of an electron in a Bohr orbit" [10]. If a charged particle follows a path P through an electromagnetic field $(\mathbf{a}, \phi)$, the Weyl scale factor (3.280) changes 'lengths' according to

$$l = l_0 \exp\left(\frac{e}{C} \int_P (\mathbf{a} \cdot \mathbf{dx} - \phi \, dt)\right), \tag{5.11}$$

where C is a constant with dimensions of action to be determined by a physical model.

For the electron in the hydrogen atom there is no magnetic field, and a simple choice of potentials is to take ϕ to be Coulombic and $\mathbf{a} = 0$; then the exponent in (5.11) evaluated over one complete orbit yields $-e\bar{\phi}\tau/C$, where τ is the period of the orbit. From Bohr's theory,

$$e\bar{\phi}\tau = 2\bar{T}\tau = \int_0^\tau 2T \, dt = S = nh, \tag{5.12}$$

---

[2] It should be noted that there is a direct quantum mechanical generalisation of (5.7) in which the action integral S, expressed in terms of either the Lagrangian (using 'coordinates') or the Hamiltonian (using 'coordinates and momenta' – phase space), is the key ingredient in Feynman's path integral formalism [8]. This can be shown to be equivalent to the conventional Hamiltonian scheme centred on the Schrödinger equation. In this perspective, de Broglie's wave (5.7) is nothing other than the semi-classical approximation to the quantum theory. Path integral methods have become very important in both condensed matter physics and high-energy physics which are based on quantum field theories; in the former case they are very convenient for formulating quantum theory at *finite* temperatures [9]. Non-relativistic QED does not seem to have taken advantage of these developments.

where T is the kinetic energy, and $\overline{X}$ is the average over the closed orbit. Thus, the scale factor (5.11) is quantised,

$$l = l_0 e^{-nh/C}. \tag{5.13}$$

Schrödinger gave other examples involving various field configurations (Zeeman effect, Stark effect, etc.) and always found the same result.[3] He pointed out that with the choice $C = -i\hbar$ the scale factor reduces to unity, and that the scale of the Weyl 'length' would reproduce itself after each quasi-period. If the Weyl 'lengths' are dropped, and the Weyl idea of 'calibration' is attached to the de Broglie wave function for a charged particle moving along a path P in an electromagnetic field so that one could write

$$\psi = \psi_0 \exp\left(\frac{ie}{\hbar}\int_P a_\mu \, \mathrm{d}x^\mu\right), \tag{5.14}$$

one has an essential part of the quantum mechanical law of electromagnetism; of course, one needs the correct 'wave function' as well.[4] Gauge invariance in a quantum mechanical setting will be discussed in Chapter 9.

Schrödinger developed these wave ideas to propose a generalisation of de Broglie's phase waves for particles, which is analogous to Kirchhoff's extension of classical geometrical optics to a fully developed physical optics that encompasses diffraction, and so was led to wave mechanics. He illustrated his new theory in a famous series of papers [14] that dealt with some exactly soluble cases: the hydrogen atom, for which he found perfect agreement with the Bohr energy formula, the harmonic oscillator and the dumbbell rotator, and he also considered the relationship of wave mechanics to the new quantum mechanics being developed independently by Dirac, and by Born, Heisenberg and Jordan. These apparently quite different formulations were soon shown to be simply alternative presentations of the same general theory – quantum mechanics.

The new quantum mechanics undermined the familiar classical description of physical systems root and branch. This was already foreshadowed by de Broglie's association of a phase wave with a particle such that the strict classical separation of the notions of particle (localised) and wave (delocalised) would become untenable in the microscopic domain. Heisenberg's penetrating analysis of the consequences of the non-commutation of pairs of conjugate operators, such as the position and momentum operators for a particle, resulted in the formulation of his celebrated uncertainty principle [15]. An immediate casualty of that fundamental result was the notion of classical trajectories, so there was no longer a picture of an atom; moreover, it identifies a particle at rest at a definite position as an unphysical situation. Clearly, pictures of molecules would also be problematic in quantum mechanics if classical description had to be given up.

---

[3] A few years later, London gave a general proof based on the relativistic HJ equation that for Coulombic systems in external fields $e\int a_\mu \, \mathrm{d}x^\mu = \int \nabla S \cdot \mathrm{d}\mathbf{x} = \int \mathbf{p} \cdot \mathrm{d}\mathbf{x}$, which evaluates to nh if the integral is taken around a Bohr orbit [11].

[4] A scholarly account of these early developments can be found in Raman and Forman [12]; the historical development of the understanding that gauge invariance is a fundamental principle in physics is presented in O'Raifeartaigh [13].

5.2 Quantisation for Particles

In quantum mechanics, the state of a physical system is associated with a vector $|\Phi\rangle$ in a separable complex Hilbert space $\mathcal{H}$. Φ is commonly called a 'state vector' (or in Dirac's terminology, a 'ket'). All observable physical quantities correspond to self-adjoint[5] operators that act on the vectors in $\mathcal{H}$. The only measurable values of a physical observable are the elements of the spectrum of the corresponding operator; in the case of an operator Λ with discrete spectrum these are determined from the eigenvalue equation

$$\Lambda|\Phi_i\rangle = \lambda_i|\Phi_i\rangle. \tag{5.15}$$

The eigenvalues of a self-adjoint operator are real, and its eigenfunctions can be taken to be orthogonal and normalised according to

$$\langle\Phi_i|\Phi_j\rangle = \delta_{ij}, \tag{5.16}$$

where $\langle a|b\rangle$ is the Hilbert space scalar product, which in general is a complex number.

If the state vector happens to be an eigenvector of the measured observable, the result of a measurement will definitely, that is with probability 1, be the corresponding eigenvalue; otherwise one has a probability distribution for finding eigenvalues. The expectation value of the result of measurement of Λ in the state described by $|\Phi\rangle$ is given by

$$\varepsilon(\Lambda, \Phi) = \frac{\langle\Phi|\Lambda\Phi\rangle}{\langle\Phi|\Phi\rangle}. \tag{5.17}$$

More generally, the notion of a mixed state in classical statistical mechanics translates into quantum theory as the use of a density operator, ρ, that acts on vectors in the Hilbert space, $\rho : \mathcal{H} \to \mathcal{H}$. For mixed states, Eq. (5.17) becomes

$$\langle\Lambda\rangle_\rho = \mathrm{Tr}(\rho\Lambda), \quad \mathrm{Tr}(\rho) = 1. \tag{5.18}$$

According to Dirac's canonical quantisation scheme, the formalism of classical Hamiltonian mechanics is to be adapted in the following way [1]. All observables for a spinless particle[6] can be constructed out of just two basic operators corresponding to position and momentum which we denote as x and p, respectively. Their classical P.B. (3.129) is replaced by a *commutation* relation,

$$[\mathsf{x}, \mathsf{p}] \equiv \mathsf{x}\mathsf{p} - \mathsf{p}\mathsf{x} = i\hbar\mathbf{1}, \tag{5.19}$$

---

[5] In infinite-dimensional spaces, this is a stronger condition than requiring an operator to be Hermitian.

[6] In the conventional Hamiltonian formulation of classical mechanics, the phase space rests on coordinates x and p that are defined globally. It is not possible to define a 'spin' variable in such a framework. However, it can be done by specifying the phase space of a classical spin-j particle as the 2-sphere, S^2, i.e. the *surface* of a sphere in the familiar Euclidean space ($\Re^3$). The radius of the sphere is the angular momentum j of the particle which can take any value ≥ 0. The P.B.s of the components of the spin angular momentum vector $\mathbf{J}$ satisfy the usual algebraic relations. Remarkably, this classical structure can *only* be quantised if j takes integer or half-integer values.

where 1 is the identity operator on the Hilbert space $\mathcal{H}$, and $\hbar$ is the reduced Planck constant $h/2\pi$. For an arbitrary number, N, of degrees of freedom one has

$$[\mathsf{x}_i,\mathsf{x}_j] = [\mathsf{p}_i,\mathsf{p}_j] = 0, \quad i,j = 1,\ldots N,$$
$$[\mathsf{x}_i,\mathsf{p}_j] = i\hbar(1)_{ij}, \quad i,j = 1,\ldots N. \tag{5.20}$$

These are the *canonical commutation relations*.

The basic observables for the particle have the formal eigenvalue equations,

$$\mathsf{x}|x'\rangle = x'|x'\rangle, \quad x' \in \mathfrak{R}, \tag{5.21}$$
$$\mathsf{p}|p'\rangle = p'|p'\rangle, \quad p' \in \mathfrak{R}, \tag{5.22}$$

and it is easy to see that the position and momentum operators also have the formal properties

$$\mathsf{x} = i\hbar\frac{\partial}{\partial\mathsf{p}}, \qquad \mathsf{p} = -i\hbar\frac{\partial}{\partial\mathsf{x}}. \tag{5.23}$$

inherited from classical mechanics. Classical functions of the canonical variables (x,p) such as the Hamiltonian, $H(x,p)$, become operators, $\mathsf{H}(\mathsf{x},\mathsf{p})$, by the direct replacements of x by x, and p by p. Furthermore, canonical quantisation postulates a direct correspondence between classical P.B.s and quantum commutators that is applied to all observables

$$\{A,B\} = C \;\Rightarrow\; [\mathsf{A},\mathsf{B}] = i\hbar\,\mathsf{C}, \tag{5.24}$$

such that quantum operators inherit the Lie group structures of the P.B.s (§3.4).

It is worth pausing here to recognise the change in the mathematical structure underlying quantum mechanics that Dirac proposed. The classical P.B.s are based on *differential calculus*, whereas the quantum commutation relations are purely *algebraic*. This move leads to far-reaching consequences in the new mechanics, the most immediate of which is that the resulting operator algebra has the possibility of non-commutativity of the quantities supposed to represent physical variables, as seen in Eq. (5.19), and this imposes limitations on what can be measured.

Although the quantum mechanical formalism is based on the vectors of an abstract Hilbert space, for many purposes it is much more convenient to introduce a representation of the Hilbert space; this is like the introduction of coordinates in algebraic geometry. Since x and p have continuous spectra, $-\infty \leq x',p' \leq +\infty$, their associated representations may be given in terms of functions on the whole real line. Thus, we may represent the abstract Hilbert space using complex-valued functions in a function space that are square integrable with respect to an appropriate measure, for example the vectors $\phi(x') \in L^2(\mathfrak{R},\mu_{x'})$ or $\psi(p') \in L^2(\mathfrak{R},\mu_{p'})$, where μ is the Lebesgue measure on the real line. If we use the functions ϕ, we speak of the coordinate representation, whereas the functions ψ describe momentum representation. The functions ϕ, ψ may be thought of as arising from an abstract vector $|\Phi\rangle$ defined by some other operator, most importantly the Hamiltonian operator and the Schrödinger equation

$$\mathsf{H}|\Phi_i\rangle = E_i|\Phi_i\rangle, \tag{5.25}$$

and we have the following relations:

$$\phi(x')_i = \langle x'|\Phi_i\rangle, \quad \psi(p')_i = \langle p'|\Phi_i\rangle. \tag{5.26}$$

Thus, the scalar product $\langle x'|\Phi_i\rangle \equiv \phi(x')_i$ is the value of the function ϕ_i at the point x'; it is in general a complex number, and is said to be the representative of the ket $|\Phi_i\rangle$ in the coordinate representation. We recognise $\phi(x')_i$ as a Schrödinger wave function for the particle.

The ket $|\Phi_i\rangle$ is associated with the particular energy E_i according to the Schrödinger equation (5.25); likewise the ket $|x\rangle$ is associated with the coordinate value x by (5.21). According to the Dirac–Jordan transformation theory [16], [17], the scalar product $\langle x|\Phi_i\rangle$ is a particular case of a transformation function or probability amplitude which has the following interpretation; the real quantity

$$|\langle x|\Phi_i\rangle|^2 \, \mathrm{d}x = |\phi(x)_i|^2 \, \mathrm{d}x \tag{5.27}$$

is the probability[7] that the position coordinate of the particle lies between x and $x + \mathrm{d}x$ when the energy is known to be E_i. The integral of this quantity, (5.27), taken over the full range of possible values of x must be 1, since the particle is certainly somewhere in the range. Thus, if $\phi(x)_i$ is to represent a physical state for a particle with coordinate x, it must be a square integrable function.

Suppose the Hamiltonian for the particle is H; then by (5.17) the energy of the particle with normalised wave function $\phi(x)_i$ is

$$E_i = \int \phi(x)_i^* \mathsf{H} \phi(x)_i \, \mathrm{d}x. \tag{5.28}$$

Since H may contain momentum operators which act by differentiation on functions of x, it follows that appropriate derivatives of $\phi(x)_i$ must also be square integrable if E_i is to be finite. In non-relativistic particle theories, the momentum operators generally appear in H with a highest power of 2 (representing kinetic energy), and so the requirement is that ϕ_i and its first *two* derivatives must be square integrable. Functions with this property belong to a subspace of L^2 called the Sobolev space H^2. This result generalises directly to many degrees of freedom.

The replacement of the classical notion of a 'state of a physical system' by a probability for the physical system to have that state is one of the most characteristic features of quantum mechanics. The probability interpretation of (5.27) is a completely general idea that applies to the relationship between the eigenvalues and eigenfunctions of every pair of observables; thus, for two observables f, g the probability that a measurement will yield the eigenvalue f of the operator f when the eigenvalue of g is definitely known to be g is given by

$$\langle f|g\rangle^* \langle f|g\rangle \equiv |\langle f|g\rangle|^2. \tag{5.29}$$

Under complex conjugation probability amplitudes have the property

$$\langle f|g\rangle^* = \langle g|f\rangle, \tag{5.30}$$

---

[7] For quantities that take continuous values, the squared modulus of the probability amplitude must be taken as a probability distribution; for variables with discrete spectra, it is a genuine probability.

and so may be considered to be elements of an infinite-dimensional Hermitian (complex symmetric) matrix indexed by the labels f, g. They are normalised and mutually orthogonal according to

$$\sum_g \langle f|g \rangle \langle g|f' \rangle = \delta_{ff'}. \tag{5.31}$$

The probability amplitude $\langle f|g \rangle$ satisfies the relation

$$|\langle f|g \rangle|^2 \leq \langle f|f \rangle \langle g|g \rangle \tag{5.32}$$

which is an instance of the *Schwartz inequality*. With the aid of (5.32), one can derive a generalised uncertainty relation for any pair of self-adjoint operators [18],

$$\langle \Delta \mathsf{f} \rangle \langle \Delta \mathsf{g} \rangle \geq \frac{1}{2} \hbar \langle [\mathsf{f}, \mathsf{g}] \rangle. \tag{5.33}$$

Furthermore, if we consider three operators $\mathsf{f}, \mathsf{g}, \mathsf{h}$ with eigenvalue equations

$$\mathsf{f}|f \rangle = f|f \rangle \quad \text{etc.,} \tag{5.34}$$

we have the fundamental composition law

$$\langle f|h \rangle = \sum_g \langle f|g \rangle \langle g|h \rangle, \tag{5.35}$$

where the 'sum' over g is taken over all possible values. Thus, the eigenfunctions $\{|g \rangle\}$ of any self-adjoint operator g provide a resolution of the identity

$$\sum_g |g \rangle \langle g| = 1. \tag{5.36}$$

These formal results can be illustrated with the properties of the basic operators for a particle, x, p. Their abstract kets $\{|x \rangle, |p \rangle\}$ satisfy the following completeness and orthonormality relations:

$$\sum_x |x \rangle \langle x| = 1, \quad \sum_p |p \rangle \langle p| = 1, \tag{5.37}$$

$$\langle x|x' \rangle = \delta_{xx'}, \quad \langle p|p' \rangle = \delta_{pp'}. \tag{5.38}$$

The representative of the eigenket $|p' \rangle$ of p in the x representation, $\langle x'|p' \rangle$, satisfies the relation

$$p' \langle x'|p' \rangle \equiv \langle x'|\mathsf{p}\, p' \rangle = -i\hbar \frac{\mathrm{d}\langle x'|p' \rangle}{\mathrm{d}x}, \tag{5.39}$$

where we have used (5.23), and so by direct integration

$$\langle x'|p' \rangle = N e^{ip'x'/\hbar}, \tag{5.40}$$

where N is an integration constant that is disposable. Application of (5.35) shows at once that the position and momentum representatives are related by Fourier transformation, for

$$\langle x|\Phi_i \rangle = \sum_p \langle x|p \rangle \langle p|\Phi_i \rangle, \tag{5.41}$$

and so

$$\phi(x)_i = \int N e^{ipx/\hbar} \psi(p)_i \, dp. \tag{5.42}$$

If ψ is normalised to unity, N is normally chosen so that ϕ has the same normalisation.

The Hamiltonian operator H for the system determines the time evolution of the system for all times, $-\infty < t < +\infty$. Using (5.24), the equation of motion for a time-independent classical variable is replaced by

$$i\hbar \frac{d\Lambda}{dt} = [\Lambda, H], \tag{5.43}$$

and for the conjugate operators x and p this leads to

$$\frac{dx}{dt} = \frac{\partial H}{\partial p}, \quad \frac{dp}{dt} = -\frac{\partial H}{\partial x}, \tag{5.44}$$

in formal correspondence with Hamilton's equations of motion (3.120). Furthermore, if $|\Phi\rangle$ is a state vector for the system at an instant in time, the state at time t later is given by

$$|\Phi(t)\rangle = U_t |\Phi\rangle, \tag{5.45}$$

where

$$U_t = \exp(-iHt/\hbar). \tag{5.46}$$

In differential form, this is the time-dependent Schrödinger equation

$$i\hbar \frac{\partial |\Phi(t)\rangle}{\partial t} = H |\Phi(t)\rangle. \tag{5.47}$$

The unitary operator U_t is commonly known as the propagator or the time evolution operator.

Notice that time appears as a parameter, unlike the position and momentum variables; there is no 'time operator', and the pair $(H, i\hbar \partial/\partial t)$ are not canonically conjugate. There has been much discussion in the literature of a 'time-energy uncertainty' relation; whatever its status is, it is not comparable to Heisenberg's uncertainty principle for the operators x and p according to which the uncertainties in position and momentum at an instant are related to Planck's constant by the inequality [15]

$$\Delta x \Delta p \geq \tfrac{1}{2}\hbar \tag{5.48}$$

which, from (5.19), can be viewed as simply a particular case of (5.33). This formulation of quantum mechanics in which the operators are constant in time and the time evolution of the system is carried by the state or wave function is known as the *Schrödinger representation*.

There is a wholly equivalent description in which the wave function is constant in time, and the time evolution is carried by time-dependent operators. This is the *Heisenberg representation*; it is generated by the unitary transformation induced by (5.46). If we denote the two representations by attaching subscripts S and H to the wave function and an operator Λ, we have the relations

$$\Lambda(t)_H = U_t^{-1} \Lambda_S U_t, \tag{5.49}$$

$$\Phi_H = U_t^{-1} \Phi_S. \tag{5.50}$$

The differential equation satisfied by the operator Λ_H is (5.43); if Λ_H involves the time explicitly, an extra term must be added to the RHS giving the general operator equation of motion for Heisenberg operators,

$$i\hbar \frac{d\Lambda_H}{dt} = [\Lambda_H, H] + i\hbar \frac{\partial \Lambda_H}{\partial t}. \tag{5.51}$$

We saw in §3.4 that a physical variable β is a constant of the motion, provided it has vanishing P.B. with the system's Hamiltonian. Such variables have a key role in the description of the system; important examples are the kinematical symmetries implied by the relativity principle and the possible occurrence of dynamical symmetries. In quantum mechanics, the corresponding statement is that an observable is a constant of the motion if, and only if, it commutes with the Hamiltonian operator. If that is the case its eigenvalues can be used to label the states of the system along with the energy. There may be more than one such observable, in which case, however, there is an important restriction implied by (5.33); the eigenvalues of two such operators can only be specified at the same instant if the operators *commute*. A complete description of a physical system according to quantum mechanics is thus given by listing the eigenvalues of a maximal set of commuting operators that also commute with the system's Hamiltonian.

The canonical quantisation prescription may not lead definitively to the quantum operators associated with a classical dynamics on phase space simply because the quantum operators x and p do not commute, and there is an ambiguity in the ordering to be adopted in polynomials in (x, p). It is known that there is no single ordering convention of the quantum operators that works for polynomials of arbitrary degree > 2. Ambiguities in the operators constructed according to the canonical quantisation rules are resolved as far as possible by the principle that the operators describing invariances of the system should have the same group theoretical properties as their classical counterparts. This means that the Lie algebras underlying the Galileo and Poincaré symmetry groups defined by the P.B. relations (3.169)–(3.175) must be reinterpreted as Lie algebras with commutator brackets according to the rule (5.24); we discuss these constructions in §5.3.1.

The original formulation of quantum theory by Heisenberg (matrix mechanics) was presented in an apparently quite different way. Instead of using a Hilbert space of square integrable functions, Heisenberg used an infinite-dimensional discrete space for which matrices play the role of operators. In matrix mechanics, one has to find a system of Hermitian matrices to represent the dynamical variables that satisfy the matrix equivalents of (5.19) and (5.43), with the Hamiltonian expressed by a diagonal matrix. As an example, consider a particle of mass m in an harmonic oscillator potential with frequency ω; Heisenberg showed that the eigenvalues of the Hamiltonian matrix were

$$E_n = (n + \tfrac{1}{2})\hbar\omega, \tag{5.52}$$

and that the role of x and p could be played by the matrices

$$\mathbf{X} = \frac{l}{\sqrt{2}} \begin{pmatrix} 0 & \sqrt{1} & 0 & 0 & 0 & \cdots \\ \sqrt{1} & 0 & \sqrt{2} & 0 & 0 & \cdots \\ 0 & \sqrt{2} & 0 & \sqrt{3} & 0 & \cdots \\ 0 & 0 & \sqrt{3} & 0 & \sqrt{4} & \cdots \\ \vdots & \vdots & \vdots & \vdots & \vdots \end{pmatrix}, \tag{5.53}$$

$$\mathbf{P} = -\frac{i\hbar}{l\sqrt{2}} \begin{pmatrix} 0 & \sqrt{1} & 0 & 0 & 0 & \cdots \\ -\sqrt{1} & 0 & \sqrt{2} & 0 & 0 & \cdots \\ 0 & -\sqrt{2} & 0 & \sqrt{3} & 0 & \cdots \\ 0 & 0 & -\sqrt{3} & 0 & \sqrt{4} & \cdots \\ \vdots & \vdots & \vdots & \vdots & \vdots \end{pmatrix}, \tag{5.54}$$

where $l = \sqrt{m\omega/\hbar}$. Direct calculation confirms that these matrices satisfy the canonical commutation relations,[8]

$$\mathbf{XP} - \mathbf{PX} = i\hbar \, \mathbf{I}. \tag{5.55}$$

The matrices are not unique since these algebraic relations are preserved under unitary transformations; that is, if

$$\mathbf{U}^{-1} = \mathbf{U}^{+}, \tag{5.56}$$

a matrix

$$\overline{\mathbf{\Lambda}} = \mathbf{U} \, \mathbf{\Lambda} \, \mathbf{U}^{-1} \tag{5.57}$$

will satisfy the defining relations if $\mathbf{\Lambda}$ does.

The underlying reason for this is easily seen. Given a basis $\{|\phi_i\rangle\}$, an operator Λ may be associated with a matrix in the following way. The action of Λ on a basis element $|\phi_i\rangle$ is another vector, $|\chi\rangle$, in the same space. Accordingly, $|\chi\rangle$ can be expanded in the basis and we may write

$$\Lambda|\phi_i\rangle = \sum_j |\phi_j\rangle \, (\mathbf{\Lambda})_{ji}. \tag{5.58}$$

The matrix of expansion coefficients $\mathbf{\Lambda}$ is the matrix representative of the operator Λ; if the basis is orthonormal, its elements can be expressed as transformation functions,

$$(\mathbf{\Lambda})_{ji} = \langle \phi_j | \Lambda \phi_i \rangle. \tag{5.59}$$

Any other orthonormal basis of the Hilbert space can be written as the unitary transform of $\{|\phi_i\rangle\}$, and so the corresponding matrix representatives of an operator Λ are related by (5.57). The eigenfunctions of the oscillator Hamiltonian in the coordinate representation are the Hermite functions, and in this basis the matrix representatives

---

[8] For matrices of finite dimension we always have $\mathrm{Tr}(\mathbf{AB}) = \mathrm{Tr}(\mathbf{BA})$, and so $\mathbf{X}$ and $\mathbf{P}$ must be infinite dimensional if they are to satisfy the commutation relation (5.55).

for x and p take the simple tri-diagonal form given above. These relations show that Heisenberg's matrix mechanics can be interpreted within the Dirac–Jordan transformation theory and confirm that Schrödinger's wave mechanics and matrix mechanics are two particular, but equivalent formulations of quantum mechanics. An obvious implication of this remark is that both formulations[9] must give identical answers to all physically relevant questions, a condition that sharply restricts the scope for 'physical interpretation' of coordinates.

5.2.1 The Continuous Spectrum

There are well-known mathematical difficulties associated with Dirac's formalism applied to operators of general physical interest if one requires the theory to be mathematically rigorous [19], [20]. These problems are connected with the passage from systems described by operators with purely discrete spectra (like Heisenberg's oscillator) to those with operators that also (or only) have continuous spectra. To see the nature of the problem we reconsider (5.17) and (5.21), choosing a coordinate representation. Equation (5.17) then takes the form

$$\mathsf{x}\, f(x) = x'\, f(x), \tag{5.60}$$

where x' is used to denote the 'eigenvalue'. x is an unbounded operator ($-\infty \leq x' \leq +\infty$) and has purely continuous spectrum. It is the operator of 'multiplication by x', and so (5.60) can be put in the form

$$(x - x')\, f(x) = 0, \tag{5.61}$$

which requires that $f(x)$ be a function that vanishes everywhere except at $x = x'$ where it is not determined. No non-zero function in Hilbert space has such properties; indeed there is no such function and x' is not a true eigenvalue. Strictly speaking, eigenvalues and eigenfunctions are associated with either operators on finite-dimensional spaces or operators with purely discrete spectra on infinite-dimensional Hilbert spaces. Nevertheless, we carry on as though the discrete space formalism can be extended to the continuous spectrum.

Dirac's ingenious solution was to introduce a set of 'improper functions', $\delta(x - x')$, with the requisite properties and set[10]

$$f(x : x') = \delta(x - x'). \tag{5.62}$$

The 'eigenvalue' equation, (5.60), then becomes

$$(x - x')\, \delta(x - x') = 0. \tag{5.63}$$

The Dirac delta function, $\delta(x)$, is now formalised properly in the theory of distributions as a functional rather than a function. The completeness relation (5.37) takes the form

$$\int_{-\infty}^{+\infty} \delta(x - x')\delta(x'' - x')\,\mathrm{d}x' = \delta(x - x''), \tag{5.64}$$

[9] This must be true of any unitarily equivalent representation.
[10] x' is the continuous label analogous to a discrete quantum number.

while the orthogonality condition (5.38) is expressed formally by

$$\int_{-\infty}^{+\infty} \delta(x-x')\delta(x-x'')\,\mathrm{d}x = \delta(x'-x''). \tag{5.65}$$

'Normalisation' is ambiguous, since

$$\int_{-\infty}^{+\infty} \delta(x-x')\delta(x-x')\,\mathrm{d}x = \delta(0), \tag{5.66}$$

and $\delta(0)$ is undefined. This reminds us that the manipulation of products of distributions is generally a delicate matter.

As discussed in Chapter 2, the δ-function can be given a rigorous meaning when it is a factor in an integrand through the relation

$$\int_{-\infty}^{+\infty} \delta(x-x')\psi(x)\,\mathrm{d}x = \psi(x'), \tag{5.67}$$

provided the function $\psi(x)$ is continuous, although this result is not much help with (5.64)–(5.66). Its use is extended to general operators that have partly or completely continuous spectra through relations of the form

$$\Lambda|\Phi(\lambda)\rangle = \lambda|\Phi(\lambda)\rangle, \tag{5.68}$$

where

$$\langle\Phi(\lambda)|\Phi(\lambda')\rangle = \delta(\lambda-\lambda'), \tag{5.69}$$

and

$$\langle\Phi(\lambda'')|\Lambda|\Phi(\lambda')\rangle = \lambda'\,\delta(\lambda''-\lambda'). \tag{5.70}$$

The functions $\{\Phi(\lambda)\}$ associated with the continuous spectrum are usually called 'generalised eigenfunctions'. Dirac was well aware of what he was doing and explicitly cautioned that the δ-function should be used advisedly [21]:

> The use of improper functions ... is, rather, a convenient notation, enabling us to express in a concise form certain fundamental formulas which we could if necessary, rewrite in a rigorous form, but only in a cumbersome way in which the parallelism with the case of discrete eigenvalues is obscured. We shall confine our use of improper functions to such elementary equations that it will be obvious that the lack of rigour associated with them will not lead to a wrong result.

A special case involving purely continuous spectrum, of great importance, is free particle motion. For example, a system of interacting particles in the absence of external fields has total momentum $\mathbf{P}$ as a conserved quantity. This is associated with the free motion of the centre of mass. It has to be taken account of, although it is of limited physical interest and it is thus advantageous to separate out the centre-of-mass contribution to the overall Hamiltonian so that the internal dynamics can be focussed on. The Hamiltonian for the centre-of-mass motion is simply that for a fictitious 'particle',

$$H = \frac{P^2}{2M}, \tag{5.71}$$

where M is the total mass of all the particles; it is easily seen that H has a purely continuous spectrum, $0 \leq E \leq \infty$. The Schrödinger equation, (5.25), in a coordinate representation becomes

$$-\nabla^2 \Theta(\mathbf{R}) = K^2 \, \Theta(\mathbf{R}), \tag{5.72}$$

where $\mathbf{R}$ is the centre-of-mass coordinate, and

$$K^2 = \frac{2ME}{\hbar^2}. \tag{5.73}$$

The solutions of (5.72) are simply plane waves,

$$\Theta_{\mathbf{K}}(\mathbf{R}) = e^{i\mathbf{K}\cdot\mathbf{R}}, \tag{5.74}$$

which are not normalisable. They are 'generalised eigenfunctions' rather than true eigenfunctions. According to the discussion in §5.2, $\Theta_{\mathbf{K}}$ does not describe a physical state with finite energy.

This apparent difficulty may be handled in various ways. One may proceed with 'δ-function normalisation', as we have discussed,

$$\int \Theta_{\mathbf{K}'}(\mathbf{R})^* \Theta_{\mathbf{K}}(\mathbf{R}) \, \mathrm{d}^3 \mathbf{R} = \delta^3(\mathbf{K}' - \mathbf{K}). \tag{5.75}$$

Another possibility is to put the system in a large box of volume Ω and impose so-called 'box normalisation',

$$\Theta_{\mathbf{K}} \rightarrow \frac{1}{\sqrt{\Omega}} e^{i\mathbf{K}\cdot\mathbf{R}}. \tag{5.76}$$

With box boundary conditions, H is self-adjoint and has a complete set of eigenfunctions with discrete energies. At the end of a calculation, for example, a scattering cross section, one has to take the $\Omega \rightarrow \infty$ limit.

A rigorous mathematical formulation can be based on so-called wave packets; these are continuous superpositions of plane waves,

$$\Phi(\mathbf{R}) = \int e^{i\mathbf{K}\cdot\mathbf{R}} \phi(\mathbf{K}) \, \mathrm{d}^3 \mathbf{K}, \tag{5.77}$$

where the weight-factor $\phi(\mathbf{K})$ is chosen so that Φ belongs to the Sobolev space $H^2(\mathbf{R})$. In this way, the results of the heuristic δ-function and box normalisation approaches can be justified; as the wave-packet formalism is rather cumbersome the latter are generally preferred for practical calculations in physics and chemistry. It is, however, important for the completeness of the quantum mechanical formalism that a method based on L^2 functions is available.

The harmonic oscillator is a paradigmatic example of a system described by a Hamiltonian with purely discrete spectrum; none of the preceding manoeuvres are then required. This fact needs to be remembered when coupled oscillators are used to model aspects of atomic and molecular systems which are described by Hamiltonian operators that generally do have discrete and continuous parts to their spectra. An important question concerns the existence of the hoped-for 'generalised eigenfunctions' for such systems; we shall return to this question in Chapter 8.

In the case of an operator Λ that has purely discrete spectrum its eigenfunctions provide a resolution of the identity that can be used to express the operator in its spectral representation,

$$\Lambda = \sum_i \lambda_i |\Phi_i\rangle \langle \Phi_i|. \tag{5.78}$$

If we define a spectral projection operator,

$$P_i \equiv |\Phi_i\rangle \langle \Phi_i|, \tag{5.79}$$

which is such that

$$P_i^2 = P_i, \quad P_i^+ = P_i, \tag{5.80}$$

the operator Λ can be expressed very simply as a direct sum of multiplication operators,

$$\Lambda = \sum_i \bigoplus \Lambda_i, \tag{5.81}$$

with

$$\Lambda_i = \lambda_i \, P_i. \tag{5.82}$$

Provided 0 is not an eigenvalue of the operator Λ, the spectral decomposition is completely equivalent to the diagonalisation of Λ. The point of this manipulation is precisely that the notion of diagonalisation does not generally apply to the continuous spectrum, whereas some form of spectral decomposition in terms of projection operators is always possible for self-adjoint operators on a Hilbert space. This is the content of the spectral theorem [22].

As an example of the spectral theorem, consider the spectral representation of the position operator x, which is easily obtained. First of all, note that

$$\langle \phi |x| \psi \rangle = \int_{-\infty}^{+\infty} \phi^*(x) x \psi(x) \; dx$$
$$= \int_{-\infty}^{+\infty} x \, d\left(\int_{-\infty}^{x} \phi^*(x') \psi(x') dx' \right), \tag{5.83}$$

with $\phi, \psi \in L^2$. Now define a family of projection operators $P(x)$, $(-\infty < x < +\infty)$ on L^2 by putting[11]

$$P(x) \, \psi(x') = \begin{cases} \psi(x') & \text{if } x \geq x' \\ 0 & \text{if } x < x' \end{cases}. \tag{5.84}$$

$P(x)$ for fixed x is obviously a projection operator; in terms of $P(x)$ we can write

$$\int_{-\infty}^{x} \phi^*(x') \psi(x') \, dx' = \int_{-\infty}^{+\infty} \phi^*(x') P(x) \psi(x') \, dx'$$
$$= \langle \phi |P(x)| \psi \rangle, \tag{5.85}$$

---

[11] In explicit form $P(x)$ acts as a Heaviside step function, $\Theta(x - x')$, in the integral over x', so its derivative in (5.86) is a delta function, $\delta(x - x')$, and one recovers Dirac's treatment of the operator x.

so that all matrix elements of x can be expressed in terms of an integral over matrix elements of the family of projection operators,

$$\langle \phi | \mathsf{x} | \psi \rangle = \int_{-\infty}^{+\infty} x \mathrm{d} \langle \phi | \mathsf{P}(x) | \psi \rangle, \tag{5.86}$$

or, as a formal operator relation,

$$\mathsf{x} = \int_{-\infty}^{+\infty} \lambda \, \mathrm{d} \mathsf{P}(\lambda). \tag{5.87}$$

More generally, one can associate a projection operator valued function (measure) P_λ with every self-adjoint operator Λ such that the operator can be represented by [23]

$$\Lambda = \int_{-\infty}^{+\infty} \lambda \, \mathrm{d} \mathsf{P}_\lambda, \tag{5.88}$$

with the spectral family P_λ being characteristic of the operator Λ. Furthermore, functions of the operator Λ, that is, $f(\Lambda)$ (like $\sqrt{\Lambda}, \Lambda^{-1}$ etc.) can be given a spectral representation as

$$f(\Lambda) = \int_{-\infty}^{+\infty} f(\lambda) \, \mathrm{d} \mathsf{P}_\lambda. \tag{5.89}$$

Spectral projection replaces the notions of eigenvalues/eigenfunctions and diagonalisation that are familiar for operators with discrete spectra. Unfortunately, there is no general practical method of finding a spectral family in any given case. A formal solution is given by the Stone formula; first we introduce the resolvent $\mathsf{R}(\lambda)$ of a self-adjoint operator Λ,

$$\mathsf{R}(\lambda) = (\Lambda - \lambda)^{-1}. \tag{5.90}$$

Because Λ is self-adjoint, its spectrum is a subset of the real axis and therefore the resolvent is defined for $\Im \lambda \neq 0$. Stone's formula relates functions of a self-adjoint operator to the discontinuity of its resolvent across the real axis [23],

$$f(\Lambda) = \frac{1}{2\pi i} \lim_{\varepsilon \to 0^+} \int_{-\infty}^{+\infty} [\mathsf{R}(\lambda + i\varepsilon) - \mathsf{R}(\lambda - i\varepsilon)] f(\lambda) \, \mathrm{d}\lambda. \tag{5.91}$$

As a particular case of Eq. (5.91), one can formally obtain the resolution of the identity:

$$\mathsf{I} = \frac{1}{2\pi i} \lim_{\varepsilon \to 0^+} \int_{-\infty}^{+\infty} [\mathsf{R}(\lambda + i\varepsilon) - \mathsf{R}(\lambda - i\varepsilon)] \, \mathrm{d}\lambda. \tag{5.92}$$

5.3 Symmetry in Quantum Mechanics

5.3.1 Relativity

We consider first the Galileo group, G_∞ (see Eqs. (3.169)–(3.174), in Chapter 3) and initially need not distinguish between the classical and quantum cases. For a system of n structureless point particles described in a laboratory fixed reference frame by variables $\mathbf{p}_n, \mathbf{x}_n, \mathbf{s}_n$ and mass m_n, the generators of the Galilean group can be formed additively as [24]

$$\mathbf{P} = \sum_n \mathbf{p}_n, \tag{5.93}$$

$$\mathbf{J} = \sum_n (\mathbf{x}_n \wedge \mathbf{p}_n + \mathbf{s}_n)$$

$$= \mathbf{R} \wedge \mathbf{P} + \mathbf{L}, \tag{5.94}$$

$$\mathbf{K} = \sum_n (\mathbf{p}_n t - m_n \mathbf{x}_n)$$

$$= \mathbf{P}t - M\mathbf{R}, \tag{5.95}$$

$$H = \sum_n \frac{p_n^2}{2m_n} + V$$

$$= \frac{P^2}{2M} + H_{\text{int}}, \tag{5.96}$$

where the centre-of-mass variables,

$$M = \sum_n m_n, \quad \mathbf{R} = \frac{1}{M} \sum_n m_n \mathbf{x}_n, \quad \mathbf{P} = \sum_n \mathbf{p}_n, \tag{5.97}$$

have been explicitly separated out. The energy V in (5.96) will be recognised as the quantity we introduced in Chapter 3 as the potential energy; provided it is independent of the centre-of-mass variables $\{\mathbf{R}, \mathbf{P}\}$ and is translation invariant, its inclusion in (5.96) does not disturb the group algebra. Likewise, the internal angular momentum $\mathbf{L}$ must be independent of the centre-of-mass variables. As we discuss in more detail in Chapter 8, it is then advantageous to make a canonical transformation from the original Hamiltonian variables $\{\mathbf{x}_n, \mathbf{p}_n\}$ to a new set comprising the conjugate variables for the centre-of-mass $(\mathbf{R}, \mathbf{P})$, and $(n-1)$ translation invariant internal variables that are independent of $\mathbf{R}$ and $\mathbf{P}$. In this way, the requirements of Galilean invariance are carried by the centre-of-mass variables and there is considerable freedom in the choice of the internal variables out of which H_{int} and $\mathbf{L}$ must be constructed.

This simple additive construction of the group generators for a collection of particles, (5.93)–(5.96), yields operators according to the canonical quantisation prescription. The commuting operator combinations,

$$\mathsf{H} - \frac{\mathsf{P}^2}{2M} = \mathsf{H}_{\text{int}} \equiv \mathsf{H}_0 + \mathsf{V}, \tag{5.98}$$

$$(\mathbf{J} - \mathbf{R} \wedge \mathbf{P})^2 = \mathbf{L}^2, \tag{5.99}$$

are the so-called Casimir invariants of the group and have eigenvalues E and $L(L+1)$, respectively, which may be used to label the states in the Hilbert space, $\mathcal{H}_n$, for the n-particle system. Physically these labels describe the energy in the centre-of-mass frame and the total angular momentum about the centre of mass.

A fundamental result of quantum theory is that the symmetry transformations of the system can be associated with a set of operators $\{\mathsf{U}_i\}$ that form a unitary representation of the symmetry group in the Hilbert space. For the kinematical symmetries, this will be the case,[12] provided the generators H, $\mathbf{P}$, $\mathbf{J}$ and $\mathbf{K}$ of G_∞ are self-adjoint operators, a condition which is easily seen to be satisfied in the case of free particles. The requirement that H should be self-adjoint leads to restrictions on the possible forms for the operator V in the case of interactions [25]. In the Galilean case, the n-particle Hilbert space $\mathcal{H}_n$ is independent of the dynamics; it is the same for free and interacting particles, and the relationship between the Hamiltonians H_0 and H_{int} is simply a unitary transformation. This is the basis of the Rayleigh–Schrödinger perturbation theory (see Chapter 6).

The unitary representations of the Galileo group are multiplier or ray representations and so involve an essential phase factor. This leads to a fundamental restriction on the scope of a Galilean invariant quantum theory [26], [27]. The general Galilean transformation, g, is the coordinate transformation

$$\mathbf{x} \to \mathbf{x}' = R\mathbf{x} + \mathbf{v}t + \mathbf{a}, \tag{5.100}$$

$$t \to t' = t + \tau, \tag{5.101}$$

where R is a rotation matrix, $\mathbf{v}$ is a constant-velocity vector, $\mathbf{a}$ is a spatial translation, and τ is a time displacement. In terms of these parameters, we describe the transformation g as $(R, \mathbf{v}, \mathbf{a}, \tau)$; the parameters are independent of $(\mathbf{x}, t)$. The unitary operator $U(g)$ that corresponds to the Galilean transformation g is such that, up to some phase factor, the transformed state may be represented by

$$|\mathbf{x}(t)\rangle' = U(g)|\mathbf{x}(t)\rangle \equiv |\mathbf{x}'(t')\rangle. \tag{5.102}$$

In terms of representatives, this is

$$\phi(\mathbf{x}, t)' = \exp(if(\mathbf{x}', t')/\hbar)\, \phi(\mathbf{x}', t'), \tag{5.103}$$

where

$$f(\mathbf{x}', t') = M\mathbf{v} \cdot \mathbf{x}' - \tfrac{1}{2}Mv^2 t' + C, \tag{5.104}$$

and C is a constant that cannot be chosen to eliminate f.

The phase f has no consequences for the operators representing spatial rotations and translations, and its constant part affects time displacement trivially through the addition of a physically unimportant constant to the Hamiltonian. Consider, however, the following sequence of translations and boosts,

$$g_I = g_4\, g_3\, g_2\, g_1 \equiv (I, -\mathbf{v}, 0, 0)\, (I, 0, -\mathbf{a}, 0)\, (I, \mathbf{v}, 0, 0)\, (I, 0, \mathbf{a}, 0), \tag{5.105}$$

that is, we start with a translation $\mathbf{a}$, and follow with a boost to a reference frame moving with constant speed $\mathbf{v}$; then we translate back by $-\mathbf{a}$, and finally reverse the boost $(-\mathbf{v})$ to restore the original frame. Let the transformation g_I be applied to a superposition state where the components refer to different masses:

$$|\Phi\rangle = |\phi(M_1)\rangle + |\phi(M_2)\rangle. \tag{5.106}$$

---

[12] The unitary representation theory for translations is summarised in the next section.

The transformed state is easily found to be

$$|\Phi\rangle' = \mathsf{U}(g_I)|\Phi\rangle = \exp(iM_1\mathbf{v}\cdot\mathbf{a}/\hbar)|\phi(M_1)\rangle + \exp(iM_2\,\mathbf{v}\cdot\mathbf{a}/\hbar)|\phi(M_2)\rangle. \qquad (5.107)$$

Thus, if $M_1 \neq M_2$, the Galilean transformation g_I (5.105), which overall amounts to the identity, produces a change in the relative phase of the superposition (5.106), and hence alters its norm.

This result leads to the conclusion that in Galilean invariant quantum mechanics consistency requires that the superposition principle is restricted to states for which $M_1 = M_2$, that is, mass is conserved; this is the law of conservation of mass in its quantum setting. The mass parameter can therefore be used as a label for the representation of the Galileo group to which the particle belongs; there are other labels that arise from more general descriptions that are not available in a purely Galilean invariant description. For example, when electromagnetic interactions are explicitly included, the charge parameter, e, of a particle plays a similar role (see Chapter 9, and also what follows). For a collection of particles with N_k of type k,

$$M = \sum_k N_k m_k, \qquad (5.108)$$

and, as noted in Chapter 2, the constancy of M implies the constancy of the individual N_k. This is a fundamental feature of atomic and molecular processes; it is not found with nuclear reactions where mass defects are the rule, and such reactions cannot be described by a Galilean invariant quantum theory.

Relativistic (Lorentz invariant) wave equations for an elementary particle, for example those bearing the names Klein–Gordon, Dirac, Proca and so on, are well known and we do not go into details. However, one very important property is that their wave functions belong to unitary representations of the Poincaré group that are characterised by the mass, m, and spin, s, of the particle [28]. Spin is a purely quantum mechanical observable with no classical analogue; the parameter s takes integer and half-integer values, $s = 0, 1/2, 1, \ldots$ and has a fundamental connection with the quantum description of identical particles which we discuss in §5.3.3. It became quite clear early on in the development of quantum theory that relativistic wave mechanics, in the sense of a relativistic quantum theory of a *fixed* number $n > 1$ of particles, is an impossibility [29], and relativistic wave mechanics gave way to quantum field theory which allows particle creation/annihilation.

The canonical construction for the Poincaré group generators associated with a classical system of n free particles ($1 < n < \infty$) can be written down in a way analogous to that used previously for the Galileo group, but cannot be extended to an interacting system because of a 'no-interaction' theorem which shows that for consistency the particles must have constant velocities, and so there are no forces [30]–[32]. The requirement that $V = 0$ for such a system described in this way can be traced to the assumption that the Poincaré group algebra should have a canonical representation with the coordinates of the particles satisfying (3.175). The familiar feature of the Galilean invariant theory that interactions only appear in H, while the other symmetry generators preserve their free particle form, is also a conventional assumption.

These insights have led to the recent development of Lorentz invariant quantum theories of interacting n-particle systems [33], [34]; for technical reasons they are restricted to short-range interactions and so are appropriate in nuclear physics where mass is not conserved in particle reactions. Nothing of this sort appears yet to be possible for atoms and molecules. It is, however, normal to describe a local Lorentz invariant quantum field theory (the $n \to \infty$ limit), including interactions, in this way using a canonical representation of the Poincaré group [29].

5.3.2 Translations and Unitary Group Representations

Both operators x and p have spectra that can be endowed with the structure of the additive group of the real line $(\Re, +)$. For definiteness, let the additive group associated with the spectrum of x be denoted by K; then for a fixed element a belonging to K, the unitary operator

$$\mathsf{U}_a = \exp\left(\frac{i}{\hbar}a\mathsf{p}\right) \tag{5.109}$$

is a true unitary representation of K since we have

$$\mathsf{U}_a\mathsf{U}_{a'} = \mathsf{U}_{a'}\mathsf{U}_a = \mathsf{U}_{a+a'} \tag{5.110}$$

$$\mathsf{U}_0 = 1, \quad \mathsf{U}_{-a} = (\mathsf{U}_a)^{-1}. \tag{5.111}$$

For every fixed p',

$$\chi(a)_{p'} = \exp\left(\frac{i}{\hbar}ap'\right) \equiv \langle a|p'\rangle \tag{5.112}$$

is a character of K, and every character may be obtained in this way. The set of characters form a group, $\tilde{K}$, called the character group, that is dual to K. A unitary representation of $\tilde{K}$ is given by

$$\mathsf{V}_b = \exp\left(\frac{i}{\hbar}b\mathsf{x}\right). \tag{5.113}$$

Now consider the effect of U_a on the basic ket $|x\rangle$; if we write

$$|X\rangle = \mathsf{U}_a|x\rangle, \tag{5.114}$$

then

$$\begin{aligned}
\mathsf{x}|X\rangle &= (\mathsf{x}\mathsf{U}_a)\,|x\rangle + \mathsf{U}_a\mathsf{x}|x\rangle \\
&= \left(i\hbar\frac{\partial\mathsf{U}_a}{\partial\mathsf{p}}\right)|x\rangle + \mathsf{U}_a\,x|x\rangle = (x-a)|X\rangle \\
&\equiv (x-a)|x-a\rangle,
\end{aligned} \tag{5.115}$$

where we have put

$$\mathsf{U}_a|x\rangle = |X\rangle = |x-a\rangle, \tag{5.116}$$

since $|X\rangle$ has the position eigenvalue $(x-a)$. The operator U_a is interpreted physically as the operator describing spatial translations, and it provides the regular representation of K, which is a locally compact Abelian group.

We can therefore write down a family of projection operators associated with this representation using the standard group theoretical result that for each $\chi \in \tilde{K}$ [35]

$$\mathsf{P}_\chi = \int \bar{\chi}(a)_p \, \mathsf{U}_a \, da, \tag{5.117}$$

where the integration is over the group parameters, and $\bar{\chi}$ is the complex conjugate of χ. This operator acts as follows on a basic ket:

$$\mathsf{P}_\chi|x\rangle = \int \bar{\chi}(a)_p \, \mathsf{U}_a|x\rangle \, da$$

$$= \int \bar{\chi}(a)_p|x-a\rangle \, da = \bar{\chi}(x)_p \int \bar{\chi}(a)_p|a\rangle \, da \equiv \bar{\chi}(x)_p|p\rangle. \tag{5.118}$$

If we choose the character labelled by the momentum eigenvalue p, we see that it projects out the component labelled by the eigenvalue p. In Dirac's notation, we have $\mathsf{P}_\chi = |p\rangle\langle p|$. Let $\mathcal{H}_\chi$ be the range of P_χ; then the $\mathcal{H}_\chi$ are mutually orthogonal invariant subspaces of $\mathcal{H}$ such that

$$\mathcal{H} = \sum_{\chi \in \tilde{K}} \bigoplus \mathcal{H}_\chi. \tag{5.119}$$

Accordingly, we may write

$$|x\rangle = \sum_\chi \mathsf{P}_\chi|x\rangle = \sum_{\chi \in \tilde{K}} \bar{\chi}(x)_p|p\rangle, \tag{5.120}$$

which shows explicitly that $|x\rangle$ and $|p\rangle$ are related by Fourier transformation [35], [36].

These results can be extended directly to the n-dimensional case ($n < \infty$) by considering n-tuples as vectors in $\mathfrak{R}^n$. The additive group K and its isomorphic dual K are then the additive group of n-tuples, so that, for example, we have to make replacements like

$$ap' \rightarrow a_1 p_1' + a_2 p_2' + \ldots a_n p_n' = \mathbf{a} \cdot \mathbf{p}', \tag{5.121}$$

using ordinary vector notation; they too are locally compact Abelian groups with a well-developed unitary representation theory. This generalisation of the notation applies equally to the position and momentum operators, and to their eigenvalues, so that, for example, spatial translation by a vector $\mathbf{a}$ is described by the operator

$$\mathsf{U}_\mathbf{a} = \exp\left(\frac{i}{\hbar}\mathbf{a} \cdot \mathbf{p}\right). \tag{5.122}$$

The requirement that the quantum theory of an isolated system should be described by a unitary representation of the symmetry group that expresses relativistic invariance (Galileo or Poincaré) leads us in a natural way to the operator $\mathsf{U}_\mathbf{a}$ and its dual $\mathsf{V}_\mathbf{b}$. Some mathematical aspects of these exponential operators are discussed in Appendix G. Here we just note that the relationship between $\mathsf{U}_\mathbf{a}$ and $\mathsf{V}_\mathbf{b}$ is the basis of the modern group theoretical approach to quantisation; quantisation can be formulated in terms of the statement of the Weyl commutation relations [37],

$$U_{\mathbf{a}}V_{\mathbf{b}} = \exp\left(\frac{i}{\hbar}\mathbf{a}\cdot\mathbf{b}\right)V_{\mathbf{b}}U_{\mathbf{a}}, \tag{5.123}$$

where

$$\mathbf{a} = \{a_1, a_2, \ldots a_n\}, \quad \mathbf{b} = \{b_1, b_2, \ldots b_n\}. \tag{5.124}$$

All irreducible representations of the Weyl commutation relations are unitarily equivalent, essentially because $U_{\mathbf{a}}$ and its dual $V_{\mathbf{b}}$ are unitary operators and hence bounded. As we have seen, coordinate systems that are related by a translation $\mathbf{a}$ cannot be distinguished, and this is a kinematical symmetry of an isolated system. The Weyl commutation relation shows that the familiar canonical commutation relation for x and p can be understood simply as a consequence of requiring translation invariance.

For systems with a finite number n of degrees of freedom, the choice of representation has a purely kinematical character since, as proved by von Neumann [19], one can pass from one representation to another by means of a unitary transformation. Clearly, the physical description will not depend on the choice of representation, which is therefore reduced purely to a matter of convenience. This is also the case for the quantum theory of the free electromagnetic field, but von Neumann's fundamental result does not hold generally in a quantum field theory if there are interactions [38]. When we relax the condition $n < \infty$, we are confronted with much more difficult mathematics, since we can no longer rely on von Neumann's theorem for a unique solution (up to unitary equivalence) for the operators x and p satisfying the Weyl commutation relations. Moreover, we have to make use of the so-called functional calculus, that is, calculus on function spaces. The summary of field quantisation in §5.4 is therefore heuristic and uses a formal analogy with the finite-dimensional case.

5.3.3 Identical Particles in Quantum Mechanics

The treatment of identical particles is a matter of fundamental difference between classical and quantum mechanics. In classical mechanics, we describe motion by specifying the orbits of the individual particles under the given classical dynamical law. Even though the particles may be identical, if the initial conditions have been set, it makes perfect sense to say that at a given instant in time, t, particle 1 is moving along orbit 1, particle 2 is moving along orbit 2 and so on. Any permutation of the identical particles will also be a solution of the dynamical problem, since the classical equations of motion are invariant under such permutations, but this solution belongs to another set of initial conditions, that is, to a distinct physical situation [36].

The characteristic features of the quantum mechanical treatment of permutation symmetry are illustrated in the simple two-particle case. As shown by Dirac [39], the permutation P_{12} which interchanges the labels of the particles can be considered to be a dynamical variable in quantum mechanics; there is no analogous quantity in classical mechanics and P_{12} is an operator which has no classical limit. If the particles are identical, all physical observables Ω_i must be symmetric functions of the variables, and so will commute with P_{12}:

$$[P_{12}, \Omega_i] = 0. \tag{5.125}$$

One of the Ω_i will be the Hamiltonian H, and thus P_{12} is a constant of the motion

$$\frac{dP_{12}}{dt} = \frac{i}{\hbar}[H, P_{12}] = 0, \tag{5.126}$$

according to (5.43). The eigenstates of H will therefore be simultaneous eigenstates of P_{12}.

Quantum mechanics requires a further remarkable property of the state vectors, as can be seen in what follows. Given a set of single-particle states $\{|\phi_k\rangle\}$, which we may take to be orthonormal, a general state of a two-particle system can be constructed as the superposition

$$|\Psi(1,2)\rangle = \sum_{k,k'} c_{k,k'} |\phi_k(1)\rangle |\phi_{k'}(2)\rangle. \tag{5.127}$$

The coefficients $\{c_{k,k'}\}$ are probability amplitudes in the sense of §5.2, $c_{k,k'} = \langle \Psi | \phi_k \phi_{k'}\rangle$ and so in principle are complex numbers. According to the usual rules of quantum mechanics, the $\{c_{k,k'}\}$ have the interpretation that $|c_{k,k'}|^2$ is the probability that particle 1 is in state k, while particle 2 is in state k'. P_{12} permutes the variables of particles 1 and 2 such that, acting on $|\Psi(1,2)\rangle$, it gives the state

$$|\Phi(1,2)\rangle = P_{12}|\Psi(1,2)\rangle = \sum_{k,k'} c_{k,k'} |\phi_k(2)\rangle |\phi_{k'}(1)\rangle. \tag{5.128}$$

Now the probability that particle 1 is in state k while particle 2 is in state k' is $|c_{k',k}|^2$ which in general is different from $|c_{k,k'}|^2$, so that the state $|\Phi(1,2)\rangle$ resulting from the action of P_{12} is a different vector in the Hilbert space. However, if the two particles are identical, $|\Phi(1,2)\rangle$ and $|\Psi(1,2)\rangle$ should be states representing the same physical situation, and hence can differ at most by a phase factor; equivalently the probabilities $|c_{k,k'}|^2$ and $|c_{k',k}|^2$ must be equal and therefore

$$c_{k,k'} = e^{i\theta} c_{k',k}, \tag{5.129}$$

where the phase angle θ must be determined from other considerations.

The missing ingredient is the notion of *spin* for quantum particles that has to be added to the non-relativistic quantum theory by hand. Spin is an intrinsic contribution to the angular momentum of a particle. It turns out that all known elementary particles fall into two groups; either they have half-integer spin or they have integer spin values (in units of $\hbar$); the former are called *fermions* and the latter are called *bosons*. Composite entities – atoms, molecules, nuclei – also have characteristic spin values obtained by combining the spins of their elementary particle constituents. When the neutron was discovered, the first idea was that it was a composite of a proton and an electron; when spin is considered, however, such a model is ruled out by observations in molecular spectroscopy. Thus, for example, the deuteron which is the bound state of 1 proton and 1 neutron, is a spin-1 particle, as may be deduced from the (molecular) band spectra of D_2. Likewise the nitrogen nucleus, N^{14}, composed of seven protons and seven neutrons, is a boson particle, as shown by the intensity pattern of the N_2 spectrum. These observations were crucial facts leading to the view that the neutron must be a spin 1/2 elementary particle [40]. To return to permutation symmetry, the important point is that integer spin values are associated with $\theta = 0$ in (5.129), while half-integer spin values are associated with $\theta = \pi$, so that the coefficient in (5.129) is either $+1$ or -1. $\theta = 0$ means that the state $|\Psi\rangle$ is symmetric in the particle variables,

$$|\Psi(1,2)\rangle = |\Psi(2,1)\rangle, \tag{5.130}$$

which is the characteristic property of wave functions for bosons, while $\theta = \pi$ is associated with an antisymmetric state,

$$|\Psi(1,2)\rangle = -|\Psi(2,1)\rangle, \tag{5.131}$$

which is characteristic of wave functions for identical fermions. This latter result is essentially a statement of the *Pauli exclusion principle* in wave mechanical terms.

Since the permutation operator is a constant of the motion, the dynamical evolution of the system cannot cause transitions from one subspace to the other; that is, they are incoherent spaces and are never mixed in superpositions. The restriction of the possible states of the system to lie in either the symmetric subspace or the antisymmetric subspace is therefore consistent with the equations of motion. The textbook description [41] of the internal states of the helium atom is a paradigmatic example of these ideas. The helium atom is the system of the He nucleus and two electrons; after separation of the centre-of-mass variables there are two internal space coordinates, $\mathbf{r}_i, i = 1, 2$ which describe the motion of the electrons relative to the nucleus. Each electron is assigned a spin variable ω and two orthonormal spin wave functions, $\alpha(\omega), \beta(\omega)$, and the single-particle states in (5.127) are of the product form:

$$|\phi(\mathbf{x}_N)_k\rangle = |\chi(\mathbf{r}_N)_j \alpha(\omega)\rangle \text{ or } |\chi(\mathbf{r}_N)_j \beta(\omega)\rangle \equiv |\phi_k(N)\rangle, \quad N = 1, 2. \tag{5.132}$$

The permutations of n identical objects form a group, S_n, called the symmetric group; it is a finite group of order $n!$. The preceding discussion has pointed to the physical significance of its two 1-dimensional representations; quite generally the symmetric representation is associated with the state vectors of boson particles, those with integer spin, for example photons, while the antisymmetric representation is associated with fermion particles, those with half-integer spin, for example electrons, neutrons, and protons.

Suppose we have a physical system of n identical particles with space–spin coordinates $\{x_n\}$, and we aim to give a quantum mechanical description in a Schrödinger representation using wave functions $\{\phi(x_1, x_2, \ldots x_n)\}$. Let P be a permutation over n objects,

$$P = \begin{pmatrix} 1 & 2 & 3 & \ldots & n \\ \sigma_1 & \sigma_2 & \sigma_3 & \ldots & \sigma_n \end{pmatrix}, \tag{5.133}$$

and let P be the operator on the n-particle wave functions corresponding to the action of P on the labels of their coordinates,

$$P\phi(x_1, x_2, \ldots x_n) = \phi(x_{\sigma_1}, x_{\sigma_2}, \ldots x_{\sigma_n}). \tag{5.134}$$

The permutation operator can be written as a product of separate space and spin variable permutation operators,

$$P = P_r P_\omega, \tag{5.135}$$

where r refers to the space variables and ω to the spin-coordinates.

Then, if ϕ is to represent a physically realisable state, we must have

$$P\phi(x_1, x_2, \ldots x_n) = \delta_P \, \phi(x_1, x_2, \ldots x_n), \tag{5.136}$$

where δ_P is either of the two one-dimensional representations of the symmetric group S_n, that is,

$$\delta_P = 1 \text{ all } P, \text{ for bosons}, \tag{5.137}$$

$$\delta_P = \text{signature of } P, \text{ for fermions}. \tag{5.138}$$

As shown by Dirac, P is a unitary operator [39],

$$P^{-1} = P^+, \tag{5.139}$$

which, just as in the two-particle case discussed in the preceding, is a constant of the motion

$$[H, P] = 0. \tag{5.140}$$

The operator P can be used to construct two Hermitian projection operators P_S, P_A, which act on a general function of the n variables to yield symmetric and antisymmetric functions, respectively. These operators are defined as

$$P_S = \frac{1}{n!} \sum_P P, \tag{5.141}$$

$$P_A = \frac{1}{n!} \sum_P \text{sgn}(P) P, \tag{5.142}$$

and have the properties

$$P_S^2 = P_S, \quad P_A^2 = P_A, \quad [P_S, P_A] = 0. \tag{5.143}$$

The common practical procedure for the construction of n-particle Schrödinger wave functions is based on forming superpositions of products of single-particle wave functions ('orbitals') that are functions of the space and spin coordinates. Given a set of single-particle wave functions, $\{\phi_1(x), \phi_2(x), \ldots \phi_k(x)\}$, a normalised antisymmetric wave function is given by the well-known Slater determinant [42]:

$$\Psi(x_1, x_2, \ldots x_n) = \frac{1}{\sqrt{n!}} \begin{bmatrix} \phi_1(x_1) & \phi_2(x_1) & \cdots & \phi_k(x_1) \\ \phi_1(x_2) & \phi_2(x_2) & \cdots & \phi_k(x_2) \\ \cdots & \cdots & \cdots & \cdots \\ \phi_1(x_n) & \phi_2(x_n) & \cdots & \phi_k(x_n) \end{bmatrix}. \tag{5.144}$$

A wave function of this form for a n-electron system is completely antisymmetric in the n electrons and so satisfies the exclusion principle, since from the properties of determinants the interchange of two rows changes the sign of the determinant, and no two of the orbitals can be the same since then the determinant would vanish. Since the single-particle states for electrons can only be of the form (5.132), we also have the familiar statement: *No two electrons can occupy the same orbital in an electronic system, and these two must have their spins opposed.* The two possible values of the spin variable for an electron are often referred to as 'up-spin' and 'down-spin' and denoted symbolically as ($\uparrow\downarrow$). The corresponding completely symmetric wave function for bosons is obtained by replacing the Slater determinant by a so-called 'positive determinant' or permanent (with a modified normalisation factor). In the boson case, there is no restriction on the number of particles that can occupy any given state.

The preceding discussion is based on the assumption of a fixed number of particles and the availability of single-particle wave functions. By making a different choice of representation of the Hilbert space, it is possible to describe situations where the number of particles is not fixed; this is essential for electromagnetic phenomena where photons are absorbed and emitted by material systems. This is the occupation number representation which will be illustrated in detail for the quantisation of the electromagnetic field (Chapter 7). It suffices here to remark that it is based on a novel choice of dynamical variables, namely a pair of adjoint operators (c^+, c) that create and annihilate particles in specified states, and an associated self-adjoint number operator

$$\mathsf{N} = \mathsf{c}^+\mathsf{c}, \quad \mathsf{N}|N\rangle = N|N\rangle, \quad N \geq 0, \tag{5.145}$$

that counts them. The integral eigenvalues of the particle number operator(s) can be used to label the state vectors of the system. In this representation, the permutation symmetry of identical particles is embodied in the fundamental algebraic properties of the annihilation and creation operators. The commutation relation,

$$\mathsf{c}\mathsf{c}^+ - \mathsf{c}^+\mathsf{c} = 1, \tag{5.146}$$

leads to a number operator for which N can be arbitrarily large.[13] This is the choice required for boson particles. On the other hand, the fermion operators are required to anticommute,

$$\mathsf{c}\mathsf{c}^+ + \mathsf{c}^+\mathsf{c} = 1, \tag{5.147}$$

since this forces the restriction $N = 0, 1$ only, which is essentially a statement of the Pauli exclusion principle [36], [43].

---

[13] N is required to be finite.

5.4 Quantisation for a Field

We introduce a real function space of square integrable test functions (Schwartz space) with elements $f, g, \ldots$ that has a scalar product

$$(f, g) = \int f(\mathbf{u}) g(\mathbf{u}) \, \mathrm{d}^3 \mathbf{u} < \infty. \tag{5.148}$$

Next, we have canonically conjugate operator valued distributions X and P which are used to describe the field,

$$\mathsf{X} = \mathsf{X}(\mathbf{u}), \quad \mathsf{P} = \mathsf{P}(\mathbf{u}), \tag{5.149}$$

and the corresponding operators $\mathsf{X}[f], \mathsf{P}[f]$ in the dual space

$$\mathsf{X}[f] = \int \mathsf{X}(\mathbf{u}) f(\mathbf{u}) \, \mathrm{d}^3 \mathbf{u}$$

$$\mathsf{P}[f] = \int \mathsf{P}(\mathbf{u}) f(\mathbf{u}) \, \mathrm{d}^3 \mathbf{u}. \tag{5.150}$$

We take X and P to be self-adjoint and define the unitary operators,

$$\mathsf{U}[f] = \exp\left(\frac{i}{\hbar} \mathsf{P}[f]\right), \ \mathsf{V}[f] = \exp\left(\frac{i}{\hbar} \mathsf{X}[f]\right). \tag{5.151}$$

The Weyl commutation relations,

$$\mathsf{U}[f]\mathsf{V}[g] = \exp\left(\frac{i}{\hbar}(f, g)\right) \mathsf{V}[g]\mathsf{U}[f], \tag{5.152}$$

are then formally equivalent to the canonical commutation relations,

$$[\mathsf{X}, \mathsf{P}] = i\hbar \mathbf{1}. \tag{5.153}$$

Using coordinates, this is

$$[\mathsf{X}(\mathbf{u}), \mathsf{P}(\mathbf{u}')] = i\hbar \delta^3(\mathbf{u} - \mathbf{u}'). \tag{5.154}$$

Formally at least, X and P define representations through their eigenvalue equations as in (5.21), (5.22) for the spinless particle,

$$\mathsf{X}|X'\rangle = X'|X'\rangle, \ \mathsf{P}|P'\rangle = P'|P'\rangle, \tag{5.155}$$

which lead to an infinite system of equations that hold at all points in the space $\mathbf{u}$, for example

$$\mathsf{X}(\mathbf{u})|X(\mathbf{u})\rangle = X(\mathbf{u})|X(\mathbf{u})\rangle. \tag{5.156}$$

The eigenvalues are functions $X(\mathbf{u})$ that also belong to the Schwartz space. The states in quantum field theory are thus quantities that depend on these functions, that is, wave functionals; for example, in the representation in which the operator X is diagonal, we have

$$\langle X | \Phi \rangle = \Phi[X], \tag{5.157}$$

where the square brackets denote functional dependence. From this, by analogy with (5.23) we can realise the commutation relations (5.153), (5.154) in terms of functional derivatives,

$$\mathsf{X} = i\hbar \frac{\delta}{\delta \mathsf{P}} \;\Rightarrow\; \mathsf{X}(\mathbf{u}) = i\hbar \frac{\delta}{\delta P(\mathbf{u})}, \tag{5.158}$$

$$\mathsf{P} = -i\hbar \frac{\delta}{\delta \mathsf{X}} \;\Rightarrow\; \mathsf{P}(\mathbf{u}) = -i\hbar \frac{\delta}{\delta X(\mathbf{u})}. \tag{5.159}$$

The completeness and orthonormality relations (5.37), (5.38) generalise formally to

$$\langle X | X' \rangle = \delta_{XX'} \equiv \delta[X - X'] = \prod_{\mathbf{u}} \delta[X(\mathbf{u}) - X(\mathbf{u})'], \tag{5.160}$$

$$\sum_{X} |X\rangle\langle X| = 1, \quad \sum_{X} \;\Rightarrow\; \int \mathrm{d}X \equiv \int \delta X(\mathbf{u}), \tag{5.161}$$

implying functional integration over all possible functions $X(\mathbf{u})$. Similarly

$$\langle P | P' \rangle = \delta_{PP'} \equiv \delta[P - P'] = \prod_{\mathbf{u}} \delta[P(\mathbf{u}) - P(\mathbf{u})'], \tag{5.162}$$

$$\sum_{P} |P\rangle\langle P| = 1, \quad \sum_{P} \;\Rightarrow\; \int \mathrm{d}P \equiv \int \delta P(\mathbf{u}). \tag{5.163}$$

We can generalise (5.39) and (5.40) to yield the representative $\langle X' | P' \rangle$ of the eigenket $|P'\rangle$ of the operator P in the 'field coordinate representation' as

$$\langle X' | P' \rangle = N \, \exp\left(\frac{i}{\hbar} (X', P') \right), \tag{5.164}$$

where

$$(X', P') = \int X(\mathbf{u})' P(\mathbf{u})' \, \mathrm{d}^3 \mathbf{u} \tag{5.165}$$

is the Schwartz space scalar product. The field coordinate and momentum representations are related by a generalised Fourier transformation (cf. (5.120)),

$$|X\rangle = \sum_{\chi \in \tilde{G}} \bar{\chi}_{P'}(X) |P'\rangle, \tag{5.166}$$

where

$$\sum_{\chi} \;\Rightarrow\; \int \mathrm{d}P' \tag{5.167}$$

implies functional integration. Functional integrations are usually done in quantum field theory using an expansion in an infinite set of discrete mode functions such as that described for the free electromagnetic field in §4.2.

Just as in the finite-dimensional case, we recognise that the functions in the Schwartz space form an Abelian group, K, under addition. The unitary operators U and V in (5.151) provide the regular representations of the group K and its dual K, respectively on the Hilbert space of states for the field; these groups, however, are not even locally compact since they describe 'translations' on an infinite-dimensional function space. An immediate illustration of this feature is the difficulty of normalisation. Formally we may write the 'group volume'

$$\Omega(K) = \int dX, \qquad (5.168)$$

in terms of which N in (5.164) should be

$$N = \frac{1}{\sqrt{\Omega(K)}}, \qquad (5.169)$$

as in the finite-dimensional case, but here $\Omega(K)$ is infinite. As a practical method of calculation the functional calculus requires the careful separation out of such infinite normalisation terms.

For fixed P and given X, the representative, (5.164),

$$\langle X|P \rangle \equiv \chi_P(X), \qquad (5.170)$$

is a character of K. We write its unitary representation on $\mathcal{H}$ as

$$U[X] = \exp\left(\frac{i}{\hbar}P[X]\right). \qquad (5.171)$$

As in (5.115), $U[X]$ acts on a basic coordinate ket by translation

$$U[X]|X'\rangle = |X' - X\rangle. \qquad (5.172)$$

The group projection operator may then be written as

$$P_{\chi_P} \equiv |P\rangle\langle P| = \int \bar{\chi}_P(X)U[X]\, dX, \qquad (5.173)$$

using (5.170) and (5.171). In terms of explicit functions this is

$$P_{\chi_P} = \int \delta X(\mathbf{u}) \exp\left[\frac{i}{\hbar}\int \left((P(\mathbf{u}) - P(\mathbf{u}))X(\mathbf{u})\right) d^3\mathbf{u}\right]. \qquad (5.174)$$

In quantum electrodynamics, the roles of the field operators X and P can be taken by the vector potential $\mathbf{a}$ and its conjugate π, for which translations such as (5.172) are important properties; for example, a gauge transformation is a translation of the vector potential by a longitudinal vector field (§2.3) and is of this type.

It is essential to recognise that in the infinite-dimensional case any two representations are in general unitarily inequivalent and hence a choice has to be made for the

Hilbert space $\mathcal{H}$. The discussion so far has been based on a formal analogy with the Schrödinger representation of quantum mechanics. Clearly, the representation ought to be chosen so that the Hamiltonian can be written as a well-defined operator in $\mathcal{H}$; this is easier said than done, however, and the infinities that have plagued quantum field theory since its inception can be understood as a reflection of the difficulty in finding a physically significant space. Our main concern is with low-energy phenomena for which a 'non-relativistic' treatment is expected to be appropriate. Consequently, everything we do really implies that a high-energy (or momentum) cut-off will be applied, and this will yield a useful regularised theory, provided the answers do not depend on the cut-off. However that may be, in practice an especially useful representation is afforded by the Fock space construction which we sketch briefly here; it will be discussed in detail for the quantised electromagnetic field in Chapter 7. We confine our attention to a field satisfying Bose–Einstein statistics.

A boson field operator $\psi(\mathbf{u})$ can be defined in terms of the basic variables, X and P, by setting

$$\psi(\mathbf{u}) = \frac{1}{\sqrt{2\hbar}} \left(X(\mathbf{u}) + iP(\mathbf{u}) \right), \tag{5.175}$$

which is such that its commutation relation with its complex conjugate ψ^* is

$$[\psi(\mathbf{u}), \psi(\mathbf{u}')^*] = \delta(\mathbf{u} - \mathbf{u}'). \tag{5.176}$$

The corresponding smeared operators are

$$\psi(f) = \int \psi(\mathbf{u}) f(\mathbf{u}) \, \mathrm{d}^3 \mathbf{u}, \tag{5.177}$$

and similarly for $\psi(\mathbf{u}')^*$. The Fock space can then be defined by the statement that there exists a vacuum vector, Φ_F, satisfying

$$\psi(f) \, \Phi_F = 0, \tag{5.178}$$

for all $f(\mathbf{u})$ in the Schwartz space. The operator $\psi(f)$ is called the field annihilation operator; its conjugate, $\psi(f)^*$ is the field creation operator. The Hilbert space is constructed by the action of polynomial combinations of $\psi(f)^*$ on the vacuum Φ_F. The number operator for a bounded volume Ω can be defined as

$$N = \int_\Omega \psi(\mathbf{u})^* \psi(\mathbf{u}) \, \mathrm{d}^3 \mathbf{u}. \tag{5.179}$$

The eigenvalues of this operator are the positive integers and zero,[14] and this leads to the particle aspect of the field. Physical states are those for which there is zero probability of finding an infinite number of particles in any finite volume Ω. In any such volume, the description is equivalent to the usual n-particle quantum mechanics (discussed in §5.2) for the n-particles in that volume. When the limit $\Omega \to \infty$ is taken, a new

---

[14] If the field satisfies Fermi–Dirac statistics, the commutation relation (5.176) must be replaced by an anti-commutator, and the eigenvalues of N reduce to 0 and 1; thereby the Pauli exclusion principle is automatically included in the formalism.

theory emerges; for example, the energy spectrum becomes entirely continuous and so genuinely irreversible processes can be discussed in this framework [44].

5.5 Quantum Mechanics and the Electromagnetic Field

Perhaps the most celebrated result of quantum mechanics is the uncertainty principle [15], according to which there is a fundamental limit to the accuracy with which the position and momentum of a particle can be known simultaneously:

$$\Delta p \, \Delta x \sim \hbar. \tag{5.180}$$

Formally, the uncertainty relation (5.180) is an immediate consequence of the fact that the observables x and p correspond to a pair of non-commuting operators. It is also instructive to consider an ideal experiment designed to measure these quantities, for example Heisenberg's γ-ray microscope. Although it is valid to use classical optics to describe the path of a convergent light beam to its focus in the microscope [45], the light beam cannot be a classical Maxwell wave; if it were, the momentum of the light could be made as small as desired so that the position of the particle could be obtained exactly without transferring momentum to it, and hence the restriction implied by (5.180) could be evaded by first measuring the momentum exactly and then using the microscope to determine the position. On the contrary, if the light beam is such that the position of the particle can be measured at the focus with an accuracy of Δx, for consistency with (5.180) the momentum of the light beam must have an uncertainty given by

$$\Delta g_x \sim \frac{\hbar}{\Delta x}. \tag{5.181}$$

The uncertainty relation (5.180) also plays a critical role in a discussion of the measurement of the field strengths **E** and **B**; although they are written as point functions, they cannot actually be measured at a specified space-time point $(\mathbf{x}, t)$. In practice, they can only be measured as average values for a small volume $\delta\Omega \sim \delta r^3$ over some time interval Δt. Such a measurement can be achieved experimentally by determining the effect of an electromagnetic field on suitable charged test bodies, for example the deflection of particle beams. The deflection can be computed classically using the Lorentz force law (3.45); however, the test particles are subject to the fundamental condition (5.180), whatever their charge and mass, and this limits the accuracy with which the deflection can be measured. This leads to an uncertainty relation for the field strengths in the same volume element,

$$\Delta E_i \, \Delta B_j \geq \frac{h}{\varepsilon_0 (\delta r)^4}, \tag{5.182}$$

with cyclic permutation of (i, j) for the other components [46]–[48]. Consequently, the classical field strengths $(\mathbf{E}, \mathbf{B})$ must be replaced by a pair of non-commuting operators (E, B) consistent with (5.182).

A general Hamiltonian description of a closed system of charged particles and electromagnetic radiation was presented in Chapter 3; according to that discussion, the full classical Hamiltonian, $H[\mathbf{g}]$, in an arbitrary gauge for the vector potential $\mathbf{a}(\mathbf{x})$ characterised by the Green's function $\mathbf{g}$, gives the correct classical equations of motion independently of the choice for $\mathbf{g}$. The formal quantisation of this scheme requires the reinterpretation of the Hamiltonian conjugate variables as operators on an Hilbert space, and the introduction of appropriate commutation relations in accordance with the discussion in this chapter.

This change has radical consequences; the celebrated Maxwell equations (Chapter 2) involve space and time derivatives of the fields $\mathbf{E}$ and $\mathbf{B}$. For simplicity let us fix the time; then the equal-time space derivatives (Curl, Div) involve the comparison of the values of the fields at a point $\mathbf{x}$ and a 'neighbouring' point $\mathbf{x} + \Delta\mathbf{x}$ as $\Delta\mathbf{x} \to \mathbf{0}$. For this to make sense, the fields must be 'smooth', that is, the fluctuations die out as $\Delta\mathbf{x}$ gets smaller. Of course, the success of Maxwell's theory for macroscopic matter is beyond dispute. But when the same relations are extended into the microscopic domain where quantisation is essential, the situation can be quite different. There are indeed 'quantum states' for which the limit $\Delta\mathbf{x} \to 0$ does not exist, because the fluctuations get larger as $\Delta\mathbf{x}$ gets smaller (notably for the 'vacuum state'), hence there are a different set of 'infinities' in QED from those already described for the classical Hamiltonian. These matters will be discussed in Part III.

Chapter 6 describes various quantum mechanical perturbation methods that seem useful for the combined system of charges and electromagnetic radiation. From the perspective of perturbation theory, the strength of the coupling between the charges and the field is expressed through the dimensionless fine-structure constant, α,

$$\alpha = \frac{e^2}{4\pi\varepsilon_0 \hbar c} \sim \frac{1}{137}. \tag{5.183}$$

Chapter 7 describes the quantisation of the electromagnetic field in the absence of sources (the free field). In Chapter 8, we describe the main results for the quantum theory of charged particles in the absence of radiation that interact through purely Coulombic forces. This is the so-called 'Coulomb Hamiltonian' which is the basis of modern atomic theory, and is the putative quantum mechanical Hamiltonian for any molecule conceived of as a collection of electrons and nuclei.

References

[1] Dirac, P. A. M. (1930), *The Principles of Quantum Mechanics*, 1st ed. (and later editions), Clarendon Press.

[2] Heisenberg, W. (1930), *The Physical Principles of the Quantum Theory*, translated by Carl Eckart and F. C. Hoyt, University of Chicago Press. Reprinted (1949) by Dover Publications Inc.

[3] Halpern, O. and Thirring, H. (1932), *The Elements of the New Quantum Mechanics*, Ch. II, Methuen & Co. Ltd.

[4] Bohr, N. (1913), Phil. Mag. Series VI **26**, 1.

[5] Birtwhistle, G. (1928), *The New Quantum Mechanics*, Cambridge University Press.

[6] Ruark, A. E. and Urey, H. C. (1930), *Atoms, Molecules and Quanta*, McGraw-Hill Book Company, Inc.

[7] de Broglie, L. (1923), Comptes Rendus **177**, 507, 548.

[8] Feynman, R. P. and Hibbs, A. R. (1965), *Quantum Mechanics and Path Integrals*, McGraw-Hill Book Company.

[9] Shankar, R. (1994), Rev. Mod. Phys. **66**, 129.

[10] Schrödinger, E. (1923), Z. Physik **12**, 13.

[11] London, F. (1927), Z. Physik **42**, 375.

[12] Raman, V. V. and Forman, P. (1969), Hist. Stud. Phys. Sci. **1**, 291.

[13] O'Raifeartaigh, L. (1997), *The Dawning of Gauge Theory*, Princeton University.

[14] Schrödinger, E. (1926), Ann. der Physik **79**, 361, 489, 734; ibid. **80**, 437.

[15] Heisenberg, W. (1927), Z. Physik **43**, 172.

[16] Dirac, P. A. M. (1927), Proc. Roy. Soc. (London) A**113**, 621.

[17] Jordan, P. (1927), Z. Physik **40**, 809.

[18] Weinberg, S. (2013), *Lectures on Quantum Mechanics*, §3.3, Cambridge University Press.

[19] von Neumann, J. (1931), Math. Annalen **104**, 570.

[20] Jauch, J. M. (1972), *On Bras and Kets* in *Aspects of Quantum Theory*, edited by A. Salam and E. P. Wigner, Cambridge University Press, Cambridge.

[21] Dirac, P. A. M. (1935), *The Principles of Quantum Mechanics*, 2nd ed., p. 74, Clarendon Press (and later editions).

[22] Roman, P. (1975), *Some Modern Mathematics for Physicists and Other Outsiders: v. 2*, Pergamon Press Ltd.

[23] Reed, M. and Simon, B. (1975), *Methods of Mathematical Physics I, Functional Analysis*, Academic Press.

[24] Schwinger, J. (1970), *Particles, Sources and Fields*, Addison-Wesley.

[25] Zhu, C. and Klauder, J. R. (1993), Amer. J. Phys. **61**, 605.

[26] Bargmann, V. (1954), Ann. Math. **59**, 1.

[27] Kaempffer, F. A. (1965), *Concepts in Quantum Mechanics*, Pure and Applied Physics **18**, Academic Press.

[28] Bargmann, V. and Wigner, E. P. (1948), Proc. Natl. Acad. Sci. USA **34**, 211.

[29] Weinberg, S. (1995), *The Quantum Theory of Fields*, vol. 1, *Foundations*, Cambridge University Press.

[30] Currie, D. G., Jordan, T. F. and Sudarshan, E. C. G. (1963), Rev. Mod. Phys. **35**, 350.

[31] Leutwyler, H. (1965), Nuovo Cimento **37**, 556.

[32] Hill, R. N. (1967), J. Math. Phys. **8**, 1756.

[33] Polyzou, W. N. (2002), J. Math. Phys. **43**, 6024.

[34] Polyzou, W. N. (2003), Phys. Rev. C**68**, 015202.

[35] Mackey, G. W. (1978), *Unitary Group Representations in Physics, Probability and Number Theory*, Benjamin/Cummings Publishing Company, Inc.

[36] Roman, P. (1965), *Advanced Quantum Theory*, Addison-Wesley.

[37] Weyl, H. (1932), *The Theory of Groups and Quantum Mechanics*, translated from the second revised German edition by H. P. Robertson and E. P. Dutton.

[38] Strocchi, F. (1985), *Elements of Quantum Mechanics of Infinite Systems*, World Scientific Publishing Co. Pte. Ltd.

[39] Dirac, P. A. M. (1929), Proc. Roy. Soc. (London) A**123**, 714.

[40] Pais, A. (1986), *Inward Bound. Of Matter and Forces in the Physical World*, pp. 409 ff., Clarendon Press.

[41] Pauling, L. and Wilson, E. B. (1935), *Introduction to Quantum Mechanics*, McGraw-Hill Book Co. Inc.

[42] Slater, J. C. (1929), Phys. Rev. **34**, 1293.

[43] Power, E. A. (1964), *Introductory Quantum Electrodynamics*, Longmans.

[44] Sewell, G. L. (1986), *Quantum Theory of Collective Phenomena*, Clarendon Press.

[45] von Weizsacker, C. F. (1931), Z. Physik **70**, 114.

[46] Bohr, N. and Rosenfeld, L. (1933), Dansk. Vid-Selsk. Mat-Fys. Medd. **12**, 8.

[47] Bohr, N. and Rosenfeld, L. (1950), Phys. Rev. **78**, 794.

[48] Heitler, W. (1954), *The Quantum Theory of Radiation*, 3rd ed., Clarendon Press, reprinted (1984) by Dover Publications, Inc.

6 Perturbation Methods

6.1 Introduction

Non-relativistic quantum mechanical theories about physical observables require the explicit or implicit solution of the Schrödinger equation for a specified physical system. Apart from a few special cases, exact solutions remain out of reach, and so approximation methods are central to applications of quantum mechanics. An important general technique is perturbation theory; it may be applied, to different ends, with time-independent and time-dependent problems. The basic idea is that, if the Hamiltonian for the problem of interest, H, can be split into two terms,

$$H = H_0 + V, \tag{6.1}$$

where the eigenfunctions of H_0 are known, it may be possible to approximate those of H by treating V as a 'small correction'. Often the perturbation operator V is proportional to a 'small' parameter, say λ, and one attempts an approximate solution of the full problem as a power series in λ. Alternatively, one may introduce λ as a formal parameter simply to classify the order of the terms, and put $\lambda = 1$ in the final answer.

Thus, for example, in the most straightforward situation in the time-independent case, an isolated eigenvalue $E_n(\lambda)$ of $H(\lambda)$, is related unambiguously to an isolated eigenvalue E_n^0 of H_0, and one may write

$$E_n(\lambda) = E_n^0 + \lambda E_n^1 + \lambda^2 E_n^2 + \ldots \tag{6.2}$$

as a convergent series within a non-trivial circle of convergence about $\lambda = 0$. In such a case the operator, $V(\lambda)$, is said to be an analytic perturbation. In other cases, an expansion such as (6.2) is at best asymptotic but may yet yield an accurate approximation to the true eigenvalue $E_n(\lambda)$; the Born–Oppenheimer expansion of molecular term values which will be discussed in Chapter 8 is an example. There are perturbation operators of physical interest that cause unperturbed eigenvalues to disappear into continuous spectra; this is not an unusual mathematical pathology. The case of most interest here, non-relativistic quantum electrodynamics, is an example. For any non-zero value of the charge parameter e, all eigenvalues of an unperturbed atomic/molecular system apart from the ground state become metastable (i.e. resonances) in the presence of electromagnetic radiation, and so one is confronted with a perturbation problem involving a continuous spectrum; this will be discussed in Chapter 11. If the perturbation is time-dependent, one speaks of transitions between the eigenstates of the unperturbed problem, and the quantity of interest is then the transition rate, or cross section, which again may be approximated as a power series in λ. It is also possible to treat scattering problems in the time-independent framework, provided the Hamiltonian H describes a *closed* system. The separation of the full Hamiltonian into an 'unperturbed' part and a 'perturbation' is a matter of choice, and different choices may significantly affect convergence of the expansion.

There is now an enormous literature devoted to perturbation theory, and in this chapter we sketch some of the different approaches which have seemed useful for thinking about atoms and molecules in electrodynamics. In electrodynamics the natural perturbation parameter is the dimensionless fine structure constant, $\alpha \approx 1/137$; however, the perturbation V involves the electromagnetic field variables, and in strong fields, for example those associated with certain types of laser radiation, in no sense may the perturbation be supposed to be small. There is also an additional feature that must be addressed due to the presence of the arbitrary Green's function $\mathbf{g}$ in the interaction potential, $V = V[\mathbf{g}]$, obtained from Eq. (3.254). In much of the literature some particular choice for $\mathbf{g}$ is made or implied, which is obviously a necessary step for computational studies, and then V is treated as a definite quantity as in textbook accounts of perturbation theory. If some other choice for $\mathbf{g}$ is made, there is no a priori guarantee that the same answer will be obtained. So it is important at the level of the basic theory to identify the condition(s) that must be met in order that independence from $\mathbf{g}$ can be assured. This will be discussed in Chapter 10.

6.2 The Rayleigh–Schrödinger Perturbation Theory

Lord Rayleigh devised a perturbation technique to study the modes of vibrating strings with small density variation, using the known results for a string of constant density [1]; later, as Schrödinger developed his wave mechanics he adapted Rayleigh's procedure to study analogous perturbation problems, and this became the first widely used version of perturbation theory in quantum chemistry for time-independent Hamiltonians (see, for example [2]). By way of introduction we restrict discussion to the simplest case of a time-independent perturbation of a non-degenerate eigenvalue in the discrete spectrum; there is a relatively straightforward modification to accommodate a degenerate initial state [3].

The Schrödinger equations for the unperturbed and perturbed systems in the notation of (6.1) are

$$H_0|\psi_n^0\rangle = E_n^0|\psi_n^0\rangle, \tag{6.3}$$

$$H|\psi_n\rangle = E_n|\psi_n\rangle. \tag{6.4}$$

The sets of unperturbed states $\{|\psi_n^0\rangle\}$ and perturbed states $\{|\psi_n\rangle\}$ are taken to be complete, orthonormal bases in the same Hilbert space and so are related by a unitary transformation. Thus we may write

$$|\psi_n\rangle = \mathsf{C}|\psi_n^0\rangle = \sum_r |\psi_r^0\rangle (\mathbf{C})_{rn}, \tag{6.5}$$

where $\mathbf{C}$ is the matrix representative of C in the unperturbed basis $\{|\psi_n^0\rangle\}$. Then we may introduce C into the Schrödinger equation for H using (6.5),

$$\mathsf{HC}|\psi_n^0\rangle = E_n \mathsf{C}|\psi_n^0\rangle, \tag{6.6}$$

so that

$$\mathsf{C}^+\mathsf{HC}|\psi_n^0\rangle = E_n \mathsf{C}^+\mathsf{C}|\psi_n^0\rangle,$$
$$= E_n |\psi_n^0\rangle, \tag{6.7}$$

since C is unitary. In terms of the diagonal eigenvalue matrices, $\mathbf{E}, \mathbf{E}^0$, and the unit matrix, $\mathbf{I}$, we have

$$(\mathbf{H})_{nm} = \langle \psi_n | \mathsf{H} | \psi_m \rangle = (\mathbf{EI})_{nm}, \tag{6.8}$$

$$(\mathbf{h})_{nm} = \langle \psi_n^0 | \mathsf{H} | \psi_m^0 \rangle = (\mathbf{E}^0 \mathbf{I} + \mathbf{V})_{nm}, \tag{6.9}$$

where the Hamiltonian matrices are related by

$$\mathbf{H} = \mathbf{C}^+ \mathbf{hC}. \tag{6.10}$$

$(\mathbf{V})_{nm}$ is the matrix representative of V in the basis of unperturbed states.

The fundamental assumption of the Rayleigh–Schrödinger (RS) perturbation theory is that the eigensolutions $\{E_n\}, \{|\psi_n\rangle\}$ can be constructed as series in powers of λ as in (6.2) with

$$|\psi_n\rangle = |\psi_n^0\rangle + \lambda |\psi_n^1\rangle + \lambda^2 |\psi_n^2\rangle + \dots . \tag{6.11}$$

The matrix $\mathbf{C}$ also has an expansion in powers of λ,

$$\mathbf{C} = \mathbf{C}_0 + \lambda \mathbf{C}_1 + \lambda^2 \mathbf{C}_2 + \dots . \tag{6.12}$$

The unitarity of C implies certain useful matrix relations between the $\{\mathbf{C}_i\}$ and their adjoints, for example to second order in V,

$$\mathbf{C}_1^+ = -\mathbf{C}_1, \quad \mathbf{C}_2^+ + \mathbf{C}_2 + \mathbf{C}_1^+ \mathbf{C}_1 = \mathbf{0}. \tag{6.13}$$

If we define

$$\varepsilon_{nr} = E_n^0 - E_r^0, \quad V_{nr} = \langle \psi_n^0 | \mathsf{V} | \psi_r^0 \rangle, \tag{6.14}$$

the terms in (6.2) are easily found through second order to be

$$E_n^1 = V_{nn}, \quad E_n^2 = \sum_{r \neq n} \frac{|V_{rn}|^2}{\varepsilon_{nr}}. \tag{6.15}$$

The unitary transformation matrices to the same order are

$$(\mathbf{C}_0)_{np} = \delta_{np}, \tag{6.16}$$

$$(\mathbf{C}_1)_{nn} = 0, \quad (\mathbf{C}_1)_{np} = \frac{V_{np}}{\varepsilon_{pn}}, \quad n \neq p, \tag{6.17}$$

$$(\mathbf{C}_2)_{nn} = -\frac{1}{2} \sum_{r \neq n} \frac{|V_{nr}|^2}{\varepsilon_{nr}^2} \tag{6.18}$$

$$(\mathbf{C}_2)_{np} = \frac{(V_{nn} - V_{pp})V_{np}}{\varepsilon_{np}^2} + \sum_{r \neq n,p} \frac{V_{nr}V_{rp}}{\varepsilon_{pn}\varepsilon_{pr}}, n \neq p. \tag{6.19}$$

The RS construction can be made for other operators that are functions, $f(H)$, of the Hamiltonian H; the matrix representative of such an operator in its eigenbasis is the diagonal matrix with elements

$$F_{nm} = \langle \psi_n | f(H) | \psi_m \rangle = f(E_n)\delta_{nm}, \tag{6.20}$$

and from (6.10) its matrix in the unperturbed basis defined by H_0 is given by

$$\mathbf{f} = \mathbf{CFC}^+. \tag{6.21}$$

This can be evaluated approximately using the perturbation series (6.2), (6.12) for each matrix factor and matrix multiplication.

6.3 The Resolvent

Convergence of these series expansions depends in part on the magnitudes of the perturbation matrix elements $\{V_{nr}\}$ relative to the energy level spacings $\{\varepsilon_{nr}^0\}$; this is not sufficient information since the perturbation expansions are infinite series. An evident difficulty with the RS perturbation theory is that in order to calculate a single perturbed eigenvalue, $E_n(\lambda)$, it is necessary to have available, in principle, the complete orthonormal basis of the unperturbed system. On the other hand intuitive arguments that can be validated by mathematical analysis suggest that there are situations where unperturbed states that are distant in energy from the one(s) of interest make 'small' contributions to the perturbation, so that accurate approximations can be developed by considering only states in a limited (finite) range of energies about the state(s) of interest.

Suppose there are k discrete unperturbed states in an energy range suggested by the problem; then we can define a projection operator on the subspace these states span,

$$P = \sum_{n=1}^{k} |\psi_n^0\rangle\langle\psi_n^0|, \tag{6.22}$$

and a complementary projector Q with the properties

$$P + Q = I, \tag{6.23}$$

$$P^2 = P, \quad Q^2 = Q, \quad PQ = QP = 0. \tag{6.24}$$

Using these projection operators the Hamiltonian H can be represented as a $2{\times}2$ matrix,

$$\begin{pmatrix} H^{PP} & H^{PQ} \\ H^{QP} & H^{QQ} \end{pmatrix}, \tag{6.25}$$

where $H^{PP} = PHP$, and similarly for the other terms. If we define

$$|\chi\rangle = P|\psi\rangle, \tag{6.26}$$

then $|\chi\rangle$ satisfies a modified, energy-dependent Schrödinger equation,

$$\left(H^{PP} + H^{PQ} \frac{1}{E I^{QQ} - H^{QQ}} H^{QP} \right) |\chi\rangle = E|\chi\rangle, \tag{6.27}$$

which formally has the same eigenvalues as the original Schrödinger equation. Here I^{QQ} is the identity operator for the subspace defined by Q.

Many different approximation schemes exist for both bound state and scattering problems based on equations like these in which the first term in (6.27) can be treated exactly, and approximations and bounds applied to the second term. In the theoretical chemistry literature such an approach is often referred to as the 'Löwdin partitioning technique'; in physics, especially with reference to scattering theory, it is associated with Feshbach. In the mathematics literature, very similar ideas in more abstract settings are associated with Grushin, Krein, Schur and others [4]–[8].

The reformulation of the Schrödinger equation just presented is a particular illustration of a more general theme. One of the most powerful tools for the study of the Hamiltonian is the resolvent; this operator is defined as (cf. §5.2.1)

$$R(E) = \frac{1}{H - E}, \tag{6.28}$$

with an analogous definition for the resolvent, R_0, of the unperturbed Hamiltonian. The spectrum of H lies on the real axis, $\Re$, and so if we regard E as a complex variable, z, the resolvent $R(z)$ is a well-defined, bounded operator over the whole complex plane other than the spectrum of H. This means that many of the powerful techniques of the analysis of functions can be carried over to the study of the Schrödinger equation [9].

It is usually convenient to separate the Hamiltonian as in (6.1) and work with the eigenstates of H_0 as a basis; there is then the useful identity

$$R = R_0 - R_0 V R, \tag{6.29}$$

which offers the obvious possibility of a perturbation series by iteration, taking R_0 as the initial value. For the particular case of $z = E \pm i\varepsilon$ where E is real and ε is infinitesimal, one usually calls the limits of the resolvents $R(E \pm i\varepsilon)$ as z approaches the real axis, Green's functions or 'propagators':

$$G^{\pm}(E) = - \lim_{\varepsilon \to \pm 0} R(E \pm i\varepsilon). \tag{6.30}$$

There is an analogous definition for the Green's functions ($G_0^{\pm}$) of the unperturbed Hamiltonian as limits of R_0. These Green's functions are central to the description of scattering experiments.

It is convenient to discuss separately the diagonal and off-diagonal matrix elements of the resolvent. Standard formal manipulations in resolvent theory [10] show that the diagonal elements may be written in the form

$$\langle \psi_i^0 | \mathsf{R}(E) | \psi_i^0 \rangle = \frac{1}{E_i^0 - E - \Sigma_i(E)}, \tag{6.31}$$

with

$$\Sigma_i(E) = \langle \psi_i^0 | \left\{ -\mathsf{V} + \mathsf{V}\mathsf{G}_0^+(E)\mathsf{V} - \dots \right\}' | \psi_i^0 \rangle, \tag{6.32}$$

where the prime means that if $\Sigma_i(E)$ is evaluated by inserting complete sets between the factors, the state ψ_i^0 must be omitted. The poles of the quantity on the LHS of (6.31) determine the energy eigenvalues of the perturbed system. It is easily seen that combining (6.31) with (6.32) reproduces both Wigner–Brillouin and Rayleigh–Schrödinger perturbation series. But there are a wide range of sophisticated approaches that lead to more powerful results than the traditional expansion in powers of V [10]–[12]. The expectation value (6.31) when combined with the mathematical technique of complex coordinate rotation [12], [13] is an important tool for the characterisation of the spectrum of a Hamiltonian; we shall meet it in Chapter 8, applied to the Coulomb Hamiltonian, and in Chapter 11, applied to the full Hamiltonian of non-relativistic quantum electrodynamics.

In the limiting case one has

$$\lim_{z \to E \pm i\varepsilon} \Sigma_i(z) = \Delta_i(E) \mp \tfrac{1}{2} i \Gamma_i(E), \tag{6.33}$$

and (6.31) leads to the condition

$$E - E_i^0 + \Delta_i(E) \mp \tfrac{1}{2} i \Gamma_i(E) = 0 \tag{6.34}$$

for the energy eigenvalues; thus Δ and Γ play the roles of level shift and level width operators respectively. Suppose now that the equation

$$E - E_i^0 - \Delta_i(E) = 0 \tag{6.35}$$

has just one root, $E = E_i$. Then there are three possibilities [9]:

1. $\Gamma_i(E_i) \neq 0$
   E_i lies on a cut of the diagonal matrix element of the resolvent, (6.31), and typically ψ_i^0 decays in the presence of the perturbation. It is called a *dissipative* state.
2. $\Gamma_i(E) \neq 0$, $E \neq E_i$
   This is the result of persistent interaction that leads to a new *dressed* state; these two cases are commonly realised when a discrete state is coupled to a quantised field.
3. $\Gamma_i(E_i) = 0$
   E_i is a simple pole of the resolvent, and there is no cut. The state ψ_i^0 is said to be *asymptotically stationary* as it is transformed into a perturbed discrete state.

Closely related to the resolvent are operators with an exponential dependence on the Hamiltonian

$$\mathsf{K}(\gamma) = e^{\gamma \mathsf{H}}, \tag{6.36}$$

where γ may be a complex number. There are two particular cases that are met in practice (and sometimes in combination):

1. γ is real: if we put $\gamma = -1/k_B T = -\beta$ where k_B is Boltzmann's constant, the operator K is proportional to the density matrix for the perturbed system in thermal equilibrium at temperature T, that is, the Gibbs state.

2. γ is pure imaginary: if we set $\gamma = i\tau/\hbar = i(t_0 - t)/\hbar$, we see that K is the unitary time evolution operator, or propagator, for the perturbed system that connects the state at t_0 to the state at time t.

The resolvent and exponential operators are formally related by standard integral transformations. For example, if H is time-independent, a formal solution of the time-dependent Schrödinger equation

$$i\hbar \frac{\partial \phi}{\partial t} = \mathsf{H}\phi \tag{6.37}$$

can be expressed in terms of the propagator K:

$$\phi(t) = e^{i\mathsf{H}(t_0 - t)/\hbar}\phi(t_0)$$
$$= \mathsf{K}(\mathsf{t_0}, \mathsf{t})\phi(t_0). \tag{6.38}$$

The propagator and the resolvent are related by a contour integral,

$$\mathsf{K}(t_0, t) = -\frac{1}{2\pi i} \oint_C \left(\frac{1}{\mathsf{H} - E} \right) e^{iE(t_0 - t)/\hbar} \, dE, \tag{6.39}$$

where the integration path C in the complex E-plane is a closed loop surrounding the real E-axis in an anticlockwise sense [10].

The probability amplitude that the state ϕ_0 has survived at time t is then, by the usual rules,

$$c_0(t) = \langle \phi_0 | \phi(t) \rangle \equiv \langle \phi_0 | \mathsf{K}(t, t_0) | \phi_0 \rangle, \tag{6.40}$$

which may be evaluated in terms of the matrix element $\langle \phi_0 | \mathsf{R}(E) | \phi_0 \rangle$ of the resolvent operator using (6.39). The evaluation of this integral by the method of residues yields the probability amplitude as a sum determined by the residues $\{R_{mk}\}$ at the poles $\{E_{mk}\}$ of $\langle \phi_0 | \mathsf{R}(E) | \phi_0 \rangle$ plus contributions, say $A(t)$, from any branch cuts. One thus has

$$c_0(t) = \sum_{E_{mk}} R_{mk} \exp(-iE_{mk}t) + A(t). \tag{6.41}$$

If there is a single pole with $\Im E_{mk} \leq 0$, the survival probability,

$$P(t)_S = |c_0(t)|^2 \rightarrow 0 \text{ as } t \rightarrow \infty, \tag{6.42}$$

that is, the state ϕ_0 shows exponential decay. In practical situations more complicated behaviour can be expected, of course; nevertheless such an approach based on complex analysis is a powerful tool for studying time evolution. An important example will be described in §10.2.4.

The exact Green's function for H, (6.30), satisfies the formal operator equation,

$$\mathsf{G}^+(E) = \mathsf{G}_0^+(E) + \mathsf{G}_0^+(E)\mathsf{V}\mathsf{G}^+(E). \tag{6.43}$$

Using $G^+(E)$, we can define a *transition operator* in closed form [10]:

$$T(E) = V + VG^+(E)V. \tag{6.44}$$

The off-diagonal matrix elements of T in the basis of unperturbed states are directly related to scattering cross-sections. T can also be given a formal operator representation in terms of the Green's function for the non-interacting system, $G_0^+(E)$,

$$T = V + V\, G_0^+(E)\, T, \tag{6.45}$$

the iteration of which leads to the familiar perturbation series

$$T = V + VG_0^+(E)V + VG_0^+(E)VG_0^+(E)V + \dots . \tag{6.46}$$

Analogous relations can be derived for the Green's functions in the time-dependent formalism; indeed, using (6.30), one easily establishes that the Fourier transform of the Green's function $G^+(E)$,

$$G^+(t) = \frac{1}{2\pi\hbar} \int_{-\infty}^{+\infty} G^+(E) e^{-(i/\hbar)Et}\, dE, \tag{6.47}$$

is a Green's function for the time-dependent Schrödinger equation in the sense that

$$\left(\frac{\hbar}{i}\frac{\partial}{\partial t} + H\right) G^+(t) = -\delta(t). \tag{6.48}$$

6.4 Time-dependent Perturbation Theory

The original formulation of time-dependent perturbation theory is due to Dirac [14], who adapted the classical perturbation methods of Euler and Lagrange in celestial mechanics to the quantum mechanical problem. Given a time-independent Hamiltonian, H_0, with eigensolutions $\{\psi_n^0, E_n^0\}$, consider the time-dependent Schrödinger equation,

$$i\hbar\frac{\partial \Phi}{\partial t} = H_0\Phi. \tag{6.49}$$

Its general solution may be expressed as an expansion over the complete set of the eigenfunctions of H_0, that is, as a wave packet,

$$\Phi(t) = \sum_n b_n e^{-iE_n^0 t/\hbar} \psi_n^0 \equiv \sum_n b_n \phi_n^0(t), \tag{6.50}$$

where the $\{b_n\}$ are constants.

Suppose that we have a perturbation problem described by the time-dependent Hamiltonian,

$$H = H_0 + V(t). \tag{6.51}$$

Unlike the time-independent case we now regard the eigensolutions of H_0 to be fixed, and interpret the modified time-dependent Schrödinger equation,

$$i\hbar \frac{\partial \Psi(t)}{\partial t} = H\Psi(t) = \left(H_0 + V(t)\right)\Psi(t), \tag{6.52}$$

in terms of transitions between the unperturbed states.

Dirac proposed to solve this equation approximately by writing its general solution in a form analogous to (6.50) but with the $\{b_n\}$ allowed to be time-dependent:

$$\Psi(t) = \sum_n b_n(t)\phi_n^0(t). \tag{6.53}$$

By substitution of this expansion into (6.52) we derive

$$i\hbar \sum_n \frac{db_n(t)}{dt}\phi_n^0 = \sum_n b_n(t)V(t)\phi_n^0, \tag{6.54}$$

or in terms of matrix elements in the unperturbed basis

$$i\hbar \frac{db_n(t)}{dt} = \sum_m e^{i\varepsilon_{nm}t/\hbar}V_{nm}(t)b_m(t), \tag{6.55}$$

where

$$V_{nm}(t) = \langle \psi_n^0 | V(t) | \psi_m^0 \rangle, \quad \varepsilon_{nm} = E_n^0 - E_m^0 \tag{6.56}$$

is (6.14) for the time-dependent case. Equation (6.55) is an infinite system of coupled first-order differential equations which must be solved subject to some initial condition specified at time $t_0 < t$. The development coefficients $\{b_n(t)\}$ are the projections of the unperturbed states on the perturbed wave packet at time t,

$$b_n(t) = \langle \phi_n^0(t) | \Psi(t) \rangle. \tag{6.57}$$

The simplest situation is where the perturbation is 'switched on' at some time t_0 when the unperturbed system is known to be in one of its eigenstates,

$$\Psi(t_0) = \psi_v^0, \quad b_n(t_0) = \delta_{nv}. \tag{6.58}$$

Then, to first order in the perturbation, we have at a later time t

$$b_n^1(t) = \delta_{nv} - \frac{i}{\hbar} \int_{t_0}^t e^{i\varepsilon_{nv}t'/\hbar}V_{nv}(t')\,dt', \tag{6.59}$$

and the first-order perturbed wave function is

$$\Psi(t)^1 = e^{-iE_v^0 t/\hbar}\psi_v^0 - \frac{i}{\hbar}\sum_n e^{-iE_n^0 t/\hbar}\int_{t_0}^t e^{i\varepsilon_{nv}t'/\hbar}V_{nv}(t')\psi_n^0\,dt'. \tag{6.60}$$

As an example, we specialise to the case where the perturbation is 'weak' and oscillates at a single frequency $\omega/2\pi$. A familiar calculation using the first-order wave function (6.60) then leads to *Fermi's golden rule* for the rate of transition to the state n,

$$\tau_{v\to n}^{-1} = \frac{2\pi}{\hbar}|V_{vn}|^2\delta(E_v - E_n \pm \hbar\omega), \tag{6.61}$$

where the $\pm$ signs denote absorption and emission of a quantum $\hbar\omega$ respectively [15].

Whatever initial condition is specified, it is evident that the probability that the system will be found in the eigenstate ψ_n^0 at time t is, by the usual rules of quantum mechanics, simply

$$P_{v \to n}(t) = |b_n(t)|^2, \quad n \neq v. \tag{6.62}$$

As before, the probability that the initial wave packet formed at time t_0 has survived at time t is

$$P(t)_S = |\langle \Psi(t_0)|\Psi(t)\rangle|^2 = \sum_n b_n^*(t_0)b_n(t) \tag{6.63}$$

in terms of the development coefficients. A perturbed wave function such as (6.60) may also be used to find expectation values of system operators at time t; we give an example in §9.5. Higher-order terms in the perturbation expansion may be obtained formally by the method of successive approximations but, as with the RS perturbation expansion, the resulting expressions rapidly become unwieldy.

Other approaches to the solution of the system (6.55) utilise integral transformations that convert the differential equations into algebraic equations; both Fourier transform and Laplace transform techniques have been proposed [16]–[19]. So, for example, a non-perturbative formal solution can be obtained by introducing a Fourier integral representation of the development coefficients, with [17]

$$b(t)_n = -\frac{1}{2\pi i} \int_{-\infty}^{+\infty} A(E)_n e^{i(E_n - E)t/\hbar} \, \mathrm{d}E. \tag{6.64}$$

The analytic properties of the coefficients $\{A(E)_n\}$ can be specified so as to incorporate the initial conditions; for example, if $A(E)_n$ is taken as a meromorphic function of E with its poles lying in only the lower half of the complex E-plane, this ensures that the initial condition $b(t)_n = 0$ for $t < 0$ is satisfied. The coefficients $\{A(E)_n\}$ are matrix elements of the resolvent for the full Hamiltonian H, closely related to (6.31).

It is important to recognise that for a quantum mechanical system with purely discrete spectrum, the superposition (6.53) leads to a quantum mechanical analogue of the Poincaré recurrence time in classical mechanics [20], [21]. The squared norm of the difference of the state vectors at times t and 0 is

$$||\Psi(t) - \Psi(0)||^2 = 2\sum_n^\infty |b_n(t)|^2(1 - \cos(\omega_n^0 t)), \tag{6.65}$$

where $E_n^0 = \hbar\omega_n^0$. The 'quantum recurrence theorem' shows that there exist intervals for t for which the RHS of (6.65) can be made arbitrarily small, so that on such intervals the state vector $\Psi(t)$ approximates the initial state $\Psi(0)$ arbitrarily closely (and infinitely often) [22]. Such a (model) system cannot show dissipation. However, if the discrete state is coupled to a continuous spectrum, the generic behaviour is that the discrete state is absorbed into the continuum.[1] This will be discussed in §10.2.4.

---

[1] The coupling of an atom or molecule to the quantised electromagnetic field is a familiar example; as we discuss in Chapter 11, all discrete states of the atomic system apart from the ground state become resonances, because of the non-zero spontaneous emission amplitude.

6.5 The Dyson Series and the S-matrix

In a scattering experiment a beam of particles comes into interaction with some target system, and as a result of the interaction, particles are scattered out of the incident beam. The scattered particles are detected at positions remote from the target where they are no longer influenced by it. Only a brief summary of scattering theory will be given here, as there are now many excellent general treatments in the literature, [15], [23], [24]. Its application to spectroscopy will be described in Chapter 10. The stationary state picture of a scattering experiment is that a state $|\Psi_i^+\rangle$ (the in-state) of the Hamiltonian H for a closed system is prepared at $t = -\infty$, and the experiment consists of measuring what this state, say $|\Psi_f^-\rangle$ (the out-state), looks like at $t = +\infty$. These states satisfy the Schrödinger equation for H with the same energy eigenvalue,

$$H\,|\Psi_i^\pm\rangle = E_i|\Psi_i^\pm\rangle. \tag{6.66}$$

The probability amplitude for the transition $i \to f$ is

$$S_{fi} = \langle\Psi_f^-|\Psi_i^+\rangle, \tag{6.67}$$

by the usual rules of quantum theory. This matrix of complex numbers is called the S-matrix; conservation of probability requires that it be unitary:

$$\mathbf{S}\mathbf{S}^+ = \mathbf{S}^+\mathbf{S} = \mathbf{I}. \tag{6.68}$$

If there were no interactions, the in- and out-states would be the same and then S_{fi} would be just δ_{fi}, so the rate of reaction is proportional to [23]

$$|S_{fi} - \delta_{fi}|^2. \tag{6.69}$$

We use the splitting of H into a free Hamiltonian H_0 and the perturbation V responsible for the interactions to introduce a set of reference states,

$$H_0\,|\psi_i^0\rangle = E_i\,|\psi_i^0\rangle. \tag{6.70}$$

The eigenvalues in (6.66) and (6.70) for the states $|\Psi_i^\pm\rangle$ and $|\psi_i^0\rangle$ are taken to be the same, so level shifts and bound states must be dealt with in the definition of H_0. The Lippmann–Schwinger equations show that the S-matrix may be expressed as

$$S_{fi} = \delta_{fi} - 2\pi i\delta(E_f - E_i)T_{fi}, \tag{6.71}$$

where the transition matrix T_{fi} is a matrix element of the operator defined by (6.45) and is related to the perturbation operator V by

$$T_{fi} \equiv \langle\psi_f^0|T|\psi_i^0\rangle = \langle\Psi_f^+|V|\psi_i^0\rangle. \tag{6.72}$$

Here $|\Psi_f^+\rangle$ is an outgoing scattering solution of the full problem with the same energy as the reference state $|\psi_i^0\rangle$. Thus physical processes are determined by matrix elements of the transition matrix T_{fi} evaluated on the energy shell; it is convenient to introduce an explicit notation for this matrix by setting

$$\mathrm{T}_{fi} = -2\pi i\,\delta(E_i - E_f)\,T_{fi}. \tag{6.73}$$

With this notation the S-matrix may be written as

$$\mathbf{S} = \mathbf{I} + \mathbf{T}, \tag{6.74}$$

and its unitarity, (6.68), requires that

$$\mathbf{T}^+\mathbf{T} = -(\mathbf{T} + \mathbf{T}^+) = \mathbf{T}\mathbf{T}^+. \tag{6.75}$$

In terms of the matrix elements (6.73) this is

$$\delta(E_i - E_f)(T_{fi} - T_{fi}^+) = -2\pi i \sum_n T_{fn}^+ T_{ni} \delta(E_f - E_n)\delta(E_n - E_i), \tag{6.76}$$

$$= -2\pi i \, \delta(E_i - E_f) \sum_n T_{fn}^+ T_{ni} \delta(E_f - E_n). \tag{6.77}$$

On the energy shell, $E_i = E_f$, and we can cancel a delta function factor to leave

$$T_{fi} - T_{fi}^+ = 2\pi i \sum_n T_{fn}^+ T_{ni} \delta(E_f - E_n). \tag{6.78}$$

Since $T_{fi}^+ = T_{if}^*$, this is also

$$T_{fi} - T_{if}^* = -2\pi i \sum_n T_{nf}^* T_{ni} \delta(E_f - E_n), \tag{6.79}$$

which is the unitarity relation for the transition matrix. For $i = f$ this reduces to the generalised optical theorem,

$$\Im T_{ii} = -\pi \sum_n |T_{ni}|^2 \delta(E_i - E_n). \tag{6.80}$$

The RHS determines the total transition rate out of the state Φ_i (and hence the total scattering cross section), while the LHS determines the elastic scattering amplitude.

Two remarks should be noted at this point. Firstly, the presentation just given deliberately introduces a 'reference system' in anticipation of a perturbation theory approach based on identification of a subsystem, described by H_0, which we suppose we have already solved. For the general theory it is important that the S-matrix can be defined as a functional of the full Hamiltonian, H, without the use of a reference Hamiltonian. Secondly, it is a $T = 0$ theory; at finite temperatures one cannot define strictly the asymptotic states, and for systems in thermal equilibrium the natural quantities to work on are the partition function and certain correlation functions (Green's functions) which can be related to observable properties (see §6.7).

The scattering theory can equally well be formulated within a time-dependent framework. At time t_0 a state $|\Psi_i\rangle$ of the full Hamiltonian H is prepared; this state then evolves in time according to the time-dependent Schrödinger equation,

$$i\hbar \frac{\partial |\Psi\rangle}{\partial t} = H|\Psi\rangle, \tag{6.81}$$

and the experiment consists of measuring what this state looks like at some later time t, say $|\Psi_f\rangle$; the probability amplitude for the transition $i \to f$ is

$$a_{fi} = \langle \Psi_f | \Psi_i \rangle. \tag{6.82}$$

As a preliminary to discussing the solution of (6.81) we first transform to a new representation in which operators carry the time dependence induced by the free Hamiltonian H_0,

$$i\hbar \frac{d\Lambda_I}{dt} = [\Lambda_I, H_0] + i\hbar \frac{\partial \Lambda_I}{\partial t}. \tag{6.83}$$

and the state in the new representation[2] satisfies a modified evolution equation driven solely by V in the new representation,

$$i\hbar \frac{\partial |\Phi\rangle_I}{\partial t} = V(t) |\Phi\rangle_I, \tag{6.84}$$

where explicitly

$$V(t) = e^{iH_0 t/\hbar} V e^{-iH_0 t/\hbar}. \tag{6.85}$$

For operators H_0 and V that do not commute, one cannot write $e^{-i(H_0 + V(t))t/\hbar}$ as a simple product of exponentials to isolate the dependence on $V(t)$; however, the relation

$$e^{-i(H_0 + V(t))t/\hbar} = e^{-iH_0 t/\hbar} U(t) \tag{6.86}$$

is an identity, provided $U(t)$ satisfies the operator differential equation [25],

$$\frac{\partial U(t)}{\partial t} = -\frac{i}{\hbar} V(t) U(t). \tag{6.87}$$

Equation (6.87) may be transformed into an equivalent integral equation that incorporates the initial condition $|\Phi(t_0)\rangle = |\Psi(t_0)\rangle$; we write

$$U(t, t_0) = 1 - \frac{i}{\hbar} \int_{t_0}^{t} V(\tau) \, U(\tau, t_0) \, d\tau, \tag{6.88}$$

which defines the unitary time development operator.

The *Dyson series* is a formal perturbation series solution for (6.88) obtained by iteration,

$$U(t, t_0) = 1 + \sum_{n=1}^{\infty} \frac{(-i/\hbar)^n}{n!} \int_{t_0}^{t} dt_1 \int_{t_0}^{t} dt_2 \ldots \int_{t_0}^{t} dt_n \, \mathcal{T}\left(V(t_1) V(t_2) \ldots V(t_n)\right). \tag{6.89}$$

The chronological operator $\mathcal{T}$ has been introduced here because the product of an operator taken at two different times generally depends on the order of the factors [10], [24]. The S-matrix defined by (6.67) can be obtained from a matrix element of the operator $U(+\infty, -\infty)$ in the basis of unperturbed states,

$$S_{fi} = \langle \psi_f^0 | U(+\infty, -\infty) | \psi_i^0 \rangle. \tag{6.90}$$

It is possible to reformulate the scattering theory so as to make explicit the division between real processes that satisfy energy conservation, and virtual processes described by the off-energy-shell matrix elements [26]. Instead of writing the T-matrix in terms of the solution of the integral equation (6.71), the problem is divided into two parts; it

[2] The new representation is now usually referred to as the *interaction* representation; formerly it was often known as the 'Dirac' representation, as it is no more than a more formal development of the time-dependent perturbation theory invented by Dirac, §6.4. This usage is uncommon now.

is convenient to use the notational convention that subscripts i, f, n will be used to designate states of the same energy $E_i = E_f = E_n = E$, and subscripts α, β denote states for which $E_\alpha, E_\beta \neq E$. The transition matrix on the energy shell, Eq. (6.73), is related to an auxiliary matrix $\mathbf{D}$ by an equation sometimes known as the Heitler integral equation,

$$T_{fi} = D_{fi} - i\pi \sum_n D_{fn} \delta(E - E_n) T_{ni}. \tag{6.91}$$

The Hermitian matrix $\mathbf{D}$ is known as the damping or reaction matrix and is only defined on the energy shell. It is easily seen from (6.91) that the S- and T- matrices on the energy-shell may be expressed in terms of the Heitler reaction matrix $\mathbf{D}$,

$$\mathbf{S} = \frac{\mathbf{I} - \frac{1}{2}\mathbf{D}}{\mathbf{I} + \frac{1}{2}\mathbf{D}}, \tag{6.92}$$

$$\mathbf{T} = \frac{\mathbf{D}}{\mathbf{I} + \frac{1}{2}\mathbf{D}}. \tag{6.93}$$

$\mathbf{T}$ and $\mathbf{D}$ commute and so can be brought to diagonal form by the same similarity transformation. If the eigenvalues of $\mathbf{D}$ are $\{d_I\}$, the matrix $\mathbf{T}$ has eigenvalues

$$t_I = \frac{d_I}{1 + i\pi d_I}, \tag{6.94}$$

which shows that an iteration solution of (6.94) is convergent for

$$(d_I^{\max})^2 < \frac{1}{\pi^2}. \tag{6.95}$$

Since $\mathbf{D}$ is Hermitian, its eigenvalues are real and the condition $d_I = d_I^*$ with (6.94) thus gives an especially simple form for the relation (6.80) that the (complex) eigenvalues of the matrix $\mathbf{T}$ must satisfy

$$\frac{1}{\pi}\Im t_I = -t_I\, t_I^*. \tag{6.96}$$

The equation that determines the damping matrix $\mathbf{D}$, however, involves general off-energy-shell matrix elements,

$$D_{\beta i} = V_{\beta i} + \left(\mathsf{V}\, \mathsf{G}_0^{\mathcal{P}}(E_i)\mathsf{D}\right)_{\beta i}, \tag{6.97}$$

with $\mathsf{G}_0^{\mathcal{P}}$ the standing-wave Green's function for H_0 defined as a principal value (ε is an infinitesimal):

$$\mathsf{G}_0^{\mathcal{P}}(E_i) = \mathcal{P}\left(\frac{1}{E_i - \mathsf{H}_0}\right) \equiv \frac{1}{2}\left(\frac{1}{E_i - \mathsf{H}_0 + i\varepsilon} + \frac{1}{E_i - \mathsf{H}_0 - i\varepsilon}\right). \tag{6.98}$$

An important property of $\mathbf{D}$ follows from iterating (6.97) and inserting complete sets of intermediate states between the operator factors. No states degenerate with the initial state $|\Phi_i\rangle$ can occur in the perturbation expansion because the principal value is always taken in the summations. Consequently, the damping matrix is built up purely from virtual processes, and the S-matrix remains unitary even when resonance is important.

The relationship of the Dyson series to the traditional RS perturbation theory is readily established. Consider the time evolution operator for the Hamiltonian H with

$(t_0 - t) = \tau$; we denote its matrix representatives in the perturbed and unperturbed bases as $\mathbf{K}(\tau)$ and $\mathbf{k}(\tau)$ respectively. We put

$$(\mathbf{k}_0(\tau))_{nm} = \langle \psi_n^0 | e^{i\mathsf{H}_0 \tau/\hbar} | \psi_m^0 \rangle, \tag{6.99}$$

and have directly

$$\mathbf{K}(\tau) = \mathbf{k}_0(\tau)\mathbf{U}(\tau), \tag{6.100}$$

where

$$(\mathbf{U}(\tau))_{nm} = \langle \psi_n^0 | \mathsf{U}(\tau) | \psi_m^0 \rangle. \tag{6.101}$$

Using the Dyson series the propagator matrix can be expressed as a perturbation expansion,

$$\mathbf{k}(\tau) = \mathbf{k}_0(\tau) + \mathbf{k}_1(\tau) + \mathbf{k}_2(\tau) + \dots, \tag{6.102}$$

with

$$(\mathbf{k}_0(\tau))_{nm} = e^{iE_n^0 \tau/\hbar} \delta_{nm}$$
$$(\mathbf{k}_1(\tau))_{nm} = e^{iE_n^0 \tau/\hbar} V_{nm} x(i\tau/\hbar)_{nm}$$
$$(\mathbf{k}_2(\tau))_{nm} = e^{iE_n^0 \tau/\hbar} \sum_k V_{nk} V_{km} y(i\tau/\hbar)_{nkm}, \tag{6.103}$$

where the matrices $\mathbf{x}(\gamma)$ and $\mathbf{y}(\gamma)$ for an arbitrary complex argument γ are defined as

$$x(\gamma)_{nm} = \frac{1 - e^{-\gamma \varepsilon_{nm}}}{\varepsilon_{nm}}, \tag{6.104}$$

$$y(\gamma)_{nkm} = \frac{x(\gamma)_{nm} - x(\gamma)_{nk}}{\varepsilon_{km}}. \tag{6.105}$$

The labels $\{k, m, n\}$ range over the complete basis without restrictions on the summations over these states, and superficially $\mathbf{x}$ and $\mathbf{y}$ appear to have singular diagonal elements. However, when two energies coincide, the elements of these matrices are of the form $\frac{0}{0}$; such terms can be evaluated by writing $E_m^0 = E_n^0 + \delta$ and using l'Hôpital's Rule for the limit $\delta \to 0$. So, for example, setting $\gamma = i\tau/\hbar$, there results

$$x(i\tau/\hbar)_{nn} = -\frac{i\tau}{\hbar}, \tag{6.106}$$

$$y(i\tau/\hbar)_{nnm} = \frac{(e^{-i\varepsilon_{nm}\tau/\hbar} - 1)}{\varepsilon_{nm}^2} + \frac{i\tau}{\hbar} \frac{1}{\varepsilon_{nm}}, \quad m \neq n \tag{6.107}$$

$$y(i\tau/\hbar)_{nmm} = -\frac{(e^{-i\varepsilon_{nm}\tau/\hbar} - 1)}{\varepsilon_{nm}^2} - \frac{i\tau}{\hbar} \frac{e^{-i\varepsilon_{nm}\tau/\hbar}}{\varepsilon_{nm}}, \quad m \neq n \tag{6.108}$$

$$y(i\tau/\hbar)_{nnn} = \frac{1}{2}\left(\frac{i\tau}{\hbar}\right)^2. \tag{6.109}$$

The RS perturbation theory determines the propagator as the unitary transformation (6.21) with $\mathbf{C}$ and $\mathbf{K}(\tau)$ given by (6.16)–(6.20). The diagonal matrix $\mathbf{K}(\tau)$ can be expanded as a perturbation series,

$$\mathbf{K}(\tau) = \mathbf{K}_0(\tau) + \mathbf{K}_1(\tau) + \mathbf{K}_2(\tau) + \dots, \tag{6.110}$$

with

$$\mathbf{K}_0(\tau) = e^{iE_n^0 \tau/\hbar} \mathbf{I}, \tag{6.111}$$

$$\mathbf{K}_1(\tau) = e^{iE_n^0 \tau/\hbar} \frac{i\tau}{\hbar} V_{nn} \mathbf{I}, \tag{6.112}$$

$$\mathbf{K}_2(\tau) = e^{iE_n^0 \tau/\hbar} \left\{ \left(\frac{i\tau}{\hbar} \right)^2 \frac{V_{nn}^2}{2} + \frac{i\tau}{\hbar} \sum_k{}' \frac{|V_{nk}|^2}{\varepsilon_{nk}} \right\} \mathbf{I}. \tag{6.113}$$

Both methods of calculation lead to the same final answer, although in any given order the resulting expressions look different because of the difference in the summation limits specified. These differences arise simply from rearranging terms in infinite series. The equivalence of this expansion with the RS perturbation theory in the case of a time-independent perturbation V may be made explicit by separating out all diagonal contributions in the Dyson series expansion, so that the intermediate state summations in (6.103) are restricted in the same way as in the traditional RS equations [27].

6.6 Quantum Mechanics and 'External Fields'

Consider a dynamical system, A, that is sufficiently isolated from its environment that its energy is sensibly constant. According to classical dynamics (Chapter 3), the energy of A can then be expressed in terms of some Hamiltonian function $H(q_k^A, p_k^A)_A$ where the label $k = 1 \dots N$ runs over the degrees of freedom. The coordinates and momenta evolve in time according to Hamilton's equations, and the integration of these equations determines the coordinates and momenta in terms of $2N$ integration constants and the time. The characterisation of the state of a classical system requires specification of values for these $2N$ integration constants. Now let system A interact with another system B; then the new, constant, total energy of the pair is described by the total Hamiltonian,

$$H = H_A + H_B + V_{AB}, \tag{6.114}$$

where the interaction V_{AB} depends on the variables of both A and B, which are coupled through the canonical equations of motion. This is the standard description of a closed system, $(A + B)$.

A quantum mechanical account of the interaction of A and B follows the same lines, using the formal rules of the canonical quantisation programme; the classical Hamiltonian variables are replaced by operators acting on the vectors of a Hilbert space. In a Schrödinger representation the Hamiltonian operator H resulting from (6.114) is time-independent, and time enters as the parameter of a one-parameter unitary group of 'time displacements' for which the Hamiltonian is the generator. Two basic situations can be distinguished:

1. The state vector at time t, $|\Psi_t\rangle$, is separable and involves a trivial phase factor,

$$|\Psi_t\rangle = e^{-iEt/\hbar} |\psi_E\rangle, \tag{6.115}$$

where $|\psi_E\rangle$ is an eigenfunction of the Hamiltonian H with eigenvalue E.

2. A state vector $|\Psi_0\rangle$ at time $t = 0$ is given, and we ask for the state vector at some later time t. This is given formally by the action of the unitary time displacement operator,

$$|\Psi_t\rangle = e^{-iHt/\hbar}|\Psi_0\rangle. \tag{6.116}$$

If $|\Psi_0\rangle$ is an eigenstate, case 2 reduces trivially to case 1; if $|\Psi_0\rangle$ is not an eigenstate it must be time-dependent, and the question arises: where does $|\Psi_0\rangle$ come from? Apparently it always arises as the result of some previous time evolution, and this leads on to the question as to how a time-dependent state vector can arise.

Although the textbooks give discussions of the effects of time-dependent perturbations, these invariably concentrate on phenomenological models (the effect of a perturbation '$V(t)$') or classical external influences (e.g. coupling to a classical electromagnetic field). Fundamentally, quantum mechanics is a time-independent formalism; it does away with the characteristic requirement for classical dynamics of a statement about initial conditions at some instant in time t; this is quite distinct from the asymptotic limits required in the definition of the S-matrix. At the same time it does away with the classical notion of a state of a physical system at an instant by replacing it, radically, with a statement about the probability of some outcome in a measurement.

The typical time-dependent problem of the quantum theory textbooks is based on a Hamiltonian operator H_B for some system B and a time-dependent perturbation V which involves some operator(s) for system B (it must affect B), the time t, and possibly classical variables that describe the perturbing agent A (e.g. the amplitude and phase of a classical radiation field). This agent is not affected by the response of system B, and so its dynamics are fixed in the specification of the problem, that is, by the integration constants arising in the solution of its classical equations of motion. In summary, we require the solutions of an evolution equation of the form

$$i\hbar\frac{\partial|\Phi\rangle}{\partial t} = H|\Phi\rangle = (H_B + V(B, t, \boldsymbol{\alpha}))|\Phi\rangle, \tag{6.117}$$

where the $\boldsymbol{\alpha}$ are the classical variables of the perturbing agent, A, which follow the classical dynamics of A. This circumstance is essential if A is to be described as an external field.

This framework may be compared with a fully quantum mechanical account based on the time-dependent Schrödinger equation for the combined Hamiltonian operator derived from the quantisation of Eq. (6.114):

$$i\hbar\frac{\partial|\Psi\rangle}{\partial t} = (H_A + H_B + V_{AB})|\Psi\rangle. \tag{6.118}$$

It may be put in a form similar to (6.117) by changing to an interaction representation based on the free Hamiltonian, H_A; this means that the A-system operators in V_{AB} carry the time-dependence of the (free) A system. We define the unitary operator

$$U(t)_A = e^{iH_A t/\hbar}, \tag{6.119}$$

and then by writing

$$|\Psi\rangle = U(t)_A|\phi\rangle, \tag{6.120}$$

we get a modified evolution equation to compare with (6.117),

$$i\hbar\frac{\partial|\phi\rangle}{\partial t} = (H_B + \mathcal{V}(t)_{AB})|\phi\rangle, \tag{6.121}$$

where

$$\mathcal{V}(t)_{AB} = U(t)_A V_{AB} U(t)_A^{-1}, \tag{6.122}$$

since $U(t)_A$ commutes with the B-system operators. It may happen that $U(t)_A$ commutes with V_{AB} so that $\mathcal{V}(t)_{AB}$ is time independent. This is the situation we describe physically as perturbation by a 'static field'; a static field is always classical. Equation (6.117) is greatly simplified since t does not appear explicitly in V, and the $\boldsymbol{\alpha}$ are simply fixed parameters. Perturbation theory is then mainly concerned with the determination of modified energy levels which can be probed in spectroscopy; the energy levels also determine the partition function and hence thermodynamic quantities at temperature T. An example of such a calculation, perturbation of an atomic/molecular fluid by a static uniform electric field, will be discussed in §12.2.

If $\mathcal{V}(t)_{AB}$ does carry time dependence, this description implies that every possible state of the free A system is involved formally in the specification of the perturbation. To see this we need only look at its matrix form using the eigenstates of the free A system,

$$H_A|\eta_{A,m}\rangle = \varepsilon_{A,m}|\eta_{A,m}\rangle. \tag{6.123}$$

Then

$$\mathcal{V}(t)_{AB}|\eta_{A,m}\rangle = \sum_k |\eta_{A,k}\rangle(V(t))_{km}, \tag{6.124}$$

where

$$(V(t))_{km} = \langle\eta_{A,k}|V_{AB}|\eta_{A,m}\rangle e^{i\Delta E_{km}t/\hbar}, \Delta E_{km} = \varepsilon_{A,k} - \varepsilon_{A,m}, \tag{6.125}$$

and the sum is taken over the whole basis. It is usually a formidable task to show how the solutions of (6.121) can be approximated by those of (6.117), and for most realistic situations it remains unfinished business. In the important case of the treatment of electromagnetic radiation there is also the complicating issue of the gauge invariance of a classical 'limit' to be considered (see Chapter 9).

6.7 Response Theory and Physical Measurements

6.7.1 The Susceptibility Tensor

Many physical measurements can be described in terms of three basic ingredients. First of all, there is the system under study which is a macroscopic sample of matter (gas, liquid, solid) prepared in some known state with the aid of heat baths, containers, pumps etc. Secondly, there is some kind of phase space probe with which we 'tickle' the system; examples of probes include electromagnetic radiation, heat and sound pulses, electric and magnetic fields, and particle beams. Finally, there is some kind of detection system with which we measure the outcome of the experiment or, as we say, the response of the system to the stimulus provided by the probe.

In the following we assume that the system is in a stable state before the probe is introduced; often this will be an equilibrium state, but long-lived metastable states are also amenable to the description of the experiment we shall give. Suppose that the probe or input is described by some function f_{in}, and the response or output is similarly described by f_{out}. A substantial part of physics is concerned with experiments in which the response is linear; the disturbance of the system caused by the probe leads to an output having a small amplitude. In such cases if two inputs are applied to the system so that

$$f_{\text{in}} \to f_{\text{out}}; f'_{\text{in}} \to f'_{\text{out}}, \qquad (6.126)$$

there is a simple linear superposition principle for constructing the output when both inputs are applied together,

$$f_{\text{in}} + f'_{\text{in}} \to f_{\text{out}} + f'_{\text{out}}. \qquad (6.127)$$

This means that we can expect to write f_{out} as a linear transformation of f_{in}; for example, restricting ourselves to a simple time-dependent probe, we generally write the time-dependent output in the form

$$f(t)_{\text{out}} = \int_{-\infty}^{+\infty} I(t - t') f(t')_{\text{in}} \, dt', \qquad (6.128)$$

where $I(\tau)$ is known as the impulse-response function [28]; it must have the obvious property $I(\tau) = 0$ for $\tau < 0$ and be integrable so that the response is causal and bounded.

Suppose that the impulse-response function possesses a Fourier transform:

$$\chi(\omega) = \int_{-\infty}^{+\infty} I(z) e^{i\omega z} \, dz, \qquad (6.129)$$

$$I(z) = \frac{1}{2\pi} \int_{-\infty}^{+\infty} \chi(\omega) e^{-i\omega z} \, d\omega. \qquad (6.130)$$

Then, (6.128) transformed to the frequency domain is a simple multiplicative relationship,

$$\overline{f}(\omega)_{\text{out}} = \chi(\omega) \overline{f}(\omega)_{\text{in}}, \qquad (6.131)$$

between the input and output Fourier transforms,

$$\overline{f}(\omega) = \int_{-\infty}^{+\infty} f(t) e^{i\omega t} \, dt. \tag{6.132}$$

The quantity χ is called the susceptibility of the system. These few natural statements already imply some strong statements about the mathematical properties of the susceptibility [10]. Although introduced as a real function of the real frequency ω, it may be analytically continued in the upper half-plane ($\Im z > 0$) to a function $\chi(z)$ that has no singularities for finite z. With some restriction on its behaviour for $|z| \to \infty$, for example, that $\chi(\omega)$ be square integrable, it follows that the real and imaginary parts of $\chi(z)$ are mutually related by a form of dispersion relation called the Krönig–Kramers transform.

Relationships of the form (6.131) are very common in physics, and we cite a few familiar examples; if we apply an AC electric field $\mathbf{E}(\omega)$ to a piece of copper wire maintained at a constant temperature, we observe an electric current $\mathbf{j}(\omega)$ given by

$$\mathbf{j}(\omega) = \boldsymbol{\sigma}(\omega)\mathbf{E}(\omega), \tag{6.133}$$

provided that $|\mathbf{E}|$ is not too large (Ohm's law); the susceptibility here is the conductivity tensor $\boldsymbol{\sigma}(\omega)$. If a solid body is subjected to a stress, the response measured is the strain and the susceptibility is the compliance. Finally, if we apply an electric field $\mathbf{E}$ to an insulating medium, we observe a dielectric polarisation $\mathbf{P}$,

$$\mathbf{P}(\omega) = \varepsilon_0 \boldsymbol{\chi}(\omega)\mathbf{E}(\omega), \tag{6.134}$$

where $\boldsymbol{\chi}$ is the electric susceptibility tensor; here $\mathbf{E}$ can just as well be associated with electromagnetic radiation, and so the frequency variation of $\boldsymbol{\chi}$ is of direct concern to spectroscopy.

In Chapter 10 we will give a detailed quantum mechanical description of this idealised experiment based on a time-independent Hamiltonian for the combined system of target and probe, regarded as a closed microscopic system. The description of the experiment focuses on the properties of the probe beam received at the detector after the interaction. It is particularly appropriate for the scattering of 'particles'. There is a complementary approach based on time-dependent perturbation theory that aims to calculate the expectation values of products of operators belonging to the target caused by the interaction. These are the susceptibilities of the target system.

This idea underlies a widely used quantum mechanical formalism, response theory, which provides a common framework for the description of a wide range of physical measurements on macroscopic matter. The number of degrees of freedom, in the sense of mechanics, in a sample of ordinary matter is of the order of Avogadro's number, N_A. Even if we believe we know the dynamics of the system, inasmuch as we can write down a Hamiltonian (e.g. based on electromagnetic interactions), it is not possible to specify the initial conditions for $\sim N_A$ degrees of freedom that correspond to any particular preparation of the system. Hence there is an essential statistical element to the approach whereby we assign probabilities to the different quantum mechanical states available to the system in accordance with the mode of preparation. For the theory to be workable the specification of the probability distribution for the states must obviously be an immensely simpler task than the specification of initial conditions for all the dynamical variables. The key new ideas here are the notions of temperature and equilibrium.

Let H_0 be the Hamiltonian for the macroscopic system conceived of as a large finite number, N, of particles in a bounded volume Ω, in the absence of the probe; the solutions of the Schrödinger equation

$$H_0|\psi_n^0\rangle = E_n|\psi_n^0\rangle \tag{6.135}$$

provide a complete orthonormal basis,

$$\langle\psi_n^0|\psi_m^0\rangle = \delta_{nm}, \quad \sum_n|\psi_n^0\rangle\langle\psi_n^0| = I, \tag{6.136}$$

for some representation of the Hilbert space $\mathcal{H}$ on which H_0 acts. The density matrix is an operator $\rho : \mathcal{H} \to \mathcal{H}$ on this space, and by virtue of the spectral theorem [29] can be represented in the form

$$\rho = \sum_n p_n|\psi_n^0\rangle\langle\psi_n^0|, \tag{6.137}$$

with

$$p_n = \langle\psi_n^0|\rho|\psi_n^0\rangle; \sum_n p_n = 1, 0 \le p_n \le 1. \tag{6.138}$$

For a system in a container with fixed walls (constant volume) there is no degeneracy and the expansion is unique. The states $\{|\psi_n^0\rangle\}$ may be taken to be real, provided the macroscopic matter has no magnetic structure and is not in the presence of an external magnetic field.

If we consider a pure fluid phase maintained at constant temperature T in a large but finite volume Ω, and in thermal equilibrium, the ensemble we use to calculate properties of the fluid is specified by the density matrix

$$\rho_0 = \frac{1}{Z_0}e^{-H_0/k_BT}, Z_0 = \mathrm{Tr}\left(e^{-H_0/k_BT}\right) = e^{-F_0/k_BT}, \tag{6.139}$$

where $F_0 = F(T,\Omega)_0$ is the Helmholtz free energy, H_0 is the Hamiltonian for the fluid, and k_B is the Boltzmann constant. The eigenvalues of H_0 increase sufficiently quickly that the Trace operation is well defined for any $T > 0$. The operator ρ_0 is known as the canonical density matrix or Gibbs state; it is uniquely characterised by the stability condition that it should minimise the free energy functional [30]

$$\overline{F}(\rho) = \mathrm{Tr}\left(\rho H_0 + k_BT\rho\ln\rho\right), \tag{6.140}$$

that is,

$$\overline{F}(\rho) = F(T,\Omega)_0 \text{ for } \rho = \rho_0$$
$$> F(T,\Omega)_0 \text{ for } \rho \ne \rho_0. \tag{6.141}$$

For a homogeneous system in a stable state, ρ_0 must be invariant under arbitrary space and time translations; we ignore any surface effects.

Symmetries of the system are expressed formally by the requirement that ρ_0 commutes with the associated generators $\{\Delta_i\}$ of the symmetries,

$$[\rho_0, \Delta_i] = \rho_0 \Delta_i - \Delta_i \rho_0 = 0. \tag{6.142}$$

Thus taking Δ_i to be the Hamiltonian, H_0, the total linear momentum $\mathbf{P}$, the total angular momentum $\mathbf{J}$, the parity operator P, and the time reversal operator Θ in (6.142) lead to, respectively, time translation invariance (stable system), space translation invariance (homogeneous system), rotational invariance (isotropic system), space-inversion invariance (absence of chirality or permanent electric polarisation), and time-reversal invariance (absence of magnetic order or permanent magnetism). Of course, if a system displays a 'broken symmetry', a more restricted density matrix than the Gibbs state is required; thus for a permanent magnet we would need to specify the direction of the magnetisation $\mathbf{M}$ in addition to the temperature since the canonical average of $\mathbf{M}$ is zero by virtue of the rotational invariance of H_0. Similarly, the canonical average of an optical rotation angle is zero because of space-inversion invariance (parity conservation) of any Hamiltonian based on electrodynamics,[3] and hence we must specify which enantiomeric substance we are concerned with.

Now suppose that the macroscopic system is perturbed by a set of external fields which in general vary in space ($\mathbf{x}$) and time (t). The fields couple to the macroscopic system through a set of extensive variables $\{\Lambda(\mathbf{x})_i\}$ that are customarily called polarisations. Here i is an index labelling the fields and their corresponding polarisations; for notational convenience we do not indicate the scalar or vector nature of these quantities. Corresponding to each polarisation $\Lambda(\mathbf{x})_i$ there is a quantum mechanical observable $\Lambda(\mathbf{x})_i$, and we write the coupling perturbation as

$$V = -\sum_{i=1}^{N} \int \Lambda(\mathbf{x})_i U(\mathbf{x}, t)_i \, d^3 \mathbf{x}, \tag{6.143}$$

where the $\{U(\mathbf{x}, t)_i, i = 1 \ldots N\}$ are the perturbing fields.[4]

The system operators $\{\Lambda_i\}$ are written in a Schrödinger representation and are time-independent; they must be assumed to be self-adjoint operators. The time dependence is carried by the density matrix $\rho(t)$ which satisfies the Liouville equation of motion,

$$i\hbar \frac{\partial \rho(t)}{\partial t} = [H, \rho(t)], \tag{6.144}$$

where the Hamiltonian H is the sum of the system Hamiltonian H_0 and the perturbation V,

$$H = H_0 + V(t). \tag{6.145}$$

---

[3] In making this statement we are ignoring any consequences of the parity-violating weak neutral currents; they are responsible for optical activity in all matter that is many orders of magnitude smaller than the usual natural optical activity of compounds like alanine.

[4] In the usual presentations of response theory the perturbing fields are taken to be classical and unaffected by the interaction with the target. However, the QED Hamiltonian (Chapter 9) in the *interaction representation* is of the form used here and can be discussed in this way.

For the further development of the response theory it is convenient to transform the polarisation operators to an interaction representation with the unitary transformation generated by H_0:

$$\Lambda(\mathbf{x},t)_i = e^{iH_0t/\hbar}\Lambda(\mathbf{x})_i e^{-iH_0t/\hbar}. \tag{6.146}$$

Observable properties of the macroscopic system can be related to expectation values of combinations of the polarisations. The mean value of the polarisation $\Lambda(\mathbf{x},t)_i$ is calculated by the usual prescription of evaluating the trace of the product of the polarisation operator, Eq. (6.146), with the density matrix ρ obtained from (6.144):

$$\Lambda(\mathbf{x},t)_i = \langle\Lambda(\mathbf{x},t)_i\rangle = \mathrm{Tr}\left(\rho(t)\Lambda(\mathbf{x},t)_i\right). \tag{6.147}$$

For the unperturbed system we have $\rho(t) = \rho_0$ which is time-independent, and write

$$\Lambda(\mathbf{x},t)_i^0 = \langle\Lambda(\mathbf{x},t)_i\rangle_0 = \mathrm{Tr}\left(\rho_0\Lambda(\mathbf{x},t)_i\right). \tag{6.148}$$

In many cases $\Lambda(\mathbf{x},t)_i^0$ will be zero; for example, an isotropic fluid has zero average electrical polarisation $\mathbf{P}(\mathbf{x},t)$ because of rotational invariance, although this would not be true of anisotropic phases such as liquid crystals. In the following we shall assume that all the $\{\Lambda(\mathbf{x},t)_i^0\}$ vanish; the correlation functions constructed as averages of products of the polarisation operators are usually non-zero, however, and, as we shall see, these quantities evaluated for the unperturbed system are related to the response of the system when it is coupled to the corresponding fields $\{U_i\}$.

The two-point fluctuation functions for the unperturbed system are defined by

$$S_{ij}^>(\mathbf{x},\mathbf{x}';t,t') = \langle\Lambda(\mathbf{x},t)_i\Lambda(\mathbf{x}',t')_j\rangle_0, \tag{6.149}$$

$$S_{ij}^<(\mathbf{x},\mathbf{x}';t,t') = \langle\Lambda(\mathbf{x}',t')_j\Lambda(\mathbf{x},t)_i\rangle_0, \tag{6.150}$$

where, as before, $\langle(\ldots)\rangle_0 = \mathrm{Tr}(\rho_0(\ldots))$. In the particular case of thermal equilibrium for which ρ_0 is the Gibbs state, such correlation functions satisfy the KMS condition [31], [32]:

$$\langle A(t)B(t')\rangle_0 = \langle B(t')A(t+i\hbar k_B T)\rangle_0. \tag{6.151}$$

The assumed space and time translation invariance of ρ_0 means that these averages can only depend upon the differences $\mathbf{x} - \mathbf{x}', t - t'$, and hence it is convenient to introduce their Fourier transforms (cf. (6.129), (6.130)) with

$$S_{ij}^\bullet(\mathbf{r},\tau) = \frac{1}{(2\pi)^4}\int_{-\infty}^{+\infty}\int e^{-i(\omega\tau - \mathbf{k}\cdot\mathbf{r})}S_{ij}^\bullet(\mathbf{k},\omega)\,\mathrm{d}\omega\,\mathrm{d}^3\mathbf{k}, \tag{6.152}$$

where $\bullet$ is $>,<$ as appropriate, so that

$$S_{ij}^>(\mathbf{k},\omega) = S_{ji}^<(-\mathbf{k},-\omega). \tag{6.153}$$

If we introduce the Fourier transforms of the polarisation operators and complete sets of states defined by (6.136), we obtain

$$S_{ij}^>(\mathbf{k},\omega) = 2\pi\sum_{m,n}p_n\left(\Lambda(\mathbf{k})_i\right)_{nm}\left(\Lambda(-\mathbf{k})_j\right)_{mn}\delta(\omega + \omega_n - \omega_m), \tag{6.154}$$

$$S_{ij}^<(\mathbf{k},\omega) = 2\pi \sum_{m,n} p_n \big(\Lambda(-\mathbf{k})_j\big)_{nm} \big(\Lambda(\mathbf{k})_i\big)_{mn} \delta(\omega - \omega_n + \omega_m). \qquad (6.155)$$

For an unperturbed system that is isotropic the two-point functions depend on the magnitude of $\mathbf{k}$ but not its direction. Comparison with Fermi's golden rule (§6.4) shows that the two-point fluctuation functions determine the transition rates for the absorption ($S^>$) and emission ($S^<$) of a quantum of energy $\hbar\omega$ and momentum $\hbar\mathbf{k}$.

It proves useful to introduce sum and difference combinations of the two-point fluctuation functions; we define the equilibrium correlation function $S_{ij}(\mathbf{k},\omega)$ by

$$S_{ij}(\mathbf{k},\omega) = \frac{1}{2}\big(S_{ij}^>(\mathbf{k},\omega) + S_{ij}^<(\mathbf{k},\omega)\big), \qquad (6.156)$$

and interpret it as the Fourier transform of the average correlation between the polarisation operators $\Lambda(\mathbf{x},t)_i$ and $\Lambda(\mathbf{x}',t')_j$. On the other hand, and anticipating the discussion that follows, we claim that the difference combination is the imaginary part of a susceptibility,

$$S_{ij}^>(\mathbf{k},\omega) - S_{ij}^<(\mathbf{k},\omega) = 2\hbar\chi_{ij}(\mathbf{k},\omega)'', \qquad (6.157)$$

where we have introduced the notation

$$\chi_{ij}(\mathbf{k},\omega) = \chi_{ij}(\mathbf{k},\omega)' + i\chi_{ij}(\mathbf{k},\omega)'' \qquad (6.158)$$

for the real and imaginary parts of the complex-valued susceptibility.

χ' and χ'' are related to each other by the Krönig–Kramers transforms,

$$\chi(\mathbf{k},\omega)'_{ij} = \frac{1}{\pi}\mathcal{P}\int_{-\infty}^{+\infty} \frac{\chi(\mathbf{k},\omega')''_{ij}}{\omega'-\omega}\,\mathrm{d}\omega'$$

$$\chi(\mathbf{k},\omega)''_{ij} = -\frac{1}{\pi}\mathcal{P}\int_{-\infty}^{+\infty} \frac{\chi(\mathbf{k},\omega')'_{ij}}{\omega'-\omega}\,\mathrm{d}\omega'. \qquad (6.159)$$

Taking (6.159) together with the prescription,

$$\zeta(\omega) = \frac{1}{\omega \pm i0} = \frac{\mathcal{P}}{\omega} \mp i\pi\delta(\omega), \qquad (6.160)$$

shows that the complex-valued susceptibility χ can be written as an integral over purely its imaginary part, χ'',

$$\chi(\mathbf{k},\omega+i0)_{ij} = \frac{1}{\pi}\int_{-\infty}^{+\infty} \frac{\chi(\mathbf{k},\omega')''_{ij}}{\omega'-(\omega+i0)}\,\mathrm{d}\omega', \qquad (6.161)$$

so that its analytic continuation for $\Im z > 0$ is

$$\chi(\mathbf{k},z)_{ij} = \frac{1}{\pi}\int_{-\infty}^{+\infty} \frac{\chi(\mathbf{k},\omega')''_{ij}}{\omega'-z}\,\mathrm{d}\omega'. \qquad (6.162)$$

For large z this can be expressed as a sum over the power moments of the susceptibility

$$\chi(\mathbf{k},\omega) = -\sum_{n=0}^{\infty} \frac{1}{z^{n+1}}\frac{1}{\pi}\int_{-\infty}^{+\infty} \omega^n \chi(\mathbf{k},\omega)''_{ij}\,\mathrm{d}\omega$$

$$= \sum_{n=0}^{\infty} \frac{1}{z^{n+1}} X(\mathbf{k})_{ij}^n, \tag{6.163}$$

where

$$X(\mathbf{k})_{ij}^n = \frac{1}{\pi} \int_{-\infty}^{+\infty} \omega^n \chi(\mathbf{k}, \omega)_{ij}^{''} d\omega \tag{6.164}$$

is the nth moment. As we have seen, $\chi(\mathbf{k}, \omega)_{ij}^{''} = -\chi(\mathbf{k}, -\omega)_{ij}^{''}$, and so all even moments must be zero; such relations are called sum-rules. If the Hermitian operators Λ_i and Λ_j are either both real or both imaginary operators, the following symmetries hold:

$$\chi(\mathbf{k}, \omega)_{ij}^{'} = \chi(\mathbf{k}, -\omega)_{ij}^{'} = \chi(-\mathbf{k}, -\omega)_{ji}^{'}$$
$$\chi(\mathbf{k}, \omega)_{ij}^{''} = -\chi(\mathbf{k}, -\omega)_{ij}^{''} = -\chi(-\mathbf{k}, -\omega)_{ji}^{''}$$
$$\chi(\mathbf{k}, \omega)_{ij} = \chi(-\mathbf{k}, -\omega)_{ji}. \tag{6.165}$$

Since the response function $\chi(\mathbf{k}, \omega)_{ij}$ is a complex number, we can write it in terms of its amplitude and a phase factor, $\phi(\mathbf{k}, \omega)$:

$$\chi(\mathbf{k}, \omega) = |\chi(\mathbf{k}, \omega)|e^{i\phi(\mathbf{k}, \omega)}. \tag{6.166}$$

The susceptibility χ will be purely real if $\phi(\mathbf{k}, \omega) = s\pi$, where s is an integer. In such a case the phase of the response Λ_i will be either equal to the phase of the driving force or π radians (180°) out of phase with it. This is characteristic of a purely dispersive (or reactive) response; if we drive an harmonic oscillator with characteristic frequency ω_0 at a frequency ω much less or much greater than ω_0, the oscillator remains in phase with the driving force if $\omega << \omega_0$ and is 180° out of phase for $\omega >> \omega_0$. For this reason, the real part of χ, that is, $\chi(\mathbf{k}, \omega)^{'}$, is called the dispersive response function or the dispersive susceptibility. On the other hand, if we drive the oscillator at a frequency ω near the natural frequency ω_0, the phase of the response is $\approx \pi/2$, and the response function is imaginary; resonance of the oscillator is accompanied by dissipation of energy (absorption of energy by the oscillator), and the imaginary part of χ, that is $\chi(\mathbf{k}, \omega)^{''}$, corresponding to $\phi = 90°$, is called the absorptive or dissipative susceptibility [28].

An explicit calculation with ρ_0 chosen as the Gibbs state, (6.139), shows that the pair of canonical averages are related by a Boltzmann factor,

$$S_{ij}^{>}(\mathbf{k}, \omega) = e^{\hbar\omega/k_B T} S_{ij}^{<}(\mathbf{k}, \omega). \tag{6.167}$$

This is an expression of the microscopic principle of detailed balance,

$$\frac{\text{emission rate}}{\text{absorption rate}} = \frac{S_{ij}^{<}(\mathbf{k}, \omega)}{S_{ij}^{>}(\mathbf{k}, \omega)} = e^{-\hbar\omega/k_B T}. \tag{6.168}$$

Using (6.167) to eliminate one of the fluctuation functions, it is easy to see that, for a system in thermal equilibrium, S_{ij} and $\chi_{ij}^{''}$ are proportional to each other,

$$S(\mathbf{k}, \omega)_{ij} = \hbar\chi(\mathbf{k}, \omega)_{ij}^{''} \coth(\hbar\omega/2k_B T), \tag{6.169}$$

which is a statement of the *fluctuation-dissipation theorem*. S_{ij} is a measure of the fluctuations in the unperturbed stable system, whereas, as we shall see, $\chi(\mathbf{k}, \omega)_{ij}^{''}$ controls the linear response of the system when coupled to the external fields $\{U_i, U_j\}$. In order to justify these statements we now make a quantum mechanical calculation of the average polarisations induced by the probe fields.

6.7.2 Linear Response Theory

Let the density matrix at time $t = -\infty$ be ρ_0, and imagine that the perturbing fields are 'switched on' (and off) smoothly and slowly ('adiabatically'). In the interaction representation the density matrix satisfies a modified equation of motion [33],

$$i\hbar \frac{\partial \rho(t)}{\partial t} = [V(t), \rho(t)], \tag{6.170}$$

subject to the initial condition,

$$\rho(-\infty) = \rho_0, \tag{6.171}$$

with the time-dependent perturbation given by

$$V(t) = -\sum_{i=1}^{N} \int \Lambda(\mathbf{x}, t)_i U(\mathbf{x}, t)_i \, \mathrm{d}^3 \mathbf{x}. \tag{6.172}$$

A formal iterative solution of Eq. (6.170) starting from the initial value ρ_0 can be written down directly:

$$\rho(t) = \rho(-\infty) - \frac{i}{\hbar} \int_{-\infty}^{t} [V(t'), \rho(-\infty)] \, \mathrm{d}t'$$

$$+ \left(\frac{i}{\hbar} \right)^2 \int_{-\infty}^{t} \int_{-\infty}^{t'} [V(t'), [V(t''), \rho(-\infty)]] \, \mathrm{d}t'' \, \mathrm{d}t' + \dots \, . \tag{6.173}$$

The linear response result for the density matrix is given by the first line of (6.173); we assume that the reaction of the system on the fields $\{U(\mathbf{x}, t)_i\}$ is negligible, so we can regard them as externally determined 'classical-like' fields, and the perturbation as being 'weak'. The higher-order terms give rise to non-linear effects associated with the use of strong perturbing fields; we shall not consider such effects here explicitly beyond noting that they can be included in this general formalism. The perturbed density matrix (6.173) is causal in character, since its value at time t depends only on values of $V(t)$ at earlier times. It is convenient to introduce the Heaviside step function Θ which has the properties

$$\Theta(z) = \begin{cases} 1 & \text{if } z \geq 0 \\ 0 & \text{if } z < 0 \end{cases},$$

so that

$$\rho(t) = \rho_0 - \frac{i}{\hbar} \int_{-\infty}^{+\infty} \Theta(t - t')[V(t'), \rho_0] \, \mathrm{d}t' \tag{6.174}$$

in the linear response approximation.

The mean value of the polarisation Λ_i can now be calculated from (6.147) with $\rho(t)$ given by (6.174). Using the invariance of the trace under cyclic permutations of the factors, we get

$$\Lambda(\mathbf{x},t)_i = \sum_{j=1}^{N} \iint_{-\infty}^{+\infty} \chi(\mathbf{x},\ \mathbf{x}';t,t')_{ij} U(\mathbf{x}',t')_j \, \mathrm{d}t' \, \mathrm{d}^3\mathbf{x}', \tag{6.175}$$

assuming $\Lambda(\mathbf{x},t)_i^0$ vanishes. Here we have introduced the complex-valued space- and time-dependent response function χ_{ij} for the polarisations i and j:

$$\chi(\mathbf{x},\mathbf{x}';t,t')_{ij} = \frac{i}{\hbar}\Theta(t-t')\big\langle [\Lambda(\mathbf{x},t)_i, \Lambda(\mathbf{x}',t')_j] \big\rangle_0. \tag{6.176}$$

As usual, we have the property that, for a homogeneous and stable system, χ can only depend on $(\mathbf{x}-\mathbf{x}')$ and $(t-t')$, and so introducing Fourier transforms, we obtain

$$\Lambda(\mathbf{k},\omega)_i = \sum_{j=1}^{N} \chi(\mathbf{k},\omega)_{ij} U(\mathbf{k},\omega)_j, \tag{6.177}$$

and the relationship to Eqs. (6.128)–(6.134) is evident. Since χ_{ij} depends only on differences in the space and time coordinates, we may write

$$\chi(\mathbf{k},\omega)_{ij} = \frac{i}{\hbar}\int_{-\infty}^{+\infty}\int e^{i\omega\tau - i\,\mathbf{k}\cdot\mathbf{r}}\Theta(\tau)\big\langle [\Lambda(\mathbf{r},\tau)_i, \Lambda(0,0)_j] \big\rangle_0 \, \mathrm{d}\tau\,\mathrm{d}^3\mathbf{r}$$

$$= \frac{i}{\hbar}\int_0^{\infty} e^{i\omega\tau}\big(\langle\Lambda(\mathbf{k},\tau)_i\Lambda(-\mathbf{k},0)_j - \Lambda(-\mathbf{k},0)_j\Lambda(\mathbf{k},\tau)_i\rangle_0\big)\,\mathrm{d}\tau. \tag{6.178}$$

The observation that the exponential dependence on the Hamiltonian H_0 shown in (6.139) and (6.146) may be combined by use of the property that the trace operation is invariant under cyclic permutation of its factors leads to the susceptibility in the form

$$\chi(\mathbf{k},\omega)_{ij} = \frac{i}{\hbar}\int_0^{\infty} e^{i\omega\tau}\big\langle \Lambda(-\mathbf{k},0)_j(\Lambda(\mathbf{k},\tau+i\hbar\beta)_i - \Lambda(\mathbf{k},\tau)_i)\big\rangle_0 \,\mathrm{d}\tau, \tag{6.179}$$

where $\beta = 1/k_B T$. Considering the integral

$$e^{i\omega z}\Lambda(z)_i, \tag{6.180}$$

taken round the rectangle with vertices at $(0,0;L,0;L,\hbar\beta;0,\hbar\beta)$ in the complex plane, and assuming that the integrand is analytic on and within the contour, yields, with the aid of Cauchy's theorem in the limit $L \to \infty$,

$$\chi(\mathbf{k},\omega)_{ij} = \frac{i}{\hbar}(e^{\beta\hbar\omega}-1)\int_0^{\infty} e^{i\omega\tau}\big\langle \Lambda(-\mathbf{k},0)_j\Lambda(\mathbf{k},\tau)_i\big\rangle_0 \,\mathrm{d}\tau$$

$$- \frac{i}{\hbar}e^{\beta\hbar\omega}\int_0^{i\hbar\beta} e^{i\omega\tau}\big\langle \Lambda(-\mathbf{k},0)_j\Lambda(\mathbf{k},\tau)_i\big\rangle_0 \,\mathrm{d}\tau. \tag{6.181}$$

If we define

$$I(t) = \int_0^t e^{i\omega\tau} S^<(\mathbf{k},\tau)_{ij} \,\mathrm{d}\tau, \tag{6.182}$$

the susceptibility may be expressed in terms of integrals over just one of the fluctuation functions:

$$\chi(\mathbf{k}, \omega)_{ij} = \frac{i}{\hbar} \left\{ (e^{\beta \hbar \omega} - 1) I(\infty) - e^{\beta \hbar \omega} I(i\hbar \beta) \right\}. \tag{6.183}$$

Perturbation by a static external field constitutes an important special case ($\omega = 0$) of the response theory. For such fields the perturbation operator (6.143) has no explicit time dependence, and thermal equilibrium in the presence of the fields can be discussed in thermodynamic terms. For a perturbed time-independent Hamiltonian, H, Eq. (6.145), the Gibbs state is

$$\rho = \frac{1}{Z} e^{-H/k_B T}, Z = \mathrm{Tr}\left(e^{-H/k_B T}\right) = e^{-F/k_B T}, \tag{6.184}$$

where the perturbed Helmholtz free energy $F = F(T, \Omega, \{U_i\})$ depends on the external fields. In these terms the thermodynamic definition of a polarisation at temperature T is

$$\Lambda_i = -\frac{\partial F}{\partial U_i} = k_B T \frac{\partial \ln(Z)}{\partial U_i}, \tag{6.185}$$

with corresponding susceptibilities,

$$\chi_{ij} = -\frac{\partial^2 F}{\partial U_i \partial U_j} = k_B T \frac{\partial^2 \ln(Z)}{\partial U_i \partial U_j}, \tag{6.186}$$

and their calculation requires the evaluation of the partition function Z. It is also of interest to study the response to a dynamical probe in the presence of a static external field. The response formalism just reviewed goes through unchanged except for the replacement of ρ_0 by ρ. For a time-independent perturbation that is 'small' in an appropriate sense, we can relate the unperturbed density matrix, ρ_0, to ρ using time-independent perturbation theory, and this provides a route to the approximate evaluation of the susceptibility (6.183) for the permanently perturbed system.

6.8 Finite Temperature Perturbation Theory

The perturbation expansions of the time development operator $U(t)$ described in the previous sections are the results of the usual perturbation theory of quantum mechanics (quantum theory at zero temperature). The operator $U(t)$ is defined for real t on the infinite line, $[-\infty \leq t \leq +\infty]$. Now consider extending t into the complex plane to give an analytic continuation, $U(z)$ of the evolution operator; if we allow time to take values on the imaginary axis,

$$t \rightarrow \frac{i\hbar}{k_B T} \equiv i\hbar \beta,$$

where k_B is Boltzmann's constant, then

$$e^{-i\hbar H/t} \rightarrow e^{-\beta H}. \tag{6.187}$$

By reworking the development in §6.5 for imaginary time, one obtains a perturbation expansion that can be used to calculate the partition function, Z, (6.184). It should be noted that the evolution now takes place on the *finite* interval $[0, \beta]$.

This is the basis for the 'finite temperature perturbation theory' originally proposed by Bloch [34] and subsequently developed as a basis for quantum statistical mechanics by Matsubara [35] as an alternative to the real-time formalism outlined in §6.7. The perturbation expansion is valid, provided the perturbation energy per particle is small compared to the thermal energy $k_B T$. The familiar expansions of $U(t)$ are valid heuristically if the perturbation energy is small compared to the separations of the energy levels between which transitions occur. Since the temperature is unrelated to the energy level spectrum of a fluid, these two conditions are not the same [36].

6.8.1 The Partition Function, Z

As before, we consider the time-independent Hamiltonian H for a fluid in thermal equilibrium at temperature T. We suppose that we can write $H = H_0 + V$ and that the energies and wave functions for H_0 are known. Using the unperturbed states as a basis for the trace operation, the partition function Z for the fluid has the Dyson series to second order,

$$
\begin{aligned}
Z &= \sum_n e^{-\beta E_n^0} U(-\beta)_{nn} \\
&= \sum_n e^{-\beta E_n^0} \left\{ 1 - \beta V_{nn} + \frac{\beta^2}{2}(V_{nn})^2 + \sum_s' \frac{|V_{ns}|^2}{\varepsilon_{ns}^2}\left(e^{-\beta \varepsilon_{sn}} + \beta \varepsilon_{sn} - 1\right) \right\},
\end{aligned}
\tag{6.188}
$$

where we have used (6.101)–(6.105) with imaginary time.

In terms of the unperturbed partition function

$$
Z_0 = \sum_n e^{-\beta E_n^0},
\tag{6.189}
$$

Eq. (6.188) may be written as

$$
Z = r(\beta)Z_0,
\tag{6.190}
$$

where

$$
r(\beta) = \left(1 - \beta \overline{V_{nn}} + \frac{\beta^2}{2}\overline{(V_{nn})^2} + \beta^2 \overline{\alpha(\beta)_{nn}}\right).
\tag{6.191}
$$

The bar notation denotes an average with respect to the unperturbed Gibbs state

$$
\overline{A_{nn}} = \sum_n p_n^0 A_{nn}.
\tag{6.192}
$$

The last term in (6.191) is

$$
\overline{\alpha(\beta)_{nn}} = \sum_{n,s}' p_n^0 |V_{ns}|^2 f(x_{sn}),
\tag{6.193}
$$

with $f(x) = (e^{-x} + x - 1)/x^2$, and

$$x_{sn} = \beta \varepsilon_{sn} = \frac{E_s^0 - E_n^0}{k_B T}. \tag{6.194}$$

In practice, the limiting cases of large and small x are often important, and $f(x)$ simplifies to

$$f(x) \approx \begin{cases} \frac{1}{x} & \text{if } x \gg 1 \\ \frac{1}{2} & \text{if } x \ll 1 \end{cases}. \tag{6.195}$$

These limits correspond physically to

$$|E_s - E_n| \to \infty \text{ and/or } T \to 0 : x \gg 1, \tag{6.196}$$
$$|E_s - E_n| \to 0 \text{ and/or } T \to \infty : x \ll 1. \tag{6.197}$$

In van Vleck's quantum theory of electric and magnetic susceptibilities [37], the limiting cases (6.195) give rise to what he termed the 'high-frequency' and the 'low-frequency' contributions. The smoothly varying function $f(x)$ is shown in Figure 6.1.

If there are non-zero matrix elements $\{V_{ns}\}$ associated with the limiting cases in (6.195), we obtain

$$\beta^2 \overline{\alpha(\beta)_{nn}} \approx \begin{cases} \beta \overline{E_n^2} & \text{if } x \gg 1 \\ \beta^2/2 \left\{ \overline{(V^2)_{nn}} - \overline{(V_{nn})^2} \right\} & \text{if } x \ll 1 \end{cases}, \tag{6.198}$$

with

$$E_n^2 = -\sum_s' \frac{|V_{ns}|^2}{\varepsilon_{ns}}, \tag{6.199}$$

which will be recognised as the usual ($T = 0$) second-order energy shift of the unperturbed state $|\psi_n^0\rangle$ due to the perturbation V (see (6.15)).

6.8.2 Thermal Averages

Using the Dyson series and working with the unperturbed states $\{|\psi_n^0\rangle\}$ throughout, (6.147) may be effectively transformed to

$$\Lambda = \langle \Lambda \rangle = \left(r(\beta) Z_0 \right)^{-1} \text{Tr}(e^{-\beta H_0} U(\beta) \Lambda) \right). \tag{6.200}$$

Thus an observable Λ is obtained as an average of matrix elements $\langle \psi_n^0 | \ldots | \psi_n^0 \rangle$ over the unperturbed Gibbs state, renormalised by the temperature-dependent factor $r(\beta)$,

$$\Lambda = r(\beta)^{-1} \sum_n p_n^0 \theta^{(n)}, \tag{6.201}$$

where

$$\theta^{(n)} = \langle \psi_n^0 | U(\beta) \Lambda | \psi_n^0 \rangle, \tag{6.202}$$

and the normalisation factor $r(\beta)$ is given explicitly by (6.191) and is also an average over the unperturbed Gibbs state. The usual situation is that H_0 describes a Galilean

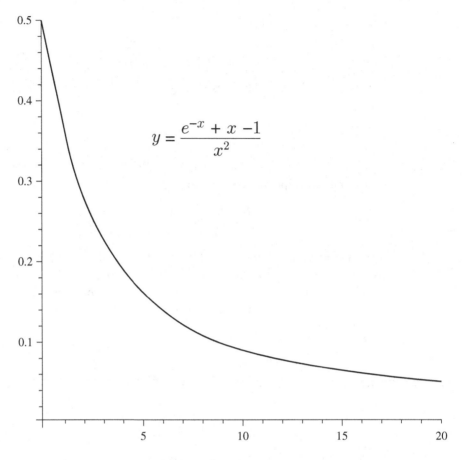

$$y = \frac{e^{-x} + x - 1}{x^2}$$

Figure 6.1 The function $f(x)$ for the calculation of the partition function, Z.

invariant system, so that the state $|\psi_n^0\rangle$ can be chosen as an eigenfunction of angular momentum, for example $|\psi_n^0\rangle = |\alpha_n JM\rangle$, where α_n represents all other appropriate quantum numbers; the intermediate states involved in $r(\beta)$ and Λ can also be written in this form. Since the unperturbed energies are independent of the M quantum number, the Wigner–Eckart theorem can be used to evaluate the sum over M required for the averages and express Λ in terms of reduced matrix elements, independent of M, and purely geometrical factors.

This quantum mechanical result for an observable has a structure similar to that given in the conventional chemical physics literature discussions, for example [38], [39], in which an average of Λ is constructed using the Boltzmann formula from classical statistical mechanics. There are, however, some differences which are important to elucidate. If the energy in state n in an external field is $u(\Lambda)_n$, the classical Boltzmann average of an observable Λ is

$$\langle \Lambda^{(n)} \rangle_{cl} = \frac{\int \Lambda(\xi) e^{-\beta u(\xi)_n} \, d\xi}{\int e^{-\beta u(\xi)_n} \, d\xi}, \qquad (6.203)$$

where Λ is now regarded as a function of a configurational variable ξ; explicitly, ξ represents the angles required to represent the orientation of a body[5] relative to a space-fixed coordinate system, for example through direction cosines [39] or a set of Euler angles Ω [40]. The exponentials can be expanded about $\beta = 0$, and the result put in the form

$$\langle \Lambda^{(n)} \rangle \approx \frac{1}{s(\beta)_{cl}} \left\{ \overline{\Lambda} - \beta \overline{\Lambda u_n} + \frac{\beta^2}{2} \overline{\Lambda u_n^2} + \dots \right\}, \tag{6.204}$$

where

$$\overline{A} = \int A(\xi) \, d\xi \tag{6.205}$$

denotes the configurational average, and

$$s(\beta)_{cl} = 1 + \beta \overline{u_n} - \frac{\beta^2}{2} \overline{(u_n)^2} + \frac{\beta^2}{2} (\overline{u_n})^2 \tag{6.206}$$

comes from the expansion of the denominator in (6.203).

The energy in the presence of the external field is given by RS perturbation theory, and from Eq. (6.15) we have, for the part depending on Λ,

$$u(\xi)_n = E_n - E_n^0$$

$$\sim V_{nn} + {\sum_r}' \frac{|V_{rn}|^2}{\varepsilon_{nr}}. \tag{6.207}$$

Noting that tracing an operator over the M quantum number using the Wigner–Eckart theorem corresponds to the classical integration over the Euler angles [37], [41], we see that Eq. (6.206) can be understood as originating from an incomplete version of the quantum trace formula (6.190). To see this, we evaluate Z using the perturbed states $\{|\psi_n\rangle\}$ for the trace; using the perturbed eigenvalues from the RS expansion, we have

$$Z = \text{Tr}(e^{-\beta H})$$

$$= \sum_n e^{-\beta(E_n^0 + E_n^1 + E_n^2 + \dots)}$$

$$\sim Z_0 \sum_n p_n^0 \{ 1 - \beta(E_n^1 + E_n^2) + \frac{\beta^2}{2}(E_n^1)^2 + \dots \}, \tag{6.208}$$

which is Eq. (6.206) if the sum over n is reduced to a sum over the M sublevels of the single state n. In this approximation, the difference between $r(\beta)$ and $s(\beta)_{cl}$ is precisely van Vleck's 'low-frequency' contribution, Eq. (6.198), which can be important when there are thermally accessible low-lying states that are coupled by non-zero off-diagonal matrix elements of V [36], [37], [42], such that a more complete evaluation of the sum over n must be made.

Similar comments apply to the numerator in (6.203) which is an approximate representation of a quantum trace formula:

$$\int \Lambda(\xi)^{(n)} e^{-\beta u(\xi)_n} \, d\xi \approx \text{Tr}(e^{-\beta H} \Lambda). \tag{6.209}$$

---

[5] The 'body' is an atom or molecule, so the passage to a description of the fluid in terms of a collection of non-interacting atoms/molecules is implied here.

The quantity $\Lambda(\xi)^{(n)}$ is identified by the following requirements:

- H must be taken in its diagonal representation in order to generate the perturbed energies in the exponent, and
- $\Lambda(\xi)^{(n)}$ should be a diagonal matrix element involving $|\psi_n^0\rangle$.

Using the perturbed states $\{|\psi_n\rangle\}$, we have

$$\text{Tr}(e^{-\beta H}\Lambda) = \sum_n \langle \psi_n|\Lambda|\psi_n\rangle e^{-\beta E_n}, \tag{6.210}$$

and with the aid of Eqs. (6.2), (6.5) and (6.207) this is

$$Z^{-1}\text{Tr}(e^{-\beta H}\Lambda) = \frac{1}{r(\beta)}\sum_n p_n^0 \langle \psi_n^0|C^+\Lambda C|\psi_n^0\rangle e^{-\beta u_n}. \tag{6.211}$$

Comparison of (6.209) and (6.211) shows that we must take

$$\Lambda(\xi)^{(n)} = \langle \psi_n^0|C^+\Lambda C|\psi_n^0\rangle, \tag{6.212}$$

with the unitary operator C constructed by perturbation theory as in §6.2, and the sum over n restricted to purely the M sublevels of a single state n.

References

[1] Lord Rayleigh (1877), *The Theory of Sound*, vol. 1, Macmillan and Co.

[2] Pauling, L. and Wilson, E. B. (1935), *Introduction to Quantum Mechanics*, McGraw-Hill Book Co. Inc.

[3] Davydov, A. S. (1976), *Quantum Mechanics*, 2nd ed., Pergamon Press.

[4] Löwdin, P.-O. (1966), in *Perturbation Theory and Its Applications in Quantum Mechanics*, ed. G. H. Wilcox, Wiley.

[5] Schur, J. (1917), J. reine u. angewandte Math., 205.

[6] Feshbach, H. (1958), Ann. Phys. (N.Y.), **5**, 357; ibid. (1962), **19**, 287.

[7] Grushin, V. V. (1970), Math. USSR Sbornik, **12**, 458.

[8] Krein, M. (1944), Comptes Rendue (Doklady) Acad. Sci. USSR (N.S.), **43**, 131; ibid. (1944), **44**, 175; ibid. (1946), **52**, 651.

[9] van Hove, L. (1955), Physica **21**, 901.

[10] Roman, P. (1965), *Advanced Quantum Theory*, Addison-Wesley.

[11] Kato, T. (1966), *Perturbation Theory for Linear Operators*, Springer-Verlag.

[12] Balslev, E. and Combes, J.-M. (1971), Comm. Math. Phys. **22**, 280.

[13] Reinhardt, W. P. (1982), Ann. Rev. Phys. Chem. **33**, 223.

[14] Dirac, P. A. M. (1930), *The Principles of Quantum Mechanics*, 1st ed. (and later editions), Oxford University Press.

[15] Weinberg, S. (2013), *Lectures on Quantum Mechanics*, §6.2, Cambridge University Press.

[16] Heitler, W. and Ma, S. T. (1949), Proc. Roy. Irish. Acad. **52**, 109.

[17] Heitler, W. (1954), *The Quantum Theory of Radiation*, 3rd ed., Clarendon Press, reprinted (1984) by Dover Publications, Inc.

[18] Blake, N. P. (1990), J. Chem. Phys. **93**, 6165.

[19] Lin, S. H. and Eyring, H. (1971), Proc. Natl. Acad. Sci. USA **68**, 76.

[20] Poincaré, H. (1890), Acta Math. **13**, 1.

[21] Carathéodory, C. (1919), Sitzungsber. Preuss. Akad. Wiss. **34**, 580.

[22] Bocchieri, P. and Loinger, A. (1957), Phys. Rev. **107**, 337.

[23] Weinberg, S. (1995), *The Quantum Theory of Fields*, vol. 1, *Foundations*, Cambridge University Press.

[24] Goldberger, M. L. and Watson, K. M. (1964), *Collision Theory*, J. Wiley.

[25] Kirzhnits, D. A. (1967), *Field Theoretical Methods in Many-Body Systems*, Pergamon Press.

[26] Goldberger, M. L. (1951), Phys. Rev. **84**, 929.

[27] Dumitru, A. G. and Woolley, R. G. (1998), Mol. Phys. **94**, 581.

[28] Pippard, A. B. (1985), *Response and Stability*, Cambridge University Press.

[29] Davies, E. B. (1976), *Quantum Theory of Open Systems*, Academic Press.

[30] Sewell, G. L. (1986), *Quantum Theory of Collective Phenomena*, Clarendon Press.

[31] Kubo, R. (1957), J. Phys. Soc. Japan **12**, 570.

[32] Martin, P. and Schwinger, J. (1959), Phys. Rev. **115**, 1342.

[33] Ben-Reuven, A. (1975), Adv. Chem. Phys. **33**, 235.

[34] Bloch, F. (1932), Z. Physik **74**, 295.

[35] Matsubara, T. (1955), Prog. Theor. Phys. **14**, 351.

[36] Landau, L. D. and Lifshitz, E. M. (1980), *Course of Theoretical Physics*, vol. 5, *Statistical Physics*, Part 1, translated by J. B. Sykes and M. J. Kearsley, Pergamon Press.

[37] van Vleck, J. H. (1932), *The Theory of Electric and Magnetic Susceptibilities*, §46, Clarendon Press.

[38] Mackrodt, W. C. (1971), Mol. Phys. **20**, 251.

[39] Barron, L. D. (1982), *Molecular Light Scattering and Optical Activity*, Cambridge University Press.

[40] Atkins, P. W. and Barron, L. D. (1968), Proc. Roy. Soc. (London) A**306**, 119.

[41] Berestetskii, V. B., Lifshitz, E. M. and Pitaevskii, L. P. (1971), *Course of Theoretical Physics*, vol. 4, *Relativistic Quantum Theory*, Part 1, translated by J. B. Sykes and J. S. Bell, Pergamon Press.

[42] Schiff, L. I. (1968), *Quantum Mechanics*, 3rd ed., McGraw-Hill.

7 Quantisation of the Electromagnetic Field

7.1 Introduction

We begin with the classical Hamiltonian for the free field in terms of the mode variables, Eq. (4.22), and in the spirit of canonical quantisation reinterpret it as an operator acting on a Hilbert space; thus

$$\mathsf{H}_{\text{rad}} = \tfrac{1}{2} \sum_{\mathbf{k},\lambda} \left(\mathsf{P}(\mathbf{k})_\lambda^2 + \omega^2 \mathsf{X}(\mathbf{k})_\lambda^2 \right), \tag{7.1}$$

where we sum over both polarisation directions, $\lambda = 1, 2$ and all wave vectors $\mathbf{k}$. The P.B.s of the mode variables are replaced by the commutation relation,

$$[\mathsf{X}(\mathbf{k})_\lambda, \mathsf{P}(\mathbf{k}')_{\lambda'}] = i\hbar \delta_{\lambda,\lambda'} \delta_{\mathbf{k},\mathbf{k}'}, \tag{7.2}$$

and all other equal-time commutators vanish. The Hamiltonian (7.1) is an infinite sum of independent harmonic oscillators, and to begin with, we consider just one term,

$$\mathsf{H}_{\mathbf{k}} = \tfrac{1}{2} \left(\mathsf{P}(\mathbf{k})^2 + \omega^2 \mathsf{X}(\mathbf{k})^2 \right), \quad \omega = kc, \tag{7.3}$$

with Schrödinger equation

$$\mathsf{H}_{\mathbf{k}} |\Psi_{n_{\mathbf{k}}}\rangle = E_{n_{\mathbf{k}}} |\Psi_{n_{\mathbf{k}}}\rangle. \tag{7.4}$$

The solution of (7.4) for the harmonic oscillator is well known and will simply be stated; the energies are

$$E_{n_{\mathbf{k}}} = (n_{\mathbf{k}} + \tfrac{1}{2})\hbar\omega, \quad n_{\mathbf{k}} = 0, 1, 2, \ldots, \tag{7.5}$$

and the self-adjoint coordinate operator $\mathsf{X}(\mathbf{k})$ has matrix elements

$$\langle \Psi_n | \mathsf{X}(\mathbf{k}) | \Psi_{n-1} \rangle = \sqrt{\frac{\hbar n}{2\omega}}, \quad \langle \Psi_n | \mathsf{X}(\mathbf{k}) | \Psi_{n'} \rangle = 0 \text{ if } n' \neq n \pm 1. \tag{7.6}$$

Since $\mathsf{P}(\mathbf{k}) = d\mathsf{X}(\mathbf{k})/dt$ for an oscillator, the matrix elements of $\mathsf{P}(\mathbf{k})$ differ from those of $\mathsf{X}(\mathbf{k})$ only by factors of $\pm i\omega$ (cf. Chapter 5). The classical field momentum is proportional to the energy, Eq. (4.35), and so the state labelled by the integer $n_{\mathbf{k}}$ has momentum eigenvalue

$$\mathbf{u}_{\mathbf{k}} = (n_{\mathbf{k}} + \tfrac{1}{2})\hbar\mathbf{k}. \tag{7.7}$$

The energy and momentum for the field are then obtained by summing (7.5) and (7.7) respectively over all the modes.

These formulae present a difficulty. The state of the field of lowest energy and momentum is obtained by putting all $n_k = 0$; this state is called the vacuum state of the electromagnetic field. In this state each mode still contributes an energy $\frac{1}{2}\hbar kc$ and momentum $\frac{1}{2}\hbar\mathbf{k}$ to the field; these are the zero-point energy and momentum of the field. Since the number of radiation oscillators for a given volume is infinite, this leads to an infinite energy and momentum for the vacuum state. Finite changes in the zero-point energy, however, are caused by changes in the cavity volume and by polarisable material bodies, as shown by, for example, the Casimir effect [1] and long-range intermolecular forces [2]–[4] which turn out to be complementary aspects of the same general quantum phenomenon. When the zero-point energy is not of interest, the difficulty is usually eliminated by the practical expedient of striking out the infinite contributions and writing

$$\mathcal{E} = \sum_{\mathbf{k},\lambda} n_{\mathbf{k},\lambda}\hbar\omega, \quad \mathbf{G} = \sum_{\mathbf{k},\lambda} n_{\mathbf{k},\lambda}\hbar\mathbf{k}. \tag{7.8}$$

It is customary, and very appealing, to introduce a new physical interpretation of these results. To begin with, suppose there is only one mode with wave vector $\mathbf{k}$ and frequency ω in the field which has the energy (7.8) identified by the integer n_k. We prefer to give a particle interpretation for (7.8) saying that the field consists of n_k quanta (particles); these quanta are called photons, and each has energy $E_{\mathrm{ph}} = \hbar\omega$. Increasing or decreasing n_k changes the energy in the mode, and we say that photons have been created or annihilated respectively. At the same time we can reinterpret the wave vector $\mathbf{k}$ of the mode in terms of the momentum of the photon, $\mathbf{p}_{\mathrm{ph}} = \hbar\mathbf{k}$.

The photon's energy and momentum behave under Lorentz transformations as the components of a 4-vector $k_\chi = (\mathbf{p}_{\mathrm{ph}}, iE_{\mathrm{ph}}/c)$ with the property

$$k_\chi k_\chi \equiv \hbar\mathbf{k} \cdot \hbar\mathbf{k} - \frac{E_{\mathrm{ph}}^2}{c^2} = 0. \tag{7.9}$$

This equation is characteristic of a particle with zero invariant mass that has speed c in all reference frames. Thus the result of quantisation is that a plane wave of light behaves as though it were a beam of n_k particles each with energy $\hbar\omega$ and momentum $\hbar\mathbf{k}$ where $|\mathbf{k}| = \omega/c$. The concept of the photon derives purely from the integer quantum numbers that appear in the quantisation formulae (7.5), (7.7) for the modes; hence photons are indistinguishable particles and their numbers are unlimited. Every mode of the field can be described in this way, and a state of the quantised field can be specified by giving the number of indistinguishable photons associated with each mode. This is exactly how a Bose–Einstein ensemble is specified in statistical mechanics, and so the photon is a boson particle.[1]

---

[1] This is also consistent with the evident requirement that in the classical limit ($\hbar \to 0$) the photons must disappear, leaving the classical Maxwell field theory. The classical limit of a collection of fermions is a system of classical particles [5].

7.2 Representation of States of the Field

So far nothing has been said about the representation of the field eigenstates $|\Psi_n\rangle$; let us initially choose a coordinate representation in which $X(\mathbf{k})$ is diagonal. The Schrödinger equation (7.4) for the mode may then be realised as an ordinary differential equation,

$$\left(-\hbar^2 \frac{d^2}{dX(\mathbf{k})^2} + \omega^2 X(\mathbf{k})^2\right) \Psi_n(X) = 2E_n \Psi_n(X), \tag{7.10}$$

and solved using the standard harmonic oscillator theory. The ground state, for example, is

$$\Psi_0(X) = N \exp\left(-\omega X(\mathbf{k})^2/2\hbar\right). \tag{7.11}$$

There is such a wave function for every $\mathbf{k}$, and so the vacuum state of the field is an infinite product of Gaussians (7.11); this can be rewritten as an exponential with an infinite sum in its exponent. In the continuum limit the sum becomes an integral over a continuous variable $\mathbf{k}$,

$$\Psi_0[X] = N \exp\left(-c/2\hbar \int k X(\mathbf{k})^2 d^3\mathbf{k}\right), \tag{7.12}$$

and so the vacuum state $|\Psi_0\rangle$ in this representation is a functional $\Psi_0[X]$ of the field variable X. The spectrum of the free electromagnetic field Hamiltonian thus consists of a simple eigenvalue at 0, corresponding to the vacuum state, Ψ_0, and absolutely continuous spectrum on the half-axis $[0, \infty)$.

Although the oscillator coordinate operator $X(\mathbf{k})$ is represented by the Hermitian matrix (7.6) and is therefore formally an observable, it is of no direct physical interest. It is much more rewarding for the development of the quantum theory to make a linear transformation from the mode operators $X(\mathbf{k}), P(\mathbf{k})$ to new variables by setting (cf. 4.24)

$$c_{\mathbf{k}} = \frac{1}{\sqrt{2\hbar\omega}}\left(\omega X(\mathbf{k}) + iP(\mathbf{k})\right), \tag{7.13}$$

$$c_{\mathbf{k}}^+ = \frac{1}{\sqrt{2\hbar\omega}}\left(\omega X(\mathbf{k}) - iP(\mathbf{k})\right), \tag{7.14}$$

or inversely,

$$X(\mathbf{k}) = \sqrt{\frac{\hbar}{2\omega}}\left(c_{\mathbf{k}} + c_{\mathbf{k}}^+\right), \tag{7.15}$$

$$P(\mathbf{k}) = -i\sqrt{\frac{\hbar\omega}{2}}\left(c_{\mathbf{k}} - c_{\mathbf{k}}^+\right). \tag{7.16}$$

Consider now the action of $c_{\mathbf{k}}$ and $c_{\mathbf{k}}^+$ on the vacuum state (7.11); we have

$$c_{\mathbf{k}} \Psi_0(X) = \frac{1}{\sqrt{2\hbar\omega}}\left(\omega X(\mathbf{k}) + \hbar\frac{d}{dX(\mathbf{k})}\right) \Psi_0(X) = 0, \tag{7.17}$$

whereas

$$c_{\mathbf{k}}^{+}\Psi_0(X) = \frac{1}{\sqrt{2\hbar\omega}}\left(\omega X(\mathbf{k}) - \hbar\frac{\mathrm{d}}{\mathrm{d}X(\mathbf{k})}\right)\Psi_0(X) \tag{7.18}$$

$$= \sqrt{\frac{2\omega}{\hbar}}X(\mathbf{k})\Psi_0(X) = \Psi_1(X). \tag{7.19}$$

More generally $c_{\mathbf{k}}$ and $c_{\mathbf{k}}^{+}$ act as ladder operators for the oscillator, stepping down and up the mode index, or photon number, $n_{\mathbf{k}}$; they are called, respectively, the annihilation and creation operators for the mode $\mathbf{k}$.

Using (7.2), (7.3), (7.13) and (7.14), we easily find

$$c_{\mathbf{k}}^{+}c_{\mathbf{k}} = \frac{1}{\hbar\omega}\left(H_{\mathbf{k}} - \frac{1}{2}\hbar\omega\right), \tag{7.20}$$

$$c_{\mathbf{k}}c_{\mathbf{k}}^{+} = \frac{1}{\hbar\omega}\left(H_{\mathbf{k}} + \frac{1}{2}\hbar\omega\right), \tag{7.21}$$

from which we infer

$$H_{\mathbf{k}} = \left(c_{\mathbf{k}}^{+}c_{\mathbf{k}} + \frac{1}{2}\right)\hbar\omega, \tag{7.22}$$

and

$$[c_{\mathbf{k}}, c_{\mathbf{k}}^{+}] = 1. \tag{7.23}$$

The matrix elements of the annihilation and creation operators are

$$\langle\Psi_{n-1}|c_{\mathbf{k}}|\Psi_n\rangle = \langle\Psi_n|c_{\mathbf{k}}^{+}|\Psi_{n-1}\rangle = \sqrt{n}, \tag{7.24}$$

and so they form a pair of adjoint operators and cannot be observables. The bilinear combination occurring in (7.22),

$$n_{\mathbf{k}} = c_{\mathbf{k}}^{+}c_{\mathbf{k}}, \tag{7.25}$$

is usually called the photon number operator; it is self-adjoint, and its eigenvalues are the integers in (7.5). This representation is so important in quantum field theory that we next describe its formal details.

7.2.1 The Fock Space Construction

According to general quantum mechanical principles, a particle is fully described by the specification of a complete set of commuting observables $\{w_i\}$. In order to describe the state of the quantised field we only need to specify the number of photons, n_i, in each state $|w_i\rangle$ since photons are non-interacting and indistinguishable particles. The photon is a boson particle, so the n_i are unrestricted integers, $(0, 1, 2, \ldots)$. The Fock space construction formalises this physical picture in terms of operators acting on a Hilbert space. We introduce a set of basis vectors $|n_1, n_2, \ldots\rangle$ that are normalised and orthogonal:

$$\langle n_{1'}, n_{2'}\ldots|n_1, n_2\ldots\rangle = \delta_{n_{1'}, n_1}\delta_{n_{2'}, n_2}\cdots. \tag{7.26}$$

The space generated by this set of vectors is a separable Hilbert space provided only that

$$\sum_i n_i < \infty. \tag{7.27}$$

We can define self-adjoint number operators n_i that are diagonal in the basis (7.26),

$$n_i|n_1,n_2,\ldots n_i,\ldots\rangle = n_i|n_1,n_2,\ldots n_i,\ldots\rangle. \tag{7.28}$$

They form a complete set of commuting operators since a basis vector is fully characterised by giving the eigenvalues of the number operators. We can also define annihilation and creation operators as linear operators that reduce by one and increase by one the number of photons in a given state

$$c_i|n_1,n_2,\ldots n_i,\ldots\rangle = \sqrt{n_i}|n_1,n_2,\ldots n_i-1,\ldots\rangle, \quad n_i > 0$$
$$= 0 \text{ for } n_i = 0, \tag{7.29}$$
$$c_i^+|n_1,n_2,\ldots n_i,\ldots\rangle = \sqrt{n_i+1}|n_1,n_2,\ldots n_i+1,\ldots\rangle. \tag{7.30}$$

Here c_i^+ is the adjoint of c_i. They have the following canonical commutation relations (CCR):

$$[c_i,c_j^+] = \delta_{ij}$$
$$[c_i,c_j] = [c_i^+,c_j^+] = 0. \tag{7.31}$$

The number operator n_i can be expressed in terms of the annihilation and creation operators,

$$n_i = c_i^+ c_i, \quad n_i+1 = c_i c_i^+, \tag{7.32}$$

since, according to (7.29), (7.30),

$$c_i^+ c_i|n_1,n_2,\ldots n_i,\ldots\rangle = n_i|n_1,n_2,\ldots n_i,\ldots\rangle, \tag{7.33}$$

in agreement with (7.28).

The Fock representation can be characterised by postulating the existence of a unique vacuum state $|\Psi_0\rangle$ which is such that

$$c_i|\Psi_0\rangle = 0, \quad \text{all } i. \tag{7.34}$$

Then given the operators defined by (7.29), (7.30), all other states of the system can be constructed by repeated action of the creation operator on the vacuum state,

$$|n_1,n_2,\ldots n_i,\ldots\rangle = \frac{1}{\sqrt{n_1!}}(c_1^+)^{n_1}\ldots\frac{1}{\sqrt{n_i!}}(c_i^+)^{n_i}\ldots|\Psi_0\rangle. \tag{7.35}$$

This can be checked by applying the number operator n_i to both sides of (7.35); use of the commutation relation,

$$[n_i,c_i^+] = c_i^+, \tag{7.36}$$

shows that the vector (7.35) satisfies (7.28). The factorial factors are required here to ensure that the boson symmetry of the photon is accounted for; the resulting many photon kets $|n_1,n_2,\ldots n_i,\ldots\rangle$ are orthonormal.

An extremely useful concept in calculations using the Fock basis is that of the normal ordered product. Every observable for the oscillator is an (operator-valued) function of X and P, and hence will be a function of sums of products of the annihilation and creation operators. Normal ordering is the process based on the use of the commutation relations (7.31) whereby all the creation operators are brought to stand to the left of all the annihilation operators. When this has been achieved, the evaluation of matrix elements is straightforward using (7.29), (7.30) and the orthogonality of the basis.

A photon in the state $|w_i\rangle$ has well-defined energy $\hbar\omega_i$, and so the field Hamiltonian must be

$$H_{rad} = \sum_i n_i \hbar \omega_i. \tag{7.37}$$

The equation of motion of the annihilation operator is then (cf. (5.43))

$$i\hbar \frac{dc_i}{dt} = [c_i, H_{rad}] = \hbar \omega_i c_i, \tag{7.38}$$

with solution

$$c(t)_i = e^{-i\omega_i t} c(0)_i. \tag{7.39}$$

This description is extremely convenient for physical processes in which the number of particles is not conserved; for example, if a particle is created in some process, it is only necessary to give the quantum numbers of the added particle, leaving the description of the original particles unchanged. The basis consisting of the vectors $|n_1, n_2, \ldots n_i, \ldots\rangle$ is called the Fock basis or the occupation number basis; it can be pictured as an inverted pyramid above the vacuum state $|\Psi_0\rangle$ [6], [7],

$$\frac{1}{\sqrt{3!}} (c_1^+)^3 |\Psi_0\rangle, \frac{1}{\sqrt{2!}} (c_1^+)^2 c_2^+ |\Psi_0\rangle \ldots, \tag{7.40}$$

$$\frac{1}{\sqrt{2!}} (c_1^+)^2 |\Psi_0\rangle, c_1^+ c_2^+ |\Psi_0\rangle \ldots, \tag{7.41}$$

$$c_1^+ |\Psi_0\rangle, c_2^+ |\Psi_0\rangle \ldots, \tag{7.42}$$

$$|\Psi_0\rangle. \tag{7.43}$$

Any such state is called a Fock state. The Hilbert space is

$$\mathcal{H} = \left\{ \sum_{n=1}^{\infty} c_n |n\rangle, \; \sum_{n=1}^{\infty} |c_n|^2 < \infty \right\}. \tag{7.44}$$

7.2.2 Coherent States

In this section we confine attention initially to a single mode of the field with definite polarisation; then we may drop the labels k and λ. We now wish to investigate further the properties of the adjoint pair of operators c and c^+ for an oscillator of frequency ω. As just discussed, the number operator for the mode is

$$n = c^+ c, \tag{7.45}$$

with eigenstates given by

$$n|n\rangle = n|n\rangle. \tag{7.46}$$

The *coherent states* of the oscillator are the eigenstates of the annihilation operator c,

$$c|\alpha\rangle = \alpha|\alpha\rangle. \tag{7.47}$$

Using (7.13), this is explicitly

$$\left(\frac{i}{\sqrt{2\hbar\omega}}P + \sqrt{\frac{\omega}{2\hbar}}X\right)|\alpha\rangle = \alpha|\alpha\rangle, \tag{7.48}$$

so that in the coordinate (X) representation for the oscillator the probability amplitude $\langle X|\alpha\rangle$ satisfies a first-order differential equation,

$$\left(\sqrt{\frac{\hbar}{2\omega}}\frac{\mathrm{d}}{\mathrm{d}X} + \sqrt{\frac{\omega}{2\hbar}}X\right)\langle X|\alpha\rangle = \alpha\langle X|\alpha\rangle. \tag{7.49}$$

Since c is not self-adjoint, the eigenvalues α will in general be complex numbers. The normalised solution is[2]

$$\langle X|\alpha\rangle = \left(\frac{\omega}{\pi\hbar}\right)^{\frac{1}{4}}\exp\left(-\left(\sqrt{\frac{\omega}{2\hbar}}X - \alpha_R\right)^2 + i\sqrt{\frac{2\omega}{\hbar}}\alpha_I X\right), \tag{7.50}$$

where we have separated α into real (α_R) and imaginary (α_I) parts.

Since $\langle\alpha|P|\alpha\rangle$ and $\langle\alpha|X|\alpha\rangle$ are both real, we can put

$$\alpha_R = \sqrt{\frac{\omega}{2\hbar}}\langle\alpha|X|\alpha\rangle \equiv \sqrt{\frac{\omega}{2\hbar}}u, \tag{7.51}$$

$$\alpha_I = \frac{1}{\sqrt{2\hbar\omega}}\langle\alpha|P|\alpha\rangle \equiv \sqrt{\frac{\hbar}{2\omega}}v. \tag{7.52}$$

Then the coherent state $\langle X|\alpha\rangle$ can be parameterised by u and v, and written as

$$\langle X|u,v\rangle = \left(\frac{\omega}{\pi\hbar}\right)^{\frac{1}{4}}\exp\left(-\frac{\omega}{2\hbar}(X-u)^2 + ivX\right). \tag{7.53}$$

The coordinate representative of the vacuum state, Ψ_0, for the oscillator in this notation is $\langle X|0,0\rangle$; the coherent state $|u,v\rangle$ can be written as a unitary transform of the vacuum state

$$|u,v\rangle = U_{u,v}|\Psi_0\rangle, \tag{7.54}$$

where the Weyl operator

$$U_{u,v} = \exp(ivX - iuP/\hbar) \tag{7.55}$$

is closely related to the Weyl commutation relation and the translation operators.[3]

---

[2] This shows that there is no eigenvalue problem in Hilbert space for the creation operator c^+; c^+ is the adjoint of c, and so the differential equation for c^+ corresponding to (7.48) must have $-i$ in the first factor, and this translates into the exponent of the Gaussian solution like (7.50) having a positive quadratic part; thus the solution cannot be normalised.

[3] Recall that $e^{-iuP/\hbar}\phi(x) = \phi(x-u)$, as discussed in §5.3.2. See Appendix G.

In terms of the annihilation and creation operators we have

$$U_{u,v} \rightarrow U_\alpha \equiv \exp\left(\alpha c^+ - \alpha^* c\right), \tag{7.56}$$

with which we easily derive

$$\begin{aligned} U_\alpha^{-1} c U_\alpha &= c + \alpha \\ U_\alpha^{-1} c^+ U_\alpha &= c^+ + \alpha^*. \end{aligned} \tag{7.57}$$

U_α is therefore commonly known as the coherent state displacement operator. We also recopy (7.54) as

$$|\alpha\rangle = U_\alpha |\Psi_0\rangle. \tag{7.58}$$

The Baker–Campbell–Hausdorff formula,

$$e^{A+B} = e^A e^B e^{-[A,B]/2}, \tag{7.59}$$

is an operator identity, provided A and B both commute with $[A, B]$. In view of (7.31) this identity can be applied to the displacement operator to yield

$$U_\alpha = e^{\alpha c^+} e^{\alpha^* c} e^{-|\alpha|^2/2}. \tag{7.60}$$

Now

$$e^{\alpha^* c}|\Psi_0\rangle = |\Psi_0\rangle, \tag{7.61}$$

from the defining property of the vacuum state (7.34), and so we can rewrite (7.58) as

$$|\alpha\rangle = e^{-|\alpha|^2/2}\, e^{\alpha c^+} |\Psi_0\rangle. \tag{7.62}$$

The equations in (7.57) can be interpreted as defining a new pair of annihilation and creation operators for the mode; the commutation relation is unchanged,

$$[c(\alpha), c(\alpha)^+] = 1, \tag{7.63}$$

but $c(\alpha)$ no longer annihilates the vacuum state $|\Psi_0\rangle$; indeed,

$$c(\alpha)|\Psi_0\rangle = \alpha|\Psi_0\rangle. \tag{7.64}$$

The coherent state $|\alpha\rangle$, (7.58), can be thought of as a new vacuum state for the transformed operators since

$$c(\alpha)|\alpha\rangle = c(\alpha)|\Psi(\alpha)_0\rangle = U_\alpha c U_\alpha^{-1} U_\alpha |0\rangle = 0. \tag{7.65}$$

The overlap between the old and new vacuum states is

$$\langle\Psi_0|\Psi(\alpha)_0\rangle = \langle\Psi_0|U_\alpha|\Psi_0\rangle = \exp(-|\alpha|^2/2) \tag{7.66}$$

by (7.69). Provided the RHS of (7.66) is > 0, (7.63) and (7.65) define a new Fock representation of the commutation relations (7.31) that is unitarily equivalent to the original representation. These relations define the *boson translation transformation*.

The relationship of the coherent states to the number states $\{|n\rangle\}$ can be established in the following way. The exponential operator is defined as

$$e^A = \sum_n^\infty \frac{A^n}{n!}, \tag{7.67}$$

and therefore the coherent state $|\alpha\rangle$ can be understood as an *infinite* superposition of the number states $\{|n\rangle\}$ in the Fock space basis:

$$\begin{aligned} |\alpha\rangle &= e^{-|\alpha|^2/2} \sum_n^\infty \frac{\alpha^n}{\sqrt{n!}} \frac{(c^+)^n}{\sqrt{n!}} |\Psi_0\rangle \\ &= e^{-|\alpha|^2/2} \sum_n^\infty \frac{\alpha^n}{\sqrt{n!}} |n\rangle. \end{aligned} \tag{7.68}$$

The expansion (7.68) can be used to characterise the formal properties of the coherent states since we know that the number states are the eigenstates of the self-adjoint operator n, and are therefore a complete set of orthonormal states. Using (7.68), we calculate for a single mode

$$\bar{n} = \langle\alpha|n|\alpha\rangle = |\alpha|^2, \; \bar{n^2} = \langle\alpha|n^2|\alpha\rangle = |\alpha|^4 - |\alpha|^2, \tag{7.69}$$

giving a dispersion

$$\bar{n^2} - (\bar{n})^2 = |\alpha|^2 = \bar{n}, \tag{7.70}$$

equal to the mean particle number. From their definition the coherent states are normalised; however, they are not orthogonal because, using (7.68), we have

$$\begin{aligned} \langle\alpha|\alpha'\rangle &= e^{-|\alpha|^2/2-|\alpha'|^2/2} \sum_n \frac{(\alpha^*\alpha')^n}{n!} \\ &= e^{-|\alpha|^2/2-|\alpha'|^2/2+\alpha^*\alpha'} \\ &= e^{-|\alpha-\alpha'|^2+i3\alpha^*\alpha'}. \end{aligned} \tag{7.71}$$

The coherent states have the valuable property that a Fock space matrix element of an operator A expressed as a normally ordered product of annihilation and creation operators can be evaluated immediately by simply substituting the coherent state parameters in place of the operators:

$$\langle\alpha|A(c^+,c)_N|\alpha'\rangle = A(\alpha^*,\alpha')\langle\alpha|\alpha'\rangle. \tag{7.72}$$

The completeness condition for the coherent states reads

$$\int |\alpha\rangle\langle\alpha| \, d\mu = 1, \tag{7.73}$$

where the measure is

$$d\mu(\alpha) = \frac{1}{\pi} d\alpha_R \, d\alpha_I, \quad -\infty \leq \alpha_R, \alpha_I \leq \infty. \tag{7.74}$$

This can be checked by inserting the LHS of (7.73) into the orthonormality condition for the number states, that is,

$$\langle n|1|n'\rangle = \delta_{nn'},\qquad(7.75)$$

and using (7.68) to verify its validity. In terms of the position and momentum parameters (u,v), the equivalent condition reads

$$\int |u,v\rangle\langle u,v|\,\mathrm{d}\mu(u,v) = 1,\qquad(7.76)$$

where

$$\mathrm{d}\mu(u,v) = \frac{1}{2\pi}\,\mathrm{d}u\,\mathrm{d}v,\quad -\infty \le u,v \le \infty.\qquad(7.77)$$

On the other hand it is evident from their relationship to the number states (7.71) that coherent states form an over-complete set; their non-orthogonality (7.71) being one manifestation of this.

7.2.3 Coherence of the Electromagnetic Field

The coherence properties of the electromagnetic field are described in terms of the properties of the mean values of polynomials of the field strength operators. These operators will involve products of the annihilation and creation operators (c_i, c_j^+) for the modes of the field which can be brought to normal ordered form. Thus the study of coherence from the theoretical point of view can be based on the evaluation of the correlation functions of the annihilation and creation operators,

$$\langle c_{i_1}^+\ldots c_{i_n}^+ c_{i_{n+1}}\ldots c_{i_{n+m}}\rangle_\rho = G^{(n,m)}(i_1,\ldots i_n, i_{n+1},\ldots i_{n+m}),\qquad(7.78)$$

where ρ is a density operator for the field and $\langle\Lambda\rangle_\rho = \mathrm{Tr}(\rho\Omega)$ according to (5.18); the annihilation and creation operators in (7.78) are all taken at the same instant in time.

The state of the electromagnetic field described by the density operator ρ is said to be fully coherent if there exists a sequence of complex numbers $\{z_i : z_1, z_2,\ldots\}$ such that for every value of n and for every set of indices $i_1,\ldots i_n, i_{n+1},\ldots i_{2n}$ we have

$$G^{(n,n)}(i_1,\ldots i_n, i_{n+1},\ldots i_{2n}) = \prod_{k=1}^{n} z_{i_k}^* \prod_{m=n+1}^{2n} z_{i_m}.\qquad(7.79)$$

If the correlation functions (7.78) possess this property only for $n \le M$, we say that the state of the field has only Mth order coherence [7]. The lowest-order correlation function, $G^{(1,1)}$, is the most familiar since it determines the intensity of scattered light in a scattering experiment, and hence cross sections.

An alternative formulation, which parallels the usual description of coherence in classical electromagnetism, utilises an average over products of the electric field operators,

$$G^{(n,m)}(t_1,\ldots t_n; t_{n+1},\ldots t_{n+m}) = \langle E(t_1^-\ldots E(t_n)^- E(t_{n+1})^+\ldots E(t_{n+m})^+\rangle\qquad(7.80)$$

such that $G^{(1,1)}(t;t)$ is again proportional to the scattered intensity. The power spectrum $S(\omega)$ of fluorescence is the Fourier transform of the two-point correlation function $G^{(1,1)}(t;t+\tau)$. These remarks apply equally to classical and quantum descriptions of the field; the difference between the two accounts lies in the interpretation of the averaging implied by $\langle\ldots\rangle$ in (7.78), (7.80). Once one goes beyond simple intensity measurements ($G^{(1,1)}$), differences arise in these correlation functions according to classical and quantum electrodynamics; measurements of the properties of the electromagnetic field in, for example, spontaneous emission and resonance fluorescence experiments confirm the quantum nature of the field [8].

Some of the higher-order correlation functions are accessible through photon-counting experiments. Let $p_{\{n\}}$ be the probability that the state with density operator ρ has n_1 photons in mode 1, n_2 photons in mode 2, ... n_i photons in mode i. This probability is the mean value of a projection operator that projects onto the Fock space vector with the occupation numbers $n_1, n_2, \ldots n_i, \ldots$. The projection operator can be developed through a group theoretical argument. The spectrum of the number operator n has the symmetry of an infinite cyclic group, the additive group of the integers; the character group of the integers is isomorphic to the group of rotations in the plane, and every character is of the form

$$\chi(\theta)_n = e^{in\theta}, \quad n = \text{integer}, \quad 0 \le \theta \le 2\pi.$$

For every θ the operator

$$U_\theta = e^{in\theta} \tag{7.81}$$

is a true unitary representation of the cyclic group; with it we may form the required projection operator on the number states (cf. §5.3.2),

$$P_n = \frac{1}{2\pi} \int_0^{2\pi} \overline{\chi}(\theta)_n \, U_\theta \, d\theta.$$

If the exponential operator (7.81), is represented using (7.67), it may be brought to normal ordered form,

$$e^{in\theta} = \sum_{k=0}^{\infty} \frac{1}{k!} \left(e^{i\theta} - 1 \right)^k (c^+)^k (c)^k,$$

and performing the θ integration yields

$$P_n = \frac{1}{n!} \sum_{k=0} \frac{(-1)^k}{k!} (c^+)^{n+k} (c)^{n+k}. \tag{7.82}$$

Equation (7.82) can be generalised to give

$$P_{\{n\}} = \prod_{i=1} \frac{1}{n_i!} \sum_{k=0} \frac{(-1)^k}{k!} (c_i^+)^{n_i+k} (c_i)^{n_i+k}. \tag{7.83}$$

According to the definitions (5.18, 7.79) a fully coherent state has

$$\langle P_{\{n\}} \rangle_\rho = \prod_{i=1} p(n_i). \tag{7.84}$$

These considerations provide the justification for the name coherent state introduced in §7.2.2. Suppose that the field is described by a coherent state vector $\Psi(\{\alpha_i\})$ labelled by the eigenvalues $\{\alpha_i\}$ of all annihilation operators $\{c_i\}$; then the density operator is

$$\rho = |\Psi(\{\alpha_i\})\rangle\langle\Psi(\{\alpha_i\})|. \tag{7.85}$$

Since all the modes are independent, and the coherent states have the property (7.72), it is evident that

$$p(n_i) = e^{-|\alpha_i|^2} \frac{|\alpha_i|^{2n_i}}{n_i!}, \tag{7.86}$$

which is a Poisson distribution about a mean $|\alpha_i|^2$, and $p_{\{n\}}$ has the product form (7.84). This is also

$$p(n_i) = e^{-\overline{n}_i} \frac{(\overline{n}_i)^{n_i}}{n_i!}, \tag{7.87}$$

and the dispersion of the distribution obtained from the coherent state is just the mean photon number as in (7.70)

$$\sigma^2 = \overline{n}_i. \tag{7.88}$$

States of the field that are not fully coherent have greater dispersions than the Poisson distribution; examples of incoherent radiation include blackbody radiation, which will be discussed in §7.4, and a pure Fock state with definite occupation numbers.

Quantum theory accounts for interference experiments in which two light beams are combined to produce interference fringes through the superposition of photon probability amplitudes. The fringe visibility, v, is defined in terms of the intensity variation

$$v = \frac{I_{\max} - I_{\min}}{I_{\max} + I_{\min}}. \tag{7.89}$$

For simplicity we restrict ourselves to just two modes of the field; their quantum properties are described by annihilation and creation operators $c_i, c_i^+, i = 1, 2$. The Fock basis will then be vectors $|n_1\rangle|n_2\rangle$ specifying the numbers of photons in each mode. We have in mind a monochromatic beam of light falling on two pinholes, as in Young's experiment, or two plane waves with wave vectors $\mathbf{k}_1$ and $\mathbf{k}_2$ ($k_1 = k_2$) that intersect at an angle θ for which interference fringes may be observed along the direction $\mathbf{r}$ perpendicular to $\mathbf{k}_1$.

The quantum theory of photodetection shows that the intensity in each mode separately, that is, when no superposition is allowed, is determined by the traces of their respective number operators

$$I_1 \propto \mathrm{Tr}(\rho n_1), \quad I_2 \propto \mathrm{Tr}(\rho n_2), \tag{7.90}$$

and v is proportional to the first-order correlation function

$$v \propto \mathrm{Tr}(\rho c_1^+ c_2). \tag{7.91}$$

It is clear that for $v \neq 0$ we require ρ to contain contributions from modes differing by one photon. For simplicity let the detected field be in a pure state with state vector $|\psi\rangle$ so that the density operator is a projector

$$\rho = |\psi\rangle\langle\psi|. \tag{7.92}$$

One way in which a non-zero v can be obtained is if $|\psi\rangle$ is a superposition of one-photon states from the two modes,

$$|\psi\rangle = a_1|1\rangle|0\rangle + a_2|0\rangle|1\rangle. \tag{7.93}$$

This result shows that the interference pattern can be built up from a sequence of 1-photon experiments, for example, by using a light source of such weak intensity that on average only one photon at a time is in the apparatus. Clearly if we take $|\psi\rangle = |n_1\rangle|n_2\rangle$ corresponding to two independent beams in Fock states, there is no correlation ($v = 0$). On the other hand a simple product of two coherent states $|\psi\rangle = |\alpha_1\rangle|\alpha_2\rangle$ can give interference fringes, provided the relative phase of the two beams is sufficiently slowly varying for the fringes to be observed. This is because the number of photons in a coherent state is indefinite with a Poisson distribution (cf. (7.86)).

Such a state can be realised with two single-mode lasers [6],[9]. The field measured at the detector belongs to the whole apparatus including the sources and the detector, such that it is not possible for a photon to contribute to the interference effect and be assigned to one or other source. The uncertainty relation (5.181) is relevant here; if the photon can be detected with an uncertainty in position Δx, it has an uncertainty in momentum Δk such that

$$\Delta x \, \Delta k \sim \hbar. \tag{7.94}$$

An analysis of the experiment then shows that if the resolution in position is better than one fringe width so that fringes can be observed, the uncertainty in momentum is such that it is no longer possible to assign the photon to a particular source. Conversely if the momentum is known sufficiently accurately to determine which beam the photon is associated with, the uncertainty in Δx is great enough to wash out the fringe pattern.

7.2.4 The Number and Phase Representation

A classical electromagnetic field is conveniently described in terms of its amplitude and phase; it is of interest, therefore, to establish the limitations imposed on such a description by the quantum theory. In §3.4 we noted that there exists a classical transformation that reduces the Hamiltonian for an oscillator to cyclic form; this involves a canonical transformation from position (x) and momentum (p) variables to action-angle variables, (J, θ). In quantum theory the analogous transformation is to the number operator, n, and a conjugate phase angle operator θ, and it is not surprising that classical electrodynamics was initially quantised in terms of this pair. The approach can still be seen (1930) in the first edition of Dirac's famous book where there is only the briefest mention of annihilation and creation operators which were adopted universally soon after. Dirac noted that the idea of conjugate number and phase operators was deeply problematic in quantum theory. By analogy with the classical discussion of a mode a 'phase operator' θ is introduced such that the operator $E(\theta) = \exp(i\theta)$ derived from it appears in the polar decomposition[4] of the annihilation operator

$$c = e^{i\theta} \frac{1}{\sqrt{n}}. \tag{7.95}$$

There is then a formal commutation relation,

$$[n\hbar, \theta] = i\hbar 1. \tag{7.96}$$

Equation (7.96) does not yield the usual uncertainty relation for conjugate variables since the eigenvalues of θ are assumed to lie in the interval $[0, 2\pi]$ and n is bounded from below with discrete spectrum. Dirac was aware that no such θ can be defined, and that its exponential is not unitary [10].

Much later it was shown that no unitary operator is available for the decomposition (7.95), and modified definitions were proposed for an adjoint pair of operators [11]

$$E(\theta) \equiv e^{i\theta} = \frac{1}{\sqrt{n+1}} c, \quad E(\theta)^+ \equiv e^{-i\theta^+} = c^+ \frac{1}{\sqrt{n+1}}, \tag{7.97}$$

together with their 'trigonometric' companions, usually, and loosely, called sin and cos,

$$\cos(\theta) = \frac{1}{2}\left(e^{i\theta} + e^{-i\theta^+}\right), \tag{7.98}$$

$$\sin(\theta) = \frac{1}{2i}\left(e^{i\theta} - e^{-i\theta^+}\right). \tag{7.99}$$

Given that the number operator does not commute with the annihilation and creation operators, it follows that E is not self-adjoint,

$$E(\theta) \neq E(\theta)^+, \tag{7.100}$$

and neither is it unitary since

$$E(\theta)\, E(\theta)^+ = 1, \tag{7.101}$$

whereas the product of the operators in reverse order yields

$$E(\theta)^+\, E(\theta) = c^+ \frac{1}{n+1}\, c. \tag{7.102}$$

In the Fock basis these operators have the simple matrix representations

$$\langle n|n|m\rangle = n\delta_{nm}, \quad \langle n|e^{i\theta}|m\rangle = \delta_{n,m-1} \tag{7.103}$$

and one infers that

$$\langle n|E(\theta)^+\, E(\theta)|m\rangle = \begin{cases} \delta_{nm}, m \neq n \\ 0, \quad m = 0 \end{cases}, \tag{7.104}$$

---

[4] The extension to operators of writing a complex variable in terms of modulus and phase, $z = re^{i\theta}$. With polar decomposition an operator A is factorised as the product of a unitary operator U and its polar partner, V, thus $A = UV$.

so that we may rewrite (7.102) as

$$e^{-i\theta^+} e^{i\theta} = 1 - \mathsf{P}_0, \tag{7.105}$$

where P_0 is the projector onto the vacuum state. The non-zero matrix elements of sin and cos in the Fock representation are

$$\langle n | \cos(\theta) | n-1 \rangle = \langle n-1 | \cos(\theta) | n \rangle = \frac{1}{2}$$

$$\langle n | \sin(\theta) | n-1 \rangle = -\langle n-1 | \sin(\theta) | n \rangle = -\frac{1}{2i}. \tag{7.106}$$

These operators provide a limited description of the phase properties of the quantised electromagnetic field in terms of their mean values and uncertainties [6]. In a Fock state $|n\rangle$ the uncertainty in the photon number is clearly zero,

$$\Delta n = 0, \tag{7.107}$$

whereas the corresponding uncertainties in $\sin(\theta)$ and $\cos(\theta)$ are given by

$$\Delta \sin(\theta) = \Delta \cos(\theta) = \frac{1}{\sqrt{2}}, \tag{7.108}$$

corresponding to a phase angle θ that is equally likely to have any value in $[0, 2\pi]$. However, they are not suitable for an account of the statistics of the quantised field; thus we have the slightly curious situation that the phase angle and its conjugate are well defined for the classical electromagnetic field, but they cannot be the 'classical limit' of corresponding quantum operators in the Schrödinger representation based on Fock space (such operators do not exist!) [12].

7.3 Quantisation of the Field Operators

If no gauge condition is imposed, the vector potential operator has a longitudinal degree of freedom in addition to the two transverse degrees of freedom that describe polarised photons; similarly its conjugate π also has three degrees of freedom. Their mutual commutation relation is canonical,

$$[\mathsf{a}(\mathbf{x})^r, \pi(\mathbf{x}')_s] = i\hbar \delta_{rs} \delta^3(\mathbf{x} - \mathbf{x}'), \tag{7.109}$$

which implies that the vector potential operator's canonical conjugate, π, may be realised as a functional derivative,

$$\pi_s = -i\hbar \frac{\delta}{\delta \mathsf{a}_s}. \tag{7.110}$$

Since the commutation relations fix the Hilbert of states, it will be 'too large', and at the outset the calculations will involve the additional degrees of freedom; an extra condition on the state space is thus required to pick out the physically significant states.

If we choose a representation in which the vector potential operator is diagonal,

$$\mathbf{a}|\mathbf{a}\rangle = \mathbf{a}|\mathbf{a}\rangle, \tag{7.111}$$

the 'eigenvalue' $\mathbf{a}$ is a classical vector potential and the 'eigenfunctions' are wave functionals $\Phi[\mathbf{a}] = \langle \mathbf{a}|\Phi\rangle$. The classical primary constraint for the free-field problem

$$\nabla \cdot \boldsymbol{\pi} \approx 0, \tag{7.112}$$

after quantisation must be expressed as a condition on the states Φ to pick out the physical Hilbert space. In close correspondence with §3.6 we define the Gauss's law operator for the free field as $\mathsf{G}_0 = \nabla \cdot \boldsymbol{\pi}$ and require that the physical states for the field be annihilated by G_0:

$$\mathsf{G}_0\Phi[\mathbf{a}] = 0. \tag{7.113}$$

We may interpret the Gauss's law operator by considering the effect of a continuous linear superposition of it with a smooth scalar field σ,

$$\mathcal{G}_0^\sigma = \int \mathsf{G}_0\sigma \, \mathrm{d}^3\mathbf{x}. \tag{7.114}$$

This operator is self-adjoint, and we may use it in the usual way to define a unitary operator,

$$\mathsf{U}_{\mathsf{G}_0}^\sigma = \exp\left(\frac{i}{\hbar}\mathcal{G}_0^\sigma\right). \tag{7.115}$$

$\mathsf{U}_{\mathsf{G}_0}^\sigma$ commutes with $\mathbf{E}$ and $\mathbf{B}$, and hence with $\mathsf{H}_{\mathrm{rad}}$ and so describes a symmetry of the system, but not with $\mathbf{a}$; we have

$$\begin{aligned}
\mathbf{a}(\mathbf{x})^s &= \mathsf{U}_{\mathsf{G}_0}^\sigma \, \mathbf{A}(\mathbf{x})^s \, (\mathsf{U}_{\mathsf{G}_0}^\sigma)^{-1} \\
&= \mathbf{A}(\mathbf{x})^s - \frac{i}{\hbar}[\mathcal{G}_0^\sigma, \mathbf{A}(\mathbf{x})^s] \\
&= \mathbf{A}(\mathbf{x})^s + (\nabla_\mathbf{x}\sigma(\mathbf{x}))^s,
\end{aligned} \tag{7.116}$$

after an integration by parts, and use of (7.109), (7.110), and $\mathsf{U}_{\mathsf{G}_0}^\sigma\Phi[\mathbf{A}] = \Phi[\mathbf{a}]$. Thus $\mathcal{G}^\sigma$ is the *generator of gauge transformations*, and $\mathsf{U}_{\mathsf{G}_0}^\sigma$ is an operator with an Abelian multiplication law that is isomorphic to $\mathcal{U}(1)$, the unitary unimodular group of complex numbers,

$$\mathsf{U}_{\mathsf{G}_0}^\sigma\mathsf{U}_{\mathsf{G}_0}^\eta = \mathsf{U}_{\mathsf{G}_0}^{\sigma+\eta} = \mathsf{U}_{\mathsf{G}_0}^\eta\mathsf{U}_{\mathsf{G}_0}^\sigma. \tag{7.117}$$

This discussion is essentially an example of the group theoretical results in §5.4, as may be seen if we make the identifications,

$$\mathbf{a} \leftrightarrow \mathbf{X}, \quad \boldsymbol{\pi} \leftrightarrow \mathbf{P}, \tag{7.118}$$

since the gauge transformations of the vector potential, (7.116), can be viewed as translations in the function space. The corresponding projection operator is easily written down from (5.174), and if the functional integration is restricted to just longitudinal vector fields, it will project out gauge-invariant states from arbitrary wave functionals depending on the vector potential.

Other classical variables for the free field that were discussed in Chapter 4 using the mode expansion, for example, the vector potential and the field strengths, can be quantised in exactly the same way as the energy and momentum. It is convenient to use the annihilation and creation operators in place of $X(\mathbf{k})$ and $P(\mathbf{k})$. For general mode and polarisation indices their commutation relation is

$$[c_{\mathbf{k},\lambda}, c_{\mathbf{q},\lambda'}^{+}] = \delta_{\lambda,\lambda'} \delta_{\mathbf{k},\mathbf{q}}, \tag{7.119}$$

and their commutation relations with the field Hamiltonian are then

$$[\mathsf{H}_{\mathrm{rad}}, c_{\mathbf{k},\lambda}] = -\hbar\omega c_{\mathbf{k},\lambda}, \tag{7.120}$$

$$[\mathsf{H}_{\mathrm{rad}}, c_{\mathbf{k},\lambda}^{+}] = \hbar\omega c_{\mathbf{k},\lambda}^{+}. \tag{7.121}$$

Quantisation of the Coulomb gauge vector potential $\mathbf{A}(\mathbf{x})$, (4.32), and its conjugate $\boldsymbol{\pi}(\mathbf{x})$, (4.33), results in the mode expansions

$$\mathbf{A}(\mathbf{x}) = \sqrt{\frac{\hbar}{2\varepsilon_0\Omega c}} \sum_{\mathbf{k},\lambda} \frac{\hat{\boldsymbol{\varepsilon}}(\mathbf{k})_\lambda}{\sqrt{k}} \left(c_{\mathbf{k},\lambda} e^{i\mathbf{k}\cdot\mathbf{x}} + c_{\mathbf{k},\lambda}^{+} e^{-i\mathbf{k}\cdot\mathbf{x}} \right), \tag{7.122}$$

$$\boldsymbol{\pi}(\mathbf{x}) = -i\sqrt{\frac{\varepsilon_0\hbar c}{2\Omega}} \sum_{\mathbf{k},\lambda} \sqrt{k}\,\hat{\boldsymbol{\varepsilon}}(\mathbf{k})_\lambda \left(c_{\mathbf{k},\lambda} e^{i\mathbf{k}\cdot\mathbf{x}} - c_{\mathbf{k},\lambda}^{+} e^{-i\mathbf{k}\cdot\mathbf{x}} \right), \tag{7.123}$$

and so the field strengths have the following operator expansions:

$$\mathbf{E}(\mathbf{x}) = i\sqrt{\frac{\hbar c}{2\varepsilon_0\Omega}} \sum_{\mathbf{k},\lambda} \sqrt{k}\,\hat{\boldsymbol{\varepsilon}}(\mathbf{k})_\lambda \left(c_{\mathbf{k},\lambda} e^{i\,\mathbf{k}\cdot\mathbf{x}} - c_{\mathbf{k},\lambda}^{+} e^{-i\mathbf{k}\cdot\mathbf{x}} \right), \tag{7.124}$$

$$\mathbf{B}(\mathbf{x}) = i\sqrt{\frac{\hbar}{2\varepsilon_0\Omega c}} \sum_{\mathbf{k},\lambda} \frac{\mathbf{k}\wedge\hat{\boldsymbol{\varepsilon}}(\mathbf{k})_\lambda}{\sqrt{k}} \left(c_{\mathbf{k},\lambda} e^{i\mathbf{k}\cdot\mathbf{x}} - c_{\mathbf{k},\lambda}^{+} e^{-i\mathbf{k}\cdot\mathbf{x}} \right). \tag{7.125}$$

The coefficient of the annihilation operator in the expansion of the vector potential, (7.122),

$$\mathbf{A}_{\mathbf{k},\lambda} = \sqrt{\frac{\hbar}{2\varepsilon_0\Omega\omega}}\,\hat{\boldsymbol{\varepsilon}}(\mathbf{k})_\lambda e^{i\mathbf{k}\cdot\mathbf{x}}, \tag{7.126}$$

can be interpreted as a wave function for the photon in momentum representation. The vector wave functions $\mathbf{A}_{\mathbf{k},\lambda}$ are orthogonal if integrated over the quantisation volume,

$$\int_{\Omega} \mathbf{A}_{\mathbf{k},\lambda}^{*} \cdot \mathbf{A}_{\mathbf{k}',\lambda'}\, d^3\mathbf{x} = \frac{\hbar}{2\varepsilon_0\omega} \delta_{\mathbf{k},\mathbf{k}'} \delta_{\lambda,\lambda'}. \tag{7.127}$$

This normalisation can be given a physical interpretation by defining analogous quantities for the electric field strength and the magnetic induction,

$$\mathbf{E}_{\mathbf{k},\lambda} = i\omega\mathbf{A}_{\mathbf{k},\lambda}, \quad \mathbf{B}_{\mathbf{k},\lambda} = i\mathbf{k}\wedge\mathbf{A}_{\mathbf{k},\lambda}, \tag{7.128}$$

in terms of which (7.124), (7.125) become

$$\mathbf{E}(\mathbf{x}) = \sum_{\mathbf{k},\lambda} \mathbf{E}_{\mathbf{k},\lambda} c_{\mathbf{k},\lambda} + \mathbf{E}_{\mathbf{k},\lambda}^{*} c_{\mathbf{k},\lambda}^{+}, \tag{7.129}$$

$$\mathbf{B}(\mathbf{x}) = \sum_{\mathbf{k},\lambda} \mathbf{B}_{\mathbf{k},\lambda} c_{\mathbf{k},\lambda} + \mathbf{B}_{\mathbf{k},\lambda}^* c_{\mathbf{k},\lambda}^+. \tag{7.130}$$

Inserting these expansions in the classical energy expression (3.258) yields the quantised Hamiltonian as

$$H_{\text{rad}} = \frac{1}{2}\varepsilon_0 \sum_{\mathbf{k},\lambda} \left(c_{\mathbf{k},\lambda} c_{\mathbf{k},\lambda}^+ + c_{\mathbf{k},\lambda}^+ c_{\mathbf{k},\lambda}\right) \int_\Omega \left(|\mathbf{E}_{\mathbf{k},\lambda}|^2 + c^2|\mathbf{B}_{\mathbf{k},\lambda}|^2\right) d^3\mathbf{x}. \tag{7.131}$$

Using the explicit forms (7.126), (7.128), the integration yields

$$\varepsilon_0 \int_\Omega \left(|\mathbf{E}_{\mathbf{k},\lambda}|^2 + c^2|\mathbf{B}_{\mathbf{k},\lambda}|^2\right) d^3\mathbf{x} = \hbar\omega, \tag{7.132}$$

in agreement with (7.22). The integral on the LHS is the mean value of the energy in the state having the wave function $\mathbf{A}_{\mathbf{k},\lambda}$ and this is equal to the energy, $\hbar\omega$, of one photon in the mode $\mathbf{k}, \lambda$; the normalisation can be interpreted as one photon per unit volume [19].

The matrix elements of the field operators in Fock space are easily obtained from those of the annihilation ($c_{\mathbf{k},\lambda}$) and creation ($c_{\mathbf{k},\lambda}^+$) operators since they are linear in $c_{\mathbf{k},\lambda}$ and $c_{\mathbf{k},\lambda}^+$. They do not commute with the number operator, and so cannot be diagonalised simultaneously with the Hamiltonian (and momentum). They are *off-diagonal* operators connecting a state $|n_{\mathbf{k}}\rangle$ only with states $|n_{\mathbf{k}} \pm 1\rangle$. We illustrate this in the following with the electric field operator defined in (7.124).

For a single mode, $\mathbf{k}$, with exactly $n_{\mathbf{k}}$ photons, we find

$$\langle n_{\mathbf{k}}|\mathbf{E}|n_{\mathbf{k}}\rangle = 0$$
$$\langle n_{\mathbf{k}}|\mathbf{E}\cdot\mathbf{E}|n_{\mathbf{k}}\rangle = \frac{\hbar kc}{\varepsilon_0\Omega}\left(n_{\mathbf{k}} + \tfrac{1}{2}\right), \tag{7.133}$$

with no time-dependence. Thus the amplitude of the field is

$$E_0 = \sqrt{\frac{2\hbar kc}{\varepsilon_0\Omega}\left(n_{\mathbf{k}} + \tfrac{1}{2}\right)}, \tag{7.134}$$

while the phase is completely uncertain. We see from (7.133) that the variance of $\mathbf{E}$ for a single mode in the Fock representation,

$$\langle n_{\mathbf{k}}|\mathbf{E}\cdot\mathbf{E}|n_{\mathbf{k}}\rangle - \langle n_{\mathbf{k}}|\mathbf{E}|n_{\mathbf{k}}\rangle \cdot \langle n_{\mathbf{k}}|\mathbf{E}|n_{\mathbf{k}}\rangle = \frac{\hbar kc}{\varepsilon_0\Omega}\left(n_{\mathbf{k}} + \tfrac{1}{2}\right), \tag{7.135}$$

is proportional to the photon number $n_{\mathbf{k}}$.

An important special case is the vacuum state Ψ_0 with no photons; at every field point $\mathbf{x}$ we have

$$\langle \Psi_0|\mathbf{E}|\Psi_0\rangle = 0, \tag{7.136}$$

$$\langle \Psi_0|\mathbf{E}\cdot\mathbf{E}|\Psi_0\rangle = \frac{\hbar c}{2\varepsilon_0\Omega}\sum_{\mathbf{k}} k\,1 = \infty, \tag{7.137}$$

so the variance in this case is infinite. Physically, the calculation is unrealistic since one cannot measure the field at a point. Suppose we probe the field with a test charge localised in a volume $\Delta V = (\Delta r)^3$ so that the field measured is the mean value averaged over the volume ΔV. Then (7.137) is replaced by

$$\langle \Psi_0 | (\mathbf{E})_{\Delta V} \cdot (\mathbf{E})_{\Delta V} | \Psi_0 \rangle \sim \frac{\hbar c}{(\Delta r)^4}, \tag{7.138}$$

showing that the finer the volume we probe, the greater the fluctuations of the field in the vacuum state [13], [14]. Equations (7.136) and (7.137) sharply circumscribe any proposed physical interpretation of the field operators. Thus, for example, the description of the polarisation state of a classical light wave can be carried by its electric field vector, $\mathbf{E}$, in terms of the phase difference between its two transverse components and their amplitudes; after quantisation, (7.136) implies that such a description is no longer possible. The infinity in (7.137) simply reflects the infinite zero-point energy of the quantised electromagnetic field that we have noted.

In contrast, the expectation values of the electric field operator in a coherent state are smooth functions. For a single-mode coherent state $|\alpha_{\mathbf{k}}\rangle$ we obtain (ignoring the polarisation state)

$$\langle \alpha_{\mathbf{k}} | \mathbf{E} | \alpha_{\mathbf{k}} \rangle = -\sqrt{\frac{2\hbar c}{\varepsilon_0 \Omega}} |\alpha_{\mathbf{k}}| \sin(\mathbf{k} \cdot \mathbf{x} - \omega t + \theta_{\mathbf{k}}), \tag{7.139}$$

where we have explicitly included the time-dependence resulting from (7.39), and set

$$\alpha_{\mathbf{k}} = |\alpha_{\mathbf{k}}| e^{i\theta_{\mathbf{k}}}. \tag{7.140}$$

Similarly,

$$\langle \alpha_{\mathbf{k}} | \mathbf{E} \cdot \mathbf{E} | \alpha_{\mathbf{k}} \rangle = \frac{\hbar k c}{2\varepsilon_0 \Omega} \left(4 |\alpha_{\mathbf{k}}|^2 \sin^2(\mathbf{k} \cdot \mathbf{x} - \omega t + \theta_{\mathbf{k}}) + 1 \right), \tag{7.141}$$

so for the coherent state mode the variance of $\mathbf{E}$ is independent of the number of photons and of α:

$$\langle \alpha_{\mathbf{k}} | \mathbf{E} \cdot \mathbf{E} | \alpha_{\mathbf{k}} \rangle - \langle \alpha_{\mathbf{k}} | \mathbf{E} | \alpha_{\mathbf{k}} \rangle \cdot \langle \alpha_{\mathbf{k}} | \mathbf{E} | \alpha_{\mathbf{k}} \rangle = \frac{\hbar k c}{2\varepsilon_0 \Omega}. \tag{7.142}$$

Equation (7.139) is a solution of the classical Maxwell wave equation for the free field (2.17); in the limit $|\alpha_{\mathbf{k}}| \to \infty$ it becomes a stable classical wave with a well-defined phase angle $\theta_{\mathbf{k}}$ and electric field amplitude

$$E_0 = \sqrt{\frac{2\hbar k c}{\varepsilon_0 \Omega}} |\alpha_{\mathbf{k}}|. \tag{7.143}$$

This rests on the fact that the uncertainty in the phase angle for the coherent state varies as $|\alpha|^{-1}$ for large $|\alpha|$. If we choose $\theta_{\mathbf{k}} = -\pi/2$, then

$$\langle \alpha_{\mathbf{k}} | \mathbf{E} | \alpha_{\mathbf{k}} \rangle = \mathbf{E}_0 \cos(\mathbf{k} \cdot \mathbf{x} - \omega t). \tag{7.144}$$

The off-diagonal matrix elements of $\mathbf{E}$ in the coherent state representation are proportional to the overlap integrals of the coherent states involved which, from (7.71), behave as

$$|\langle \alpha_{\mathbf{k}} | \beta_{\mathbf{k}} \rangle|^2 = \exp\left(-|\alpha_{\mathbf{k}} - \beta_{\mathbf{k}}|^2 \right), \tag{7.145}$$

and so only become negligible for $|\alpha_{\mathbf{k}} - \beta_{\mathbf{k}}| \to \infty$.

Later on (Chapter 9) we will meet examples of the boson translation transformation, (7.57), applied to the operators of the quantised electromagnetic field. The modes of the field are labelled by their wave vectors $\mathbf{k}$ and polarisations λ, and the annihilation and creation operators for different modes all commute, so we may apply the transformation to the individual modes. If we write the transformation operator

$$\mathsf{U} = \exp(i\mathsf{S}) = \exp\left(i\sum_{\mathbf{k},\lambda}\mathsf{S}_{\mathbf{k},\lambda}\right), \tag{7.146}$$

the generator of the transformation for the mode $\mathbf{k}, \lambda$ is

$$\mathsf{S}_{\mathbf{k},\lambda} = -i(\alpha_{\mathbf{k},\lambda}^{*}\,\mathsf{c}_{\mathbf{k},\lambda} - \alpha_{\mathbf{k},\lambda}\,\mathsf{c}_{\mathbf{k},\lambda}^{+}), \tag{7.147}$$

with $\{\alpha_{\mathbf{k},\lambda}\}$ a set of complex-valued scalars. Let $\mathbf{V}(\mathbf{x})$ stand for any one of the field variables, (7.122)–(7.125), and $\mathbf{T}(\mathbf{x})$ be some vector field operator (to be specified by the physical context). Then provided their functional scalar product

$$\int \mathbf{T}(\mathbf{x}) \cdot \mathbf{V}(\mathbf{x})\, \mathrm{d}^3\mathbf{x} \, < \, \infty, \tag{7.148}$$

it is easy to see using the mode expansion of $\mathbf{V}(\mathbf{x})$ that the integral (7.148) can be put in the form of S in (7.146). Thus such expressions, which will be met in Chapter 9, can be interpreted as generators of boson translations. In the infinite volume, continuum limit, the $\{\alpha_{\mathbf{k},\lambda}\}$ are replaced by a function, $\alpha(\mathbf{k})_{\lambda}$, and in (7.66) we make the replacement

$$|\alpha|^2 \, \to \, \sum_{\lambda} \int |\alpha(\mathbf{k})_{\lambda}|^2 \mathrm{d}^3\mathbf{k}. \tag{7.149}$$

Provided $\alpha(\mathbf{k})_{\lambda}$ is square integrable (belongs to $\mathcal{L}^2$), the boson translation is unitary and simply generates a unitarily equivalent representation of the original Hilbert space. However, there are physically important situations where this condition is not met and the mean number of quanta in the new state is infinite; in such a case the transformation leads to a different Hilbert space, and the possibility of new physics [15].

We have emphasised the photon picture of the quantised electromagnetic field because it provides a powerful basis for calculations. By using the mode expansion, the functional calculus can be kept in the background. It is, however, only one possible representation, and there are others that emphasise the wave aspects of radiation, for example one can use the fields themselves as coordinates. We conclude this section with a brief discussion of how this may be done.

In §5.5 we saw that a consistent quantum theory of charges and radiation requires that the field strengths be replaced by operators that satisfy (5.182). Using (7.119), (7.124), (7.125) we compute their commutator directly,

$$[\mathsf{E}(\mathbf{x})_i, \mathsf{B}(\mathbf{x}')_j] = \frac{\hbar}{2\varepsilon_0\Omega} \sum_{\mathbf{k},\lambda} \hat{\boldsymbol{\varepsilon}}(\mathbf{k})_{\lambda i}(\mathbf{k} \wedge \hat{\boldsymbol{\varepsilon}}(\mathbf{k})_{\lambda})_j \left(e^{i\mathbf{k}\cdot(\mathbf{x}-\mathbf{x}')} + e^{-i\mathbf{k}\cdot(\mathbf{x}-\mathbf{x}')}\right). \tag{7.150}$$

The sum over the polarisation vectors can be performed with the relation (2.26). Now

$$(\mathbf{k} \wedge \hat{\boldsymbol{\varepsilon}}(\mathbf{k})_{\lambda})_j = \varepsilon_{jmn}k_m\hat{\varepsilon}(\mathbf{k})_{\lambda n}, \tag{7.151}$$

and so

$$\sum_\lambda \hat{e}(\mathbf{k})_{\lambda i}(\mathbf{k} \wedge \hat{\mathbf{e}}(\mathbf{k})_\lambda)_j = \varepsilon_{jmi}\, k_m, \tag{7.152}$$

since $\mathbf{k} \wedge \mathbf{k} = 0$. The factor of $\mathbf{k}$ may be removed from the sum by replacing it with $i\partial/\partial\mathbf{x}'$, so that the exponentials can be combined, and the commutator reduces to

$$[E(\mathbf{x})_i, B(\mathbf{x}')_j] = \frac{i\hbar}{\varepsilon_0}\varepsilon_{ijm}\frac{\partial}{\partial x_m'}\frac{1}{\Omega}\sum_\mathbf{k} e^{i\mathbf{k}\cdot(\mathbf{x}-\mathbf{x}')}. \tag{7.153}$$

With periodic boundary conditions for the volume Ω we have

$$\frac{1}{\Omega}\sum_\mathbf{k} e^{i\mathbf{k}\cdot(\mathbf{x}-\mathbf{x}')} = \delta_{\mathbf{x},\mathbf{x}'}. \tag{7.154}$$

In the continuum limit the Fourier series for the field variables (7.122)–(7.125) must be written as integrals with the replacement

$$\frac{1}{\sqrt{\Omega}}\sum_\mathbf{k} \rightarrow \frac{1}{\sqrt{(2\pi)^3}}\int d^3k, \tag{7.155}$$

and (7.154) becomes

$$\frac{1}{\Omega}\sum_\mathbf{k} e^{i\mathbf{k}\cdot(\mathbf{x}-\mathbf{x}')} \rightarrow \frac{1}{(2\pi)^3}\int e^{i\mathbf{k}\cdot(\mathbf{x}-\mathbf{x}')}\,d^3k = \delta^3(\mathbf{x}-\mathbf{x}'). \tag{7.156}$$

All other commutators of the field strength operators vanish; this result is consistent with (5.182) [5], [16].

Suppose we choose a representation in which the electric field strength operator is diagonal, so that its eigenvalues are ordinary (classical) electric fields,

$$\mathbf{E}(\mathbf{x})|\mathbf{E}\rangle = \mathbf{E}(\mathbf{x})|\mathbf{E}\rangle. \tag{7.157}$$

The commutation relation (7.153) with (7.156) implies that the field operators themselves can be represented using functional differentiation, and in this representation the magnetic induction operator is

$$B(\mathbf{x})_i = -\frac{i\hbar}{\varepsilon_0}\varepsilon_{ijk}\nabla_k\frac{\delta}{\delta E(\mathbf{x})_j}. \tag{7.158}$$

Now consider the Hamiltonian for the free field,

$$H_{rad} = \tfrac{1}{2}\varepsilon_0\int (\mathbf{E}\cdot\mathbf{E} + c^2\mathbf{B}\cdot\mathbf{B})\,d^3x, \tag{7.159}$$

where both fields are transverse,

$$\nabla\cdot\mathbf{E} = \nabla\cdot\mathbf{B} = 0. \tag{7.160}$$

Once again we can factorise the Hamiltonian into what are effectively annihilation and creation operators; the obvious factorisation of $x^2 + y^2 = (x + iy)(x - iy)$ used earlier is not useful here because we have to accommodate the Curl operator in the representation (7.158). Instead, we write the Hamiltonian as [17],

$$H_{rad} = \tfrac{1}{2}\varepsilon_0 \int \mathbf{F}^+ \cdot \mathbf{F} \, d^3\mathbf{x}, \qquad (7.161)$$

where

$$\mathbf{F}(\mathbf{x}) = c\mathbf{B}(\mathbf{x}) + \frac{i}{2\pi^2}\nabla \wedge \int \frac{\mathbf{E}(\mathbf{y})}{|\mathbf{x}-\mathbf{y}|^2} \, d^3\mathbf{y}. \qquad (7.162)$$

This odd-looking formula is easily understood if we introduce Fourier transforms; a straightforward calculation using the expansions of the fields (7.124), (7.125) yields

$$\mathbf{F}(\mathbf{k}) = \int \mathbf{F}(\mathbf{x}) \, e^{-i\mathbf{k}\cdot\mathbf{x}} \, d^3\mathbf{x} \qquad (7.163)$$

$$= i\sqrt{\frac{\hbar c(2\pi)^3}{2\varepsilon_0 k}} \, \hat{\mathbf{k}} \wedge \sum_{\lambda} \hat{\boldsymbol{\varepsilon}}(\mathbf{k})_\lambda c(\mathbf{k})_\lambda, \qquad (7.164)$$

where $c(\mathbf{k})_\lambda$ is the annihilation operator for the mode with wave vector $\mathbf{k}$ and polarisation λ, with $\mathbf{k}$ now treated as a continuous variable. In momentum space, the Fourier transforms of the Curl operation $(\nabla\wedge)$ and of $1/|\mathbf{x}|^2$ are just what are required to equalise the coefficients of $c(\mathbf{k})_\lambda$ in the expansions of $\mathbf{E}(\mathbf{x})$ and $\mathbf{B}(\mathbf{x})$ (see (7.128)).

In this formulation the vacuum state of the field is the vector that is annihilated by $\mathbf{F}$,

$$\mathbf{F}(\mathbf{x})|\Psi_0\rangle = 0. \qquad (7.165)$$

In the $\{|\mathbf{E}\rangle\}$ representation this is a functional differential equation,

$$\left(-\frac{i\hbar c}{\varepsilon_0}\varepsilon_{ijk}\nabla_k \frac{\delta}{\delta E(\mathbf{x})_j} - \frac{i}{2\pi^2}\varepsilon_{ijk}\nabla_k \int \frac{\mathbf{E}(\mathbf{y})}{|\mathbf{x}-\mathbf{y}|^2} \, d^3\mathbf{y} \right) \Psi[\mathbf{E}]_0 = 0, \qquad (7.166)$$

with solution

$$\Psi[\mathbf{E}]_0 = N\exp\left(-\frac{\varepsilon_0}{4\pi^2\hbar c} \int\int \frac{\mathbf{E}(\mathbf{x})\cdot\mathbf{E}(\mathbf{y})}{|\mathbf{x}-\mathbf{y}|^2} \, d^3\mathbf{y} \, d^3\mathbf{x} \right). \qquad (7.167)$$

This vacuum state wave functional $\Phi[\mathbf{E}]_0$, and states constructed by the polynomial action of $\mathbf{F}^+$ on it, obviously meet the criterion (7.113), since $\mathbf{F}$ and its adjoint commute with G_0. The Hamiltonian for the free field, (7.159), involves purely transverse field variables representing the two physical degrees of freedom, and so its gauge-invariant eigenstates may also be expressed as functionals, $\Phi[\mathbf{A}]$, of the *Coulomb gauge* vector potential.

7.4 Electromagnetic Radiation in Thermal Equilibrium

The quantum mechanical picture of blackbody radiation in a cavity of volume Ω is that it consists of a 'gas' of photons in thermal equilibrium with their surroundings. The photon has no electric charge, and at non-relativistic thermal energies there can be no photon–photon interactions; consequently, the photons cannot equilibrate among themselves, and thermal equilibrium is achieved through continuous absorption and emission of photons by matter in the cavity. The distinguishing feature of this description compared to the familiar thermodynamics of the 'Ideal Gas' (a non-interacting atomic gas) is that the particle number is not a conserved quantity; hence we must take the chemical potential μ to be zero, and the Gibbs free energy of the photon gas is zero. The (Helmholtz) free energy at constant volume is, however, non-trivial and can be used in the usual way to calculate all the customary thermodynamic parameters.

In thermal equilibrium at temperature T, the probability that the mode with wave vector $\mathbf{k}$ is occupied with $n_{\mathbf{k}}$ photons is given by Boltzmann's formula,

$$p(n_{\mathbf{k}}) = \frac{e^{-E_{n_{\mathbf{k}}}/k_B T}}{Z}. \tag{7.168}$$

where $E_{n_{\mathbf{k}}}$ is the quantised mode energy, (7.5), and Z is the partition function for the gas,

$$Z = \sum_{n_{\mathbf{k}}} e^{-E_{n_{\mathbf{k}}}/k_B T}. \tag{7.169}$$

Writing

$$x = \exp(-\hbar\omega/k_B T), \tag{7.170}$$

we obtain

$$p(n_{\mathbf{k}}) = (1-x)x^{n_{\mathbf{k}}}, \tag{7.171}$$

since Z is a simple geometric series in x with sum $(1-x)^{-1}$.[5]

The average energy of the mode at temperature T calculated with this probability can then be put in the form [6]

$$\overline{E}_{\mathbf{k}} = \overline{n}_{\mathbf{k}} \hbar\omega, \tag{7.172}$$

where the mean photon number is

$$\overline{n}_{\mathbf{k}} = \frac{1}{\exp(\hbar\omega/k_B T) - 1}. \tag{7.173}$$

The average energy of the mode $\mathbf{k}$ at temperature T, (7.172), tends to the classical value given by the equipartition theorem, $k_B T$, for those modes satisfying

$$\hbar\omega \ll k_B T, \tag{7.174}$$

---

[5] Since $\omega = 0$ is excluded (cf. Chapter 4), $x < 1$ is assured.

which corresponds to the limiting case,

$$\overline{n}_{\mathbf{k}} \gg 1. \tag{7.175}$$

Thus the mean energy density at temperature T of modes with frequencies in the interval $[\omega, \omega + d\omega]$ is obtained from the Rayleigh–Jeans law by replacing the classical value $k_B T$ with (7.172); this yields the Planck formula which describes the spectral distribution of blackbody radiation in the cavity correctly:

$$\overline{\mathcal{E}(\omega)} \, d\omega = \left(\frac{\omega^2}{\pi^2 c^3} \right) \left(\frac{\hbar \omega \, d\omega}{\exp(\hbar \omega / k_B T) - 1} \right). \tag{7.176}$$

The average total energy density in the cavity, u, is obtained by summing up the contributions from all the modes; this is an integral over the Planck distribution (7.176) for all positive frequencies,

$$u = \int_0^\infty \overline{\mathcal{E}(\omega)} \, d\omega = AT^4, \tag{7.177}$$

where

$$A = \frac{\pi^2 k_B^4}{15 \hbar^3 c^3}. \tag{7.178}$$

Hence the internal energy U of the radiation[6] in the cavity is

$$U(\Omega, T) = A \Omega T^4. \tag{7.179}$$

Experimentally, it is the intensity, I, of blackbody radiation leaving a small area of the surface of the cavity that is more directly accessible, rather than the energy density in the cavity; the two are, however, closely related since

$$I = \frac{1}{4} c u = \sigma T^4, \tag{7.180}$$

where $\sigma = cA/4$ is the Stefan–Boltzmann constant expressed in terms of fundamental constants.

It is now straightforward to characterise the fluctuations in the blackbody radiation. Accepting the ergodic theorem, we can think of the mean photon number (7.173) as either the long time average of the number of photons in a cavity mode with wave vector $\mathbf{k}$, or as an ensemble average. The thermal distribution (7.168) can be expressed in terms of $\overline{n}_{\mathbf{k}}$ using (7.170), (7.171) as

$$p(n_{\mathbf{k}}) = \frac{(\overline{n}_{\mathbf{k}})^{n_{\mathbf{k}}}}{(\overline{n}_{\mathbf{k}} + 1)^{n_{\mathbf{k}}+1}}. \tag{7.181}$$

The extent of the fluctuations in the photon numbers is given by the dispersion $\sigma_{\mathbf{k}}$ of this distribution; this is easily found to be

$$\sigma_{\mathbf{k}} = \overline{n}_{\mathbf{k}} (\overline{n}_{\mathbf{k}} + 1), \tag{7.182}$$

so that the fluctuation, $\Delta n_{\mathbf{k}} = \sqrt{\sigma_{\mathbf{k}}}$, is always greater than the average photon number, $\overline{n}_{\mathbf{k}}$ [6], [7]. As regards fluctuations of the energy in the cavity, thermodynamics gives a

---

[6] The radiation constant $A = 7.556 \, 10^{-16} \, \text{J m}^{-3} \, \text{K}^{-4}$.

general relation between the dispersion in $\mathcal{E}(\omega)$ and the temperature derivative of its average value [18],

$$\overline{\Delta \mathcal{E}(\omega)^2} = k_B T^2 \frac{\overline{d\mathcal{E}(\omega)}}{dT}. \tag{7.183}$$

With the aid of (7.176) this gives

$$\overline{\Delta \mathcal{E}(\omega)^2} = \hbar\omega \, \overline{\mathcal{E}(\omega)} + \frac{\left(\overline{\mathcal{E}(\omega)}\right)^2}{\rho_\omega \, d\omega}. \tag{7.184}$$

The first term, which disappears in the classical limit, can be associated with the photons; the second appears in Maxwell's theory, and the occurrence of both types of term in the quantum theory result (7.184) reminds us of the limitations of the classical particle and wave concepts [16].

Another form of the condition for the field to be quasi-classical can be obtained as follows. As noted previously (§7.1), in practice we can measure values of the field averaged over some time interval Δt which must be less than the time over which the field varies appreciably. Only frequencies such that $\omega \Delta t \leq 1$ will contribute to the average value of $\mathbf{E}$ (or $\mathbf{B}$), so we only need consider modes with frequencies less than $\omega_m = 1/\Delta t$. The total energy density, which is proportional to $|\mathbf{E}|^2$, may be written as

$$W = \int_0^{\omega_m} \rho_\omega n_\omega \hbar\omega \, d\omega \approx \bar{n} \int_0^{\omega_m} \rho_\omega \hbar\omega \, d\omega = \frac{\bar{n}\hbar\omega_m^4}{\pi^2 c^3}. \tag{7.185}$$

Then in order of magnitude

$$\bar{n} \gg 1 \quad \text{requires} \quad |\mathbf{E}| \; \gg \; \frac{\sqrt{\hbar c}}{(c\Delta t)^2}. \tag{7.186}$$

Thus for classical behaviour of the time-averaged field, $|\mathbf{E}|$ must reach a minimum value which increases as Δt decreases. A time-varying field, if sufficiently weak, can never be classical; on the other hand, a static electromagnetic field is always classical since we can take $\Delta t \to \infty$ [19].

More generally, the Helmholtz free energy is related to the partition function (cf. Chapter 6) by

$$F = -k_B T \log_e Z. \tag{7.187}$$

Using the evaluation of Z above and (7.173), the free energy can be expressed in the same way as the internal energy in terms of mode contributions,

$$f_\omega \, d\omega = -k_B T \left(\frac{\omega^2}{\pi^2 c^3} \right) \log_e(1 + \bar{n}_k), \tag{7.188}$$

so that the total free energy density is

$$f = \int_0^\infty f_\omega \, d\omega. \tag{7.189}$$

An integration by parts shows that (7.189) is a simple multiple of (7.177), and there results

$$F(\Omega, T) = -\frac{1}{3} U(\Omega, T).$$ (7.190)

The equation of state is then

$$P = -\left(\frac{\partial F}{\partial \Omega}\right)_T = \frac{1}{3} u(\Omega, T),$$ (7.191)

which shows that, unlike the usual 'Ideal Gas', the pressure is *not* an independent variable. Similarly the entropy is given by

$$S = -\left(\frac{\partial F}{\partial T}\right)_\Omega = \frac{4}{3} \frac{U(\Omega, T)}{T},$$ (7.192)

which vanishes at $T = 0$.

7.5 Angular Momentum and Polarisation

In Chapter 4 we derived expansions of the vector potential $\mathbf{A}(\mathbf{x})$ in terms of a complete orthogonal set of transverse vector functions with definite angular momentum (JM) and definite parity, (4.126), (4.127). These functions are composed of vector spherical harmonics and appropriately chosen radial functions; such expansions can be quantised in exactly the same way as the expansion of the field in terms of plane waves, §7.3. Instead of a box of volume V we quantise the field in a large sphere of radius R. The coefficients of the orthogonal vector functions can be interpreted as annihilation and creation operators for photons with definite angular momentum and parity. For each frequency ω and fixed parity we have a commutation relation that replaces (7.119),

$$[c^+_{\omega JM}, c_{\omega J'M'}] = \delta_{JJ'}\, \delta_{MM'},$$ (7.193)

with all other commutators vanishing. For a definite value of J the energy E_J of the field is determined by an integral over the radial function, (4.135), and this can be used to fix the normalisation of the expansion functions by setting E_J equal to the energy $\hbar\omega$ of a photon.

We thus arrive naturally at the idea that the photon can carry a certain angular momentum. There are, however, some special features in such a description which arise because the photon is a particle with zero mass. Such a particle must be considered to travel at the speed of light in all reference frames and hence has no rest frame. The conventional identification of the quantum mechanical spin of a particle as its angular momentum in its rest frame is therefore inapplicable. Furthermore, for such a relativistic particle there is no consistent definition of a (local) position operator, and so the customary description of orbital angular momentum based on the classical formula $\mathbf{L} = \mathbf{r} \wedge \mathbf{p}$ is not available. The momentum of a photon is, however, a well-defined quantity and can serve equally well for the formulation of an account of photon angular momentum. The orbital angular momentum operator in this 'momentum' representation can be taken to be

$$\mathbf{L} = -i\hbar \mathbf{k} \wedge \nabla_{\mathbf{k}}, \tag{7.194}$$

and therefore differs from the orbital angular momentum operator in the position representation only in that $\mathbf{x}$ is replaced by $\mathbf{k}$ [19].

The expansion functions for the angular momentum decomposition of the vector potential $\mathbf{A}(\mathbf{x})$ are the vector spherical harmonics, $\mathbf{Y}(\mathbf{x})_{JlM}$. As shown in Chapter 4, the transformation properties of such functions under rotations can be labelled by two integral parameters which we denote by S and l, and can interpret as 'spin' and 'orbital' angular momentum labels respectively for the photon. The value $S = 1$ is required because these functions are vectors, and the value of l describes the order of the spherical harmonic, Y_{lm}, used to build the vector spherical harmonic $\mathbf{Y}_{JlM}$. For a description of the angular momentum properties of the photon it therefore suffices to take over the account of vector spherical harmonics in Chapter 4 with the simple change of replacing the unit vector $\mathbf{x}$ by the unit vector $\mathbf{n}$ derived from the photon momentum vector, $\mathbf{k}$, as the argument of the vector spherical harmonic functions which can be regarded as photon wave functions. A photon with angular momentum J and parity $(-1)^J$ is usually referred to as an electric 2^J-pole (or EJ) photon; one with parity $(-1)^{J+1}$ is called a magnetic 2^J-pole or (MJ) photon. Thus a parity even state with $J = 1$ corresponds to a magnetic dipole photon, and one with $J = 2$ corresponds to an electric quadrupole photon. The boson operators in (7.193) for EJ photons commute with those for MJ photons. The angular momentum representation is particularly convenient for describing multipole radiation [19], [20].

In a relativistic theory the total angular momentum, $\mathbf{J} = \mathbf{L} + \mathbf{S}$, is conserved, while its components, $\mathbf{L}$ and $\mathbf{S}$, are not separately conserved. The component of $\mathbf{J}$ along the direction of the photon momentum, $\mathbf{n}$, is also a conserved quantity, and since by (7.194) $\mathbf{L} \cdot \hat{\mathbf{n}} = 0$, this is

$$\mathbf{J} \cdot \hat{\mathbf{n}} = (\mathbf{L} + \mathbf{S}) \cdot \hat{\mathbf{n}} = \mathbf{S} \cdot \hat{\mathbf{n}}. \tag{7.195}$$

Thus there is a conserved quantity that can be associated with the photon 'spin'; this operator quantity is called the helicity of the photon.[7] Since a scalar product is invariant under rotations, the helicity operator commutes with J^2 and J_z, and so there is a representation for photons with well-defined angular momentum and helicity since these are a complete set of commuting observables.

The polarisation properties of the classical electromagnetic field are completely described by the Stokes parameters discussed in §4.3; they were expressed in terms of the oscillator variables for a single mode $\mathbf{k}$. After quantisation of the field the Stokes parameters $\{S_i\}$ can be reinterpreted as corresponding Stokes operators, $\{\Sigma_i\}$, [21]–[23]. Using (7.15), (7.16) we can write down directly these operators for the mode[8] in terms of annihilation and creation operators for the two orthogonal polarisation

---

[7] The helicity is well defined for any relativistic particle.

[8] Thus we have an infinite set of operators to describe the field.

states, $\lambda = 1, 2$, thus (cf. (4.49)–(4.47))

$$\Sigma(\mathbf{k})_0 = \mathsf{n}(\mathbf{k})_1 + \mathsf{n}(\mathbf{k})_2, \tag{7.196}$$

$$\Sigma(\mathbf{k})_1 = \mathsf{c}(\mathbf{k})_1^+ \mathsf{c}(\mathbf{k})_2 + \mathsf{c}(\mathbf{k})_2^+ \mathsf{c}(\mathbf{k})_1, \tag{7.197}$$

$$\Sigma(\mathbf{k})_2 = -i \left(\mathsf{c}(\mathbf{k})_1^+ \mathsf{c}(\mathbf{k})_2 - \mathsf{c}(\mathbf{k})_2^+ \mathsf{c}(\mathbf{k})_1 \right), \tag{7.198}$$

$$\Sigma(\mathbf{k})_3 = \mathsf{n}(\mathbf{k})_1 - \mathsf{n}(\mathbf{k})_2. \tag{7.199}$$

The only non-zero commutator of the mode operators is

$$[\mathsf{c}(\mathbf{k})_\lambda^+, \mathsf{c}(\mathbf{k})_{\lambda'}] = \delta_{\lambda\lambda'}, \ \lambda, \lambda' = 1, 2, \tag{7.200}$$

and so the commutation relations of the Stokes operators are

$$[\Sigma(\mathbf{k})_i, \Sigma(\mathbf{k})_j] = 2i\varepsilon_{ijn}\Sigma(\mathbf{k})_n, \quad i, j, n = 1, 2, 3. \tag{7.201}$$

Thus the Stokes operators $\{\Sigma(\mathbf{k})_i\}$ are isomorphic to the operators that represent an angular momentum $j = \frac{1}{2}$. Since

$$[\Sigma(\mathbf{k})_0, \Sigma(\mathbf{k})_i] = 0, \quad i = 1, 2, 3, \tag{7.202}$$

they may also be thought of as the generators of the Lie group SU(2) that leaves the operator $\Sigma(\mathbf{k})_0$ invariant. The eigenvectors of $\Sigma(\mathbf{k})_0$ must therefore transform according to some irreducible representation of this symmetry group.

Let $n = 2j$ be an eigenvalue of $\Sigma(\mathbf{k})_0$, and let $|\phi_n\rangle$ be an eigenvector,

$$\Sigma(\mathbf{k})_0 |\phi_n\rangle = n |\phi_n\rangle \quad n = 0, 1, 2, \ldots . \tag{7.203}$$

In general this is a degenerate eigenvalue, and in the subspace spanned by the associated eigenvectors $\{|\phi_n\rangle\}$, the Casimir operator, V_1, of the group satisfies

$$V_1 \equiv \Sigma(\mathbf{k})_1^2 + \Sigma(\mathbf{k})_2^2 + \Sigma(\mathbf{k})_3^2 = j(j+1)\mathsf{I}. \tag{7.204}$$

The Stokes operators $\{\Sigma(\mathbf{k})_i : i = 1, 2, 3\}$ thus belong to the representation D^j of the rotation group, and the eigenvalues of the operators $\frac{1}{2}\Sigma(\mathbf{k})_i$ in a given subspace are just $+j, j-1, \ldots, -j$. Since the Stokes operators do not mutually commute, the simultaneous measurement of this set of eigenvalues is not possible, but as we have noted, they all commute with $\Sigma(\mathbf{k})_0$, and so the photon states for a given wave vector $\mathbf{k}$ can be labelled by the eigenvalues of $\Sigma(\mathbf{k})_0$ and of one component of $\boldsymbol{\Sigma}(\mathbf{k})$.

In terms of Fock space states based on the orthogonal polarisation vectors $\{\hat{\boldsymbol{\varepsilon}}_\lambda, \lambda = 1, 2\}$ we have directly

$$\Sigma(\mathbf{k})_3 |\mathbf{k}, \hat{\boldsymbol{\varepsilon}}_1\rangle = +1 |\mathbf{k}, \hat{\boldsymbol{\varepsilon}}_1\rangle, \tag{7.205}$$

$$\Sigma(\mathbf{k})_3 |\mathbf{k}, \hat{\boldsymbol{\varepsilon}}_2\rangle = -1 |\mathbf{k}, \hat{\boldsymbol{\varepsilon}}_2\rangle. \tag{7.206}$$

Elementary calculations then show that the matrix representatives of the Stokes operators in this basis are just the familiar Pauli matrices,

$$\Sigma(\mathbf{k})_i |\mathbf{k}, \hat{\boldsymbol{\varepsilon}}_\lambda\rangle = \sum_{\lambda'} |\mathbf{k}, \hat{\boldsymbol{\varepsilon}}_{\lambda'}\rangle \, \sigma_{\lambda,\lambda'}^i, \tag{7.207}$$

with

$$\sigma^1 = \begin{bmatrix} 0 & 1 \\ 1 & 0 \end{bmatrix}, \quad \sigma^2 = \begin{bmatrix} 0 & i \\ -i & 0 \end{bmatrix}, \quad \sigma^3 = \begin{bmatrix} 1 & 0 \\ 0 & -1 \end{bmatrix}. \tag{7.208}$$

Thus another complete specification of the photon is afforded by specifying the momentum of the photon and its polarisation state; this is the representation assumed earlier in this chapter.

The helicity and polarisation of the photon are simply related, for if we take $\mathbf{n}$ to define the z-direction, the spherical polarisation unit vectors $\{\hat{\mathbf{e}}_\lambda, \lambda = \pm 1, 0\}$ can be recognised as the eigenfunctions of the helicity operator $\mathbf{S} . \mathbf{n}$ with eigenvalues $\pm 1, 0$ (in units of $\hbar$), as may be seen from (4.97)–(4.99). Equally, from the properties of the Pauli matrices, we infer that the Fock space states $\{|\mathbf{k}, \hat{\mathbf{e}}_\lambda\rangle, \lambda = 1, 2\}$ are eigenstates of the Stokes operator $\Sigma(\mathbf{k})_2$ with eigenvalues ± 1. In other words, the helicity states with eigenvalues $+1$ and -1 of a photon with definite momentum $\mathbf{k}$ correspond to right-handed and left-handed circular polarisation respectively for the photon.

According to the general principles of quantum mechanics we can consider super-positions of helicity states, so that, for example, the state

$$|\phi\rangle = a_{+1}|\hat{\mathbf{e}}_{+1}\rangle + a_{-1}|\hat{\mathbf{e}}_{-1}\rangle, \quad |a_{+1}|^2 + |a_{-1}|^2 = 1 \tag{7.209}$$

describes a photon state for which $|a_{+1}|^2$ and $|a_{-1}|^2$ are the probabilities for measuring right-handed and left-handed circular polarisation respectively. The generic case is one of elliptic polarisation with $|a_{\pm 1}|$ both non-zero and unequal; linear polarisation arises when $|a_{+1}| = |a_{-1}|$. The overall phase of these coefficients has no physical significance, but the relative phase is still important. For linear polarisations with $a_{-1} = a_{+1}^*$ the phase of a_{+1} may be identified as the angle between the plane of polarisation and a fixed direction that is orthogonal to the photon wave vector $\mathbf{k}$ [24]. The superposition $|\phi\rangle$ can just as well be referred to the orthogonal polarisation base states $|\hat{\mathbf{e}}_\lambda\rangle$ and then, by virtue of the isomorphism between the Pauli matrices and the Stokes operators just described, different superpositions correspond to rotations in the polarisation state space induced by the Stokes operators.

In order to explore the quantum mechanical description of polarisation further it is convenient to make a simple change of basis for the polarisation state space and express a general polarisation state in terms of its projections on the orthogonal polarisation Fock space states $\hat{\mathbf{e}}_\lambda$, $\lambda = 1, 2, 3$:

$$|\mathbf{k}, \boldsymbol{\pi}\rangle = \sum_{\lambda=1,2,3} |\mathbf{k}, \hat{\mathbf{e}}_\lambda\rangle\langle\hat{\mathbf{e}}_\lambda|\boldsymbol{\pi}\rangle, \tag{7.210}$$

where

$$|\mathbf{k}, \hat{\mathbf{e}}_\lambda\rangle = c_{\mathbf{k},\lambda}^+ |[0]\rangle. \tag{7.211}$$

The transformation amplitudes follow from the geometry of the polarisation ellipse, Figure 4.1, and the transversality condition, [25]

$$\langle\hat{\mathbf{e}}_1|\boldsymbol{\pi}\rangle = \cos(\eta)\cos(\theta) - i\sin(\eta)\sin(\theta) \equiv b_1, \tag{7.212}$$

$$\langle\hat{\mathbf{e}}_2|\boldsymbol{\pi}\rangle = \cos(\eta)\sin(\theta) + i\sin(\eta)\cos(\theta) \equiv b_2, \tag{7.213}$$

$$\langle \hat{\varepsilon}_3 | \boldsymbol{\pi} \rangle = 0, \tag{7.214}$$

where θ and η are the azimuth and ellipticity.

We have noted that the classical description of polarisation using the electric field vector, $\mathbf{E}$, is not possible in quantum theory because the quantised field operator, $\mathbf{E}$, is linear in the photon annihilation and creation operators, and so has vanishing diagonal elements. Bilinear combinations of these operators, however, have non-zero expectation values in general and can provide a description of the polarisation properties of the photon. To this end we define the set of real Stokes parameters as the expectation values of the Stokes operators just described. The one-photon state $|\mathbf{k}, \boldsymbol{\pi}\rangle$, (7.210), yields

$$S_\mu = \langle \mathbf{k}, \boldsymbol{\pi} | \Sigma(\mathbf{k})_\mu | \mathbf{k}, \boldsymbol{\pi} \rangle, \quad \mu = 0, 1, 2, 3, \tag{7.215}$$

with

$$S_0 = b_1 b_1^* + b_2 b_2^* = 1, \tag{7.216}$$
$$S_1 = b_1 b_2^* + b_1 b_2^* = \cos(2\eta)\sin(2\theta), \tag{7.217}$$
$$S_2 = i(b_1 b_2^* - b_2 b_1^*) = \sin(2\eta), \tag{7.218}$$
$$S_3 = b_1 b_1^* - b_2 b_2^* = \cos(2\eta)\cos(2\theta), \tag{7.219}$$

in terms of the quantities in (7.212) and (7.213).

Photon states with definite momentum and polarisation are pure states in quantum mechanics; so-called mixed states photons are also possible and may be described by a density-matrix formalism. For simplicity the discussion is restricted to a one-photon state with definite momentum but with a polarisation state that in general must be described by assigning probabilities to possible pure polarisation states. In this situation the reduced one-photon polarisation density matrix is a 2×2 Hermitian matrix, ρ_{ij}; it is normalised by the usual condition

$$\mathrm{Tr}(\boldsymbol{\rho}) = 1 \rightarrow \rho_{11} + \rho_{22} = 1. \tag{7.220}$$

Hermiticity requires the diagonal elements to be real, and $\rho_{12} = \rho_{21}^*$. A two-dimensional density matrix can always be expanded in terms of the identity matrix and the Pauli matrices, and so be specified by three real parameters $\{P_1, P_2, P_3\}$ which are conventionally thought of as the components of a polarisation vector, $\mathbf{P}$, [26]:

$$\boldsymbol{\rho} = \frac{1}{2}(\mathbf{I} + \mathbf{P} \cdot \boldsymbol{\sigma}). \tag{7.221}$$

We may express the polarisation vector directly in terms of the Stokes parameters $\{S_1, S_2, S_3\}$, and taking the Pauli matrices in the familiar representation in which σ^3 is diagonal, (7.208), obtain

$$\boldsymbol{\rho} = \frac{1}{2} \begin{bmatrix} 1 + S_3 & S_1 - iS_2 \\ S_1 + iS_2 & 1 - S_3 \end{bmatrix}. \tag{7.222}$$

This density matrix may be transformed to its diagonal representation

$$\boldsymbol{\rho} = \frac{1}{2} \begin{bmatrix} 1 + P & 0 \\ 0 & 1 - P \end{bmatrix}, \tag{7.223}$$

where $P^2 = S_1^2 + S_2^2 + S_3^2$; evidently $P_1 = P_2 = 0$ and so $P = S_3$. The parameter P, which takes the values $0 \leq P \leq 1$, is called the degree of polarisation. The values $P = +1$ and $P = -1$ correspond to complete polarisation along the directions $\hat{\boldsymbol{\varepsilon}}_1$ and $\hat{\boldsymbol{\varepsilon}}_2$ respectively, that is, to pure linear polarisation along these directions. In the case of an unpolarised photon we have $P = 0$ and $S_1 = S_2 = S_3 = 0$; the density matrix is simply

$$\rho_{\alpha\beta} = \frac{1}{2}\,\delta_{\alpha\beta}, \tag{7.224}$$

which tells us that all directions of polarisation are equally likely. This does not mean that an unpolarised photon can be thought of as the superposition of two completely polarised pure states with opposite directions of polarisation, for there are an infinite number of ways the density matrix can be decomposed as a superposition.

The construction (7.210) is readily generalised. The state vector $|\mathbf{k}, \boldsymbol{\pi}[n]\rangle$ which describes the state with n photons in the mode with wave vector $\mathbf{k}$ and polarisation $\boldsymbol{\pi}$ may be constructed by the n-fold application of the creation operator $c^+_{\mathbf{k},\boldsymbol{\pi}}$ to the vacuum state for this mode, $|\mathbf{k}, [0]\rangle$. In terms of the creation operator $c^+_{\mathbf{k},\lambda}$ for a plane-polarised photon, we may write

$$c^+_{\mathbf{k},\boldsymbol{\pi}} = \sum_{\lambda=1,2} b_\lambda\, c^+_{\mathbf{k},\lambda}, \tag{7.225}$$

and so

$$|\mathbf{k}, \boldsymbol{\pi}[n]\rangle = \frac{1}{\sqrt{n!}}\left(c^+_{\mathbf{k},\boldsymbol{\pi}}\right)^n |\mathbf{k}, [0]\rangle \tag{7.226}$$

$$= \frac{1}{\sqrt{n!}}\left(\sum_{\lambda=1,2} b_\lambda\, c^+_{\mathbf{k},\lambda}\right)^n |\mathbf{k}, [0]\rangle \tag{7.227}$$

$$= \sum_r \sqrt{\binom{n}{r}}\, b_1^{n-r}\, b_2^r\, |\mathbf{k}_1[n-r]\,\mathbf{k}_2[r]\rangle, \tag{7.228}$$

where $\binom{n}{r}$ is a binomial coefficient, and $|\mathbf{k}_1[n-r]\mathbf{k}_2[r]\rangle$ is a fully symmetrised boson state. The n-photon density matrix is described in [25].

7.6 Other Field Quantisation Schemes

The quantisation of the free electromagnetic field described in this chapter is equivalent to the usual one in covariant QED which starts from a continuum model of the field, whereas here we have used the device of box quantisation which is eventually removed by changing the sum over (discrete) wave vectors $\mathbf{k}$ to integration over a continuous $\mathbf{k}$, and the replacement of the quantisation volume Ω with $(2\pi)^3$. The expansion in either plane waves or spherical waves leads to the energy eigenstates of the quantised electromagnetic field, and the recognition that there is a Hilbert space represented as a Fock space based on the photon number states $\{n_i\}$, where i is either $\mathbf{k}$ for plane waves or

ωJM for spherical waves. Although not of direct concern here, it is the case that there are other possibilities that are applicable in non-relativistic QED; these modern developments are central now to both fundamental research and applications in 'quantum optics', 'cavity QED' and allied areas of physics.

The essential condition for free-space quantisation, alluded to in Chapter 4, is that the quantisation box has dimensions much greater than any wavelengths of interest; the results of calculations are then independent of any details of the box. For example, we assumed that the fact the box has conducting walls, used to fix the boundary conditions for the plane wave expansion, could be ignored; however, a charge near a conducting wall is associated with an 'image charge' and, if the 'box' is of microscopic dimensions, the image charge is relevant. In particular, the commutation relations for the field involve the image charge [27]. This situation can be realised in solid-state materials, and this has led to a dynamic area of modern physics research – 'cavity QED' [28].

Another possibility is the recognition that in some experiments there is nothing recognisable as a 'cavity', and yet an infinite plane wave does not capture the physical situation, for example light in a waveguide or optical fibre. Moreover, the plane wave expansion is incapable of capturing the highly engineered structure that can be achieved with laser radiation so that the use of box quantisation may lead to spurious results. The simplest form of a better description could be based on a single light beam travelling in a straight line taken to be of infinite length with a constant finite beam cross-sectional area $\mathcal{A}$. If we choose the z-axis parallel to the beam direction and ignore transverse excitations ($k_x = k_y = 0$), the mode expansion of the Coulomb gauge vector potential, Eq. (7.122), is replaced by

$$\mathbf{A}(z) = \mathbf{A}(z)^+ + \mathbf{A}(z)^-, \tag{7.229}$$

where

$$\mathbf{A}(z)^+ = \sum_{\lambda} \int_{-\infty}^{+\infty} dk \sqrt{\frac{\hbar}{4\pi\varepsilon_0 ck\mathcal{A}}} \,\hat{\boldsymbol{\varepsilon}}(k)_\lambda \mathsf{c}(k)_\lambda e^{ikz} \tag{7.230}$$

and $\mathbf{A}^-$ is the Hermitian conjugate of (7.230). The construction of a Fock space, coherent states and so on then follows similar lines to that discussed in this chapter [29].

An even more refined description of a beam has been developed in modern quantum optics since laser radiation typically has novel properties that are not found with conventional light sources. In particular the assumption we just made of a constant cross-sectional area $\mathcal{A}$ can be improved upon. The basic ideas of the wave approach to optics can be based on the scalar Helmholtz equation for an amplitude factor $\Psi(\mathbf{x})$,[9]

$$(\nabla^2 + k^2)\Psi(\mathbf{x}) = 0. \tag{7.231}$$

---

[9] The amplitude will be modulated by the time-dependent factors $e^{\pm i\omega t}$, where $\omega = kc$ in the full solution to the wave equation.

The idea is to construct a complete set of modes other than the plane waves we have generally used so as to make mode expansions of the field variables that can be quantised to give a Fock space structure. This is the route to 'quantum optics'. The scalar wave equation is related to the vector wave equations for the field variables (e.g. (2.62), (2.63)) in the following way. First they can be rewritten in a more symmetrical form through the introduction of two vector fields known as the electric and magnetic Hertz potentials:

$$\phi = -\frac{1}{\varepsilon_0} \mathbf{\nabla} \cdot \mathbf{\Pi}_e \tag{7.232}$$

$$\mathbf{a} = \frac{1}{\varepsilon_0 c^2} \frac{\partial \mathbf{\Pi}_e}{\partial t} + \mathbf{\nabla} \wedge \mathbf{\Pi}_m. \tag{7.233}$$

In free space in the absence of sources, the Hertz potentials can be expressed in terms of two scalar functions and a fixed direction denoted by a unit vector which we will call $\hat{\mathbf{\varepsilon}}_3$,

$$\mathbf{\Pi}_e = \chi \hat{\mathbf{\varepsilon}}_3, \quad \mathbf{\Pi}_m = \psi \hat{\mathbf{\varepsilon}}_3. \tag{7.234}$$

Commonly, $\hat{\mathbf{\varepsilon}}_3$ will be associated with the propagation direction of a light beam which we take to be the z-axis ($\hat{\mathbf{\varepsilon}}_3 = \hat{\mathbf{z}}$); the remarkable and varied nature of the light that can be produced with laser devices is seen in the plane transverse to the propagation axis. For monochromatic fields in homogeneous, isotropic and optically linear media, both χ and ψ are solutions of (7.231) with $k = 2\pi/\lambda$.

Given Ψ, one can work back through the Hertz potentials and construct the electromagnetic fields $(\mathbf{E}, \mathbf{B})$, for example

$$\mathbf{E} = a_{\mathrm{TE}} (\mathbf{\nabla} \wedge \hat{\mathbf{z}} \Psi) + a_{\mathrm{TM}} k^{-1} (\mathbf{\nabla} \wedge \mathbf{\nabla} \wedge \hat{\mathbf{z}} \Psi), \tag{7.235}$$

where the (complex) coefficients a specify the transverse electric and magnetic components of the light beam. A widely used approximation is to attempt a solution for Ψ in terms of a plane wave of infinite extent along the propagation direction, z, multiplied into a complex amplitude u:

$$\Psi(\mathbf{x}) = u(\mathbf{x}) e^{ikz}. \tag{7.236}$$

The paraxial approximation is to take $u(\mathbf{x})$ as the solution of the *paraxial* equation

$$\left(2ik \frac{\partial}{\partial z} + \nabla_T^2 \right) u(x, y) = 0 \tag{7.237}$$

in terms of the two-dimensional Laplacian for the transverse variables. This is justified under the assumption that the variation of u along z is slow within the distance of a wavelength $\lambda = 2\pi/k$; that is, within a distance $\Delta z = \lambda$, the change Δu is much smaller than u itself. The paraxial equation has important applications in optics, where it provides solutions that describe the propagation of electromagnetic waves in the form of, for example, paraboloidal waves or Gaussian beams.

The scalar wave equation has a fundamental symmetry group describing invariance under translations and rotations – the Euclidean group E(3) [30]. If we define the differential operators

$$P_i = \frac{\partial}{\partial x_i}, \quad J_i = (\mathbf{x} \wedge \mathbf{P})_i, \quad i = 1, 2, 3, \tag{7.238}$$

these six quantities satisfy a closed Lie algebra,[10]

$$[P_i, P_k] = 0, \quad [J_i, J_k] = \varepsilon_{ikj} J_j, \quad [J_i, P_k] = \varepsilon_{ikj} P_j, \tag{7.239}$$

where $[.,.]$ is the commutator of the two first-order differential operators.[11] Of special importance are the Casimir invariants of the group, those quadratic combinations of the generators that commute with all individual generators; these are

$$\mathbf{P} \cdot \mathbf{P} = \sum_i^3 P_i^2 \quad \mathbf{J} \cdot \mathbf{P} = \sum_i^3 J_i P_i, \tag{7.240}$$

that is, the square of the 'momentum' and the projection of the 'angular momentum' on the 'momentum' vector. The eigenvalues of the Casimir invariants can be used to index the solutions of the wave equation.

Separation of variables is an important route to the solution of a partial differential equation like (7.231). It is associated with the existence of other quadratic combinations of the generators (7.240) that commute with the Casimir operators. A very elementary example is the observation that the squares of the components of $\mathbf{P}$ in rectangular Cartesian coordinates individually commute with $|\mathbf{P}|^2$, and if we choose two of them, say P_x and P_y, there are two more eigenvalue equations,

$$P_x^2 \Psi(\mathbf{x}) = -k_x^2 \Psi(\mathbf{x}), \quad P_y^2 \Psi(\mathbf{x}) = -k_y^2 \Psi(\mathbf{x}), \tag{7.241}$$

and the solution of (7.231) is a simple product of plane waves,

$$\Psi(\mathbf{x}) = e^{-ik_x x} e^{-ik_y y} e^{-ik_z z}, \quad k^2 = k_x^2 + k_y^2 + k_z^2. \tag{7.242}$$

More interesting is the observation that in circular cylindrical coordinates (r, ϕ, z) there are two commuting operators, P_z^2 and J_z^2; their eigenvalue equations are

$$P_z^2 \Psi = -k_z^2 \Psi, \quad J_z^2 \Psi = -m \Psi, \quad \text{integer } m \tag{7.243}$$

and the modes can be expressed using Bessel functions,

$$\Psi_m(\mathbf{x}, k_\perp, k_z) = J_m(k_\perp r) e^{im\phi} e^{-ik_z z}, \tag{7.244}$$

where

$$k_\perp = \sqrt{k_x^2 + k_y^2} \tag{7.245}$$

is the magnitude of the momentum $\mathbf{k}_\perp$ transverse to the z-direction. Any solution of the scalar wave equation that is regular at the origin can be expressed as a superposition of these modes. The quantisation of so-called 'Bessel beams' is described in [31]. These ideas can be extended to other sets of orthogonal functions. So the expansion in

---

[10] A subgroup of the relativity groups described in Chapter 5.
[11] Although no dimensions are specified, they are usually known as 'momentum' (**P**) and 'angular momentum' (**J**).

terms of plane waves in the traditional route to the quantisation of the electromagnetic field can be generalised to encompass the structured radiation that laser devices can produce.

The plane wave expansions of the quantised field variables $\{\mathbf{A}, \mathbf{E}^{\perp}, \mathbf{B}\}$ can thus be transformed so as to make apparent this structure; one can define field operators which can create or annihilate a single photon in a defined spatial mode (e.g. in a Hermite–Gauss or Laguerre–Gauss mode) by action on a suitably defined vacuum state [32]–[36]. Such states are said to have an orbital angular momentum associated with a wavefront with a helical phase structure – 'twisted light'; in terms of cylindrical coordinates such beams have a phase $e^{il\phi}$, where l is an integer which is interpreted as a photon having orbital angular momentum $\hbar l$; since the photon has zero mass, this is a challenging interpretation! However that may be, structured laser beams have been the subject of much recent research and are of particular interest for molecular chiroptical light scattering. In the notation we have just used an appropriate vector potential can be represented as

$$\mathbf{A}^{+} = \sum_{\mathbf{k}, \lambda, l, p} \sqrt{\frac{\hbar}{2\varepsilon_0 \Omega \mathcal{E}_{\mathbf{k}}}} \hat{\boldsymbol{\varepsilon}}(\mathbf{k}, l, p)_{\lambda} \mathrm{c}(\mathbf{k}, l, p)_{\lambda} F(r)_{l,p} e^{i(kz+l\phi)}, \qquad (7.246)$$

where $F(r)_{l,p}$ is a normalised radial distribution describing the beam's profile transverse to the propagation direction (e.g. Laguerre–Gaussian) and l, p are further quantum numbers required to characterise the beam [37].

References

[1] Casimir, H. B. G. (1948), Proc. Kon. Nederl. Akad. Wet **51**, 793.

[2] Casimir, H. B. G. and Polder, D. (1948), Phys. Rev. **73**, 360.

[3] Verwey, E. J. W. and Overbeek, J. T. G. (1948), *Theory of the Stability of Lyophobic Colloids*, Elsevier.

[4] Power, E. A. (1974), Phys. Rev. **A10**, 756.

[5] Heitler, W. (1954), *The Quantum Theory of Radiation*, 3rd ed., Clarendon Press, reprinted (1984) by Dover Publications, Inc.

[6] Loudon, R. (1983), *The Quantum Theory of Light*, 2nd ed., Clarendon Press.

[7] Bialynicki-Birula, I. and Bialynicki-Birula, Z. (1975), *Quantum Electrodynamics*, PWN and Pergamon.

[8] Wódkiewicz, K. (1980), *Foundations of Radiation Theory and Quantum Electrodynamics*, ed. A. O. Barut, p. 109, Plenum Press.

[9] Walls, D. F. (1977), Amer. J. Phys. **45**, 952.

[10] Dirac, P. A. M. (1930), *The Principles of Quantum Mechanics*, 1st ed., Clarendon Press.

[11] Susskind, L. and Glogower, J. (1964), Physics **1**, 49.

[12] Ozawa, M. (1997), Ann. Phys. **257**, 65.

[13] Bohr, N. and Rosenfeld, L. (1933), Dansk. Vid-Selsk. Mat-Fys. Medd. **12**, 8.

[14] Bohr, N. and Rosenfeld, L. (1950), Phys. Rev. **78**, 794.

[15] Strocchi, F. (1985), *Elements of the Quantum Mechanics of Infinite Systems*, World Scientific Press.

[16] Heisenberg, W. (1930), *The Physical Principles of the Quantum Theory*, translated by Carl Eckart and F. C. Hoyt, University of Chicago Press. Reprinted (1949) by Dover Publications Inc.

[17] Menikoff, R. and Sharp, D. H. (1977), J. Math. Phys. **18**, 471.

[18] Hill, T. L. (1960), *An Introduction to Statistical Thermodynamics*, Addison-Wesley Publishing Co. Inc., reprinted (1986) by Dover Publications, Inc.

[19] Berestetskii, V. B., Lifshitz, E. M. and Pitaevskii, L. P. (1971), *Course of Theoretical Physics*, vol. 4, *Relativistic Quantum Theory*, Part 1. Translated from the Russian by J. B. Sykes and J. S. Bell, Pergamon Press.

[20] Hamilton, J. (1959), *The Theory of Elementary Particles*, Clarendon Press.

[21] Kaempffer, F. A. (1965), *Concepts in Quantum Mechanics*, Pure and Applied Physics **18**, Academic Press.

[22] Jauch, J. M. and Rohrlich, R. (1976), *The Theory of Photons and Electrons*, 2nd expanded edition, Springer-Verlag.

[23] Akhiezer, A. I. and Berestetskii, V. B. (1965), *Quantum Electrodynamics*. Translated from the Russian second edition by G. M. Volkoff. Interscience Monographs and Texts in Physics and Astronomy, **XI**, Interscience Publishers.

[24] Weinberg, S. (1995), *The Quantum Theory of Fields*, vol. 1, *Foundations*, Cambridge University Press.

[25] Atkins, P. W. and Barron, L. D. (1969), Mol. Phys. **16**, 453; ibid. **18**, 721, 729.

[26] Roman, P. (1965), *Advanced Quantum Theory*, Addison-Wesley.

[27] Power, E. A. and Thirunamachandran, T. (1990), Phys. Rev. **A25**, 2473.

[28] Haroche, S. and Raimond, J.-M. Sci. Am. (1993), **268**, 54; accessed at www.scientificamerican.com/article/cavity-quantum-electrodynamics.

[29] Blow, K. J., Loudon, R., Phoenix, S. J. D. and Shephered, T. J. (1990), Phys. Rev. A **42**, 4102.

[30] Rodríguez-Lara, B. H., El-Ganainy, R. and Guerrero, J. (2018), Science Bulletin, **63**, 244.

[31] Jáuregui, R. and Hacyan, S. (2005), Phys. Rev. **A71**, 033411.

[32] Dávila Romero, L. C., Andrews, D. L. and Babiker, M. (2002), J. Opt. B: Quantum Semiclass. Opt. **4**, S66.

[33] Deutsch, I. H. and Garrison, J. C. (1991), Phys. Rev. **A43**, 2498.

[34] Aiello, A. and Woerdman, J. P. (2005), Phys. Rev. **A72**, 060101.

[35] Garrison, J. C. and Chiao, R. Y. (2008), *Quantum Optics*, Oxford University Press.

[36] Rubinsztein-Dunlop, H. et al. (2017), J. Optics, **19**, 013001 (51 pages).

[37] Forbes, K. A. and Salam, A., (2019), Phys. Rev. **A100**, 053413 (13 pages).

8 The Coulomb Hamiltonian

8.1 Introduction

In this chapter we review the properties of the Hamiltonian operator obtained from the canonical quantisation of the classical Hamiltonian for particles interacting through electrostatic forces only. If all the terms involving the electromagnetic field are omitted from the general Hamiltonian for electrodynamics obtained in Chapter 3, *and* the choice $\mathbf{P}(\mathbf{x})^\perp = 0$ is made, there remains

$$H_{\text{charges}} = \sum_i^n \frac{p_i^2}{2m_i} + \sum_{i<j}^n \frac{e_i e_j}{4\pi\varepsilon_0 |\mathbf{x}_i - \mathbf{x}_j|} = T + V, \tag{8.1}$$

with Dirac bracket

$$[x_i^r, p_j^s]^* = \delta_{ij}\,\delta_{rs}, \tag{8.2}$$

for the canonical particle position and momentum variables. T accounts for the total kinetic energy of the particles, while V gives the total potential energy due to the pairwise interactions of all the charges through the static classical Coulomb force.

According to a theorem due to Schwarzschild the classical dynamics under a potential V that is bounded from below can be characterised as follows. Almost all orbits $(x(t), p(t))$ fall into two classes: either $x(t)$ remains bounded for all t, or $x(t)$ becomes unbounded in both time directions, $t \to \pm\infty$. If the two-body interaction $v(\mathbf{r})$ has a Fourier transform $\tilde{v}(\mathbf{k})$, the total potential energy can be expressed as

$$V = \sum_{i<j}^n e_i e_j v(|\mathbf{x}_i - \mathbf{x}_j|) = -\frac{n}{2}v(0) + \frac{1}{(2\pi)^3}\int \tilde{v}(\mathbf{k}) \left|\sum_i e_i\, e^{i\mathbf{k}\cdot\mathbf{x}_i}\right|^2 d^3\mathbf{k}. \tag{8.3}$$

In the case of the Coulomb interaction $\tilde{v}(\mathbf{k}) = 4\pi/k^2 > 0$ and so the potential energy V is bounded from below by $-nv(0)/2$; however, for point charges as $r \to 0, v(r) \to \pm\infty$. Thus the r^{-1} singularity in the Coulomb potential energy can lead to pathological dynamics in which a particle is neither confined to a bounded region, nor escapes to infinity for good [1]. Another problem was encountered already by Newton, who solved the classical problem with the inverse square force law by reducing the orbit equation to quadrature; for $n > 2$ the equations of motion are not separable and lead to a very complicated dynamics. Newton struggled famously to account quantitatively for the orbit of the moon in the earth-moon-sun problem ($n = 3$). We now know that the underlying reason for his difficulties is the existence of solutions carrying the signature of chaos [2].

Canonical quantisation replaces the variables in the classical system (8.1), (8.2) with formal quantum mechanical operators. Thus the classical Hamiltonian becomes

$$H = \sum_i^n \frac{\mathsf{p}_i^2}{2m_i} + \sum_{i<j}^n \frac{e_i e_j}{4\pi\varepsilon_0 |\mathbf{x}_i - \mathbf{x}_j|} = \mathsf{T} + \mathsf{V}, \tag{8.4}$$

acting on the Hilbert space of square integrable functions $\omega = \Re^{3n}$, customarily denoted as $\mathcal{H} = L^2(\omega)$. Similarly, in place of (8.2) the particle position and momentum operators have the commutation relation

$$[\mathsf{x}_i^r, \mathsf{p}_j^s] = i\hbar \delta_{ij}\delta_{rs}. \tag{8.5}$$

In the modern literature (8.4) is usually referred to as the Coulomb Hamiltonian and we have followed this practice. Specialised to the case of one nucleus, it is the basis for the quantum theory of the atom; in the multinuclear case it provides the starting point for the quantum theory of molecules and of condensed matter (the 'many-body problem') [3]. Two points need to be recognised; firstly, the choice $\mathbf{P}(\mathbf{x})^\perp = 0$ is the choice of a definite gauge for the vector potential – the *Coulomb gauge*, so everything that follows is dependent on that choice, and lurking in the background is the question of the *gauge invariance* of conclusions. Secondly, it is taken for granted that now the charge and mass parameters in (8.1) are the correct empirical values rather than the formal values introduced in Chapter 3. Although explicit solutions can only be found in the two-body case (as in classical dynamics), much can be said about qualitative aspects of its spectral and symmetry properties. In this chapter we consider how much and what is known about it, concentrating particularly on the molecular case [4]–[9]. From the point of view of practical calculations for molecules the celebrated Born–Oppenheimer approximation is probably the most important idea in molecular physics and therefore is given a detailed treatment in §8.5.

8.2 Fundamental Properties of the Coulomb Hamiltonian

The pathologies of the classical many-body problem with inverse square law forces are swept away by quantisation thanks to a celebrated proof by Kato [10] that the Coulomb Hamiltonian is essentially self-adjoint on the same domain as that for free particles, that is, the domain of T, the Sobolev space $\mathcal{H}^2(\chi)$ (see (8.19)). This is true for arbitrary numbers of particles, with any values for their mass and charge parameters. An important aspect of Kato's proof is that it does not require the explicit solutions of the Schrödinger equation; it amounts to the demonstration that the Coulomb interaction, although unbounded, is infinitely small in comparison with the kinetic energy; this is expressed more precisely by the Kato–Rellich theorem [11]:

$$||V\psi|| \leq \alpha\,||T\psi|| + \beta(\alpha)\,||\psi||, \quad \alpha > 0. \tag{8.6}$$

A corollary of Kato's theorem shows that H is bounded from below [1]; there is thus a lowest energy $E_0 > -\infty$, and we have the operator relation

$$H \geq E_0 1. \tag{8.7}$$

For normalised wave functions in the L^2-Hilbert space one has the Rayleigh–Ritz variational principle,

$$\langle H \rangle = \frac{\langle \psi | H | \psi \rangle}{\langle \psi | \psi \rangle} \geq E_0, \tag{8.8}$$

and by using a scaling argument originally due to Hylleraas, it can be shown that [12]

$$\langle H \rangle \leq 0. \tag{8.9}$$

However, there can be no presumption that the normalisable states achieve the minimum energy.

Kato's result guarantees several properties that are essential in any quantum theory, for example, the time evolution of a Schrödinger wave function given by

$$\psi(t) = e^{-iHt/\hbar}\,\psi(0) \equiv U_t\psi(0) \tag{8.10}$$

is unitary, and so conserves probabilities. Moreover, H has a spectral representation,

$$H = \int \lambda\,dE_\lambda, \tag{8.11}$$

so that the energy distribution $d\langle \psi | E_\lambda | \psi \rangle$ for any normalised state ψ is a probability measure on the spectrum $\sigma(H)$ of H. The spectrum $\sigma(H) = |E_0, \infty)$ lies on the real E-axis.

It is easily seen that the Coulomb interaction is translation invariant and so does not disturb the algebra of the Galilean group (cf. §5.3.1). Thus the total momentum operator, which in the present notation is

$$\mathbf{P_R} = \sum_{i}^{n} \mathbf{p}_i, \tag{8.12}$$

commutes with H, and H has a purely continuous spectrum. Physically the centre of mass of the whole system, with position operator

$$\mathbf{R} = \frac{1}{M}\sum_{i}^{n} m_i\mathbf{x}_i, \quad M = \sum_{i}^{n} m_i, \tag{8.13}$$

behaves like a free particle. It is desirable then to introduce $\mathbf{R}$ and its conjugate $\mathbf{P_R}$, together with appropriate internal coordinates, $\{\mathbf{t}\}$, into H to effect the separation of the centre of mass and the internal dynamics; this will be described explicitly in §8.4. Formally H may be written as a direct integral,

$$H = \int_{\Re^3}^{\oplus} H(P)\,dP, \tag{8.14}$$

where

$$H(P) = \frac{P^2}{2M} + H' \tag{8.15}$$

is the Hamiltonian at fixed total momentum P. H' is independent of P and acts on $L^2(\mathfrak{R}^{3(n-1)})$ [13]; it is translationally invariant.

The focus of the discussion can then be shifted to the internal Hamiltonian H'. There are infinitely many possible choices of the internal coordinates $\{t\}$ that are unitarily equivalent, so that the form of H' is not determined; but whatever coordinates are chosen, the essential point is that H' is the same operator specified by the decomposition (8.15). Since the total angular momentum is constant and can be written in the form

$$\mathbf{L} = \mathbf{L_R} + \mathbf{L}_{int}, \tag{8.16}$$

where both terms are separately conserved, one could further transform H' by making its rotational invariance explicit. The angular momentum operators have discrete spectra, however, and so do not alter the qualitative description of the spectrum.

The centre-of-mass variables do not appear in the internal molecular Hamiltonian, and the wave functions associated with H through the Schrödinger equation

$$H\psi = E\psi, \tag{8.17}$$

can therefore be written in product form,

$$\psi = \Theta(\mathbf{R})_k \Psi(\mathbf{t})_i, \tag{8.18}$$

where $\Psi(\mathbf{t})_i$ is an eigenfunction for the internal Hamiltonian $H'(\mathbf{t})$. If they are to belong to $L^2(\omega)$, Θ must be taken as an appropriate wave packet constructed like (5.77), and Ψ_i must be associated with the energy eigenvalue E_i of a bound state of the molecule.

The Schrödinger equation for H' defined by (8.15) in position representation,

$$H'\Psi(\mathbf{t}) = E\Psi(\mathbf{t}), \tag{8.19}$$

is formally an elliptic partial differential equation (PDE) in the internal coordinates $\{t\}$ on the reduced configuration space $\mathcal{X} = \mathfrak{R}^{3(n-1)}$. The occurrence of the Coulomb singularities in H' and the physical interpretation of Ψ require that (8.19) must be placed in a mathematical setting involving the notion of *distributional derivatives* if it is to be given a precise meaning. One has to view (8.19) in the following way.

We define the Sobolev space $\mathcal{H}^2(\mathcal{X})$ as the space of $\mathcal{L}^2(\mathcal{X})$-functions such that their distributional derivatives up to second order all belong to $\mathcal{L}^2(\mathcal{X})$. For $\Psi \in \mathcal{H}^2(\mathcal{X})$, each term in (8.19) makes sense as a $\mathcal{L}^2(\mathcal{X})$-function and the equality takes place in this space $\mathcal{L}^2(\mathcal{X})$; thus if we denote the kinetic energy part of H' as T, the term $T\Psi$ is a $\mathcal{L}^2(\mathcal{X})$-function that satisfies, for all smooth functions h on $\mathcal{X}$ with bounded support,

$$\langle T\Psi | h \rangle_{\mathcal{L}^2(\mathcal{X})} = \langle \Psi | Th \rangle_{\mathcal{L}^2(\mathcal{X})}, \tag{8.20}$$

where Th is now computed in the usual way. This idea applied to all terms in H' means that for all smooth functions h on $\mathcal{X}$ with bounded support,

$$\langle \Psi | H'h \rangle_{L^2(\mathcal{X})} = E \langle \Psi | h \rangle_{L^2(\mathcal{X})}. \tag{8.21}$$

One should therefore see an eigenfunction $\Psi \in L^2(\mathcal{X})$ as a *distributional solution* to (8.19).

Essentially what is done here is the differentiations in H′ are transferred from the wave function to suitably smooth functions h using integration by parts as required. This point of view is already necessary in the simplest case, the hydrogen atom. After removal of the centre-of-mass motion, the internal Coulomb Hamiltonian involves the electron-proton relative coordinate $\mathbf{r}$; its ground state is given by $\Psi_0 = c\exp(-|\mathbf{r}|)$, in appropriate units. This function is continuous everywhere, and differentiable outside the collision at 0. But it is not differentiable at 0 and so (8.19) cannot be understood as a classical PDE. By contrast, if the potential energy terms are smooth functions, for example Hooke's Law for coupled oscillators, the generalised formalism just described would yield (smooth) solutions everywhere and (8.19) would be recognised as just such a classical PDE [14].

8.3 Spectral Properties

There are various ways in which the spectrum $\sigma(A)$ of a self-adjoint operator A may be classified. From the point of view of measure theory the natural decomposition is into pure point, absolutely continuous and singular continuous parts. The sets are closed but need not be disjoint. The absolutely continuous spectrum describes states in which the particles leave a bounded region in a finite time; these are the scattering states of the system. The discrete spectrum, $\sigma_d(A)$, is the subset of the pure point spectrum that consists of isolated eigenvalues of finite multiplicity; it describes bound states in which the particles stay infinitely long in a bounded region. The essential spectrum, $\sigma_{ess}(A)$, is the complement of the discrete spectrum and is the infinite dimensional part

$$\sigma_{ess}(A) = \sigma(A)\backslash\sigma_d(A). \tag{8.22}$$

The discrete spectrum and the essential spectrum are, by definition, disjoint; however, although the essential spectrum is always closed, the discrete spectrum need not be. $\sigma_{ess}(A)$ includes the absolutely continuous spectrum, $\sigma_{ac}(A)$, and the singular continuous spectrum, $\sigma_{sc}(A)$, this last consisting of infinitely degenerate eigenfunctions.

The quantum analogue of Schwarzschild's classical theorem for a Coulombic system is due to Ruelle. Since H is self-adjoint, the Hilbert space $\mathcal{H}$ of states $\{\psi\}$ decomposes into two U_t-invariant subspaces:

$$\mathcal{H} = \mathcal{H}_b \oplus \mathcal{H}_c. \tag{8.23}$$

$\mathcal{H}_b$ is the subspace of bound states spanned by the eigenfunctions of H′. Such states have the property that for any $\sigma > 0$, x stays with probability $1 - \sigma$ in a finite ball $|x| \le R(\sigma)$ for all σ. $\mathcal{H}_c$ is the complement of $\mathcal{H}_b$; it describes the scattering states, for which the probability to find x in any finite ball $|x| \le R(\sigma)$ at time t vanishes in the time average over both time directions, $-\infty \le t \le 0$ and $0 \le t \le +\infty$ [4], [15]. Thus for the Coulomb Hamiltonian the singular continuous spectrum is empty, $\sigma_{sc}(H) = \emptyset$.

A theorem of Weyl shows that if A is self-adjoint and an operator B differs from it by only a compact perturbation, $\sigma_{ess}(A) = \sigma_{ess}(B)$ [16]. The essential spectrum of the kinetic energy operator, $-\nabla^2$, is easily seen to be $\sigma_{ess}(-\nabla^2) = [0,\infty]$; Weyl's theorem suffices to prove that $\sigma_{ess}(-\nabla^2 + V) = \sigma_{ess}(-\nabla^2) = [0, \infty)$ for a large class of two-body Schrödinger operators (after removal of the centre-of-mass contribution) with $V \to 0$ at infinity, including the case of the Coulomb interaction $1/r$. Suppose we write

$$H = H_0 + V, \tag{8.24}$$

and define the resolvents of H and H_0 as in Chapter 6. Then by standard manipulations we have [17]

$$R(E) = R_0(E) - R_0(E)VR(E). \tag{8.25}$$

In the two-body Coulombic case with the choice of position representation one has $H_0 = -\nabla^2$, $V = 1/r$, and the crucial result is that $R_0(E)V$ is a compact operator, and so is bounded [18].

This property fails in the n-body case where V is a sum over two-particle interactions $V = \sum_{i<j} V_{ij}(\mathbf{x}_i - \mathbf{x}_j)$ which are such that $V_{ij}(\mathbf{r}) \to 0$ as $r \to \infty$, where $r = |\mathbf{x}_i - \mathbf{x}_j|$ is the ij relative distance coordinate. Then even though the two-body potential is well behaved at infinity, V fails to go to zero at infinity in tubes where

$$\sum_{i=1}^{n-1} |\mathbf{x}_i - \mathbf{x}_n|^2 \to \infty, \tag{8.26}$$

while some $|\mathbf{x}_i - \mathbf{x}_j|$ remain finite. As a result, $R_0(E)V$ is not compact, and Weyl's theorem is no longer applicable. During the 1960s and 1970s a different decomposition of the resolvent was discovered and utilised in scattering theory, and this culminated finally in the proof of the Hunziker–van Winter–Zhislin (HVZ) theorem [19]–[21] (Theorem XIII.17 in [22]) which demonstrates that the essential spectrum can be written as $\sigma_{ess}(H') = [\Sigma, \infty)$, where Σ is the energy of the lowest two-body[1] cluster decomposition of the n-particle system. Subsequently a rather different method of proof based on geometrical ideas was developed; a valuable discussion of the HVZ theorem from this point of view can be found in [18].

Even without recourse to detailed mathematics, it is clear that the essential spectrum of the hydrogen atom begins at zero energy; it describes the scattering states of a single electron and a nucleus. It is absolutely continuous and does not contain any pure point members. For all other atoms the first ionisation energy is such that the essential spectrum begins at some energy E below zero energy. It contains states describing the scattering of an electron from a singly ionised atom, two electrons from a doubly ionised atom and so on. These states occur at energies below zero. Thus in the case of the helium atom $\sigma_d(H') = [E_0, E)$ and $\sigma_{ess}(H') = [E, \infty)$. At energies above zero, the spectrum is absolutely continuous and describes the scattering of the electrons by the nucleus.

The Coulomb interaction belongs to a special class of two-body potentials $V(\mathbf{r})$, where $\mathbf{r}$ is the interparticle separation, that are called dilatation analytic; suppose that V has the property that the operator

$$J(\theta) = V(e^\theta \mathbf{r})(-\nabla^2 + 1)^{-1} \tag{8.27}$$

---

[1] It is only necessary to consider the two-body cluster threshold because the three-body cluster threshold can be shown to be above the two-body cluster one and so on (see result (20b) of XIII.5 in [22]).

is compact and has an analytic continuation from real θ to $\{\theta, |\Im \theta| < \varepsilon\}$ for some $\varepsilon > 0$. This suffices to guarantee that any eigenvalues of the internal Hamiltonian H' are of finite multiplicity and can only accumulate at thresholds [23]; physically, the thresholds correspond to fragmentation energies. The exploitation of the dilatation analyticity of the Coulombic interaction is commonly referred to as the complex coordinate rotation technique [23]–[25] because it involves a unitary transformation of the original Hamiltonian that introduces complex coordinates and momenta. This remarkable idea due to Balslev and Combes [23] leads to a transparent picture of the whole spectrum.

Let $\mathbf{x}$ and $\mathbf{p}$ stand collectively for all the coordinate and momentum operators of the n particles described by H, and write a position space wave function as $\psi^n(\mathbf{x})$ with $\mathbf{x} = (\mathbf{x}_1, \mathbf{x}_2, \ldots \mathbf{x}_n)$, where, as discussed previously, §8.2, only $(n-1)$ of the $\mathbf{x}_i$ are independent of the centre-of-mass coordinate $\mathbf{R}$. Now introduce a dilatation transformation with a unitary operator parameterised by an angle θ such that

$$U(\theta)\psi^n(\mathbf{x}) = e^{i3(n-1)\theta/2}\psi^n(\mathbf{x}e^{i\theta}). \tag{8.28}$$

The effect of $U(\theta)$ is to transform $\mathbf{x}$ and $\mathbf{p}$ as

$$\mathbf{x} \to \mathbf{x}e^{i\theta}, \quad \mathbf{p} \to \mathbf{p}e^{i\theta}. \tag{8.29}$$

Then we can construct the unitary transform of the Hamiltonian H under $U(\theta)$

$$H(\theta) = U(\theta) \, H \, U(\theta)^{-1}. \tag{8.30}$$

Complex coordinate rotation allows a particularly vivid visualisation of the entire spectrum and generalises the description of the helium atom just given. The absolutely continuous spectrum of H' lies above a certain number[2] of discrete states which are bound stationary states of the n-particle cluster (atom or molecule). They are embedded in the continuous spectrum of H, a fact that explains why it is advantageous to separate out the centre-of-mass contribution. The singular continuous spectrum of H' is empty, $\sigma_{sc}(H') = \emptyset$. In the complex energy plane the continuous spectrum is a branch cut on the real E axis starting at some value Σ and extending to $+\infty$; it is rich in structure, having numerous sets of other bound states embedded in it. These are associated with the fragments that the n-particle cluster can break up into. A new cluster fragment does not appear until the energy has exceeded a characteristic threshold value. This description of the spectrum is what one would expect intuitively from the facts of mass spectrometry.

With the assumption of Coulombic interactions between the particles the following precise conclusions can be stated [25]:

(i)   Bound state eigenvalues of $H(\theta)$ are independent of θ, and identical to those of H, for $|\theta| < \pi/2$;

(ii)   scattering thresholds corresponding to the possibility of fragmentation into all possible cluster fragments in different states of excitation are also independent of θ for $|\theta| < \pi/2$;

---

[2] This number can range from 0 to ∞.

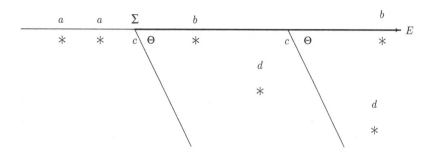

Figure 8.1   The spectrum of the rotated Hamiltonian, $H(\theta)$; a: discrete eigenvalues of H', b: continuum embedded eigenvalues, c: thresholds, d: resonances. Σ is the start of the essential spectrum; $\Theta = 2\theta$.

(iii)   the segments of continua beginning at each scattering threshold rotate by an angle 2θ into the lower half plane ($\theta \geq 0$), each about its own threshold;

(iv)   in addition, discrete eigenvalues of $H(\theta)$ may appear in the sector $0 > \arg(z - \Sigma) > -2\theta$, where z is the complex energy and Σ is the bottom of $\sigma_{ess}(H)$; these are associated with bound states with finite lifetimes (resonances). These features are shown in Figure 8.1.

The extent of the discrete spectrum is by no means obvious, and for a Coulomb Hamiltonian describing a given collection of electrons and nuclei the difficult technical problem is to find out if there is any discrete spectrum at all before the start of the essential spectrum. For the moment let us assume that the nuclei are *fixed* at definite locations defined by (classical) position vectors $\{\mathbf{R}_k\}$. Then the main result is the demonstration of a so-called 'binding condition', first obtained by Zhislin [19]; for a collection of N electrons and A fixed nuclei with total (positive) charge Ze, the bottom of the spectrum of H' is a genuine N-particle bound state that satisfies the Schrödinger equation with some energy E_0 for each choice of the locations of the nuclei, provided $N < Z + 1$. The binding condition is based on a comparison between the ground state energies of the N-electron Hamiltonian with and without the electron-nuclear attraction potential. For fixed nuclei, the internuclear repulsion energy is a constant and can be dropped for the purposes of the comparison. Let

$$H^V(N) = \sum_i^N \left(\frac{p_i^2}{2m} - \frac{e^2}{4\pi\varepsilon_0} \sum_g^A \frac{Z_g}{r_{ig}} \right) + \frac{e^2}{4\pi\varepsilon_0} \sum_{i<j}^N \frac{1}{r_{ij}}, \tag{8.31}$$

$$H^0(N) = \sum_i^N \frac{p_i^2}{2m} + \frac{e^2}{4\pi\varepsilon_0} \sum_{i<j}^N \frac{1}{r_{ij}}, \tag{8.32}$$

and define

$$Z = \sum_g^A Z_g. \tag{8.33}$$

Let $E^V(N)$ and $E^0(N)$ be the bottom of the spectra of $H^V(N)$ and $H^0(N)$ respectively; then Zhislin proved, for $N < Z + 1$ [19],

$$E^V(N) < \min\{E^V(N') + E^0(N - N') : 0 \le N' < N\}. \tag{8.34}$$

It was also shown by Zhislin [19], and later by Uchiyama [26], that in the single nucleus case there are an infinite number of bound states if the atom is electrically positive or neutral. Proofs of this are accessible in Thirring [27] and Simon [28]. A given nucleus with charge Ze can only bind a finite number, $n(Z)$, of electrons [29], [30], although $n(Z)$ is not known explicitly.

If the system has an overall negative charge, then it has at most a finite number of bound states as, again, was first shown by Zhislin [31]. For example, Nyden Hill [32] showed explicitly that the H^- ion has only one bound state. If the system has more than one nucleus, the Hamiltonian H', with the nuclei treated as fixed, has spectral properties just like those of an atom, but the form of the spectrum depends upon the choice of the nuclear geometry; at one extreme the spectrum will be that of the united atom, and at the other, that of the separated atoms [27].

There is the following fundamental theorem due to Weyl: if one has a trial wave function Φ with an expectation value $\langle H'\rangle_\Phi$ which is below the bottom of the essential spectrum, then H' has at least one discrete negative eigenvalue. A proof of this result can be based on the min-max principle pioneered by Weyl and Courant [33]; this characterises the (exact) eigenvalues in terms of an extremum property of the operator. A modern formulation is Theorem XIII.1 in [22]; a self-adjoint operator A that is bounded from below, that is $A \ge cI$ for some finite c, has eigenvalues $\lambda_k(A)$ that can be expressed as

$$\lambda_k(A) = \sup U_A(\phi_1, \phi_2, \ldots \phi_{k-1}), \tag{8.35}$$

where

$$U_A(\phi_1, \phi_2, \ldots \phi_m) = \inf\langle \psi|A|\psi\rangle, \tag{8.36}$$

and

$$\langle \psi|\psi\rangle = 1, \psi \in [\phi_1, \ldots \phi_m]^\perp,$$
$$[\phi_1, \ldots \phi_m]^\perp = \{\psi|\langle \psi|\phi_i\rangle = 0, i = 1, 2, \ldots m\}. \tag{8.37}$$

Then for each fixed k, either

(a)   there are k eigenvalues below $\sigma_{ess}(A)$, and λ_k is the kth eigenvalue

or

(b) λ_k is the bottom of $\sigma_{ess}(A)$, and there are at most $(k - 1)$ eigenvalues below $\sigma_{ess}(A)$. The significance of this result is that if $\sigma_{ess}(A) = [a, \infty)$ for some a and we know that $\lambda_k < a$, there must be at least k eigenvalues [22].

On the other hand if the variation in (8.35)–(8.37) is restricted to a k-dimensional subspace of the domain of the operator A described by an orthogonal projection P, then the min-max principle can be used to characterise the eigenvalues $\tilde{\lambda}_k$ of PAP, and this leads to

$$\lambda_k \le \tilde{\lambda}_k, \tag{8.38}$$

which is the Rayleigh–Ritz principle. Thus if a variational calculation with some trial function yields an expectation value $\langle A \rangle$ below the bottom of $\sigma_{ess}(A)$, there must be at least one eigenvalue below $\langle A \rangle$, by (8.38).

Weyl's theorem justifying the Rayleigh–Ritz principle and the HVZ theorem taken together provide the requisite mathematical justification for a conjecture on which quantum chemistry was founded originally; namely, if one can find an expectation value $\langle H' \rangle$ which is a lower energy than the lowest energy of all possible separated clusters, then there exists a bound ground state. The studies of the energy of the He atom by Hylleraas and of the H_2 molecule by James and Coolidge are typical examples in early quantum chemistry [12], [34]. To make practical use of this result in any particular system, however, one must determine the start of the essential spectrum and find a trial variational function that bounds this starting energy from below; both are very difficult to do.

At present the most that has been proved is that the hydrogen molecule has at least one bound state (Richard et al. [35]). Ordinary chemical experience makes it seem likely that there are some atomic combinations that do not have any bound states but, so far, there are no rigorous results that enable it to be said that a particular kind of neutral system has no bound states. It is known that if a system gets either too positive or too negative, it does not have any bound states at all [36], [37]. Simon [38] argued very persuasively that a neutral system will have an infinite number of bound states only if the position of the bottom of the essential spectrum is determined by breakup into a pair of oppositely charged ionic clusters; if the clusters are neutral, there will be, at most, only a finite number of bound states. Subsequently Vugal'ter and Zhislin [39] showed rigorously that Simon's conjecture about the spectrum in the case of neutral clusters was well founded, and Evans et al. [40] were able to show that his belief about the charged clusters was too. An examination of tables of experimental values of electron affinities and ionisation energies leads to the conclusion that it is very unlikely that any diatomic molecule has an infinite number of bound states. This observation is not inconsistent with spectroscopic experience. A comprehensive review of the technical means used to characterise the full spectrum, discrete and continuous, for the N-body Hamiltonian can be found in [4].

As in the classical case the question of stability is an important one. We have seen that quantisation prevents collapse due to the singularity of the Coulomb potential, and that is sufficient to put the quantum theory of the atom on a sound basis. However, there is no prior limitation on the number of particles that the theory can describe, and it is reasonable to ask whether the quantum mechanical theory applies to macroscopic matter. The energy of ordinary matter is extensive, that is, proportional to the number of particles, n, and the laws of thermodynamics are valid. The thermodynamic limit ($n \rightarrow \infty$) exists, provided the particle density is bounded. For these basic results to be reproduced by quantum mechanics it is necessary that the ground state energy of a 'large' system should be proportional to the number of particles n,

$$\langle \psi_0 | H | \psi_0 \rangle \geq -a\,n, \tag{8.39}$$

where a is independent of n. The $-$ sign indicates a positive binding energy per particle. (8.39) is a highly non-trivial result about a many-body problem; it is significantly stronger than the earlier result that the energy is bounded from below.

To see this, consider the following elementary line of argument. For a system of N electrons (charge $-e$) and A nuclei (charges $Z_g e : g = 1, \dots A$) the sums in (8.4) can be separated into four terms,[3] with the following classical interpretations: the kinetic energy of the nuclei, the internuclear repulsion, the interelectron repulsion and the kinetic energy of the electrons + the electron–nuclear attraction. The first, second and third sums are positive, so let's concentrate on the fourth term which may have either sign. It can be rewritten as a double sum over a collection of one-electron Hamiltonians with an effective nuclear charge of $A Z_g$,

$$\sum_i^N \sum_g^A A^{-1} \left(\frac{p_i^2}{2m} - \frac{e^2}{4\pi\varepsilon_0} \frac{A Z_g}{r_{ig}} \right). \tag{8.40}$$

The inter-particle separation operator $|\mathbf{x}_u - \mathbf{x}_v|$ is denoted by r_{uv}. The ground state energy of the hydrogenic atom with nuclear charge z is

$$e_0 = -\frac{m e^4 z^2}{2\hbar^2 (4\pi\varepsilon_0)^2} = -z^2 \, \varepsilon_H. \tag{8.41}$$

This energy is a lower bound for the corresponding one-electron Hamiltonian, and so we get the crude estimate [34]

$$\sum_i^N \left(\frac{p_i^2}{2m} - \frac{e^2}{4\pi\varepsilon_0} \sum_g^A \frac{Z_g}{r_{ig}} \right) \geq -N A \varepsilon_H \sum_g^A Z_g^2. \tag{8.42}$$

Consider a gas of hydrogen for which $N = A = n/2$, and all $Z_g = 1$; then the RHS of (8.42) varies roughly as n^3. It is evident that the equality in this bound is much too low since it would correspond to putting all the electrons into the lowest energy state, whereas the Pauli exclusion principle (§5.3.3) tells us that no more than two electrons ($\uparrow\downarrow$) can occupy any electron energy level, and that the minimum energy of the N-electron system is found in accordance with Bohr's Aufbauprinzip; hence particle statistics are a vital ingredient in the theory. We note in passing that if the relativistic kinetic energy $\sqrt{p^2 + m^2}$ is substituted for the non-relativistic form $p^2/2m$, the bound (8.39) does not necessarily hold; electrons collapse into the nuclei if the latter have sufficiently high charges $Z_g e$ [1]. If the temperature of the system $T > 0$, we need to discuss the free energy via the partition function Z; stability of the second kind implies an upper bound on Z and so a lower bound on f, the free energy per particle. If f can be shown to have a limit as $|\Omega| \to \infty$ at constant particle density, the thermodynamic limit exists; these matters are discussed in considerable detail in [41]. It is now known that the inclusion of the nuclear kinetic energy operators does not disturb the demonstrations of the existence of a stable ground state for a system of electrons and

---

[3] This is done explicitly in (8.48) in §8.4.

nuclei interacting via purely Coulombic potentials, and of the stability of matter. As an example of the kind of argument involved we just record one elegant demonstration of this result.[4]

Without approximation one can write the full Coulomb Hamiltonian in the form

$$H = T_N + H_0, \tag{8.43}$$

where T_N represents all the nuclear kinetic energy operators. Let φ be any *electronic* wave function with nuclear positions as parameters, and Ψ be any wave function for the full Coulomb Hamiltonian (square integrable and of finite energy). Now, it is shown in [41] that φ satisfies a lower bound with some constant C that does *not* depend on the nuclear positions,

$$\langle \varphi | H_0 | \varphi \rangle_{el} \geq - C \langle \varphi | \varphi \rangle_{el}; \tag{8.44}$$

the subscript indicates that the expectation value is taken over purely the electronic variables. Since C doesn't involve the positions of the nuclei, we also have

$$\langle \Psi | H_0 | \Psi \rangle_{el} \geq - C \langle \Psi | \Psi \rangle_{el}, \tag{8.45}$$

and this implies

$$\langle \Psi | H_0 | \Psi \rangle \geq - C \langle \Psi | \Psi \rangle, \tag{8.46}$$

where the integrations are now over all electronic and nuclear variables. Since the nuclear kinetic energy operator T_N is *non-negative* we can also write

$$\langle \Psi | T_N + H_0 | \Psi \rangle \equiv \langle \Psi | H | \Psi \rangle$$
$$\geq - C \langle \Psi | \Psi \rangle, \tag{8.47}$$

which proves stability for moving electrons and nuclei. We shall return to this discussion in Chapter 11 when the quantised electromagnetic field is included in the Hamiltonian; then much more delicate questions have to be answered.

8.4 Symmetry and the Coulomb Hamiltonian

Let the position variables in the problem be designated as $\mathbf{x}_i$ which will be treated as a column matrix of three Cartesian components $x_{i\alpha}$, $\alpha = x, y, z$; we regard the $\mathbf{x}_i$ collectively as the 3-by-n matrix $\mathbf{x}$. When it is necessary to distinguish between electrons and nuclei, the variables may be split up into two sets, one set consisting of N variables, $\mathbf{x}_i^e$, describing the electrons with charge $-e$ and mass m, and the other set of A variables, $\mathbf{x}_g^n$, describing the nuclei with charges $+Z_g e$ and masses m_g, $g = 1, \ldots A$; then $n = N + A$.

---

[4] I am most grateful to Prof Thierry Jecko for showing me this.

With this notation the Hamiltonian operator for a system of N electrons and A atomic nuclei may be written as

$$H = \sum_{g}^{A} \frac{p_g^2}{2m_g} + \frac{e^2}{4\pi\varepsilon_0} \sum_{g<h}^{A} \frac{Z_g Z_h}{r_{gh}} + \frac{e^2}{4\pi\varepsilon_0} \sum_{i<j}^{N} \frac{1}{r_{ij}} + \sum_{i}^{N} \left(\frac{p_i^2}{2m} - \frac{e^2}{4\pi\varepsilon_0} \sum_{g}^{A} \frac{Z_g}{r_{ig}} \right), \tag{8.48}$$

where each sum has the classical interpretation referred to in §8.3. As noted in §8.2 the symmetries of the Coulomb Hamiltonian are the symmetries associated with the Galilean group, and it is convenient to separate off the centre-of-mass motion so that attention can be transferred to the internal Hamiltonian H′. In addition, there is the permutation symmetry of identical particles. Although it is the case that the internal Hamiltonian is the same operator irrespective of any choices made about the translationally invariant coordinates, for the purposes of practical calculations some choice must be made. We start with a general discussion; for molecules in the perspective of the Born–Oppenheimer approximation (§8.5), being able to clearly identify electrons and nuclei while maintaining the overall symmetry requirements is paramount [42].

The following discussion of the required coordinate transformations is put in classical terms; since only linear relations are involved, the quantum mechanical version is the same with classical variables replaced by operators. We use the centre-of-mass coordinate **R**, (8.13), to write

$$\mathbf{x}_i = \mathbf{R} + \mathbf{r}_i, \quad i = 1, \ldots n, \tag{8.49}$$

and then define a set of $(n-1)$ linearly independent translationally invariant internal coordinates $\{\mathbf{t}_k, k = 1, \ldots n-1\}$ by a linear transformation of the $\{\mathbf{r}_i\}$,

$$\mathbf{t}_k = \sum_i C_{ki} \mathbf{r}_i, \quad i = 1, \ldots n; \, k = 1, \ldots n-1$$

$$0 = \sum_i C_{ki}, \tag{8.50}$$

where the zero row-sum for the matrix **C** ensures the translation invariance of the $\{\mathbf{t}_k\}$. The coordinates $\{\mathbf{r}_i\}$ are not linearly independent since

$$\sum_i m_i \mathbf{r}_i = 0, \tag{8.51}$$

but may be thought of as internal coordinates relative to the centre of mass.

In a similar way we may decompose the momentum variables into a centre-of-mass contribution involving $\mathbf{P_R}$, and a set of internal momenta,

$$\mathbf{p}_i = \frac{m_i}{M} \mathbf{P_R} + \boldsymbol{\sigma}_i, \quad i = 1, \ldots, n, \tag{8.52}$$

with

$$\sum_i \boldsymbol{\sigma}_i = 0. \tag{8.53}$$

The $\{\boldsymbol{\sigma}_i\}$ can be linearly transformed into a set of independent momenta $\{\boldsymbol{\pi}_k\}$ that are conjugate to the translationally invariant coordinates $\{\mathbf{t}_k\}$.

All that is needed to carry this through explicitly is a coordinate transformation symbolised by

$$(\mathbf{t}\,\mathbf{R}) = \mathbf{x}\,\mathbf{V}. \tag{8.54}$$

In (8.54) $\mathbf{t}$ is a 3-by-$n-1$ matrix and $\mathbf{R}$ is a 3-by-1 matrix, so that the combined (bracketed) matrix on the left of (8.54) is 3-by-n. $\mathbf{V}$ is an n-by-n matrix which, from the structure of the left side of (8.54), has a special last column whose elements are

$$V_{in} = M_T^{-1} m_i, \quad M_T = \sum_{i=1}^{n} m_i. \tag{8.55}$$

As the coordinates $\mathbf{t}_j, j = 1, 2, \dots n-1$ are to be translationally invariant, we require the condition

$$\sum_{i=1}^{n} V_{ij} = 0, \qquad\qquad j = 1, 2, \dots n-1, \tag{8.56}$$

on each remaining column of $\mathbf{V}$ and it is easy to see that (8.56) forces $\mathbf{t}_j \to \mathbf{t}_j$ as $\mathbf{x}_i \to \mathbf{x}_i + \mathbf{a}$, all i.

The $\mathbf{t}_i$ are independent if the inverse transformation,

$$\mathbf{x} = (\mathbf{t}\,\mathbf{R})\mathbf{V}^{-1}, \tag{8.57}$$

exists. The structure of the right side of (8.57) shows that the bottom row of $\mathbf{V}^{-1}$ is special and, without loss of generality, we may require its elements to be

$$(\mathbf{V}^{-1})_{ni} = 1, \qquad\qquad i = 1, 2, \dots n. \tag{8.58}$$

The inverse requirement on the remainder of $\mathbf{V}^{-1}$ implies that

$$\sum_{i=1}^{n} (\mathbf{V}^{-1})_{ji} m_i = 0, \qquad\qquad j = 1, 2, \dots n-1. \tag{8.59}$$

When we write the column matrix of the Cartesian components of the partial derivative operator as $\partial / \partial \mathbf{x}_i$, the coordinate change (8.54) gives

$$\frac{\partial}{\partial \mathbf{x}_i} = \sum_{j=1}^{n-1} V_{ij} \frac{\partial}{\partial \mathbf{t}_j} + m_i M_T^{-1} \frac{\partial}{\partial \mathbf{R}}, \tag{8.60}$$

and when it seems more convenient, this column matrix of derivative operators will also be denoted as the vector grad operator $\vec{\nabla}(\mathbf{x}_i)$. If a second set $\mathbf{t}'$ of translationally invariant coordinates is constructed, it may be related to the original set by

$$(\mathbf{t}'\,\mathbf{R}) = (\mathbf{t}\,\mathbf{R})\overline{\mathbf{V}}, \quad \overline{\mathbf{V}} = \mathbf{V}^{-1}\mathbf{V}', \tag{8.61}$$

where $\mathbf{V}'$ is the transformation matrix that defines $\mathbf{t}'$. The matrix $\overline{\mathbf{V}}$ is

$$\begin{pmatrix} \mathbf{G} & \mathbf{0} \\ \mathbf{0} & 1 \end{pmatrix}, \quad G_{ij} = (\overline{\mathbf{V}})_{ij}, \quad i, j = 1, 2, \dots n-1. \tag{8.62}$$

It is easily seen that the form of (8.60) is preserved under a change from $\mathbf{t}$ to $\mathbf{t}'$. It is thus the case that any set of translationally invariant coordinates can be related to any other set by means of a linear transformation.

The Hamiltonian (8.4) in a position representation now separates into two parts as in (8.15):

$$H(\mathbf{t}, \mathbf{R}) = -\frac{\hbar^2}{2M_T}\nabla^2(\mathbf{R}) - \frac{\hbar^2}{2}\sum_{i,j=1}^{n-1}\frac{1}{\mu_{ij}}\boldsymbol{\nabla}(\mathbf{t}_i).\boldsymbol{\nabla}(\mathbf{t}_j) + \frac{e^2}{4\pi\varepsilon_0}\sum_{i>j=1}^{n}\frac{Z_iZ_j}{r_{ij}(\mathbf{t})}$$

$$= -\frac{\hbar^2}{2M_T}\nabla^2(\mathbf{R}) + H'(\mathbf{t}). \tag{8.63}$$

Here

$$1/\mu_{ij} = \sum_{k=1}^{n} m_k^{-1}V_{ki}V_{kj}, \qquad\qquad i, j = 1, 2, \ldots n-1. \tag{8.64}$$

The $n-1$ dimensional square matrix composed of all the $1/\mu_{ij}$ is denoted as $\boldsymbol{\mu}^{-1}$. The operator r_{ij} is the interparticle distance operator expressed as a function of $\mathbf{t}_i$. Thus

$$r_{ij}(\mathbf{t}) = \sqrt{\sum_{\alpha}\left(\sum_{k=1}^{n-1}((V)_{kj}^{-1} - (V)_{ki}^{-1})t_{\alpha k}\right)^2}. \tag{8.65}$$

In (8.63) the first term represents the centre-of-mass kinetic energy operator, and the $\vec{\nabla}(\mathbf{t}_i)$ are grad operators expressed in the Cartesian components of $\mathbf{t}_i$.

It is straightforward to enumerate the other symmetries of the translationally invariant Hamiltonian. The total angular momentum operator may be written as

$$L(\mathbf{x}) = \frac{\hbar}{i}\sum_{i=1}^{n}\check{\mathbf{x}}_{\mathbf{i}}\frac{\partial}{\partial \mathbf{x}_i}, \tag{8.66}$$

where $L(\mathbf{x})$ is a column matrix of Cartesian components and the skew-symmetric matrix $\check{\mathbf{x}}_{\mathbf{i}}$ is

$$\check{\mathbf{x}}_{\mathbf{i}} = \begin{bmatrix} 0 & -x_{iz} & x_{iy} \\ x_{iz} & 0 & -x_{ix} \\ -x_{iy} & x_{ix} & 0 \end{bmatrix}. \tag{8.67}$$

The matrix $\check{\mathbf{x}}_{\mathbf{i}}$ can also be written in terms of the infinitesimal rotation generators (4.89) so that

$$\check{\mathbf{x}}_i = \sum_{r} x_{ir}\mathbf{M}^{rT}. \tag{8.68}$$

A variable symbol with a $\check{}$ sign over it will be used to denote a skew-symmetric matrix as defined by (8.68).

Transforming to coordinates $\mathbf{R}$, $\{\mathbf{t}_i\}$ gives the total angular momentum operator as the sum of two commuting terms

$$L(\mathbf{x}) \rightarrow \frac{\hbar}{i}\sum_{i=1}^{n-1}\check{\mathbf{t}}_i\frac{\partial}{\partial\mathbf{t}_i} + \frac{\hbar}{i}\check{\mathbf{R}}\frac{\partial}{\partial\mathbf{R}}. \tag{8.69}$$

The first term will be denoted as $L(t)$ and called the translationally invariant angular momentum. Similarly the inertia tensor takes the separable form

$$\mathbf{I}(\mathbf{t}, \mathbf{R}) = \sum_{i=1}^{n-1} \mu_{ij} \check{\mathbf{t}}_i^T \check{\mathbf{t}}_j + M_T \check{\mathbf{R}}^T \check{\mathbf{R}}, \qquad (8.70)$$

where

$$\mu_{ij} = \sum_{k=1}^{n} m_k (V)_{ki}^{-1} (V)_{kj}^{-1}, \qquad i, j = 1, 2, \dots n - 1, \qquad (8.71)$$

and is a matrix inverse to (8.64). The square of the operator $L(t)$ and its z-component commute, and both commute with the translationally invariant Hamiltonian.

The three angular momentum components constitute the Lie algebra from which the invariance group SO(3) of the Coulomb Hamiltonian may be constructed. The invariance group O(3) arises from this Lie algebra with the addition of the inversion operator. The translationally invariant Hamiltonian will therefore have eigenfunctions which provide a basis for irreducible representations (irreps) of the orthogonal group in three dimensions O(3). Thus the eigenfunctions are expected to be of two kinds classified by their parity; each kind consists of eigenfunction sets, each with degeneracy $2J + 1$, according to the irrep $J = 0, 1, 2, \dots$ of SO(3) to which the eigenfunctions belong. The representations of O(3) are distinct for each parity, and so there is no group theoretical reason to expect eigenfunctions with different parity p to be degenerate. It would be possible to construct directly, in terms of translationally invariant coordinates, states $\{|EJMp\rangle\}$ which were simultaneously angular momentum eigenfunctions with definite parity under inversion. This is the usual way of constructing approximate solutions for the atomic Coulomb Hamiltonian.

The general permutation of identical particles can be written as

$$\mathcal{P}(\mathbf{x}^e \mathbf{x}^n) = (\mathbf{x}^e \mathbf{x}^n) \begin{pmatrix} \mathbf{P}^e & \mathbf{0} \\ \mathbf{0} & \mathbf{P}^n \end{pmatrix} \equiv \mathbf{x}\mathbf{P}, \qquad (8.72)$$

where $\mathbf{P}^e$ and $\mathbf{P}^n$ are standard permutation matrices. They are orthogonal with determinant ± 1 according to whether the permutation is of even or odd parity. The matrix $\mathbf{P}^n$ will have non-zero entries only for each group of identical atomic nuclei and is most conveniently visualised as having block diagonal form, one block for each group of identical nuclei. The effect of this permutation on the translationally invariant coordinates is

$$\mathcal{P}(\mathbf{t}\,\mathbf{R}) = (\mathbf{t}\,\mathbf{R}) \begin{pmatrix} \mathbf{F} & \mathbf{0} \\ \mathbf{0} & 1 \end{pmatrix}, \qquad (8.73)$$

where

$$(\mathbf{F})_{ij} = (\mathbf{V}^{-1} \mathbf{P} \mathbf{V})_{ij}, \qquad i, j = 1, 2, \dots..n - 1. \qquad (8.74)$$

The matrix $\mathbf{F}$ is not necessarily in standard permutational form, neither is it orthogonal, even though it has determinant ± 1 according to the sign of $|\mathbf{P}|$. Thus the translationally invariant coordinates will transform under any permutation of like particles into linear combinations of themselves. Any chosen transformed coordinate will, generally, involve both $\mathbf{P}^e$ and $\mathbf{P}^n$ in its definition. If it is desired to identify electrons with a particular set of translationally invariant coordinates, specialised coordinate choices must be made to avoid $\mathbf{P}^n$ becoming involved in the definition of their transformed forms, and even more specialised choices must be made to ensure that members of the chosen set transform only into each other under $\mathbf{P}^e$.

The eigenfunctions of the translationally invariant Hamiltonian will necessarily provide irreducible representations for the symmetric group S_n of the system. This group comprises the direct product of the permutation group S_N for the electrons with the permutation groups S_{A_i} for each set of identical nuclei i comprising A_i members. The physically realisable representations of this group are restricted by the requirement that, when spin is properly incorporated into the eigenfunctions, the eigenfunctions form a basis only for the totally symmetric representation, if bosons (spin 0, 1, 2 etc.), or for the antisymmetric representation, if fermions (spin 1/2, 3/2, 5/2 etc.). Both of these representations are one-dimensional. Irreducible representations of the translationally invariant Hamiltonian which correspond to physically realisable states are said to be permutationally allowed. In general such representations will be many-dimensional and so we would expect to have to deal with degenerate sets of eigenfunctions in attempting to identify a molecule in the solutions to the translationally invariant problem.[5]

Let a particular irreducible representation of the symmetric group for the N electronic coordinates be denoted as $[\boldsymbol{\lambda}]^N$ and let the conjugate representation be denoted as $[\widetilde{\boldsymbol{\lambda}}]^N$. For spin $\frac{1}{2}$ particles like electrons the representation of the symmetric group carried by the spin eigenfunctions $\Theta_{S,M_S,i}$ must be one described by no more than a two-rowed Young diagram, that is, $[\boldsymbol{\lambda}]^N \equiv [\lambda_1, \lambda_2]$, where

$$\lambda_1 = N/2 + S, \quad \lambda_2 = N/2 - S. \tag{8.75}$$

The representations are independent of the choice of M_S, and i labels the rows (columns) of the representation. The dimension[6] of the representation is given by the Wigner number

$$f_S^N = \frac{(2S+1)N!}{(N/2+S+1)!(N/2-S)!}. \tag{8.76}$$

Assuming that the translationally invariant part of the Coulomb Hamiltonian for the chosen system has eigenfunctions and discrete spectrum, then, among them, there will be a degenerate set that provides a basis for the representation conjugate to that for the chosen spin eigenfunctions. The representation and the conjugate representation have the same dimension, and a basis of space-spin products can be formed which belongs to the antisymmetric representation of the symmetric group and hence satisfies the Pauli exclusion principle.

This sort of argument could be extended to particles with spins other than $\frac{1}{2}$ and with Bose rather than Fermi statistics. To do so is, however, much more difficult because it is much harder in the general case than it is with particles of spin $\frac{1}{2}$ to associate the spin functions with their space parts to produce functions of appropriate symmetry.

---

[5] See the discussion of Theorem XIII.46 in [22].
[6] For most polyatomic molecules the Wigner number f_0^N for typical singlet ground states, $S = 0$, is huge.

Although it is true that particles of spin s can provide a basis for representations of the symmetric group corresponding to Young diagrams with, at most, only $2s + 1$ rows, it is not in general possible to determine the lengths of these rows simply from the total S and N values in the problem. One cannot then make any simple association between particular spin eigenfunctions and a matched spatial function. To achieve the required results it is necessary to deploy the formal machinery of the symmetric group in constructing a Clebsch–Gordan series from the full set of space-spin products in order to isolate in the direct product space either antisymmetric irreps for fermions, or symmetric irreps for bosons [43].

Some of these results may be demonstrated straightforwardly by the special case of the translationally invariant Hamiltonian for a diatomic molecule. In this case we can choose one of the translationally invariant coordinates as the internuclear separation vector,

$$\mathbf{r} = \mathbf{x}_0 - \mathbf{x}_1. \tag{8.77}$$

The translationally invariant electronic variables can be taken as

$$\mathbf{t}_i^e \equiv \mathbf{t}_i = \mathbf{x}_i^e + v_0 \mathbf{x}_0 + v_1 \mathbf{x}_1, \quad i = 1, 2, \ldots N, \quad v_0 = -\frac{m_0}{M}, \ v_1 = -\frac{m_1}{M}, \tag{8.78}$$

where $M = m_0 + m_1$. If we introduce the momentum $\mathbf{p}$ conjugate to $\mathbf{r}$, and the electronic momenta $\{\boldsymbol{\pi}_i\}$, the transformation from the original laboratory frame variables to the new space-fixed coordinates is given explicitly by the linear relations:

$$\mathbf{x}_0 = \mathbf{R} + \frac{m_1}{m_0 + m_1} \mathbf{r} - \frac{m}{M} \sum_j \mathbf{t}_j \tag{8.79}$$

$$\mathbf{p}_0 = \frac{m_0}{M} \mathbf{P_R} + \mathbf{p} - \frac{m_0}{m_0 + m_1} \sum_j \boldsymbol{\pi}_j \tag{8.80}$$

$$\mathbf{x}_1 = \mathbf{R} - \frac{m_0}{m_0 + m_1} \mathbf{r} - \frac{m}{M} \sum_j \mathbf{t}_j \tag{8.81}$$

$$\mathbf{p}_1 = \frac{m_1}{M} \mathbf{P_R} - \mathbf{p} - \frac{m_1}{m_0 + m_1} \sum_j \boldsymbol{\pi}_j \tag{8.82}$$

$$\mathbf{x}_i = \mathbf{R} + \mathbf{t}_i - \frac{m}{M} \sum_j \mathbf{t}_j, \quad i = 2, \ldots n \tag{8.83}$$

$$\mathbf{p}_i = \frac{m}{M} \mathbf{P_R} + \boldsymbol{\pi}_i, \quad i = 2, \ldots n. \tag{8.84}$$

Here and in what follows the sums over i, j are from $2, \ldots n$. The coordinate axes in the new space-fixed frame are parallel to those chosen for the original laboratory frame. The appropriate forms for the $\{\mathbf{r}_i, \boldsymbol{\sigma}_i\}$ variables can be read off directly by comparing (8.79)–(8.84) with (8.49) and (8.52).

The quantum mechanical transformation,

$$(\mathbf{x}_n; \mathbf{p}_n) \rightarrow (\mathbf{R}, \mathbf{P_R}; \mathbf{t}_k, \boldsymbol{\pi}_k), \tag{8.85}$$

is a unitary transformation that preserves commutation relations. In terms of these variables the internal Hamiltonian, H', for a diatomic molecule takes the well-known form

$$H' = \frac{p^2}{2\mu} + \sum_j \frac{\pi_j^2}{2m_j} + \frac{1}{2(m_0 + m_1)} \sum_{i \neq j} \pi_i \cdot \pi_j + V(\{t_i\}, r), \tag{8.86}$$

where $\mu = m_0 m_1 / (m_0 + m_1)$. The potential energy $V(\{t_i\}, r)$ depends on all pairs of interparticle separation vectors $\{x_i - x_n\}$ which can be written down directly from (8.79)–(8.84).

The transformation to a set of space-fixed coordinates can be carried through for all other observables. Thus, for example, the electric and magnetic multipole moment operators may be defined in terms of the translationally invariant coordinates and their conjugate momenta. The electric dipole moment (**d**) and electric quadrupole moment (**Q**) operators are

$$\mathbf{d} = \sum_i e_i \mathbf{r}_i$$

$$\mathbf{Q} = \sum_i e_i \mathbf{r}_i \mathbf{r}_i, \tag{8.87}$$

and similarly the magnetic dipole moment operator can be defined as

$$\mathbf{m} = \sum_i \frac{e_i}{2m_i} (\mathbf{r}_i \wedge \sigma_i), \tag{8.88}$$

where the $\{\sigma_i\}$ are defined in (8.52).

These definitions [44] are different from the traditional ones that contain an unnecessary origin dependence (§2.4.1). In the special case of the diatomic molecule use of the coordinates just defined yields for the electric dipole moment operator of a neutral molecule, **d**,

$$\mathbf{d} = \left(\frac{e_0 m_1 - e_1 m_0}{m_0 + m_1} \right) \mathbf{r} + \sum_j e_j \mathbf{t}_j. \tag{8.89}$$

This formula suggests an important distinction between homo- and hetero-nuclear diatomic molecules; for the former the dipole operator can be expressed purely in terms of the electronic coordinates $\{t_k\}$, whereas for the latter, which include isotopic species like HD, the internuclear vector **r** also appears. Consequently dipole matrix elements within a given electronic state of a homonuclear diatomic molecule can be expected to be tiny, or zero by virtue of the orthogonality of vibration-rotation wave functions, within an adiabatic approximation.

For the electric quadrupole and magnetic dipole operators the corresponding formulae in terms of this choice of internal coordinates are

$$\mathbf{Q} = \left(\frac{e_0 m_1^2 + e_1 m_0^2}{2(m_0 + m_1)^2} \right) \mathbf{rr} + \frac{1}{2} \sum_j e_j \mathbf{t}_j \mathbf{t}_j - \mathbf{d}\delta, \tag{8.90}$$

$$
\begin{aligned}
\mathbf{m} = {} & \left(\frac{e_0 m_1^2 + e_1 m_0^2}{2(m_0 + m_1)^2} \right) \mathbf{r} \wedge \mathbf{p} - \left(\frac{e_0}{2m_0} - \frac{e_1}{2m_1} \right) \delta \wedge \mathbf{p} \\
& - \frac{e_0 + e_1}{2(m_0 + m_1)} \delta \wedge \gamma - \frac{m_0 m_1}{2(m_0 + m_1)^2} \left(\frac{e_0}{m_0} - \frac{e_1}{m_1} \right) \mathbf{r} \wedge \gamma \\
& + \sum_i \frac{e_i}{2m_e} \mathbf{t}_i \wedge \pi_i - \sum_i \frac{e_i}{2m_e} \delta \wedge \pi_i,
\end{aligned}
\tag{8.91}
$$

where

$$\gamma = \sum_j \pi_j, \quad \delta = \frac{m_e}{M} \sum_j \mathbf{t}_j. \tag{8.92}$$

For the general polyatomic case a somewhat different approach is necessary when considering the rotation-reflection symmetry of the molecule. The translationally invariant coordinates are not immediately adapted to describe such symmetry. Descriptions of this kind of symmetry are usually offered in terms of three angular coordinates and a parity specification, together with $3n$–6 internal coordinates, which are invariant under all rotation-reflections. The process of making these choices is often called fixing or embedding a frame in the body. To construct the frame fixed in the body it is supposed that the three orientation variables are specified by means of an orthogonal matrix $\mathbf{C}$, the elements of which are expressed as functions of three Eulerian angles $\{\phi_m, m = 1, 2, 3\}$ which are orientation variables. We require that the matrix $\mathbf{C}$ is specified in terms of the translationally invariant coordinates $\mathbf{t}$. Thus the Cartesian coordinates $\mathbf{t}$ are considered related to a set $\mathbf{z}$ by

$$\mathbf{t} = \mathbf{Cz}, \tag{8.93}$$

so the matrix $\mathbf{C}$ may be thought of as a direction cosine matrix, relating the laboratory frame to the frame fixed in the body.

The laboratory frame may always be chosen as a right-handed frame, but it is not always the case that there is the freedom to choose the frame fixed in the body as a right-handed one. Since $\mathbf{z}$ are fixed in the body, not all their $3n$–3 components are independent, for there must be three relations between them. Hence components of $\mathbf{z}_i$ must be writable in terms of $3n$–6 independent internal coordinates $\{q_i, i = 1, 2, \ldots, 3n-6\}$. Some of the q_i may be components of $\mathbf{z}_i$, but generally the q_i are expressible in terms of scalar products of the $\mathbf{t}_i$ (and equally of the $\mathbf{z}_i$) since scalar products are the most general constructions that are invariant under orthogonal transformations of their constituent vectors. There are many rather delicate topological problems raised by such a choice. The space defined is a quotient manifold that is only locally Euclidean and so internal coordinates can be defined only locally; in fact it always takes more than one coordinate system to cover the quotient manifold. For this reason the transformation to the body fixed coordinates and momenta is not a unitary transformation of the original canonical variables.

8.5 The Born–Oppenheimer Approximation

The conventional account of the operator (8.48) in the multinuclear case is dominated by the famous work *Quantum Theory of Molecules* by M. Born and J. R. Oppenheimer, [45], and we begin with a summary account of that historic paper. Much of the groundwork for Born and Oppenheimer's treatment of the energy levels of molecules was laid down in an earlier attempt by Born and Heisenberg [46]. The basic idea of both calculations is that the low-lying excitation spectrum of a molecule can be obtained by regarding the nuclear kinetic energy as a 'small' perturbation of the energy of the electrons for stationary nuclei *in an equilibrium configuration*. The physical basis for the idea is the large disparity between the mass of the electron and the masses of the nuclei; classically the light electrons undergo motions on a 'fast' timescale ($\tau_e \approx 10^{-16} \leftrightarrow 10^{-15}$s), while the vibration-rotation dynamics of the much heavier nuclei are characterised by 'slow' timescales ($\tau_N \approx 10^{-14} \leftrightarrow 10^{-12}$s).

Born and Heisenberg started from the usual classical non-relativistic Hamiltonian (8.1) for a system comprised of N electrons and A nuclei interacting via Coulombic forces. They assumed there is an arrangement of the nuclei which is a stable equilibrium (a molecular structure) that can be used as a reference configuration for the calculation. Formally the rotational motion of the system can be dealt with by requiring the coordinates for the reference structure to satisfy[7] what were later to become known as the Eckart conditions [47]. Then with a suitable set of internal variables the Hamiltonian was transformed to

$$H = H_0 + \lambda^2 H_2 + \dots, \tag{8.94}$$

where the expansion parameter is

$$\lambda = \left(\frac{m}{M}\right)^{\frac{1}{2}}. \tag{8.95}$$

This is in an appropriate form for the application of the action-angle perturbation theory Born had developed.

The 'unperturbed' Hamiltonian H_0 is the full Hamiltonian for the electrons with the nuclei fixed at the equilibrium structure; the next term in the expansion H_1 disappears because of the equilibrium condition. H_2 is quadratic in the nuclear variables (harmonic oscillators) and also contains the rotational energy,[8] while ... stands for higher-order anharmonic vibrational terms. With considerable effort there follows the usual separation of molecular energies, although, of course, no concrete calculation was possible within the Old Quantum Theory framework. A feature of this calculation is that it gives the electronic energies at a *single* configuration because the perturbation calculation requires the introduction of the (assumed) equilibrium structure. This is different from the earlier *adiabatic* approach Nordheim had tried unsuccessfully using the Old Quantum Theory to get the electronic energy at *any* separation of the nuclei [48] (cf. Chapter 1).

With the discovery of Quantum Mechanics, Born and Oppenheimer proposed to reconsider this system of electrons and nuclei. It will be useful to review a few key points in their discussion. Let the physical variables of the former be denoted by lower-case letters (mass m, coordinates x, momenta p) and of the latter by capital letters (mass M, coordinates X, momenta P). The small parameter for the perturbation expansion

---

[7] This also deals with the uninteresting overall translation of the molecule.

[8] The rotational and vibrational energies occur together because of the choice of the parameter λ; as is well known, Born and Oppenheimer later showed that using the quarter power of the mass ratio separates the vibrational and rotational energies in the orders of the perturbation expansion [45].

must clearly be some power of m/M_o, where M_o can be taken as any one of the nuclear masses or their average. In contrast to the earlier calculation Born and Oppenheimer preferred to take

$$\kappa = \left(\frac{m}{M_o}\right)^{\frac{1}{4}}, \tag{8.96}$$

instead of Born and Heisenberg's $\lambda = \kappa^2$. In an obvious shorthand notation using a coordinate representation, the kinetic energy of the electrons is then[9]

$$T_e = T_e\left(\frac{\partial}{\partial x}\right), \tag{8.97}$$

and the Coulomb energy is simply $U(x,X)$; the dependence on κ is contained entirely in the nuclear kinetic energy term

$$T_N = \kappa^4 H_1\left(\frac{\partial}{\partial X}\right). \tag{8.98}$$

The 'unperturbed' Hamiltonian is thus

$$T_e + U = H_0\left(x, \frac{\partial}{\partial x}, X\right), \tag{8.99}$$

in terms of which the total Hamiltonian is

$$H = H_0 + \kappa^4 H_1. \tag{8.100}$$

This splitting of the Hamiltonian into an 'unperturbed' part ($\kappa = 0$) and a 'perturbation' is essentially the same as in the earlier Old Quantum Theory version [46]. The difference here is that the action-angle perturbation theory of the Old Quantum Theory is replaced by the Rayleigh–Schrödinger wave mechanical perturbation theory (Chapter 6) using κ as the perturbation parameter. The Schrödinger equation is

$$(H - E)\psi(x, X) = 0 \tag{8.101}$$

with the expansions[10]

$$\psi = \psi^{(0)} + \kappa\,\psi^{(1)} + \kappa^2\,\psi^{(2)} + \dots, \tag{8.102}$$

$$E = E^{(0)} + \kappa E^{(1)} + \kappa^2 E^{(2)} + \dots . \tag{8.103}$$

It remains to discuss the dependence, if any, of the unperturbed Hamiltonian on κ.

First, something must be done about the translational and rotational degrees of freedom. For a molecule with A nuclei, Born and Oppenheimer chose $3A - 6$ functions,

$$\xi_i = \xi_i(X), \tag{8.104}$$

---

[9] The details can be found in the original paper [45], and in various English language presentations, for example [49]–[51].

[10] This is now recognised as a case of singular perturbation theory since the small parameter κ appears as a coefficient of the highest-order differential operator H_1; as a consequence the perturbation series must be an asymptotic expansion, at best.

to describe the relative positions of the nuclei with respect to each other, and 6 functions,

$$\theta_i = \theta_i(X), \tag{8.105}$$

to determine the position and orientation of the nuclear configuration in space. We can identify three of the $\{\theta_i\}$ as the Euler angles that parameterise the matrix $\mathbf{C}$, (Eq. 8.93); the other three are the components of the centre-of-nuclear mass coordinate, $\mathbf{X}$.

The crucial step in their argument now comes; they wrote [49]

> We consider, as the unperturbed system, the electronic motion for an arbitrary but henceforth fixed nuclear configuration, ξ_o. We then develop all quantities with respect to small changes of the ξ_i, which we designate by $\kappa\zeta_i$; we presume then that the 'domain' of oscillation is such that κ is close to zero, an assumption which is only justified by its success.

Thus an essential feature of the 'Born–Oppenheimer approximation' is that the electrons and nuclei are treated in quite different ways. The nuclei are initially taken as fixed, classical sources of a potential experienced by the electrons; only subsequently are the nuclei treated dynamically according to quantum theory. Moreover the nuclei, even if identical, are regarded as distinguishable particles localised in their own potential wells.

The Schrödinger equation for the unperturbed Hamiltonian H_0 can be solved for any choice of the nuclear parameters ξ, and yields an unperturbed energy $E^0(X)$ for the configuration X; the crucial observation that makes the calculation successful is the choice of ξ_o. For the consistency of the whole scheme it turns out that ξ_o cannot be arbitrarily chosen, but must correspond to a *minimum* of the electronic energy. That there is such a point is assumed to be self-evident for the case of a stable molecule. The result of the calculation was a triumph; the low-lying energy levels of a stable molecule can be written in the form

$$E_{\text{Mol}} = E_{\text{Elec}} + \kappa^2 E_{\text{Vib}} + \kappa^4 E_{\text{Rot}} + \dots, \tag{8.106}$$

in agreement with a considerable body of spectroscopic evidence. The eigenfunctions that correspond to these energy levels are simple products of an electronic wave function obtained for the equilibrium geometry and suitable vibration-rotation wave functions for the nuclei.

Many years after his work with Heisenberg and Oppenheimer, Born returned to the subject of molecular quantum theory and developed a different account of the separation of electronic and nuclear motion [51], [52]. As in the original Born–Oppenheimer calculation, a crucial step is to assign the nuclear coordinates the role of parameters in the Schrödinger equation (8.107) for the electronic Hamiltonian; it differs from the earlier approach of Born and Oppenheimer because now the values of X range over the whole nuclear configuration space. Born's argument can be summarised as follows. Consider the unperturbed electronic Hamiltonian (8.99) at a fixed nuclear configuration X that corresponds to some molecular structure (not necessarily an equilibrium structure). Its Schrödinger equation is

$$\left(H_0(x, \frac{\partial}{\partial x};X) - E^o(X)_m\right)\varphi(x,X)_m = 0, \tag{8.107}$$

where m stands conventionally for both discrete and continuous labels. The *bound state* eigenvalues considered as functions of the X are the molecular potential energy surfaces.

More abstractly, H_o is an operator on the electronic Hilbert space, $L^2(x)$; the $\{\varphi_m\}$ are supposed complete in $L^2(x)$, so there is a resolution of the identity

$$1(X)_x = \sum_m |\varphi(x,X)_m\rangle\langle\varphi(x,X)_m|, \tag{8.108}$$

at fixed X. Born proposed to solve the full molecular Schrödinger equation, (8.101), by an expansion using the fixed-nuclei eigenfunctions as a complete set of states,

$$\psi(x,X) = \sum_m \Phi(X)_m \, \varphi(x,X)_m, \tag{8.109}$$

with coefficients $\{\Phi(X)_m\}$ that play the role of nuclear wave functions. Substituting this expansion into (8.101), multiplying the result by $\varphi(x,X)_n^*$ and integrating over the electronic coordinates x leads to an infinite-dimensional system of coupled classical differential equations for the nuclear functions $\{\Phi\}$,

$$\left(T_N + E^o(X)_n - E\right)\Phi(X)_n + \sum_{nn'} B(X,P)_{nn'}\Phi(X)_{n'} = 0, \tag{8.110}$$

where the coupling coefficients $\{B(X,P)_{nn'}\}$ have a well-known form which we need not record here [51].

In this formulation the adiabatic approximation consists of retaining only the diagonal terms in the coupling matrix $\mathbf{B}(X,P)$, for then a state function can be written as

$$\psi(x,X) \approx \psi(x,X)_n^{\text{AD}} = \varphi(x,X)_n \, \Phi(X)_n, \tag{8.111}$$

and a product wave function corresponds to additive electronic and nuclear energies. The special character of the electronic wave functions $\{\varphi(x,X)_m\}$ is, by (8.107), that they diagonalise the electronic Hamiltonian H_o; they are said to define an 'adiabatic' basis (cf. the approximate form (8.111)) because the electronic state label n is not altered as X varies. The Born approach does not really require the diagonalisation of H_o; it is perfectly possible to define other representations of the electronic expansion functions through unitary transformations of the $\{\varphi\}$, with concomitant modification of the coupling matrix $\mathbf{B}$. This leads to so-called 'diabatic' bases; the freedom to choose the representation is very important in practical applications to spectroscopy and atomic/molecular collisions [53], [54]. For many years now these equations have been regarded in the theoretical molecular spectroscopy/quantum chemistry literature as defining the 'Born–Oppenheimer approximation', the original perturbation method being relegated to the status of historical curiosity. Commonly they are said to provide an exact (in principle) solution [55]–[58] for the stationary states of the molecular Schrödinger equation (8.101), it being recognised that drastic truncation of the infinite set of equations (8.110) is required for any practical application.

With the benefit of hindsight it is clear that Born and Oppenheimer [45] piloted their way with remarkable physical insight through what is actually a singular perturbation problem so as to classify the molecular term energies in powers of κ. The later argument of Born [51], [52] cannot be justified because his assumption that the electronic functions $\{\varphi(x,X)\}$ at fixed X are a 'complete set of states' for an expansion of the molecular wave functions, $\{\Psi(x,X)\}$, is not valid. To make the arguments mathematically precise and to estimate errors, it is necessary firstly to find a way of dealing with the mathematical problems raised by the symmetries of the Hamiltonian, while preserving a plausible distinction between electronic and nuclear variables, and secondly to characterise the singular perturbation expansion. This latter task requires recognition of the proper description of the continuous spectrum of self-adjoint operators.

In view of the discussion in §8.4 and the arguments advanced by Born and Oppenheimer [45], we start by expressing the internal Hamiltonian $H'(\mathbf{t})$, (8.15), in terms of two sets of coordinates. For generality,[11] the systems that we shall consider must have $A \geq 4$. One set consists of $A-1$ translationally invariant coordinates $\mathbf{t}_i^n$ expressed entirely in terms of the original $\mathbf{x}_i^n$:

$$\mathbf{t}_i^n = \sum_{j=1}^A \mathbf{x}_j^n V_{ji}^n, \quad i = 1,2,\ldots,A-1. \tag{8.112}$$

Here $\mathbf{V}^n$ is a non-singular matrix whose last column is special, with elements

$$V_{iA}^n = M^{-1} m_i, \quad M = \sum_{i=1}^A m_i, \tag{8.113}$$

so that the coordinate $\mathbf{X}$, defined by its last column, is the coordinate of the centre-of-nuclear mass. The overall centre of mass $\mathbf{R}$ and the centre-of-nuclear mass $\mathbf{X}$ are related by

$$\mathbf{R} = \mathbf{X} + \frac{m}{M_T} \sum_{i=1}^N \mathbf{t}_i^e. \tag{8.114}$$

The elements in the first $A-1$ columns of $\mathbf{V}^n$ each sum to zero, exactly as in the general case, to ensure translational invariance. The other set comprises N translationally invariant coordinates of the form

$$\mathbf{t}_i^e = \mathbf{x}_i^e + \sum_{j=1}^A v_j \mathbf{x}_j^n, \quad i = 1,2,\ldots N. \tag{8.115}$$

For definiteness we shall choose $v_i = -m_i/M$ so that the translationally invariant electronic coordinates are the original electronic coordinates expressed relative to the centre-of-nuclear mass. The inverse relations are

$$\mathbf{x}_i^e = \mathbf{X} + \mathbf{t}_i^e, \tag{8.116}$$

$$\mathbf{x}_i^n = \mathbf{X} + \sum_{j=1}^{A-1} \mathbf{t}_j^n ((\mathbf{V}^n)_{ji}^{-1}, \tag{8.117}$$

[11] The particular cases $A=1$, $A=2$ and $A=3$ (the nuclear configurations that define, respectively, a point, a line and a plane) must be given individual treatments.

with

$$((\mathbf{V}^n)^{-1}_{Ai} = 1, \quad i = 1, 2, \ldots.A, \tag{8.118}$$

while the inverse requirement on the remaining rows gives

$$\sum_{i=1}^{A}((\mathbf{V}^n)^{-1}_{ji}m_i = 0 \qquad j = 1, 2, \ldots.A - 1. \tag{8.119}$$

With these coordinate definitions the translationally invariant Coulomb Hamiltonian takes the following form:

$$H'(\mathbf{t}) \to H^e(\mathbf{t}^e) + H^n(\mathbf{t}^n) + H^{en}(\mathbf{t}^n, \mathbf{t}^e). \tag{8.120}$$

The part of the Hamiltonian which can be associated with the electronic variables is

$$H^e(\mathbf{t}^e) = -\frac{\hbar^2}{2\mu}\sum_{i=1}^{N}\nabla^2(t_i^e) - \frac{\hbar^2}{2M}\sum_{ij=1}^{N}{}'\nabla(t_i^e)\cdot\nabla(t_j^e) + \frac{e^2}{4\pi\varepsilon_0}\sum_{i>j=1}^{N}\frac{1}{|t_j^e - t_i^e|}, \tag{8.121}$$

with

$$1/\mu = 1/m + 1/M, \tag{8.122}$$

while the part that can be associated with the nuclei is

$$H^n(\mathbf{t}^n) = -\frac{\hbar^2}{2}\sum_{i,j=1}^{A-1}\frac{1}{\mu_{ij}^n}\nabla(t_i^n).\nabla(t_j^n) + \frac{e^2}{4\pi\varepsilon_0}\sum_{i>j=1}^{A}\frac{Z_iZ_j}{r_{ij}(\mathbf{t}^n)}, \tag{8.123}$$

where $r_{ij}(\mathbf{t}^n)$ is defined just as in the general case (8.65) but using the t_i^n and $(\mathbf{V}^n)^{-1}$, that is,

$$r_{ij}(\mathbf{t}^n) = \sqrt{\sum_{\alpha}\left(\sum_{k=1}^{A-1}((\mathbf{V}^n)^{-1}_{kj} - (\mathbf{V}^n)^{-1}_{ki})t_{\alpha k}^n\right)^2}, \tag{8.124}$$

and the inverse mass matrix is similarly specialised as

$$1/\mu_{ij}^n = \sum_{k=1}^{A}m_k^{-1}V_{ki}^n V_{kj}^n, \qquad i, j = 1, 2, \ldots A - 1. \tag{8.125}$$

The electronic and nuclear variables are coupled only via a potential term:

$$H^{en}(\mathbf{t}^n, \mathbf{t}^e) = -\frac{e^2}{4\pi\varepsilon_0}\sum_{i=1}^{A}\sum_{j=1}^{N}\frac{Z_i}{r_{ij}'(\mathbf{t}^n, \mathbf{t}^e)}, \tag{8.126}$$

and the electron–nucleus distance expression becomes

$$|\mathbf{x}_i^n - \mathbf{x}_j^e| \equiv r_{ij}' = \left|\sum_{k=1}^{A-1}t_k^n(\mathbf{V}^n)^{-1}_{ki} - t_j^e\right|. \tag{8.127}$$

It follows from (8.73) that this choice of translationally invariant coordinates is such that the t_i^e transform under a permutation of the $\mathbf{x}_i^e$ exactly as do the $\mathbf{x}_i^e$ and remain unchanged under any permutation of the $\mathbf{x}_i^n$. The t_i^n are invariant under any permutation of the $\mathbf{x}_i^e$, while a permutation of the $\mathbf{x}_i^n$ produces a transformation among the

t_i^n just as in (8.73), but with the variables restricted to the t_i^n, the permutations only among the identical nuclei and with transformation matrices involving only the nuclear variables. This choice makes $H^e(t^e)$ trivially invariant under permutations of the original electronic coordinates and independent of any particular choice of translationally invariant nuclear coordinates. Similarly, $H^n(t^n)$ is independent of any particular choice of translationally invariant electronic coordinates and is also invariant under any permutation of the original coordinates of identical nuclei. The interaction operator $H^{en}(t^n, t^e)$ is obviously invariant under a permutation of the original electronic coordinates and is also invariant under a permutation of the original coordinates of identical nuclei.

The internal molecular Hamiltonian, Eq. (8.120), expressed in terms of the translationally invariant internal coordinates defined by (8.112), (8.115), may now be separated into two parts as required by the Born–Oppenheimer analysis. The full electronic Hamiltonian arising from (8.120) on ignoring the nuclear kinetic energy operators is obtained as

$$H^{elec} = H^e(t^e) - \frac{e^2}{4\pi\varepsilon_0}\sum_{i=1}^{A}\sum_{j=1}^{N}\frac{Z_i}{r'_{ij}(t^n, t^e)} + \frac{e^2}{4\pi\varepsilon_0}\sum_{i>j=1}^{A}\frac{Z_i Z_j}{r_{ij}(t^n)}, \qquad (8.128)$$

with domain $\mathcal{L}^2[\mathfrak{R}^{3N}] \oplus \mathfrak{R}^{3(A-1)}$. It is the sum of (8.121) and (8.126) together with the last term from (8.123). This Hamiltonian is invariant under all orthogonal transformations of coordinates so that its eigenfunctions, if any, are angular momentum eigenstates with definite parity. It is also invariant under the permutation of the variables of any set of identical particles; it can play the role of H_0 in (8.100). The nuclear kinetic energy is given by the first term in (8.123); as expected, it is proportional to κ^4.

Although, for a given choice of boundary conditions, the spectrum of the translationally invariant Hamiltonian is precisely the same with this choice of coordinates as it would be for any other choice, we might hope to recognise an eigenfunction corresponding to a molecule rather more easily than we would in some other choices. It might be reasonably hoped that for some set of nuclei and electrons there were bound state solutions of the Schrödinger equation

$$H'(t)\Psi_h(t) = E_h\Psi_h(t), \qquad (8.129)$$

among which molecules could be identified. Here h is used to denote a set of quantum numbers $(J M p \mathbf{r} i)$: J and M for the angular momentum state, p specifying the parity of the state, $\mathbf{r}$ specifying the permutationally allowed irreps within the groups of identical particles and i to specify a particular energy value. For a given J such solutions will be degenerate for all $2J + 1$ values of M, and the permutational irreps can be, as we have seen, extensively degenerate too.

Let us now go over again the arguments that Born and his collaborators used. As we have seen, the idea that the kinetic energy of the massive nuclei could be treated as a perturbation of the electronic motion was first formulated in the framework of the Old Quantum Theory. Recall from §8.1 that for particles with classical Hamiltonian variables $\{\mathbf{x}_i, \mathbf{p}_i\}$ the Hamiltonian for a collection of charged particles is

$$H = \sum_i^n \frac{p_i^2}{2m_i} + \sum_{i<j}^n \frac{e_i e_j}{4\pi\varepsilon_0 |\mathbf{x}_i - \mathbf{x}_j|}, \tag{8.130}$$

with the non-zero Dirac bracket

$$[\mathbf{x}_i, \mathbf{p}_j]^* = \delta_{ij}. \tag{8.131}$$

We denote the classical dynamical variables for the electrons collectively as $\mathbf{x}, \mathbf{p}$, and those for the nuclei by $\mathbf{X}, \mathbf{P}$ and write the classical Hamiltonian as $H(\mathbf{x}, \mathbf{p}, \mathbf{X}, \mathbf{P})$. It may be separated into two parts to isolate the nuclear momentum variables:

$$H(\mathbf{x}, \mathbf{p}, \mathbf{X}, \mathbf{P}) = H_0(\mathbf{x}, \mathbf{p}, \mathbf{X}) + \kappa^4 H_1(\mathbf{P}). \tag{8.132}$$

According to Hamilton's equations for the unperturbed problem ($\kappa = 0$),

$$\frac{d\mathbf{X}}{dt} = [\mathbf{X}, H_0]^* = 0, \tag{8.133}$$

using Dirac-bracket notation, and H_0 is interpreted (correctly) as describing the dynamics of the electrons in the field of stationary nuclei. This was the starting point of Born and Heisenberg's calculations [46].

Now we move to quantum theory and recast (8.132) as an operator relation; after the customary canonical quantisation the Hamiltonian variables become time-independent operators in a Schrödinger representation, and we may write the internal molecular Hamiltonian operator in an obvious shorthand as

$$H'(\mathbf{t}^e, \boldsymbol{\nabla}^e, \mathbf{t}^n, \boldsymbol{\nabla}^n) = H^{elec}(\mathbf{t}^e, \boldsymbol{\nabla}^e, \mathbf{t}^n) + \kappa^4 H_1(\boldsymbol{\nabla}^n). \tag{8.134}$$

Since there are no nuclear momentum operators in the Hamiltonian (8.128), the nuclear position operators commute with H^{elec},

$$[H^{elec}, \mathbf{t}^n] = 0, \tag{8.135}$$

from which we infer the nuclear position operators $\mathbf{t}^n$ are constants of the motion under H^{elec}.

What to do now is the crucial point in any argument for the 'Born–Oppenheimer approximation'; in the original presentation, Born and Oppenheimer had to make an interpretation of the unperturbed Hamiltonian, H_0, and wrote[12]

> If one sets $\kappa = 0$....one obtains a differential equation in the x_k alone, the X_l appearing as parameters
>
> $$\left[H_0\left(x, \frac{\partial}{\partial x}, X \right) - E \right] = 0. \tag{8.136}$$
>
> Evidently, this represents the electronic motion for stationary nuclei.

Of course, it is not 'evident'! It is a far-reaching assumption which allows the discussion to join up with the classical structural account of molecules; making the interpretation that follows from (8.133) is problematic since specifying precisely the positions $\{\mathbf{t}^n\}$

---

[12] English translation from [49].

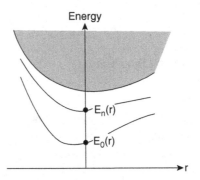

Figure 8.2   The spectrum $\sigma(r)$ for a diatomic molecule [18].

for stationary nuclei violates Heisenberg's uncertainty principle. If one continues with quantum mechanics per se, a quite different picture emerges, as we now discuss.

In view of (8.135), H^{elec} and the $\{\mathbf{t}^n\}$ may be simultaneously diagonalised, and we may search for the eigenvectors of H^{elec} corresponding to definite values of the $\mathbf{t}^n$. Let $\mathbf{b}$ be some eigenvalue of the $\mathbf{t}^n$ corresponding to choices $\mathbf{x}_g = \mathbf{a}_g$ in the laboratory-fixed frame; then the $\mathbf{a}_g$ describe a classical nuclear geometry. The set, X, of all $\mathbf{b}$ is $\mathfrak{R}^{3(A-1)}$.

We denote the electronic Hamiltonian evaluated at the eigenvalue $\mathbf{b}$ as $H^{elec}(\mathbf{b}, \mathbf{t}^e)$. $H^{elec}(\mathbf{b}, \mathbf{t}^e)$ is very like the usual clamped-nuclei Hamiltonian, but it is explicitly translationally invariant and has an extra term, the second term in (8.121), which is often called either the Hughes–Eckart or the mass polarisation term. It can be analysed with the HVZ theorem, and has both discrete[13] and continuous parts to its spectrum [18]–[27]:

$$\sigma(\mathbf{b}) \equiv \sigma(H^{elec}(\mathbf{b}, \mathbf{t}^e)) = \{E(\mathbf{b})_0, \dots E(\mathbf{b})_m\} \bigcup [\Lambda(\mathbf{b}), \infty), \qquad (8.137)$$

where the $\{E(\mathbf{b})_m\}$ are isolated eigenvalues of finite multiplicities. $\Lambda(\mathbf{b})$ is the bottom of the essential spectrum; it is the lowest eigenvalue of an $(N-1)$ electron clamped-nuclei Hamiltonian, and this is still true [59] even if $H^{elec}(\mathbf{b}, \mathbf{t}^e)$ is reduced with respect to its symmetries.[14] In the case of a diatomic molecule the electronic eigenvalues depend only on the internuclear separation r, and have the form of the familiar potential curves shown in Figure 8.2.

The corresponding wave functions are the solutions of the Schrödinger equation

$$H^{elec}(\mathbf{b}, \mathbf{t}^e) \, \varphi(\mathbf{b}, \mathbf{t}^e)_m = E(\mathbf{b})_m \, \varphi(\mathbf{b}, \mathbf{t}^e)_m. \qquad (8.138)$$

For every $\mathbf{b}$, $H^{elec}(\mathbf{b}, \mathbf{t}^e)$ is self-adjoint on the electronic Hilbert space $\mathcal{H}(\mathbf{b})$. This Hamiltonian's natural domain $\mathcal{D}$ is the set of square integrable electronic wave functions

---

[13] Cf. Zhislin's binding condition, §8.3.

[14] In general the symmetry of $H^{elec}(\mathbf{b}, \mathbf{t}^e)$ considered as a function of the electronic coordinates, $\mathbf{t}^e$, is much lower than that of H^{elec} since they are determined by the (point group) transformations that leave the geometrical figure defined by the $\mathbf{a}_g$ invariant. In a chiral structure there is no symmetry operation other than the identity, so even space-inversion symmetry is lost.

$\{\varphi_m\}$ with square integrable first and second derivatives; $\mathcal{D}$ is independent of $\mathbf{b}$. We may suppose the $\{\varphi_m\}$ are orthonormalised independently of $\mathbf{b}$:

$$\int d\mathbf{t}^e \; \varphi(\mathbf{b}, \mathbf{t}^e)_n^* \; \varphi(\mathbf{b}, \mathbf{t}^e)_m \; = \; \delta_{nm}. \tag{8.139}$$

The $\{\varphi(\mathbf{b}, \mathbf{t}^e)\}$ are defined only up to a phase factor of the form $\exp[iw(\mathbf{b})]$, where w is any single-valued real function of the $\{\mathbf{b}_i\}$ which can be different for different electronic states. In the absence of degeneracies ('curve-crossing') the $\{\varphi_m\}$ may be chosen to be real; otherwise the phase factor must be considered. Specific phase choices may therefore be needed when tying this part to the nuclear part of the product wave function so as to make the electronic wave function a continuous function of the formal nuclear variables, $\mathbf{b}$, and the complete product function single valued. This is the origin of the Berry phase in clamped-nuclei calculations involving intersecting potential energy surfaces; for a discussion of these matters see [60]. It is worth noting explicitly that notions of molecular Berry phases and conical intersections of PE surfaces are tied to the clamped-nuclei viewpoint. They are only 'observable' to the extent that experimental data are interpreted within that framework.

We thus have a family of electronic Hilbert spaces $\{\mathcal{H}(\mathbf{b})\}$ that are the 'eigenspaces' of the family of self-adjoint operators $\{\mathrm{H}^{elec}(\mathbf{b}, \mathbf{t}^e)\}$; they are parameterised by the nuclear position vectors $\mathbf{b} \in X$. From them we can construct a 'big' Hilbert space as a direct integral over all the $\mathbf{b}$ values,

$$\mathcal{H} \; = \; \int_X^\oplus \mathcal{H}(\mathbf{b}) \; d\mathbf{b}, \tag{8.140}$$

and this is the Hilbert space for H^{elec}. Each $\mathcal{H}(\mathbf{b})$ is a subspace of $\mathcal{H}$, and pairs of distinct subspaces $\mathcal{H}(\mathbf{b}), \mathcal{H}(\mathbf{b}')$ are orthogonal [61]. The Hamiltonian H^{elec} has the same invariance under the rotation-reflection group O(3) as does the full translationally invariant Hamiltonian (8.63) and it has a somewhat extended invariance under nuclear permutations, since the nuclear masses appear only in symmetrical sums.

In view of (8.140) we see that, since H^{elec} commutes with all the $\mathbf{t}^n$, (8.128) has the direct integral decomposition

$$\mathrm{H}^{elec} \; = \; \int_X^\oplus \mathrm{H}^{elec}(\mathbf{b}, \mathbf{t}^e) \; d\mathbf{b}, \tag{8.141}$$

and an abstract result [22] shows that H^{elec} as a direct integral of self-adjoint operators is itself self-adjoint, even though the potential energy operator cannot be regarded as small in the sense of Kato. The relation (8.141) implies that the spectrum of H^{elec} is purely continuous,

$$\sigma \; = \; \sigma(\mathrm{H}^{elec}) \; = \; \bigcup_\mathbf{b} \sigma(\mathbf{b}) \; \equiv \; [V_0, \infty), \tag{8.142}$$

where V_0 is the minimum value of $E(\mathbf{b})_0$; in the diatomic molecule case this is the minimum value of $E_0(r)$ (see Figure 8.2). Although H^{elec} is bounded from below there are no discrete molecular energy levels and it has no normalisable eigenfunctions. The conventional response has been to suppose that there corresponds to every solution of (8.138) a 'generalised eigenvector' of H^{elec} which may be written in terms of Dirac's delta function notation (cf. §5.2.1) as a distribution,

$$\Phi(\mathbf{t}^e, \mathbf{t}^n) = \varphi(\mathbf{b}, \mathbf{t}^e)_m \, \delta(\mathbf{t}^n - \mathbf{b}), \tag{8.143}$$

in the $(\mathbf{t}^e, \mathbf{t}^n)$ position representation [55]. One cannot hope to calculate a bound state wave function as an expansion in κ with a basis of unnormalisable functions such as (8.143) [62].

This analysis is quite independent of any particular coordinate choice. Allowing the nuclear masses to increase without limit would not produce an operator with a discrete spectrum since this would just cause the mass polarisation term to vanish and the effective electronic mass to become the rest mass. An extra choice of fixed nuclear positions, ignoring effectively the integration in (8.141), must be made to give any discrete spectrum and normalisable L^2 eigenfunctions. It is thus not possible to reduce without approximation the molecular Schrödinger equation to a system of coupled differential equations of classical type as suggested by Born (the 'Born–Huang' coupled equations). The essential difficulty is that it is not possible to give a precise interpretation of 'eigenvalues' and 'eigenfunctions' for the continuous part of the spectrum of the internal molecular Hamiltonian; this vagueness means that one cannot check their properties, and one cannot give a precise meaning to the formulae in which they occur. If a formalism cannot be checked, how can one possibly declare it to be 'exact'?

Both time-independent and time-dependent approaches to more mathematically rigorous accounts of the Born–Oppenheimer perturbation series for molecular energy levels have been developed in the mathematical physics literature. They make clear that the 'Born–Oppenheimer approximation' is an approximation closely related to the classical limit of quantum mechanics. The occurrence of identical nuclei is a characteristic feature of the generic molecular formula, but so far the permutation symmetry this implies is absent from these studies. In the time-independent approach approximate calculations of low-lying eigenvalues of the full Coulomb Hamiltonian are described. An important feature is that the total energy of the molecular system is fixed at the outset, and then one constructs an effective Hamiltonian that depends on this energy using projection operator techniques. The accuracy of the approximation is defined by the quality of the replacement of the true Hamiltonian by the effective one; this can be expressed in terms of the usual expansion parameter κ [9]. The time-dependent approach aims to study the dynamics of certain initial states under the action of the time-development operator for the Coulomb Hamiltonian in the limit $\kappa \to 0$; the wave functions of interest are products of electronic clamped-nuclei eigenfunctions with wave packets that are generalisations of the usual harmonic oscillator eigenfunctions (coherent states) for the nuclei [63]. The recent reviews by Hagedorn and Joye [64], and by Jecko [9] include extensive reference lists for the modern literature.

The expansion in powers of κ, (8.106), is an asymptotic expansion that can be obtained by a WKB (semiclassical) analysis of the effective Hamiltonian for the nuclear dynamics. This requires a more complete treatment than the adiabatic model using the partitioning technique to project the full Coulomb Hamiltonian, H', onto the adiabatic subspace. A normalised electronic eigenvector $|\varphi(\mathbf{b})_j\rangle$ is associated with a projection operator by the usual correspondence:

$$P(\mathbf{b})_j = |\varphi(\mathbf{b})_j\rangle\langle\varphi(\mathbf{b})_j|. \tag{8.144}$$

In view of our earlier discussion of the 'big' Hilbert space $\mathcal{H}$, we can form a direct integral over all nuclear positions,

$$P_j = \int_X^{\oplus} P(\mathbf{b})_j \, d\mathbf{b}, \tag{8.145}$$

to yield a projection operator on the adiabatic subspace. If we want to include m discrete electronic levels, we can form a direct sum of the contributing $\{P_j\}$:

$$P = \bigoplus_{j=0}^{m} P_j. \tag{8.146}$$

This is an Hermitian projection operator, and it and its complement, Q, have the usual properties:

$$P + Q = 1, \quad P^2 = P, \quad Q^2 = Q, \quad PQ = QP = 0. \tag{8.147}$$

Using these projection operators, the original molecular Schrödinger equation

$$H'|\Psi\rangle = E|\Psi\rangle, \tag{8.148}$$

can be partitioned as described in Chapter 6. Further progress depends crucially on developing a tractable approximation to the energy-dependent operator in the subspace spanned by Q, and for this to be achieved the range of energies to be considered must be specified. The difficulty is that an 'exact' representation of the projection operator Q involves a description of the continuous part of the electronic spectrum in terms of its spectral family (cf. §5.2.1) which is unknown; this is the underlying reason why the Born–Huang expansion (8.110) is purely formal.

The microscope transformation used by Combes and Seiler [18], [66], [67] to give a rigorous version of the Born–Oppenheimer theory of a diatomic molecule results in an essentially semiclassical theory. It is applicable if there is a minimum in the potential $V_0 = V(\{\mathbf{x}_n^0\})$ associated with a particular configuration of the nuclei that is deep enough for the lowest-energy eigenstates to be localised about $\{\mathbf{x}_n^0\}$. Multiple wells can also be treated in this way. Consider a unitary transformation of the form

$$U_\lambda = \exp(iS_\lambda/\hbar), \tag{8.149}$$

where λ is a real parameter to be determined,[15] and S_λ depends on all the nuclear position and momentum operators. The usual rules yield

$$\Psi_\lambda = U_\lambda^{-1}\psi, \quad H_\lambda = U_\lambda^{-1} H U_\lambda, \tag{8.150}$$

$$\mathbf{x}(\lambda)_n = \mathbf{x}_n - \frac{\partial S_\lambda}{\partial \mathbf{p}_n}, \quad \mathbf{p}(\lambda)_n = \mathbf{p}_n + \frac{\partial S_\lambda}{\partial \mathbf{x}_n}. \tag{8.151}$$

---

[15] U_1 is the identity operator.

The operator S_λ is chosen so that the transformation produces a translation to make $\{x_n^0\}$ the origin of the coordinates, and a simultaneous dilation (scale transformation) of the position and momentum operators

$$\frac{\partial S_\lambda}{\partial p_n} = (1-\lambda)(x_n - x_n^0), \quad \frac{\partial S_\lambda}{\partial x_n} = \frac{(1-\lambda)}{\lambda} p_n \tag{8.152}$$

so that

$$x(\lambda)_n = x_n^0 + \lambda(x_n - x_n^0), \quad p(\lambda)_n = \frac{1}{\lambda} p_n. \tag{8.153}$$

In this way one can look at the low-lying states with a 'microscope' with a certain resolving power that depends on Planck's constant. These relations show that S_λ is a non-integrable function of the phase space variables $\{x_n, p_n\}$, since

$$\frac{\partial^2 S_\lambda}{\partial x_n \partial p_n} \neq \frac{\partial^2 S_\lambda}{\partial p \partial x_n}, \quad \lambda \neq 1. \tag{8.154}$$

Thus it doesn't have a definite value at (x_n, p_n), but it does have well-defined derivatives given by Eq. (8.152) except at $\lambda = 0$; at this point the transformation equations have a singularity.

Under this transformation a Hamiltonian of the form

$$H = \sum_n \frac{p_n^2}{2m_n} + V(\{x_n\}) \tag{8.155}$$

is transformed to

$$H_\lambda = \sum_n \frac{p(\lambda)_n^2}{2m_n} + V(\{x(\lambda)_n\}), \tag{8.156}$$

which may be written as

$$H_\lambda = V(\{x_n^0\}) + K(\lambda) \tag{8.157}$$

where

$$K(\lambda) = -\frac{\hbar^2}{\lambda^2} \sum_n \frac{\nabla_n^2}{2m_n} + V(x_n^0 + \lambda(x_n - x_n^0)) - V(\{x_n^0\}). \tag{8.158}$$

With the choice $\lambda = \sqrt{\hbar}$, unitary equivalence of the spectrum implies that the eigenvalues of the original Hamiltonian H are related to those of $K(\lambda)$ by

$$E_n = V(\{x_n^0\}) + \hbar \mu_n(\lambda). \tag{8.159}$$

Provided V is analytic in λ, it can be expanded about $\lambda = 0$, and this puts $K(\lambda)$, in lowest order, into the form of a sum of coupled oscillators so that the first approximation for the eigenvalue function μ_n is

$$\mu_n \approx \sum_k \left(n_k + \tfrac{1}{2} \right) \omega_k, \tag{8.160}$$

and the wave functions are the familiar Hermite functions for oscillators. In the Born–Oppenheimer calculation for the diatomic molecule analyticity of the effective potential V in λ could be proven, and the role of $\sqrt{\hbar}$ was taken by the usual BO expansion parameter $\kappa = (m_e/M_N)^{\frac{1}{4}}$. In this way the molecular energy level formula (8.106) is recovered as an asymptotic expansion [18].

This sort of discussion can also be extended to solid-state physics; if the minimum energy configuration $\{\mathbf{x}_n^0\}$ is assumed to be a regular lattice, the description is precisely the usual Debye theory of *phonons* in a large system [3], [69] with wave functions $\Psi_{p,\mathbf{k}}$ labelled by wave vectors $\mathbf{k}$ just as in the Band theory of electrons in a crystalline solid. The variables in the Hermite functions can be expressed in terms of pairs $(\mathbf{x}_n, \mathbf{x}_m)$ of the original nuclear variables, so the $\Psi_{p,\mathbf{k}}$ are strongly correlated states. The phonons in this description are collective excitations in which the nuclei always remain attached to their 'own' potential wells.

The singular nature of the microscope transformation can be demonstrated explicitly for the diatomic molecule by the modification of the spectrum associated with the limit $\kappa \to 0$. The spectrum of the Coulomb Hamiltonian for a molecule was discussed in §8.3; for the diatomic molecule, $\sigma_{ess}(\mathsf{H})$ starts at the lowest two-body threshold $\Sigma = \lambda_A(m_A) + \lambda_B(m_B)$ given by the minimal value of the sums of pairs of binding energies for atoms A and B with finite masses m_A and m_B respectively. On the other hand the spectrum of the electronic Hamiltonian, H^{elec}, is purely continuous, $\sigma(\mathsf{H}_0) = [\mathsf{V}_0, \infty)$. In the limit $m_A, m_B \to \infty$, Σ does not generally converge to V_0; instead the missing part of the continuous spectrum $[\mathsf{V}_0, \lambda_A(\infty) + \lambda_B(\infty)]$ is provided by an accumulation of bound states in this interval [68].

While the microscope transformation is formally applicable to the polyatomic case, it may not be sufficient to control the asymptotic behaviour. The Hamiltonian (8.128) is the one used by Klein et al. [70] in their consideration of the precise formulation of the Born–Oppenheimer approximation for polyatomic systems. Their work was based on a powerful abstract operator method, the pseudodifferential calculus [9], [70], [71] and a formalism related to the partitioning technique described previously. In [70] it is assumed that (8.128) has a discrete eigenvalue which has a minimum as a function of the $\mathbf{t}^{\mathrm{n}}$ in the neighbourhood of some values $\mathbf{t}_i^{\mathrm{n}} = \mathbf{b}_i$. Because of the rotation-inversion invariance, such a minimum exists on a three-dimensional sub-manifold for all $\mathbf{b}_i$ such that

$$\mathbf{b}_i \to \mathcal{R}\mathbf{b}_i, \quad \mathcal{R} \in O(3). \tag{8.161}$$

The $\mathbf{b}_i$ therefore define the geometrical shape of the minimum in the usual way. If the minimum figure is a plane, then the potential well is diffeomorphic to $SO(3)$, while if it is non-planar, then it is diffeomorphic to $O(3)$, and so the well is actually a symmetric double well. In either case, Klein et al. show that the eigenvalues and eigenfunctions of the full problem can be obtained as WKB-type expansions to all orders of the expansion parameter, the square root of the ratio of the electronic to a typical nuclear mass $(= \kappa^2)$. Because of the way the Hamiltonian is formulated, its invariance under permutations of the electronic variables is readily considered, and the electronic wave function can easily be chosen in permutationally allowed form, no matter what the nuclear geometry happens to be. However, permutational invariance when some of the nuclei are identical was not considered in [70].

8.5.1 Quantum Chemistry and the Coulomb Hamiltonian

There are various approaches to the solution of the molecular Schrödinger equation in the quantum chemistry/chemical physics literature starting from the Coulomb Hamiltonian. Firstly, the functions in (8.143) can be used as the basis of a Rayleigh–Ritz calculation being, hopefully, well-adapted to the construction of appropriate trial functions. Several different lines have been developed; in the adiabatic model the trial function is written as the continuous linear superposition

$$\Psi(\mathbf{t}^e, \mathbf{t}^n)_m = \int F(\mathbf{b}) \, \varphi(\mathbf{b}, \mathbf{t}^e)_m \, \delta(\mathbf{t}^n - \mathbf{b}) \, d\mathbf{b}$$
$$= F(\mathbf{t}^n) \, \varphi(\mathbf{t}^n, \mathbf{t}^e)_m, \tag{8.162}$$

where the square integrable weight factor $F(\mathbf{t}^n)$ may be determined by reducing (8.129) to an effective Schrödinger equation for the nuclei in which $F(\mathbf{t}^n)$ appears as the eigenfunction [72]. Since the $\{\varphi_m\}$ are orthonormal we have

$$\langle \Psi_m | \Psi_m \rangle = \iint |\Psi(\mathbf{t}^e, \mathbf{t}^n)_m|^2 \, d\mathbf{t}^e \, d\mathbf{t}^n = \int |F(\mathbf{t}^n)|^2 \, d\mathbf{t}^n. \tag{8.163}$$

On the other hand the (mm) matrix element of the translationally invariant Hamiltonian H' can be written as

$$\langle \Psi_m | H' | \Psi_m \rangle = \iint \Psi_m^* \left(H' \, \Psi_m \right) d\mathbf{t}^e \, d\mathbf{t}^n = \int F(\mathbf{t}^n)^* \left(H_m \, F \right) (\mathbf{t}^n) \, d\mathbf{t}^n, \tag{8.164}$$

where we have defined the effective nuclear Hamiltonian,

$$\left(H_m \, F \right) (\mathbf{t}^n) = \int \varphi(\mathbf{t}^e, \, \mathbf{t}^n)_m \left[H' \, \varphi(\mathbf{t}^e, \mathbf{t}^n)_m \, F(\mathbf{t}^n) \right] d\mathbf{t}^e. \tag{8.165}$$

The Rayleigh–Ritz quotient,

$$E[\Psi_m] = \frac{\langle \Psi_m | H' | \Psi_m \rangle}{\langle \Psi_m | \Psi_m \rangle}, \tag{8.166}$$

is stationary for those functions that are solutions of the effective nuclear 'Schrödinger equation',

$$H_m \, F_s = E_{ms} \, F_s. \tag{8.167}$$

In particular, using the ground electronic state φ_0, the Rayleigh–Ritz quotient leads to an upper bound to the ground state energy E_0 of H'. This calculation amounts to the diagonalisation of H' in the one-dimensional subspace spanned by Ψ_0. The subspace may be enlarged, and the accuracy thereby improved, by using the subspace spanned by a set of trial functions $(\Psi_0, \Psi_1, \cdots, \Psi_m)$ of the form of (8.162). So far no symmetries of the Coulomb Hamiltonian have been discarded.

In practice the adiabatic model is implemented as follows; a collection of energies $E(\mathbf{b}_i)$ is found through standard quantum chemical computations for different geometries $\{\mathbf{b}_i\}$ and fitted to produce a function $V(\mathbf{t}^n)$ that is treated as a potential energy contribution to (8.167), it being assumed that the fitted function somehow approximates the assumed electronic 'eigenvalue' $E(\mathbf{t}^n)$ in the equation

$$H^{\text{elec}}(\mathbf{t}^n, \mathbf{t}^e) \Phi_p(\mathbf{t}^n, \, \mathbf{t}^e) = E_p(\mathbf{t}^n) \Phi_p(\mathbf{t}^n, \mathbf{t}^e). \tag{8.168}$$

From the discussion in §8.5 this is problematic, even without considering nuclear permutations, since H^{elec} has purely continuous spectrum and no 'eigenvalues'. Having
set up the variational calculations with square integrable functions, the approximate
ground state is naturally a discrete state; the discussion, however, yields no information
about the bottom of the essential spectrum.

   The unnormalisable product wave function of the adiabatic model, (8.143), can be
regularised by replacing the Dirac delta function by a suitable approximation to the
identity (cf. §2.4.1); for example, if $\xi(x)_a$ is defined as

$$\xi(x)_a = \frac{1}{(2\pi a)^{n/2}} e^{-x\cdot x/2a}, \tag{8.169}$$

we have

$$\lim_{a \to 0} \xi(x)_a = \delta^n(x) \tag{8.170}$$

in the sense of distributions. Here x is a vector with n components, and a is a parameter
that will have to be chosen in the light of results obtained when we do *not* take the limit.

   As in §8.5 it is convenient to use the collective notations x and X for all the electronic and nuclear position variables respectively, and P for all the nuclear momentum
variables; we assume there are A nuclei, so the nuclear vectors have $3A$ components.[16]
The electronic wave function can be taken, as before, as a solution of the clamped-
nuclei Schrödinger equation for some nuclear configuration b formed by the A nuclei,
$\varphi(b,x)$. Thinking about molecular structures, we choose the nuclear wave function to
be strongly peaked about the same configuration b; it can be obtained by appropriate
translations of $\xi(X)_a$, which is peaked about the origin, that is,

$$\xi(X,b)_a = e^{-ibP/\hbar}\xi(X)_a. \tag{8.171}$$

Furthermore, a coherent state for the nuclei can be formed as the Weyl transform of
$\xi(X)_a$ which is (cf. Appendix G)

$$\psi(X)_{u,b} = e^{iuX-ibP/\hbar}\xi(X)_a. \tag{8.172}$$

The coherent state ψ has the property that the average position is $\langle X \rangle = b$, and the
average momentum is $\langle P \rangle = \hbar u$.

   We define the molecular 'intrinsic state' as the product

$$\chi(x,X:u,b) = \varphi(b,x)\psi(X)_{u,b}, \tag{8.173}$$

which we assume is normalised. A trial wave function for a Rayleigh–Ritz variational
calculation of the energy of the molecular Hamiltonian, analogous to the adiabatic
model, is then the continuous superposition

$$\Psi(x,X) = \int G(u,b)\chi(x,X:u,b)\,d\mu(u,b). \tag{8.174}$$

This variational calculation of the energy requires the solution of an integral equation
to determine the value of the weight function $G(u,b)$, and the result would be the optimal way to dispose the classical quantities u and b. It is a non-adiabatic formalism since

---

[16] Thus b and u below are also $3A$-dimensional vectors.

we no longer have a simple Hartree product of electronic and nuclear wave functions. Wave functions Ψ of this form belong to a subspace $\mathcal{H}_\chi$ of the Hilbert space $L^2(\omega)$ characterised in part by the dimensionality of the vectors u and b.

The simplest calculation one can do with an intrinsic function χ is to compute the expectation value of the Hamiltonian and appeal to the Rayleigh–Ritz principle to find the 'optimal' value of (u, b) by calculating the minimum value of $E(u, b)$:

$$E(u,b) = \langle \chi(u,b)|\mathsf{H}|\chi(u,b)\rangle$$
$$0 = \frac{\partial E(u,b)}{\partial u}\Big|_{u=u_0}, \quad \frac{\partial E(u,b)}{\partial b}\Big|_{b=b_0} = 0. \tag{8.175}$$

The expectation value in (8.175) implies integration over *all* the variables for the electrons and nuclei. This will be the best estimate of an eigenvalue of H that a trial function of this form can support,[17] while values of $E(u, b)$ away from the minimum define an 'energy surface' in the phase space (u,b).

The approximation

$$G(u,b) \sim \delta(u)F(b) \tag{8.176}$$

reduces (8.174) to

$$\Psi(x,X) = \int F(b)\chi(x,X:b)\,\mathrm{d}\mu(b), \tag{8.177}$$

which is the ansatz of the Generator Coordinate Method (GCM) [73]–[75]. In the GCM the effective Schrödinger equation for the weight function $F(b)$ becomes an integral equation (the Hill–Wheeler equation). The trial wave function may be improved, in the sense of a variational calculation, by forming linear superpositions of the wave functions $\{\Psi\}$; this has been done for diatomic molecules for which a fairly complete GCM account that maintains all the symmetries of the Coulomb Hamiltonian has been developed [76]. In the applications that were made to polyatomic molecules, however, the dependence on the nuclear variables $\{\mathbf{t}^\mathrm{n}\}$ was not expressed through functions adapted to nuclear permutation symmetry, nor was parity explicitly involved. It would seem the computational burden is still too great to make this a mainstream approach to polyatomic molecules; detailed accounts were given in [77]–[79].

An important property of intrinsic functions is specified by the Brink–Weiguny condition [80]. Let U be a unitary representation of some symmetry of the system and consider its action on a wave function, Ψ, of the type above. Brink and Weiguny proved that

$$\mathsf{U}\Psi \in \mathcal{H}_\chi \text{ if } \Psi \in \mathcal{H}_\chi, \text{ iff } \mathsf{U}\chi(\mathbf{z},\mathbf{b}) = \chi(\mathbf{z},\sigma(\mathbf{b})), \tag{8.178}$$

where $\sigma(\mathbf{b})$ is a transformation in the GC-parameter space having the same effect as U. As an example, let $\mathsf{T_a}$ be the operator of coordinate translation by an amount $\mathbf{a}$. Then for instrinsic states of the form

$$\chi(\mathbf{z},\mathbf{b}) = \chi(\mathbf{z}-\mathbf{b}), \tag{8.179}$$

---

[17] The intrinsic function may contain other parameters and in principle they too can be optimised.

we have

$$\mathsf{T}_{\mathbf{a}}\chi(\mathbf{z},\mathbf{b}) = \chi(\mathbf{z}-\mathbf{a},\mathbf{b}) \equiv \chi(\mathbf{z},\mathbf{b}+\mathbf{a}). \tag{8.180}$$

Thus by appropriate choice of the intrinsic states all of the symmetries of the physical system can be referred to the GC-parameter space. In this way we can make an important distinction between

1. dynamical aspects of the Hamiltonian, and all other observables, and
2. purely geometric (and possibly topological) aspects of the GC-parameter space.

As we have seen, the Coulomb Hamiltonian, H, commutes with the operators of overall rotation, translation, space inversion and any applicable permutation symmetries; the corresponding degeneracy of the wave functions can be incorporated in weight functions $F(\mathbf{b})$ which factorise in the form

$$F(\mathbf{b}) = F_{\text{trans}} \, F_{\text{rot}} \, F_{\text{int}}. \tag{8.181}$$

Thus part of the structure of $F(\mathbf{b})$ can be determined purely by symmetry arguments. The important point is that the trial wave functions $\{\Psi\}$ can be properly symmetry adapted even if the intrinsic functions are not. On the other hand the wave function in the generator coordinate approach is a continuous superposition over products of 'electronic' and 'nuclear' wave functions and is *not* separable, so this is a non-adiabatic formalism. Obviously one recovers the usual adiabatic Born–Oppenheimer wave function if one reverts to the singular limit (8.162); it should be noted that it is the weight function F that passes over to the BO nuclear wave function rather than ξ_a.

References

[1] Thirring, W. (1987), in *Schrödinger, Centenary Celebrations of a Polymath*, ed. C. W. Kilmister, p. 65, Cambridge University Press.
[2] Gutzwiller, M. (1998), Rev. Mod. Phys. **79**, 589.
[3] Anderson, P. W. (1984), *Basic Notions of Condensed Matter Physics*, Frontiers of Physics **55**, Benjamin/Cummings: Addison-Wesley.
[4] Hunziker, W. and Sigal, I. M. (2000), J. Math. Phys. **41**, 3448.
[5] Woolley, R. G. and Sutcliffe, B. T. (2003), Ch. 3, vol. 1 of *Fundamental World of Quantum Chemistry: A Tribute Volume to the Memory of Per-Olov Löwdin*, edited by E. J. Brändas and E. S. Kryachko, Kluwer Academic Publishers.
[6] Sutcliffe, B. T. and Woolley, R. G. (2005), Phys. Chem. Chem. Phys. **7**, 3664.
[7] Sutcliffe, B. T. and Woolley, R. G. (2012), J. Chem. Phys. **137**, 22A544.
[8] Sutcliffe, B. T. and Woolley, R. G. (2014), J. Chem. Phys. **140**, 037101.
[9] Jecko, T. (2014), J. Math. Phys. **55**, 053504; doi:10.1063/1.4870855.
[10] Kato, T. (1951), Trans. Am. Math. Soc. **70**, 212.
[11] Simon, B. (2000), J. Math. Phys. **41**, 3523.

[12] Löwdin, P.-O. (1988), in *Molecules in Physics, Chemistry and Biology*, vol. 2, ed. J. Maruani, Kluwer Academic Publishers.

[13] Loss, M., Miyao, T. and Spohn, H., (2007), J. Funct. Anal. **243**, 353.

[14] Jecko, T., Sutcliffe, B. T. and Woolley, R. G. (2015), J. Phys. A: Mathematical and Theoretical **48**, 445201 (20 pages).

[15] Ruelle, D. (1969), Nuovo Cimento A**61**, 655.

[16] Weyl, H. (1909), Rend. Circ. Mat. Palermo **27**, 373.

[17] Roman, P. (1965), *Advanced Quantum Theory*, Addison-Wesley Publishing Co.

[18] Combes, J.-M. and Seiler, R. (1980), in *Quantum Dynamics of Molecules*, ed. R. G. Woolley, NATO ASI **B57**, Plenum Press, 435.

[19] Zhislin, G. M. (1960), Trudy. Mosk. Mat. Obsc. **9**, 82.

[20] van Winter, C. (1964), Kgl. Danske Vid. Selskab. **1**, 1.

[21] Hunziker, H., (1966), Helv. Phys. Acta **39**, 451.

[22] Reed, M. and Simon, B. (1978), *Methods of Modern Mathematical Physics, IV, Analysis of Operators*, Academic Press.

[23] Balslev, E. and Combes, J.-M. (1971), Commun. Math. Phys. **22**, 280.

[24] Aguilar, J. and Combes, J.-M. (1971), Commun. Math. Phys. **22**, 269.

[25] Reinhardt, W. P. (1982), Ann. Rev. Phys. Chem. **33**, 223.

[26] Uchiyama, J. (1967), Pub. Res. Inst. Math. Sci. Kyoto A**2**, 117.

[27] Thirring, W. (1981), *A Course in Mathematical Physics, 3, Quantum Mechanics of Atoms and Molecules*, translated by E. M. Harrell, Springer-Verlag.

[28] Simon, B. (1971), *Quantum Mechanics for Hamiltonians Defined as Quadratic Forms*, Princeton University Press.

[29] Ruskai, M. B. (1982), Commun. Math. Phys. **85**, 325.

[30] Sigal, I. M. (1982), Commun. Math. Phys. **85**, 309.

[31] Zhislin, G. M. (1971), Theor. Math. Phys. **7**, 571.

[32] Nyden Hill, R. (1977), J. Math. Phys. **18**, 2316.

[33] Courant, R. and Hilbert, D. (1953), *Methods of Mathematical Physics*, vol. 1, Interscience Publishers, Inc.

[34] Löwdin, P.-O. (1989), Pure Appl. Chem. **61**, 2065.

[35] Richard, J-M., Fröhlich, J., Graf, G-M. and Seifert, M. (1993), Phys. Rev. Lett. **71**, 1332.

[36] Ruskai, M. B. (1990), Ann. Inst. Henri Poincaré **52**, 397.

[37] Ruskai, M. B. (1991), Commun. Math. Phys. **137**, 553.

[38] Simon, B. (1970), Helv. Phys. Acta **43**, 607.

[39] Vugal'ter, S. A. and Zhislin, G. M. (1977), Theor. Math. Phys. **32**, 602.

[40] Evans, W. D., Lewis, R. T. and Saito, Y. (1992), Phil. Trans. Roy. Soc. (London) A**338**, 113.

[41] Lieb, E. H. and Seiringer, R. (2010), *The Stability of Matter and Quantum Mechanics*, Cambridge University Press.

[42] Sutcliffe, B. T. (1993), J. Chem. Soc. Faraday Trans. **84**, 2321.

[43] Katriel, J. (2001), J. Mol. Struct. Theochem **547**, 1.

[44] Dumitru, A. G., and Woolley, R. G. (1998), Mol. Phys. **94**, 595.

[45] Born, M. and Oppenheimer, J. R. (1927), Ann. der Physik **84**, 457.

[46] Born, M. and Heisenberg, W. (1924), Ann. der Physik **74**, 1.

[47] Eckart, C. (1935), Phys. Rev. **47**, 552.

[48] Nordheim, L. (1923), Z. Physik **19**, 69.

[49] Blinder, S. M. accessed at `www.ulb.ac.be/cpm/people/bsutclif/main.html`.

[50] Lathouwers, L. and van Leuven, P. (1982), Adv. Chem. Phys. **49**, 115.

[51] Born, M. and Huang, K. (1955), *Dynamical Theory of Crystal Lattices*, Oxford University Press.

[52] Born, M. (1951), Gött. Nachr. Math. Phys. Kl. 1.

[53] O'Malley, T. F. (1971), Adv. At. Mol. Phys. **7**, 223.

[54] Hall, G. G. (1987), Int. J. Quant. Chem. **31**, 383.

[55] Messiah, A. (1960), *Quantum Mechanics*, John Wiley & Sons, Inc.

[56] Mead, C. A. and Moscowitz, A. (1967), Int. J. Quant. Chem. **1**, 243.

[57] Monkhorst, H. (1987), Phys. Rev. A**36**, 1544.

[58] Lodi, L. and Tennyson, J. (2010), J. Phys. B At. Mol. Opt. Phys. **43**, 133001.

[59] Balslev, E. (1972), Ann. Phys. (N.Y.) **73**, 49.

[60] Mead, C. A. (1992), Rev. Mod. Phys. **64**, 51.

[61] Roman, P. (1975), *Some Modern Mathematics for Physicists and Other Outsiders*, vol. 2, Pergamon Press Inc.

[62] Weinberg, S. (2013), *Lectures on Quantum Mechanics*, Ch. 5, Cambridge University Press.

[63] Hagedorn, G. A. (1980), Commun. Math. Phys. **77**, 1.

[64] Hagedorn, G. A. and Joye, A. (2007), *Proceedings of Symposia in Pure Mathematics*, **76**, *Spectral Theory and Mathematical Physics. A Festschrift in Honor of Barry Simon's 60th Birthday*, edited by F. Gesztezy, P. Deift, C. Galvez, P. Perry and W. Schlag, p. 230, American Mathematical Society.

[65] Löwdin. P. O. (1966), in *Perturbation Theory and Its Application in Quantum Mechanics*, Proceedings of Madison Symposium, ed. C. H. Wilcox, John Wiley and Sons, Inc.

[66] Combes, J.-M., Duclos, P. and Seiler, R. (1981), in *Rigorous Atomic and Molecular Physics*, edited by G. Velo and A. Wightman, p. 185 Plenum.

[67] Combes, J.-M., Duclos, P. and Seiler, R. (1983), J. Funct. Anal. **52**, 257.

[68] Combes, J.-M. (1977), Acta Phys. Austr. **17**, Suppl., 139.

[69] Ashcroft, N. and Mermin, N. W. (1976), *Solid State Physics*, Holt, Rinehart, Winston.

[70] Klein, M., Martinez, A., Seiler, R. and Wang, X. P. (1992), Commun. Math. Phys. **143**, 607.

[71] Feffermann, C. L. (1983), Bull. Am. Math. Soc. **9**, 129.

[72] Messiah, A. (1960), *Quantum Mechanics*, John Wiley & Sons, Inc.

[73] Peierls, R. and Yoccoz, J. (1957), Proc. Phys. Soc. A**70**, 381.

[74] Lathouwers, L., van Leuven, P. and Bouten, M. (1977), Chem. Phys. Letters **52**, 479.

[75] Lathouwers, L. and van Leuven, P. (1980), Chem. Phys. Letters **70**, 410.

[76] Lathouwers, L. and van Leuven, P. (1982), Adv. Chem. Phys. **49**, 115.

[77] Deumens, E., Ohrn, Y., Lathouwers, L. and van Leuven, P. (1986), J. Chem. Phys. **84**, 3944.

[78] Lathouwers, L., van Leuven, P., Deumens, E. and Ohrn, Y. (1987), J. Chem. Phys. **86**, 6352.

[79] Broeckhove, J., Lathouwers, L. and van Leuven, P. (1991), J. Math. Chem. **6**, 207.

[80] Brink, D. M. and Weiguny, A. (1968), Nuclear Phys. A**120**, 59.

PART III

NON-RELATIVISTIC QUANTUM ELECTRODYNAMICS

9 The Quantisation of Electrodynamics

9.1 Introduction

Rather than proceeding directly to quantum electrodynamics (QED), it may be helpful to summarise first the actual historical development since this still exerts a powerful influence on the way the theory of the interaction of electromagnetic radiation with matter in the 'non-relativistic' (low-energy) regime is presented. The return to electrodynamics that followed the success of quantum mechanics in accounting for the stability of atoms in terms of electrostatic forces was based on a very different conception from the earlier classical one; henceforth the electromagnetic field was to be treated as a weak perturbation of the atomic states, and so the concept of an isolated atom became a central feature of the new mechanics. The pathologies due to self interaction ('radiation reaction') that had plagued classical electrodynamics were put to one side with the assumption that one could use the experimental charge and mass parameters of the electron and the atomic nucleus. By treating the electromagnetic field as an external, classical perturbation of an atom, Schrödinger was able to calculate the Einstein B-coefficient for stimulated absorption and emission, and the cross section for linear light scattering [1] which turned out to be equivalent to the formula obtained earlier by Kramers and Heisenberg using the correspondence principle. These calculations were the prototypes for what has become known as the semiclassical radiation model which we shall describe in modern terms in §9.5.

Shortly afterwards, quantum electrodynamics for the atom-radiation system was developed by Dirac, who discovered the boson quantisation of the free radiation field and used it to represent the Coulomb gauge vector potential as a quantum mechanical operator [2]. Dirac was able to reproduce Schrödinger's results for stimulated absorption and emission and linear light scattering, but he also calculated directly Einstein's A-coefficient for spontaneous emission. A particularly important result of Dirac's calculation is that the relationship between the A- and B-coefficients for a transition at frequency ω,

$$A = \frac{\hbar \omega^3}{\pi^2 c^3} \, B,\tag{9.1}$$

is quite general and is not limited to radiation in thermal equilibrium with its surroundings as originally assumed by Einstein [3]. On the other hand it quickly became apparent that the ugly pathologies due to self-interaction would also have to be revisited in the quantum mechanical theory [4], [5].

In this chapter we shall discuss some general features of both classical and quantum mechanical descriptions of the electromagnetic field, paying particular attention to the freedom to make gauge transformations of the field potentials, and sketch briefly a few ideas about the relationship of the semiclassical model to QED. Within the perturbation approach the question of the stability of atoms and molecules is no longer a question for QED, as it had been for classical physics, and so an extensive quantum theory of atoms and molecules has developed over many years in which electromagnetic radiation plays only the subsidiary role of causing transitions between their states in absorption, emission and scattering processes. This is the case for both classical and quantum mechanical descriptions of the electromagnetic field. The ground state of an atom[1] cannot decay through the spontaneous emission mechanism, and so its stability is not in question in the perturbation theory framework. More recently the existence of a stable ground state for an atom and the fate of Bohr's excited 'stationary states' in the presence of the quantised radiation field has been considered in a non-perturbative framework using the methods of functional analysis; we shall take this up in Chapter 11.

We saw in Chapter 7 that quantisation of the field implies the use of operator-valued field variables, for example the photon annihilation and creation operators and the vector potential operator, together with the specification of the states of the field in terms of a representation of an Hilbert space, for example, the Fock space for the modes that contribute. The classical description of the field collapses these two aspects into the classical field variables. As noted previously, the semiclassical radiation model is based on quantising the atomic system as usual, while the electromagnetic field is treated as an external classical time-dependent perturbation in the Schrödinger equation for the atom. This approach has turned out to be very popular; historically, it was developed in parallel with non-relativistic QED and was mainly used in low-order perturbation theory to describe conventional light scattering processes [6], [7]. The results obtained generally parallel those of the S-matrix theory in QED. In practice the semiclassical model is often based on the further assumption that the vector potential is restricted to the Coulomb gauge; alternatively it is also commonly supposed that the coupling to the classical electromagnetic field is effectively only through an electric field that is spatially uniform over the volume of the atom. This is called the electric dipole approximation since the interaction operator is reduced to

$$V = -\mathbf{d} \cdot \mathbf{E}^{\perp}, \tag{9.2}$$

where $\mathbf{d}$ is the atomic electric dipole moment, and $\mathbf{E}^{\perp}$ is the uniform field.

In §2.3 we saw that the field potentials $(\mathbf{a}, \phi)$ are introduced in classical electromagnetism as auxiliary quantities from which the physical fields $(\mathbf{E}, \mathbf{B})$ can be calculated

---

[1] Here and later on we sometimes use 'atom' as a shorthand for 'atom or molecule'; the context should make clear whether the remark is restricted to just the atomic case (one nucleus).

according to

$$\mathbf{B} = \mathbf{\nabla} \wedge \mathbf{a}, \tag{9.3}$$

$$\mathbf{E} = -\frac{\partial \mathbf{a}}{\partial t} - \mathbf{\nabla}\phi. \tag{9.4}$$

These two equations imply that equally good potentials $(\mathbf{a}', \phi')$ can be defined by a gauge transformation:

$$\mathbf{a} \rightarrow \mathbf{a'} = \mathbf{a} - \mathbf{\nabla}\chi, \tag{9.5}$$

$$\phi \rightarrow \phi' = \phi + \frac{\partial \chi}{\partial t}. \tag{9.6}$$

It is evident that, provided the function χ is differentiable and integrable, the transformed potentials $(\mathbf{a}', \phi')$ yield the same physical field $(\mathbf{E}, \mathbf{B})$; thus the potentials are not determined uniquely by Eqs. (9.3), (9.4). On the other hand the use of potentials ensures that the homogeneous Maxwell equations

$$\mathbf{\nabla} \cdot \mathbf{B} = 0, \quad \mathbf{\nabla} \wedge \mathbf{E} = -\frac{\partial \mathbf{B}}{\partial t} \tag{9.7}$$

are satisfied automatically. Gauge invariance is the requirement that physical observables calculated with the aid of the potentials be independent of the transformation function $\chi(\mathbf{x}, t)$, so that making a choice of gauge in a calculation carries with it the requirement of ensuring gauge invariance of the results obtained.

The basic laws of classical electrodynamics are the Lorentz force law for the charges and the Maxwell equations for the electromagnetic field, both of which are gauge invariant. Quantum electrodynamics is formulated in terms of either Lagrangian or Hamiltonian dynamics, and the fundamental law is of a quite different, geometrical character. In infinitesimal form, it is based on a differential 1-form $\mathrm{d}v = \mathbf{a}(\mathbf{x}) \cdot \mathrm{d}\mathbf{x}$ involving the vector potential, $\mathbf{a}(\mathbf{x})$, and the phase of the wave function for a charge e, located at position $\mathbf{x}$. The finite, integrated version of the QED law is a line integral over $\mathrm{d}v$ on some path ending at the position $\mathbf{x}$. It amounts to a reworking in the quantum context of Weyl's failed attempt at unifying classical electromagnetism and gravitation (§3.7.1). This is what is meant when QED is described as a 'gauge theory'.

The main features of the gauge invariance of electrodynamics in a quantum mechanical description were established very early on by the pioneers [10]–[12]. The essential point is that a gauge-invariant description requires a wave function for a charged particle interacting with the electromagnetic field to contain a phase factor that involves the ratio $(e/\hbar)$ and a line integral over the potentials describing the field. This is true for both classical and quantum descriptions of the field; line integrals of this form are familiar from Dirac's theory of magnetic monopoles [12], the Bohm–Aharonov effect [13], and in the Power–Zienau–Woolley (PZW) transformation theory in atomic and molecular physics [14]–[20]. More generally, if charges are regarded as the quanta of a quantised field, essentially the same phase factor must be associated with their annihilation and creation operators to give a gauge-invariant formalism [21]–[23]. These ideas will be introduced in §9.2 and developed in §9.3, where it will become apparent that the introduction of the electric polarisation field operator in the quantum theory through the action integral

$$F = \int \mathbf{P}(\mathbf{x}) \cdot \mathbf{a}(\mathbf{x}) \, d^3\mathbf{x} \tag{9.8}$$

is really only a disguised form of the fundamental line integral over the vector potential.

With the advent of the laser in the 1960s, high-intensity radiation sources became widely available and their use revealed a host of non-linear optical processes. In the language of QED these phenomena are 'multiphoton' processes and are much harder to describe in the photon picture because they require high-order perturbation theory. Accordingly researchers turned to the semiclassical radiation model and began to investigate non-perturbative solutions to the time-dependent Schrödinger equation with the aid of computers. A physical rationale for taking the electromagnetic field to be 'classical' can be seen from the estimate in §7.1 and the recognition that lasers can deliver large electric fields in short pulse times. More precisely, the radiation from a single-mode laser operating well above threshold is well described by a coherent state $|\alpha_k\rangle$ with a large parameter α; the expectation value of the electric field operator approximates the classical Maxwell form (7.144), and the simplest statistical properties of the radiation are classical-like [8], [9] (cf. Chapter 7). As just suggested, however, it is still a far from trivial matter to show that these results justify treating the Hamiltonian in an arbitrary gauge as a classical-quantum hybrid since they are properties of the *state* of the field. We shall pursue this further in §9.5 where we comment on the relationships between classical, semiclassical and quantum electrodynamics.

9.2 Wave Mechanics and Electromagnetism

By way of introduction we consider the case of a particle in an external classical electromagnetic field according to wave mechanics. The wave function for a particle in a coordinate representation $\psi(x)$ is a complex-valued function of the space-time coordinates, $x = (\mathbf{r}, t)$. The phase of the wave function at a particular point has no physical meaning, and its value can be assigned arbitrarily; thus if we put

$$\psi(x) \rightarrow \exp(-ib/\hbar)\psi(x), \tag{9.9}$$

where b is a free parameter with the dimensions of action, both wave functions describe the same physical state. Only the relative phase between two different space-time points is significant; more precisely we may assume that the relative phase is definite only if the two points are neighbouring [12]. In this framework one can repeat more or less verbatim Weyl's argument described in §3.7.1 with the real vector in space-time replaced by the complex-valued wave function for the particle $\psi(x)$; the ramifications of this approach were described by Weyl in 1929 in a classic paper that showed how electromagnetism in quantum mechanics should be understood as a gauge theory [11].

In general b can be taken to be a local function, $b = b(x)$, since the phase may vary with the position of the particle. Let us suppose that the phase of the wave function at a point has been chosen according to some specific calibration. A change in the phase of the wave function can be thought of as a unitary transformation, provided $b(x)$ satisfies appropriate (and fairly mild) analytical conditions,

$$\psi(x) \;\rightarrow\; \psi(x)' = \mathsf{U}_b\,\psi(x), \tag{9.10}$$

with

$$\mathsf{U}_b = \exp\left(-ib(x)/\hbar\right). \tag{9.11}$$

In infinitesimal form the unitary transformation (9.10) is

$$\psi(x) \;\rightarrow\; \psi(x)' = \left(1 - \frac{i}{\hbar}b(x)\right)\psi(x). \tag{9.12}$$

The complex numbers $\{\mathsf{U}_b\}$ for suitable $b(x)$ provide a unitary representation of the unitary unimodular group $\mathcal{U}(1)$, and (9.10) describes the action of a group element at the point x on the wave function ψ; $\mathcal{U}(1)$ has the unit circle in the complex plane as its group space. If we equip the circle with axes, for example, the real and imaginary parts of $\psi(x)$, (9.10) can be viewed as a rotation with respect to these axes, and it formalises the freedom we have to assign a different $\mathcal{U}(1)$ rotation at each point x. Only in the special case $b =$ constant are the axes fixed; in general, we cannot assume that the 'internal' axes will stay the same if we move from x to another point x'. Equivalently we may say that in moving the wave function along some path $\mathcal{C}$ between these points parameterised by, say, λ, the basis vectors in the internal space will depend on λ. A characteristic feature of electromagnetism is that the interaction between a charged particle and the field is proportional to the charge e. This fact can be brought into the description by specialising the unitary operators $\{\mathsf{U}_b\}$ to the form

$$\mathsf{U}_f = \exp\left(-ief(x)/\hbar\right), \tag{9.13}$$

so that the electric charge parameter e determines the character of the representation of the abelian group $\mathcal{U}(1)$ to which the particle is assigned.

In order to compare wave functions at points that differ by an infinitesimal displacement, we make use of its derivative which involves the difference of its value at two neighbouring points,

$$d\psi(x) = \psi(x+dx) - \psi(x), \tag{9.14}$$

and so provides information about both the variation of the wave function with position, and the rotation of the axes of the internal space on moving from x to $x+dx$. Under the transformation (9.12) the derivative changes according to

$$\frac{\partial \psi(x)}{\partial x^{\alpha}} \rightarrow \frac{\partial \psi(x)}{\partial x^{\alpha}} - \frac{ie}{\hbar} f(x) \frac{\partial \psi(x)}{\partial x^{\alpha}} - \frac{ie}{\hbar} \left(\frac{\partial f(x)}{\partial x^{\alpha}} \right) \psi(x). \qquad (9.15)$$

It is apparent that the wave function and its derivative do not transform in the same way under the phase change (compare (9.12) and (9.15)); this is a consequence [24] of the rotation of the axes in the internal space. A covariant derivative of $\psi(x)$ is one which *does* transform in the same way as $\psi(x)$ itself under the transformation induced by U_f; such a derivative may be constructed by parallel transport as in general relativity. Instead of (9.14) we must compare $\psi(x+\mathrm{d}x)$ with the value $\psi(x)$ would have if it were parallel transported to $x+\mathrm{d}x$,

$$\delta \psi(x) = \psi(x+\mathrm{d}x) - \psi(x)^{\|}. \qquad (9.16)$$

The quantum mechanical law for electromagnetism, which replaces the classical Lorentz force law for a particle with charge e in an electromagnetic field, can be stated [25] in terms of (9.16) as follows; $\delta \psi(x)$ is proportional to $\psi(x)$ and to a differential 1-form $\mathrm{d}v$ that depends on a vector field, $a(x)_{\alpha}$, and the charge e,

$$\delta \psi(x) = \frac{i}{\hbar} \mathrm{d}v \ \psi(x), \qquad (9.17)$$

where

$$\mathrm{d}v = e a(x)_{\alpha} \mathrm{d}x^{\alpha}. \qquad (9.18)$$

The vector field $a(x)_{\alpha}$ is identified with the electromagnetic field potentials through $a(x)_{\alpha} = (\frac{1}{c} \phi(\mathbf{x},t), -\mathbf{a}(\mathbf{x},t))$. Finite displacements may be described by integration of the 1-form $\mathrm{d}v$; as it is not usually a perfect derivative, the result will depend on the integration path. Note especially that the quantum law is expressed in terms of the field potentials rather than the physical field strengths.

We now have two different expressions for displaced wave functions at the point $x+\mathrm{d}x$: $\psi(x) + \mathrm{d}\psi(x)$ and $\psi(x) + \delta \psi(x)$. The covariant derivative of the wave function is obtained from the difference between these two quantities:

$$D\psi(x) = (\psi(x) + \mathrm{d}\psi(x)) - (\psi(x) + \delta \psi(x))$$
$$= \mathrm{d}\psi(x) - \frac{i}{\hbar} \mathrm{d}v \ \psi(x), \qquad (9.19)$$

where we have used (9.17) for $\delta \psi(x)$. The covariant derivative operator D_{α} is then defined by

$$\frac{D\psi(x)}{\mathrm{d}x^{\alpha}} = \frac{\partial \psi(x)}{\partial x^{\alpha}} - \frac{ie}{\hbar} a_{\alpha} \psi(x) = D_{\alpha} \psi(x). \qquad (9.20)$$

In the presence of an electromagnetic field the covariant derivative replaces the ordinary derivative as the generator of translations for the particle; the quantum law (9.17) is thus precisely the rule (in geometrical terms, the connection) for the parallel transport of the charged particle (and its associated wave function) to neighbouring points. Moreover, it implies that the wave function $\psi(x)$ for a charged particle in the presence of an electromagnetic field is not single-valued. This may be seen as follows.

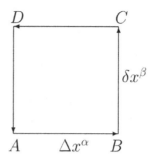

Figure 9.1   Parallel transport of a charge round a closed path in an electromagnetic field.

Suppose we are given a wave function $\psi(x)_{A,0}$ for a charge in an electromagnetic field at a space-time point x which we denote as A; we may take the wave function round a closed path by infinitesimal displacements so that it returns to its original position x, as shown in Figure 9.1. To allow for a possible change we denote the final result after one complete circuit by $\psi(x)_{A,1}$. To compute the effect of the displacements we use parallel transport and the covariant derivative [22], [24]; thus transporting $\psi(x)_{A,0}$ from A to B yields

$$\psi(x)_{\text{at B}} = (1 + \Delta x^\alpha \mathsf{D}_\alpha)\, \psi(x)_{A,0}, \tag{9.21}$$

and continuing in this way yields, after one complete circuit,

$$\psi(x)_{A,1} = \left(1 + \Delta x^\alpha\, \delta x^\beta\, [\mathsf{D}_\alpha, \mathsf{D}_\beta]\right) \psi(x)_{A,0}. \tag{9.22}$$

The commutator of the covariant derivative operator can be evaluated using (9.20),

$$
\begin{aligned}
[\mathsf{D}_\alpha, \mathsf{D}_\beta] &= \left[\frac{\partial}{\partial x^\alpha} - \frac{ie}{\hbar} a_\alpha, \frac{\partial}{\partial x^\beta} - \frac{ie}{\hbar} a_\beta \right] \\
&= -\frac{ie}{\hbar} \left(\frac{\partial a_\beta}{\partial x^\alpha} - \frac{\partial a_\alpha}{\partial x^\beta} \right) \equiv -\frac{ie}{\hbar} f(x)_{\alpha\beta},
\end{aligned}
\tag{9.23}
$$

where $f(x)_{\alpha\beta}$ is the Faraday field tensor for the electromagnetic field evaluated at the position of the particle, x, and so

$$\psi(x)_{A,1} = \left(1 - \frac{ie}{\hbar} \sigma^{\alpha\beta}\, f(x)_{\alpha\beta}\right) \psi(x)_{A,0}, \tag{9.24}$$

for a non-zero electromagnetic field. From the figure we see that $\Delta x^\alpha \delta x^\beta = \sigma^{\alpha\beta}$ is the area of the infinitesimal surface bounded by $ABCD$. Thus the space is not simply connected because the closed path $ABCD$ cannot be shrunk to a point. Eq. (9.24) says that the space about the charge e has a curvature measured by the field tensor. This calculation tells us that if we transport the wave function for a charge in the presence of an electromagnetic field along some path, the result will depend on the chosen path [22] which is exactly the content of the Bohm–Aharonov effect predicted on the basis of quite different considerations [13], [26], [27].

Considered as an infinitesimal transformation, (9.17) may be put in the form of (9.12):

$$\psi(x) \ \rightarrow \ \psi(x)' = \left(1 + \frac{i}{\hbar}\,d\omega\right)\psi(x). \tag{9.25}$$

For a finite displacement along a curve C to the point x we take the infinite product of the factors on the RHS of (9.25) for each of the infinitesimal segments comprising the curve C to calculate the modified phase of $\psi(x)$:

$$\psi(x) \rightarrow \psi(x)' = \prod^x\left(1 + \frac{i}{\hbar}\,d\omega\right)\psi(x) \equiv \exp\left(\frac{ie}{\hbar}\int_C^x a(y)_\alpha\,dy^\alpha\right)\psi(x). \tag{9.26}$$

The phase factor in (9.26) does not have a definite value at the point x, but it does have a definite derivative, since[2]

$$\frac{\partial}{\partial x_\mu}\int_C^x a(y)_\alpha\,dy^\alpha \ = \ a(x)_\mu, \tag{9.27}$$

the field potential. It is easily seen that if we regard the RHS of (9.26) as a path-dependent operator, U_C, acting on the wave function $\psi(x)$, then in view of (9.27), the ordinary derivative d_α and the covariant derivative D_α are related by the intertwining formula

$$U_C\,d_\alpha = D_\alpha U_C. \tag{9.28}$$

Introducing the classical gauge transformations, (9.6), into (9.26), we automatically reproduce the arbitrary character of the phase of the wave function described in (9.13), since putting

$$a_\alpha \ \rightarrow \ a_\alpha - \frac{\partial\chi}{\partial x_\alpha} \tag{9.29}$$

into (9.26) yields, after an integration by parts,

$$\exp\left(\frac{ie}{\hbar}\int_C^x a(y)_\alpha\,dy^\alpha\right) \rightarrow \exp\left(-ie\chi(x)/\hbar\right)$$
$$\times \ \exp\left(\frac{ie}{\hbar}\int_C^x a(y)_\alpha\,dy^\alpha\right). \tag{9.30}$$

Suppose now we chose a path C' which is formed by a finite displacement of the original path C with the same end points; then the path-dependent wave functions for these two paths are related by [28]

$$\psi(x:C') = \exp\left(-\frac{ie}{\hbar}\oint a(x)_\alpha\,dx^\alpha\right)\psi(x:C), \tag{9.31}$$

where the integral is taken over the closed path $C' - C$. An application of Stokes's theorem to the exponent in (9.31) as in (3.282) expresses it in terms of an integral of the field tensor over the surface S bounded by the two paths.

---

[2] Some specifics of the calculus of path-dependent functionals are given in Appendix E.

Just as physical observables are required to be independent of the gauge of the 4-potential $a(x)_\alpha$, so too must the choice of path C be one solely of convenience. The path-dependent wave function $\psi(x : C)$ is not an observable; on the other hand its squared modulus is interpreted in quantum mechanics as a physically meaningful probability density, and for this we have

$$|\psi(x : C')|^2 = |\psi(x : C)|^2. \tag{9.32}$$

If we choose another path, C'', (9.31) will again hold with C' replaced by C'', and the integration is taken over another surface, S', that is bounded by $C'' - C$. The difference in the integrals of the field tensor over these two surfaces can be expressed as an integral over the volume enclosed by the two surfaces (Gauss's theorem), and this vanishes if the homogeneous Maxwell equations are assumed to hold. Provided the arbitrary phase function $\chi(x)$ is integrable, the change in the 4-potential, (9.29), means the wave function ψ with its non-integrable phase is *gauge invariant* [29].

As we saw in §3.7.1, Weyl hoped the introduction of the group of dilations on the real line affecting the lengths of vectors at different space-time points as a modification of the (Riemann) geometry of space-time would unify gravitation and electromagnetism. That change implies observable consequences which are definitively not compatible with physics. The differential 1-form dv also appears naturally in the Hamiltonian formulation of classical electrodynamics as a modification of the symplectic geometry of phase space (cf. §3.7.2); there the field potentials are just auxiliary variables and the geometry does not lead to observable consequences. The classical canonical transformations with the generator F become formally unitary transformations in the quantum theory. Quantum mechanics, in which complex numbers are intrinsic, transfers dv to the unitary group $\mathcal{U}(1)$ – the phase transformations of the wave function just discussed. These give rise to observable effects such as the magnetic field Bohm–Aharonov effect. The change in phase of a wave function and the simultaneous transformation of the potentials together define a gauge transformation in wave mechanics. Thus it is the requirement for calibration (gauging) of the arbitrary phase of the wave function $\psi(x)$ that makes possible the gauge transformation of the field potential. So finally a physically significant mathematical context for the differential 1-form dv has been found.

An elementary example may be helpful here; the Hamiltonian for a free spin-zero non-relativistic charge e with mass m is

$$\mathsf{H} = \frac{\mathsf{p}^2}{2m}, \tag{9.33}$$

where $\mathbf{p} = -i\hbar\nabla$ is the usual expression for the momentum operator in position representation; time evolution is determined by the time-dependent Schrödinger equation,

$$i\hbar\frac{\partial\psi(\mathbf{x})}{\partial t} = \mathsf{H}\psi(\mathbf{x}). \tag{9.34}$$

The charge only enters the description when electromagnetic interactions are invoked; it is customary to make an appeal to the principle of minimal coupling to make the following substitutions in (9.34) to represent the interaction

$$\mathbf{p} \rightarrow \mathbf{p} - e\mathbf{a}(\mathbf{x}), \quad \frac{\partial}{\partial t} \rightarrow \frac{\partial}{\partial t} + \frac{ie}{\hbar}\phi(\mathbf{x}). \tag{9.35}$$

The preceding discussion, however, implies that even when the electromagnetic field is introduced, one may keep the 'free' Hamiltonian (9.33), but one must accept that the wave function $\psi(\mathbf{x})$ is a multi-valued quantity because of the non-integrable phase. Eq. (9.35) is nothing other than the differential form of the (integral) relationship, (9.26). In other words 'interaction' in the sense of the Newtonian and Maxwell–Lorentz descriptions is replaced by a geometrical description in the manner of Einstein and Weyl with the introduction of the non-integrable phase of the wave function (9.26); this was first demonstrated in a justly famous paper by Dirac, who showed that the formalism of quantum mechanics can accommodate the notion of a non-integrable phase, provided the same change in phase round a closed path is associated with all possible wave functions of the particle [12]. This geometrical framework is capable of considerable generalisation; the abelian group $\mathcal{U}(1)$ can be replaced by the non-abelian Lie groups $\mathcal{SU}(N), N = 2, 3$, and this leads to the Yang–Mills theory that underlies the Standard Model of modern particle physics [30]. In that context the (open) line integral in (9.30) is usually known as a 'Wilson line', while the closed line integral in (9.31) is a 'Wilson loop' [31].

We know from Chapter 3 that in the general Hamiltonian description of electrodynamics the scalar potential is eliminated, as it is a redundant variable. The preceding considerations for the wave function can then be expressed entirely in terms of the vector potential, $\mathbf{a}(\mathbf{x})$, using three-dimensional vector notation. So, for example, the exponential operator in (9.26) becomes

$$\mathsf{U}_C = \exp\left(\frac{ie}{\hbar}\int_C^{\mathbf{X}} \mathbf{a}(\mathbf{z}) \cdot d\mathbf{z}\right), \tag{9.36}$$

where $\mathbf{X}$ is the position variable for the charge e. If now we define the formal quantity,

$$\mathbf{P}(\mathbf{x}) = e\int_C^{\mathbf{X}} \delta^3(\mathbf{x} - \mathbf{z}) \, d\mathbf{z}, \tag{9.37}$$

we recover from the exponent of (9.36) the familiar form of the action integral F, introduced in §3.3:

$$e\int_C^{\mathbf{X}} \mathbf{a}(\mathbf{z}) \cdot d\mathbf{z} = \int \mathbf{P}(\mathbf{x}) \cdot \mathbf{a}(\mathbf{x}) \, d^3\mathbf{x} \equiv F. \tag{9.38}$$

We will see in the next section that the operator form of F plays a central role in the formalism of quantum electrodynamics. The occurrence of the path-dependent operator defined by (9.26) has a long history reaching back to the early years of quantum theory [18], [22], [32]–[36]. A cautionary word is in order, however; we may regard (9.26) as a transformation to a new representation that defines *gauge-invariant* states as the transforms of states in some gauge,

$$|\phi_C\rangle = \mathsf{U}_C|\psi\rangle, \tag{9.39}$$

and it is natural to enquire as to the overlap between the states in the two representations. If the field potential is treated as a quantised field operator, it has long been known [37] that the overlap vanishes for point particles,

$$\langle \psi | \phi_C \rangle = 0, \tag{9.40}$$

which implies that the two Hilbert spaces are orthogonal, and that U_C is *not* unitary, as it takes the states ψ out of the original Hilbert space. This will be explored further in Chapter 11.

9.3 Quantum Electrodynamics

9.3.1 Introduction

The full quantisation of the classical Hamiltonian formalism for electrodynamics described in Chapter 3 can be approached in several different ways, depending on how the classical second class constraint for the interacting system of charges and field,

$$\Omega_2 = \nabla \cdot \boldsymbol{\pi} + \rho \approx 0, \tag{9.41}$$

is dealt with, and on the choice of variables.

1. The classical gauge-invariant Hamiltonian (3.258) and its P.B. algebra (3.259) – (3.262) are interpreted in terms of Hilbert space operators, and the Maxwell equations for the field operators,

$$\nabla \cdot \mathbf{B} = 0 \tag{9.42}$$

$$\varepsilon_0 \nabla \cdot \mathbf{E} = \rho, \tag{9.43}$$

hold as initial conditions.

2. Classical variables such as $\boldsymbol{\pi}$ and ρ are interpreted as Hilbert space operators, and physical states of the system, $\{\Psi_k\}$, are selected by the requirement that they are annihilated by Ω_2; that is, a physical state must satisfy the relation

$$\left(\nabla \cdot \boldsymbol{\pi} + \rho\right) \Psi_k = 0. \tag{9.44}$$

In this case, canonical P.B.s are still valid, and so the corresponding quantum operators satisfy canonical commutation relations. The Hamiltonian operator is given by the canonical quantisation of Eq. (3.225).

3. The canonical P.B.s are redefined as Dirac brackets by the imposition of a gauge condition for the vector potential so that $\Omega_2 = 0$ is valid as an ordinary equation (one of the Maxwell equations). The reduced Hamiltonian and the Dirac brackets given by (3.254)–(3.257) are then reinterpreted as operator relations on a Hilbert space that is fixed by the commutation relations for the chosen gauge.

Only the third possibility has been developed sufficiently for practical calculations involving atoms/molecules and radiation, simply because of the unique significance of the instantaneous Coulomb interaction when there is more than one charged particle.

Moreover, there is the difficulty of realising the full set of gauge-invariant operators in a usable form. Approaches 1 and 2 do, however, provide useful insights into the relationship of QED with classical electrodynamics and gauge invariance.

The Hamiltonian in the explicitly gauge-invariant formalism is the generalisation of (9.33) to the electrodynamics of charges and the electromagnetic field considered as a closed system:

$$H = \sum_n \frac{1}{2m_n}|\bar{\mathbf{p}}_n|^2 + \tfrac{1}{2}\varepsilon_0 \int \left(|\mathbf{E}|^2 + c^2|\mathbf{B}|^2 \right) d^3\mathbf{x}. \qquad (9.45)$$

The charge-field coupling parameter e_n appears in the non-zero commutators of the $\{\bar{\mathbf{p}}_n\}$ operators which are not canonical variables,

$$[\bar{p}_n^t, \bar{p}_m^r] = i\hbar e_n \delta_{nm} \varepsilon_{rts} B(\mathbf{x}_n)^s, \qquad (9.46)$$

$$[\bar{p}_n^r, E(\mathbf{x})^s] = i\hbar e_n \varepsilon_0^{-1} \delta_{rs} \delta^3(\mathbf{x}_n - \mathbf{x}). \qquad (9.47)$$

To these relations must be conjoined the conditions

$$[x_n^s, \bar{p}_m^r] = i\hbar \delta_{nm} \delta_{sr}, \qquad (9.48)$$

$$[E(\mathbf{x})^r, B(\mathbf{x}')^s] = i\hbar \varepsilon_0^{-1} \varepsilon_{rst} \nabla_{\mathbf{x}'}^t \delta^3(\mathbf{x} - \mathbf{x}'), \qquad (9.49)$$

where the superscripts label the components of the three-dimensional vectors. The $\{\bar{\mathbf{p}}_n\}$ operators provide the covariant differentiation described in the previous section.

The commutation relations (9.47) and (9.48) suggest we may write

$$\varepsilon_0 \nabla_{\mathbf{x}_n} \cdot \mathbf{E}(\mathbf{x}, \mathbf{x}_n) = -e_n \delta^3(\mathbf{x}_n - \mathbf{x}). \qquad (9.50)$$

Since the Dirac delta function has the representation

$$-\delta^3(\mathbf{x}_n - \mathbf{x}) = \nabla_{\mathbf{x}_n}^2 \left(\frac{1}{4\pi|\mathbf{x}_n - \mathbf{x}|} \right), \qquad (9.51)$$

the separation

$$\mathbf{E}(\mathbf{x}) = \mathbf{E}(\mathbf{x})^{\|} + \mathbf{E}(\mathbf{x})^{\perp} \qquad (9.52)$$

with

$$\mathbf{E}(\mathbf{x})^{\|} = \sum_n e_n \nabla_{\mathbf{x}_n} \left(\frac{1}{4\pi\varepsilon_0|\mathbf{x}_n - \mathbf{x}|} \right) \qquad (9.53)$$

is consistent with Gauss's law, (9.43), and with (9.49). We may then write

$$\tfrac{1}{2}\varepsilon_0 \int \left(|\mathbf{E}|^2 + c^2|\mathbf{B}|^2 \right) d^3\mathbf{x} = H_{\text{rad}} + V_{\text{Coulomb}} \qquad (9.54)$$

with

$$H_{\text{rad}} = \tfrac{1}{2}\varepsilon_0 \int \left(|\mathbf{E}^{\perp}|^2 + c^2|\mathbf{B}|^2 \right) d^3\mathbf{x}$$

$$V_{\text{Coulomb}} = \sum_{n,m} \frac{e_n e_m}{4\pi\varepsilon_0|\mathbf{x}_n - \mathbf{x}_m|}, \qquad (9.55)$$

so that (9.45) becomes

$$H = \sum_n \frac{1}{2m_n} |\overline{\mathbf{p}}_n|^2 + H_{\text{rad}} + V_{\text{Coulomb}}. \tag{9.56}$$

This argument does not exclude the possibility that there is a contribution to the transverse electric field that depends on the charges since (9.47) provides no information about the transverse field, $\mathbf{E}^\perp$.

There is still the difficulty of how to represent the $\{\overline{\mathbf{p}}_n\}$ operators such that (9.46) is satisfied. The customary response is to introduce a *gauge field* through the separation

$$\overline{\mathbf{p}}_n = \mathbf{p}_n - e_n \mathbf{a}(\mathbf{x}_n), \tag{9.57}$$

where $\mathbf{p}_n$ is the usual canonical conjugate to $\mathbf{x}_n$ and $\mathbf{a}(\mathbf{x}_n)$ is a vector potential that is required to satisfy

$$\mathbf{\nabla} \wedge \mathbf{a}(\mathbf{x}) = \mathbf{B}(\mathbf{x}). \tag{9.58}$$

This leads to the second possible form of quantum electrodynamics; we recognise that if no gauge condition is imposed, the vector potential operator has a longitudinal degree of freedom in addition to the two transverse degrees of freedom that describe polarised photons. Using (9.47) again and writing $\varepsilon_0 \mathbf{E} = -\boldsymbol{\pi}$, we are led to a canonical commutation relation,

$$[a(\mathbf{x})^r, \pi(\mathbf{x}')^s] = i\hbar \delta_{rs} \delta^3(\mathbf{x} - \mathbf{x}'), \quad r, s = 1, 2, 3 \tag{9.59}$$

which implies that the conjugate, $\boldsymbol{\pi}$, may be realised as the functional differentiation operator:

$$\pi_s = -i\hbar \frac{\delta}{\delta a_s}. \tag{9.60}$$

Since the commutation relations fix the Hilbert space of states, the Hilbert space will be 'too large' and at the outset the calculations will involve the extra degrees of freedom; an extra condition on the state space is thus required to pick out the physically significant states, and this is provided by (9.44).

We choose a representation which is diagonal in the field and particle 'coordinates', $(\{\mathbf{x}_n\}, \mathbf{a}(\mathbf{x}))$; then we can replace the operator ρ by the c-number $\rho(\mathbf{x})$, the classical charge density, and as in Chapter 2, introduce the classical electric polarisation field, $\mathbf{P}(\mathbf{x})$, so that (9.44) becomes

$$\mathbf{\nabla} \cdot \big(\boldsymbol{\pi}(\mathbf{x}) - \mathbf{P}(\mathbf{x})\big) \Psi_k(\mathbf{x}) = 0. \tag{9.61}$$

Ψ_k is a functional of the vector potential $\mathbf{a}(\mathbf{x})$ and a function of the particle position variables which we suppress to simplify the notation. There is no requirement for the physical states to be eigenfunctions of $\boldsymbol{\pi}$ with eigenvalues $\mathbf{P}(\mathbf{x})$; however, we may write[3]

$$\boldsymbol{\pi}(\mathbf{x}) \Psi_k(\mathbf{x}) = \mathbf{P}(\mathbf{x}) \Psi_k(\mathbf{x}) + \mathbf{T}_k(\mathbf{x}), \tag{9.62}$$

---

[3] In the following, $\mathbf{T}_k$ and χ_k are also functionals of the vector potential and functions of the particle coordinates.

where $\mathbf{T}_k$ is transverse,

$$\boldsymbol{\nabla} \cdot \mathbf{T}_k(\mathbf{x}) = 0. \tag{9.63}$$

We may view (9.62) as a first-order functional differential equation, using (9.60) to represent $\boldsymbol{\pi}$, and rewrite it as

$$\frac{\delta}{\delta \mathbf{a}(\mathbf{x})} \left\{ \exp\left(-\frac{i}{\hbar} \int \mathbf{P}(\mathbf{x}') \cdot \mathbf{a}(\mathbf{x}') \, d^3 \mathbf{x}' \right) \Psi_k \right\} = \boldsymbol{\eta}_k(\mathbf{x}), \tag{9.64}$$

where

$$\boldsymbol{\eta}_k(\mathbf{x}) = \frac{i}{\hbar} \mathbf{T}_k(\mathbf{x}) \exp\left(-\frac{i}{\hbar} \int \mathbf{P}(\mathbf{x}') \cdot \mathbf{a}(\mathbf{x}') \, d^3 \mathbf{x}' \right). \tag{9.65}$$

Since $\boldsymbol{\eta}_k$ as expressed by (9.64) is a functional derivative, it must satisfy the integrability condition,

$$\frac{\delta \boldsymbol{\eta}_k(\mathbf{x})}{\delta \mathbf{a}(\mathbf{y})} = \frac{\delta \boldsymbol{\eta}_k(\mathbf{y})}{\delta \mathbf{a}(\mathbf{x})}, \tag{9.66}$$

and this requires that

$$\mathbf{T}_k = \mathbf{t}_k(\mathbf{a}(\mathbf{x})) \exp\left(\frac{i}{\hbar} \int \mathbf{P}(\mathbf{x}') \cdot \mathbf{a}(\mathbf{x}') \, d^3 \mathbf{x}' \right), \tag{9.67}$$

where $\mathbf{t}_k(\mathbf{a})$ is a transverse vector *function* of the vector potential, $\mathbf{a}(\mathbf{x})$, [38]. Thus a physical state can be put in the form

$$\Psi_k(\mathbf{x}) - \exp\left(\frac{i}{\hbar} \int \mathbf{P}(\mathbf{x}') \cdot \mathbf{a}(\mathbf{x}') \, d^3 \mathbf{x}' \right) \int^{\mathbf{a}(\mathbf{x})} \delta \mathbf{a}'(\mathbf{x}) \cdot \mathbf{t}_k(\mathbf{a}'(\mathbf{x}))$$

$$= \exp\left(\frac{i}{\hbar} \int \mathbf{P}(\mathbf{x}') \cdot \mathbf{a}(\mathbf{x}') \, d^3 \mathbf{x}' \right) \psi_k(\{\mathbf{x}_n\}, \mathbf{a}(\mathbf{x})). \tag{9.68}$$

As for the free field (cf. §7.3), we may define a Gauss's law operator for the interacting system, $\mathsf{G} = \boldsymbol{\nabla} \cdot \boldsymbol{\pi} + \rho$; the physical states Ψ of the interacting system will then, as just shown, be required to satisfy the condition (9.61) which we write as

$$\mathsf{G}\Psi_k[\mathbf{a}, \{\mathbf{x}_n\}] = 0. \tag{9.69}$$

The generator of gauge transformations for the interacting system is now

$$\mathcal{G}^f = \int \mathsf{G} f \, d^3 \mathbf{x}$$

$$= -\int (\boldsymbol{\nabla} f) \cdot \boldsymbol{\pi} d^3 \mathbf{x} + \int f \rho \, d^3 \mathbf{x}, \tag{9.70}$$

with associated unitary operator

$$\mathsf{U}_\mathsf{G}^f = \exp\left(\frac{i}{\hbar} \mathcal{G}^f \right). \tag{9.71}$$

Under the action of U_G^f we have

$$\begin{aligned} \mathsf{a}^s &\to \mathsf{a}^s - (\boldsymbol{\nabla} f)^s \\ \mathsf{p}_n^s &\to \mathsf{p}_n^s + e(\boldsymbol{\nabla} f)^s, \end{aligned} \tag{9.72}$$

which confirms again that the combination $\bar{\mathbf{p}}_n \equiv \mathbf{p}_n - e\mathbf{a}(\mathbf{x}_n)$ is *gauge invariant*. A new feature is the transformation rule for the states,

$$\Psi_k[\mathbf{a}, \{\mathbf{x}_n\}]' = U_G^f \Psi_k[\mathbf{a}, \{\mathbf{x}_n\}]$$

$$= \exp\left(\frac{i}{\hbar}\int f\rho\, \mathrm{d}^3\mathbf{x}\right)\Psi_k[\mathbf{a}', \{\mathbf{x}_n\}]. \tag{9.73}$$

Once again we have a unitary representation of the group $\mathcal{U}(1)$; the difference here is that in the presence of charged particles we have a multiplier representation, since a gauge transformation of the vector potential is associated with a concomitant change in the *phase* of the states determined by the charges present, as described in the previous section. Provided that charge is conserved, there is unrestricted validity for the quantum mechanical superposition principle because the phase factor is the same for all possible states. Conversely one cannot have superpositions of states associated with different charge densities since their relative phases could be changed by a gauge transformation; this is the quantum mechanical formulation of the principle of charge conservation in terms of the states of the system.

The relationship between G_0 and G parallels that found for their classical counterparts; if $\Phi[\mathbf{a}]$ is a physical state of the free field, there is a corresponding state of the interacting system, Ψ, given by

$$\Psi = W^{-1}\Phi, \tag{9.74}$$

such that Ψ satisfies (9.69), and

$$G = WG_0W^{-1}. \tag{9.75}$$

It is readily verified that the (formally) unitary operator W is

$$W = W[\mathbf{P}] = \exp\left\{-\frac{i}{\hbar}\int \mathbf{P}\cdot\mathbf{a}\, \mathrm{d}^3\mathbf{x}\right\}, \tag{9.76}$$

where $\mathbf{P}$ is the electric polarisation field operator for the charges satisfying, as usual,

$$\nabla\cdot\mathbf{P} = -\rho. \tag{9.77}$$

The phase factor involving the electric polarisation field that appears in (9.76) will be recognised as the quantised form of the action F discussed in the previous section, §9.2, and (9.68) is of the form of (9.74) in a 'coordinate' representation. According to the discussion in Chapter 5 the operator W considered as a functional of the field 'coordinate' $\mathbf{a}$ induces translations of the conjugate field momentum,

$$W\boldsymbol{\pi}W^{-1} = \boldsymbol{\pi} + \mathbf{P}. \tag{9.78}$$

It commutes with the particle position operators $\{\mathbf{x}_n\}$ and the field coordinate, $\mathbf{a}$, and adds a gradient proportional to its exponent to the particle momentum operators, $\{\mathbf{p}_n\}$,

$$\mathbf{p}_n \rightarrow \mathbf{p}_n + \nabla_{\mathbf{x}_n}F. \tag{9.79}$$

Suppose now that we have two static charges, say e_1 at $\mathbf{x}_1$ and e_2 at $\mathbf{x}_2$, and the electromagnetic field which mediates their mutual interaction. The charge density for two point particles is

$$\rho(\mathbf{x}) = e_1\delta^3(\mathbf{x} - \mathbf{x}_1) + e_2\delta^3(\mathbf{x} - \mathbf{x}_2). \tag{9.80}$$

The unitary operator (9.76) in an arbitrary gauge can be expressed in terms of the *Coulomb gauge vector potential* as [39]

$$W[\mathbf{P}] = \exp\left\{-\frac{i}{\hbar}\int f\rho\,d^3\mathbf{x}\right\}\exp\left\{-\frac{i}{\hbar}\int \mathbf{P}\cdot\mathbf{A}\,d^3\mathbf{x}\right\}, \tag{9.81}$$

where f is identified with χ given by (2.52). For the particular case of an overall neutral system with $e_1 = -e$ and $e_2 = +e$, the electric polarisation field may be written as [40] (cf. §2.4.3)

$$\mathbf{P}(\mathbf{x}:C) = e\int_{\mathbf{x}_1}^{\mathbf{x}_2}\delta^3(\mathbf{z}-\mathbf{x})\,d\mathbf{z}, \tag{9.82}$$

where the path C starts at particle 1 at $\mathbf{x}_1$ and ends at particle 2 at $\mathbf{x}_2$. Thus

$$W[\mathbf{P}] = \exp\left\{-\frac{ie}{\hbar}\left(f(\mathbf{x})_2 - f(\mathbf{x})_1\right)\right\}\exp\left\{-\frac{ie}{\hbar}\int_{\mathbf{x}_1}^{\mathbf{x}_2}d\mathbf{z}\cdot\mathbf{A}(\mathbf{z})\right\}. \tag{9.83}$$

Now if $\Phi[\mathbf{a}]$ satisfies (7.113), then so does the gauge-invariant state

$$\overline{\Phi}[\mathbf{a}] = \exp\left\{\frac{ie}{\hbar}\int_{\mathbf{x}_1}^{\mathbf{x}_2}d\mathbf{z}\cdot\mathbf{A}(\mathbf{z})\right\}\Phi[\mathbf{a}] \tag{9.84}$$

because $\nabla\cdot\mathbf{A} = 0$. This means that the state $\Psi[\mathbf{a}]$ that satisfies (9.69) can be decomposed into the form

$$\Psi[\mathbf{a}] = \exp\left\{\frac{ie}{\hbar}\left(f(\mathbf{x})_2 - f(\mathbf{x})_1\right)\right\}\overline{\Phi}[\mathbf{a}]. \tag{9.85}$$

To complete the specification of the quantum mechanical Hamiltonian one must also define its domain; this is left unspecified by the canonical quantisation algorithm. This is not a trivial matter, and unfortunately is not really solved. The conventional perturbation theory assumes that the Hilbert space of the full system is the same as that for the 'free' reference system (atoms/molecules and EM field without coupling), as in ordinary quantum mechanics, so that the diagonalisation of the full Hamiltonian expressed in the reference system basis can be expressed as a certain unitary transformation to be constructed approximately as a perturbation series. The assumption is that the field and particle variables are independent and so there are no non-zero commutators between them. But in quantum electrodynamics if one takes the charges to be 'point particles' this is never the case; the usual remedy is to smooth out point charges, which is physically plausible for nuclei, but not at all evident for electrons. In the gauge-invariant scheme 1, the commutators that follow from (9.46) and (9.47) involve the field and are proportional to the particle charge, suggesting that a gauge-invariant description of charged particles cannot be based on free particles and the free field as a reference system.

9.3.2 The QED Hamiltonian

Canonical quantisation of the classical Hamiltonian scheme (3.254)–(3.257) leads to the usual form of non-relativistic quantum electrodynamics. The Hamiltonian operator for a closed system of charged particles and electromagnetic radiation, with the vector potential in an arbitrary gauge, is

$$\mathsf{H} = \tfrac{1}{2} \sum_{n}^{N} \frac{1}{m_n} (\mathbf{p}_n - e_n \mathbf{a}(\mathbf{x}_n))^2 + \tfrac{1}{2}\varepsilon_0 \int \left(\varepsilon_0^{-2} \boldsymbol{\pi} \cdot \boldsymbol{\pi} + c^2 \mathbf{B} \cdot \mathbf{B} \right) d^3\mathbf{x}. \qquad (9.86)$$

The equal time commutation relations for the particle and field variables are

$$[\mathsf{x}_n^r, \mathsf{p}_m^s] = i\hbar \delta_{nm} \delta_{rs}, \qquad (9.87)$$

$$\left[\mathsf{a}(\mathbf{x})^r, \boldsymbol{\pi}(\mathbf{x}')^s \right] = i\hbar \left(\delta_{rs} \delta^3(\mathbf{x} - \mathbf{x}') - \nabla_{\mathbf{x}}^r g(\mathbf{x}', \mathbf{x})^s \right), \qquad (9.88)$$

$$\left[\boldsymbol{\pi}(\mathbf{x})^r, \mathsf{p}_n^s \right] = i\hbar e_n \nabla_n^s g(\mathbf{x}, \mathbf{x}_n)^r, \qquad (9.89)$$

where $g(\mathbf{x}, \mathbf{x}')$ is a Green's function for the divergence operator, (2.64).

As in the classical theory the constraint (9.41) becomes an ordinary equation, but now between operators,

$$\boldsymbol{\nabla} \cdot \boldsymbol{\pi} + \rho = 0, \qquad (9.90)$$

which is Gauss's law. In a Schrödinger representation, the operators are time independent, and the time evolution is carried by the quantum states according to the Schrödinger equation

$$i\hbar \frac{\partial \Psi_S}{\partial t} = \mathsf{H}\Psi_S. \qquad (9.91)$$

In order to cast the Hamiltonian (9.86) into a form appropriate for quantum mechanical perturbation theory through the partition

$$\mathsf{H} = \mathsf{H}_0 + \mathsf{V}, \qquad (9.92)$$

the first term in (9.86), which overall is gauge invariant, must be multiplied out as

$$\sum_n \left(\frac{|\mathbf{p}_n|^2}{2m_n} - \frac{e_n}{2m_n} \mathbf{p}_n \cdot \mathbf{a}(\mathbf{x}_n) - \frac{e_n}{2m_n} \mathbf{a}(\mathbf{x}_n) \cdot \mathbf{p}_n + \frac{e_n^2}{2m_n} \mathbf{a}(\mathbf{x}_n) \cdot \mathbf{a}(\mathbf{x}_n) \right). \qquad (9.93)$$

The first term contributes to H_0, while the remainder belongs to V. The division of the Hamiltonian into parts ('system' + 'perturbation') in (9.92) is conventional and must not affect the final results of any calculation. We choose to locate the dependence on the arbitrary Green's function $\mathbf{g}$ in the 'perturbation' part as a step towards the customary methods of quantum theory. It will be convenient in the following, however, to regard the complete Hamiltonian (9.86) as a functional of $\mathbf{g}$, and we denote it by $\mathsf{H} = \mathsf{H}[\mathbf{g}]$. Since $\mathbf{g}^\perp$ can be chosen at will, $\mathsf{V}[\mathbf{g}]$ has an arbitrary character which can be identified with the occurrence of gauge transformations in electrodynamics.

In the canonical scheme developed so far, the commutators (9.88)–(9.89) carry an explicit dependence on the arbitrary function $\mathbf{g}(\mathbf{x}, \mathbf{x}')$, while the form of the Hamiltonian remains fixed, whatever gauge the vector potential is chosen in. It is possible to redefine the variables so that the dependence on **g** is made explicit in the Hamiltonian, while working with one particular set of commutators which remains fixed as **g** is changed. As a first step towards rewriting the Hamiltonian with its dependence on **g** explicit we use the fact that (9.89) can also be written as

$$[\mathbf{p}_n^r, \pi(\mathbf{x})^s]^* = -e_n \nabla_n^r g(\mathbf{x}, \mathbf{x}_n)^s \equiv [\mathbf{p}_n^r, \mathsf{P}(\mathbf{x})^s], \tag{9.94}$$

in terms of the electric polarisation field **P** associated with the charges. We can then separate π in the form (cf. (9.78))

$$\boldsymbol{\pi} = \boldsymbol{\pi}^\perp + \mathbf{P}. \tag{9.95}$$

This is consistent with (9.88), provided $\pi^\perp$ is interpreted as the conjugate of the Coulomb gauge vector potential **A**, and so is associated purely with electromagnetic radiation;[4] if we adopt (9.95), the commutator (9.94) is redundant and can be dropped.

The second term in (9.86) can now be written as

$$\tfrac{1}{2}\varepsilon_0 \int \left(\mathbf{E}\cdot\mathbf{E} + c^2\mathbf{B}\cdot\mathbf{B}\right) d^3\mathbf{x} = \tfrac{1}{2}\varepsilon_0 \int \left(\mathbf{E}^\perp\cdot\mathbf{E}^\perp + c^2\mathbf{B}\cdot\mathbf{B}\right) d^3\mathbf{x}$$
$$- \int \mathbf{E}^\perp\cdot\mathbf{P}\, d^3\mathbf{x} + \frac{1}{2\varepsilon_0}\int \mathbf{P}\cdot\mathbf{P}\, d^3\mathbf{x}. \tag{9.96}$$

The first term on the RHS of (9.96) is the Hamiltonian for free electromagnetic radiation,

$$H_{\text{rad}} = \tfrac{1}{2}\varepsilon_0 \int \left(\mathbf{E}^\perp\cdot\mathbf{E}^\perp + c^2\mathbf{B}\cdot\mathbf{B}\right) d^3\mathbf{x}, \tag{9.97}$$

which was discussed in Chapter 7. The term linear in $\mathbf{E}^\perp$ couples charges to electromagnetic radiation since **P** depends on the $\{e_n\}$ according to (2.142). At this stage in the discussion it has always been conventional to separate the polarisation field **P** into its longitudinal and transverse parts so as to make the Coulombic interaction energy of the charges explicit using the known longitudinal component [16]. We write

$$\mathbf{P}(\mathbf{x}) = \mathbf{P}(\mathbf{x})^\| + \mathbf{P}(\mathbf{x})^\perp, \tag{9.98}$$

so that

$$\mathbf{P}(\mathbf{x})\cdot\mathbf{P}(\mathbf{x}) = \mathbf{P}(\mathbf{x})^\|\cdot\mathbf{P}(\mathbf{x})^\| + \mathbf{P}(\mathbf{x})^\perp\cdot\mathbf{P}(\mathbf{x})^\perp, \tag{9.99}$$

with (using (2.65, 2.142))

$$\mathbf{P}(\mathbf{x})^\| = \sum_n e_n \nabla_\mathbf{x} \left(\frac{1}{4\pi|\mathbf{x} - \mathbf{x}_n|}\right). \tag{9.100}$$

For a collection of n charges $\{e_n\}$ we recover the familar Coulomb interaction energy:

$$\frac{1}{2\varepsilon_0}\int |\mathbf{P}(\mathbf{x})^\||^2 d^3\mathbf{x} = \tfrac{1}{2}\sum_{n,m} \frac{e_n e_m}{4\pi\varepsilon_0|\mathbf{x}_n - \mathbf{x}_m|}. \tag{9.101}$$

---

[4] Cf. the remark after (9.56).

The first term in (9.93) together with (9.101) yields the Coulomb Hamiltonian (Chapter 8) for the charged particles:

$$H_{\text{charges}} = \sum_n \frac{|\mathbf{p}_n|^2}{2m_n} + \frac{1}{2}\sum_{n,m}\frac{e_n e_m}{4\pi\varepsilon_0 |\mathbf{x}_n - \mathbf{x}_m|}. \tag{9.102}$$

Thus the remainder of the Hamiltonian after (9.97) and (9.102) have been accounted for is the interaction between the charges and the radiation, and (9.92) explicitly is

$$H_0 = H_{\text{charges}} + H_{\text{rad}}$$

$$V[\mathbf{g}] = -\sum_n \frac{e_n}{2m_n}\left(\mathbf{p}_n \cdot \mathbf{a}(\mathbf{x}_n) + \mathbf{a}(\mathbf{x}_n)\cdot\mathbf{p}_n - e_n\mathbf{a}(\mathbf{x}_n)\cdot\mathbf{a}(\mathbf{x}_n)\right)$$

$$-\int \mathbf{P}(\mathbf{x})\cdot\mathbf{E}^\perp(\mathbf{x})\,d^3\mathbf{x} + \frac{1}{2\varepsilon_0}\int \mathbf{P}(\mathbf{x})^\perp\cdot\mathbf{P}(\mathbf{x})^\perp\,d^3\mathbf{x}, \tag{9.103}$$

where

$$\mathbf{a}(\mathbf{x}') = \mathbf{A}(\mathbf{x}') - \boldsymbol{\nabla}_{\mathbf{x}'}\int \mathbf{A}(\mathbf{x}'')\cdot\mathbf{g}(\mathbf{x}'',\mathbf{x}')\,d^3\,\mathbf{x}'' \tag{9.104}$$

is the general vector potential expressed in terms of the Coulomb gauge vector potential operator, $\mathbf{A}$, and the Green's function $\mathbf{g}(\mathbf{x},\mathbf{x}')$, Eq. (2.64).

Implicit in this wholly conventional discussion is the expectation that the last term in (9.103) is only a 'small' perturbation of H_0, amounting to 'contact terms' and 'self-interactions' to be dealt with by renormalisation and so on [14], and so it is mostly ignored. Such a discussion is really only possible within the multipole approximation for the polarisation fields. We saw in Chapter 2 that the full line integral form for the electric polarisation field leads to the complete cancellation of the static Coulomb interaction (9.101) required for the atomic Hamiltonian, and replaces it with an interaction mediated by the transverse electromagnetic fields, as well as singular and contact interactions. However that may be, Eqs. (9.97), (9.102), (9.103) supplemented by the commutation relations,

$$[x_n^r, p_m^s] = i\hbar\delta_{nm}\delta_{rs}, \tag{9.105}$$

$$[A(\mathbf{x},t)^r, E(\mathbf{x}',t)^{\perp s}]^* = -i\hbar\varepsilon_0^{-1}\delta_{rs}^\perp(\mathbf{x}-\mathbf{x}'), \tag{9.106}$$

constitute the conventional Hamiltonian theory of a closed system of charged particles interacting with electromagnetic radiation in quantum mechanical form.

In Chapter 3, we saw that the Coulomb gauge theory is obtained by choosing $\mathbf{g}(\mathbf{x},\mathbf{x}')^\perp = \mathbf{0}$; this reduces (9.104) to $\mathbf{a}(\mathbf{x}') = \mathbf{A}(\mathbf{x}')$ and removes the terms in $\mathbf{P}^\perp$ from (9.103). We denote the Coulomb gauge Hamiltonian by $H[0]$; it is given explicitly by Eqs. (9.97), (9.102), (9.103) with

$$V[0] = -\sum_n \frac{e_n}{m_n}\mathbf{p}_n\cdot\mathbf{A}(\mathbf{x}_n) + \sum_n \frac{e_n^2}{2m_n}\mathbf{A}(\mathbf{x}_n)\cdot\mathbf{A}(\mathbf{x}_n), \tag{9.107}$$

where we have explicitly used the transversality of $\mathbf{A}$ to simplify the first term. The relationship between the Coulomb gauge Hamiltonian and the Hamiltonian $H[\mathbf{g}]$ for any other non-zero **g** may be expressed by a formally unitary transformation of $H[0]$,

$$H' = \Lambda\, H[0]\, \Lambda^{-1}. \tag{9.108}$$

The operator Λ is precisely the unitary operator W introduced in the last section with the vector potential **a** specialised to the Coulomb gauge operator, **A**,

$$\Lambda = \exp(-iF/\hbar), \quad F = \int \mathbf{P}(\mathbf{x}) \cdot \mathbf{A}(\mathbf{x})\, d^3\mathbf{x}. \tag{9.109}$$

F commutes with the canonical 'position' operators for the field and particles, so the transformation yields

$$\mathbf{A} \;\rightarrow\; \mathbf{A}' = \mathbf{A}, \tag{9.110}$$

$$\boldsymbol{\pi} \;\rightarrow\; \boldsymbol{\pi}' = \boldsymbol{\pi} + \mathbf{P}, \tag{9.111}$$

$$\mathbf{x}_n \;\rightarrow\; \mathbf{x}'_n = \mathbf{x}_n, \tag{9.112}$$

$$\mathbf{p}_n \;\rightarrow\; \mathbf{p}'_n = \mathbf{p}_n + \boldsymbol{\nabla}_{\mathbf{x}_n} F, \tag{9.113}$$

$$H[0] \rightarrow H'. \tag{9.114}$$

Obviously this is not a gauge transformation since F commutes with the Coulomb gauge vector potential, **A**; since it is unitary, the commutation relation between **A** and its conjugate $\boldsymbol{\pi}$ is unchanged. F also commutes with Gauss's law (9.90); we will continue to work with the *original* canonical variables $(\{\mathbf{x}_n\}, \{\mathbf{p}_n\}, \mathbf{A}, \boldsymbol{\pi})$ so that the form of the Hamiltonian is modified. Using the power series expansion of the exponential operator, we obtain the transformed Hamiltonian as

$$H' = H[0] + \frac{i}{\hbar}\,[H[0], F] + \frac{1}{2!}\left(\frac{i}{\hbar}\right)^2 [\,[H[0], F]\,, F\,]. \tag{9.115}$$

The series terminates because the evaluation of the double commutator in (9.115) yields terms which commute with F. To see this, we note that the non-zero contributions to the expansion (9.115) arise from commutation of F with the particle momenta and the electric field intensity operator in $H[0]$; when all the terms are evaluated we find

$$H' = H[\mathbf{g}]. \tag{9.116}$$

Consequently the general Hamiltonian, given by (9.92) and (9.103), can be understood as the unitary transform by the operator Λ of the Coulomb gauge Hamiltonian. This calculation mirrors perfectly the classical canonical transformation theory summarised in §3.6 with Dirac brackets replaced by commutators.

The foregoing can be related to the discussion in §9.2 quite directly. We put the line integral form (2.69) for $\mathbf{g}(\mathbf{x}, \mathbf{x}')$ into (9.104) to make explicit the vector potential in (9.103) as

$$\mathbf{a}(\mathbf{x}') = \mathbf{A}(\mathbf{x}') - \boldsymbol{\nabla}_{\mathbf{x}'} \int_C^{\mathbf{x}'} \mathbf{A}(\mathbf{z}) \cdot d\mathbf{z}. \tag{9.117}$$

If we regard the integral as a multi-valued function, leaving the path C unspecified, then as in (9.27)

$$\boldsymbol{\nabla}_{\mathbf{x}'} \int_C^{\mathbf{x}'} \mathbf{A}(\mathbf{z}) \cdot \mathrm{d}\mathbf{z} = \mathbf{A}(\mathbf{x}') \Rightarrow \mathbf{a}(\mathbf{x}') = 0. \qquad (9.118)$$

All the terms in $\mathsf{V}[\mathbf{g}]$ involving the vector potential are eliminated and $\mathsf{H} = \mathsf{H}_0 + \mathsf{V}[\mathbf{g}]$ becomes the sum of the 'free' Hamiltonians for charges and radiation, that is, the gauge-invariant Hamiltonian (9.45). From §9.2 we learnt that if there is interaction between the charges and the radiation, we must then let H act on a space of multi-valued wave functions characterised by a non-integrable phase factor. Equivalently, we can regard the phase factor as a formally unitary operator and achieve the same result by 'unitary transformation' with the operator Λ, (9.109). A (single-valued) Coulomb gauge wave function $\Psi[0]$ is thereby transformed to a multi-valued wave function φ which is gauge invariant:

$$\varphi = \Lambda\Psi[0]. \qquad (9.119)$$

Unfortunately this does not offer a straightforward way to gauge-invariant calculation, and in practice non-relativistic QED has followed a quite different strategy. One can envisage a two-stage approach in which we first fix on a definite path in $\mathbf{g}(\mathbf{x},\mathbf{x}')$ so that the integral in (9.103) is single-valued, and subsequently decide what can be calculated that is *independent* of the choice of path. Such quantities will be gauge invariant and hence candidates for physical observables. This leads to the Power–Zienau–Woolley (PZW) transformation theory to which we now turn.

9.3.3 The Power–Zienau–Woolley Transformation

The functional scalar product of the electric polarisation field, $\mathbf{P}(\mathbf{x})$, and the classical vector potential, $\mathbf{a}(\mathbf{x})$, is defined as the integral over all space of their scalar product:

$$F = \int \mathbf{P}(\mathbf{x}) \cdot \mathbf{a}(\mathbf{x})\,\mathrm{d}^3\mathbf{x}. \qquad (9.120)$$

We first met it in Chapter 3, where it appeared in the discussion of the gauge invariance of the Lagrangian for classical electrodynamics; as noted there, it has the same dimensions as Planck's constant, h, that is, dimensions of action. In the classical Hamiltonian scheme for electrodynamics it is the generator of a canonical transformation that displays the relationship between the Coulomb gauge Hamiltonian ($\mathbf{g}^\perp = 0$) and the Hamiltonian in an arbitrary gauge parameterised by some non-zero $\mathbf{g}^\perp$. In quantum theory F becomes a formally self-adjoint operator after quantisation of either the particle variables (F_{sc} in the semiclassical radiation model, §9.5) or both the particle and field variables (F in QED, §9.3.2) and, as we have just seen, it acts, in close analogy with the classical theory, as the generator of an important unitary transformation that plays the same role as the classical canonical transformation.

In the original formulations of the unitary transformation (9.108), (9.109), definite choices were made for the polarisation field; Power and Zienau expressed it using the leading terms of the multipole series development obtained from the atomic charge density operator (cf. the discussion in §2.4.3),

$$\mathbf{P}(\mathbf{x}) \approx (\mathbf{d} + \mathbf{Q} \cdot \boldsymbol{\nabla} + \ldots) \, \delta^3(\mathbf{x}), \tag{9.121}$$

$$\boldsymbol{\nabla} \cdot \mathbf{P} = -\rho, \tag{9.122}$$

clearly reflecting a prior conception of an atom as a bound collection of charges centred on the origin [14], [41]. The semiclassical form of the generator of the transformation had been given much earlier in a simplified form by Goeppert–Mayer [42]; only the electric dipole operator in the multipole series was retained, and the classical vector potential had no spatial variation:

$$\mathsf{F}_{\mathrm{sc}} \approx \mathbf{d} \cdot \mathbf{A}(t). \tag{9.123}$$

Later, and independently of Power and Zienau, Fiutak showed that the complete multipole series representation of the action F_{sc} could be summed up into an integral; if we choose a fixed vector $\mathbf{O}$ as the atomic origin about which the multipole expansion is made and set

$$\mathbf{x}_n = \mathbf{q}_n + \mathbf{O} \tag{9.124}$$

for the particle coordinates, the integral representation for the semiclassical case is

$$\mathsf{F}_{\mathrm{sc}} = \sum_n e_n \mathbf{q}_n \cdot \int_0^1 \mathbf{A}(\sigma \mathbf{q}_n, t) \, \mathrm{d}\sigma. \tag{9.125}$$

Fiutak also discussed briefly the QED case and concluded that the transformed QED Hamiltonian could only be equivalent to the original Coulomb gauge Hamiltonian for first-order radiative processes [43]. This cannot be right since even if the vector potential is quantised, (9.125) defines a formally self-adjoint operator, and so the transformation should lead to a unitarily equivalent Hamiltonian. He seems to have been concerned about the appearance of the term involving $|(\mathbf{P}^\perp)|^2$ in H[g] which did not arise in the semiclassical transformation he considered (cf. §9.5). That this is the correct result was first demonstrated by Atkins and Woolley [15].

Irrespective of whether $\mathbf{A}$ is chosen as a classical field or as an operator-valued quantity, the action F defined by (9.125) can be put in the standard form [15], [16],

$$\mathsf{F} = \int \mathbf{P}(\mathbf{x}) \cdot \mathbf{A}(\mathbf{x}) \, \mathrm{d}^3\mathbf{x}, \tag{9.126}$$

where

$$\mathbf{P}(\mathbf{x}) = \sum_n e_n \int_{C_n}^{\mathbf{x}_n} \delta^3(\mathbf{z} - \mathbf{x}) \, \mathrm{d}\mathbf{z}. \tag{9.127}$$

In non-relativistic QED the transformation of the Coulomb gauge Hamiltonian with the unitary operator

$$\Lambda_{\mathrm{pzw}} = \exp\left(-\frac{i}{\hbar} \sum_n e_n \int_{C_n}^{\mathbf{x}_n} \mathbf{A}(\mathbf{z}) \cdot \mathrm{d}\mathbf{z} \right) \tag{9.128}$$

came to be known as the Power–Zienau–Woolley (PZW) transformation [9], [15]–[18], [20]. The path C_n was specified as a straight line between the charge at position $\mathbf{x}_n$ and the origin about which the multipole expansion is made,

$$\mathbf{z}_n = \mathbf{x}_n - \sigma\mathbf{q}_n, \quad 0 \le \sigma \le 1. \tag{9.129}$$

The result of the PZW transformation is easily obtained by substituting the particular polarisation field (9.127) into the general Hamiltonian H[$\mathbf{g}$] discussed in §9.3.2. Explicit formulae require the evaluation of the derivative of the action F for the calculation of the transformed particle momentum variables since

$$\Lambda_{\mathrm{pzw}}\mathbf{p}_n\Lambda_{\mathrm{pzw}}^{-1} = \mathbf{p}_n + \nabla_{\mathbf{x}_n}\mathsf{F}, \tag{9.130}$$

using the PZW form (9.127). Direct calculation yields (see Appendix E)

$$\nabla_{\mathbf{x}_n}\int_{C_n}^{\mathbf{x}_n}\mathbf{A}(\mathbf{z})\cdot\mathrm{d}\mathbf{z} = \mathbf{A}(\mathbf{x}_n) + \int_0^1(1-\sigma)\mathbf{q}_n\wedge\mathbf{B}(\mathbf{z}(\sigma)_n)\,\mathrm{d}\sigma, \tag{9.131}$$

where $\mathbf{B}$ is the magnetic induction operator. Thus the vector potential in V[$\mathbf{g}$], (9.103), can be expressed in terms of the magnetic induction operator as

$$\mathbf{a}(\mathbf{x}) = -\int_0^1(1-\sigma)\mathbf{q}\wedge\mathbf{B}(\mathbf{z}(\sigma))\,\mathrm{d}\sigma, \tag{9.132}$$

where $\mathbf{q}$ is the vector displacement of the field point $\mathbf{x}$ from the fixed origin, $\mathbf{x} = \mathbf{q} + \mathbf{O}$ (see Appendix E).

The interaction term linear in $\mathbf{a}$,

$$-\sum_n\frac{e_n}{2m_n}\left(\mathbf{p}_n\cdot\mathbf{a}(\mathbf{x}_n) + \mathbf{a}(\mathbf{x}_n)\cdot\mathbf{p}_n\right), \tag{9.133}$$

can be put in the classical form,

$$-\int\mathbf{M}(\mathbf{x})\cdot\mathbf{B}(\mathbf{x})\,\mathrm{d}^3\mathbf{x}, \tag{9.134}$$

where the 'magnetisation density' operator is defined by

$$\mathbf{M}(\mathbf{x}) = \sum_n\frac{e_n}{m_n}\int_0^1(1-\sigma)\delta^3(\mathbf{z}(\sigma)_n - \mathbf{x})\mathbf{q}_n\wedge\mathbf{p}_n\,\mathrm{d}\sigma, \tag{9.135}$$

which must be symmetrised to make it Hermitian. The interaction quadratic in the vector potential can be expressed as a kind of generalised diamagnetic term,

$$\sum_n\frac{e_n^2}{2m_n}\left(\int_0^1(1-\sigma)\mathbf{q}_n\wedge\mathbf{B}(\mathbf{z}(\sigma)_n)\,\mathrm{d}\sigma\right)^2$$

$$= \tfrac{1}{2}\iint\mathcal{X}(\mathbf{x},\mathbf{x}')_{\alpha\beta}\mathsf{B}(\mathbf{x})_\alpha\mathsf{B}(\mathbf{x}')_\beta\,\mathrm{d}^3\mathbf{x}'\,\mathrm{d}^3\mathbf{x}, \tag{9.136}$$

where

$$\mathcal{X}(\mathbf{x}, \mathbf{x}')_{\alpha\beta} = \sum_n \frac{e_n^2}{m_n}(\delta_{\alpha\beta}q_n^2 - q_n^\alpha q_n^\beta)$$

$$\times \int_0^1 \int_0^1 (1-\sigma)(1-\sigma')\delta^3(\mathbf{z}(\sigma)_n - \mathbf{x})\delta^3(\mathbf{z}(\sigma')_n - \mathbf{x}')\,d\sigma'\,d\sigma. \quad (9.137)$$

The multipole expansion of this expression yields an interaction quadratic in the magnetic field in multipolar form – see, for example, [18]. Essentially the same computations lead to the corresponding forms for $\mathbf{M}$ and $\mathcal{X}$ derived from the paired-particle representation of $\mathbf{P}$, (2.164), in which there is no arbitrary origin [40].

For the case of a uniform magnetic field we can ignore the dependence of the delta functions on σ, σ' and (9.135), (9.137) reduce to

$$\mathbf{M}(\mathbf{x}) \sim \mathbf{m}\,\delta^3(\mathbf{x}), \quad (9.138)$$

$$\mathcal{X}(\mathbf{x},\mathbf{x}')_{\alpha\beta} \sim \chi_{\alpha\beta}\,\delta^3(\mathbf{x})\delta^3(\mathbf{x}'), \quad (9.139)$$

where

$$\mathbf{m} = \sum_n \frac{e_n}{2m_n}\,\mathbf{q}_n \wedge \mathbf{p}_n, \quad (9.140)$$

$$\chi_{\alpha\beta} = \sum_n \frac{e_n^2}{4m_n}(\delta_{\alpha\beta}q_n^2 - q_n^\alpha q_n^\beta) \quad (9.141)$$

are the familiar magnetic dipole operator and the diamagnetic susceptibility tensor for an atomic system. The final result of the PZW transformation based on (9.127) therefore is the Hamiltonian for a closed system of charges and particles expressed in terms of polarisation densities and the transverse components of the electromagnetic field,

$$H_{\text{pzw}} = \sum_n \frac{|\mathbf{p}_n|^2}{2m_n} + H_{\text{rad}} - \int \mathbf{P}(\mathbf{x})\cdot\mathbf{E}(\mathbf{x})^\perp d^3\mathbf{x} - \int \mathbf{M}(\mathbf{x})\cdot\mathbf{B}(\mathbf{x})\,d^3\mathbf{x}$$

$$+ \tfrac{1}{2}\iint \mathcal{X}(\mathbf{x}, \mathbf{x}')_{\alpha\beta}B(\mathbf{x})_\alpha B(\mathbf{x}')_\beta\, d^3\mathbf{x}'\,d^3\mathbf{x}$$

$$+ \frac{1}{2\varepsilon_0}\int \mathbf{P}(\mathbf{x})\cdot\mathbf{P}(\mathbf{x})\,d^3\mathbf{x}, \quad (9.142)$$

which is of the form presented originally, where the polarisation densities were expressed in terms of the complete multipolar series [15], [16]. It should be noted that the static Coulomb interaction conveyed by the longitudinal component of the polarisation field has been combined with the transverse component to give the last term in (9.142).

Two polarisation fields related by

$$\mathbf{P}' = \mathbf{P} + \nabla \wedge \mathbf{U} \quad (9.143)$$

cannot be differentiated between; in particular the equations of motion are independent of $\mathbf{U}$. Now considering the transformations with these polarisation fields based on

$$\Lambda_{\mathbf{P}} = \exp\left(-\frac{i}{\hbar}\int \mathbf{P}\cdot\mathbf{A}\,d^3\mathbf{x}\right),\tag{9.144}$$

we have

$$\Lambda_{\mathbf{P'}} = \Lambda_{\mathbf{P}}\Lambda_{\mathbf{B}},\tag{9.145}$$

where

$$\Lambda_{\mathbf{B}} = \exp\left(-\frac{i}{\hbar}\int \mathbf{U}\cdot\mathbf{B}\,d^3\mathbf{x}\right).\tag{9.146}$$

This operator is gauge invariant but depends on the arbitrary field $\mathbf{U}$.

Choosing a different path corresponds to choosing a $\mathbf{U}$ field. We can use (2.169) to write the transformation operator for path 1 as

$$\Lambda_1 = \exp\left(-\frac{ie}{\hbar}\int_{C_1}d\mathbf{z}\cdot\mathbf{A(z)}\right) = \Lambda_{\mathrm{pzw}}\Lambda_{\mathbf{B}_{12}},\tag{9.147}$$

where

$$\Lambda_{\mathbf{B}_{12}} = \exp\left(\frac{ie}{\hbar}\int_{\Sigma_{12}}d\mathbf{S}\cdot\mathbf{B}\right)\tag{9.148}$$

is a Wilson loop.

The result of transforming the Coulomb gauge Hamiltonian $H[0]$ with the modified operator Λ_1 can therefore be put in the form

$$\begin{aligned}H_1 &= \Lambda_1 H[0]\Lambda_1^{-1}\\&= \Lambda_{\mathbf{B}_{12}}H_{\mathrm{pzw}}\Lambda_{\mathbf{B}_{12}}^{-1}.\end{aligned}\tag{9.149}$$

Note that $\Lambda_{\mathbf{B}_{12}}$ is *independent* of the particle variables although it depends on the charge parameter e and the boundary Σ_{12}, so that the PZW forms for the polarisation fields $\mathbf{P},\mathbf{M}$ and the susceptibility $\mathcal{X}$ are unchanged. The transformed Hamiltonian H_1 for the altered path therefore differs from H_{pzw} because of the non-commutation of $\Lambda_{\mathbf{B}_{12}}$ with $\mathbf{E}^\perp$, and the changes amount to changes in the energy of the field (i.e. to changes in the second and third terms in (9.142)).

As discussed in §2.4.1, (9.127) is a particular case of the general form of solution for (9.122) given by

$$\mathbf{P(x)} = \int \mathbf{g(x,x')}\rho(\mathbf{x'})\,d^3\mathbf{x'},\tag{9.150}$$

with

$$\mathbf{g(x,x')} = \int_C^{\mathbf{x'}}\delta^3(\mathbf{z}-\mathbf{x})\,d\mathbf{z}\tag{9.151}$$

taken over a straight-line path C. There is no physical or mathematical requirement however, for the lower end point of the integration path to be at the centre of a charge distribution, and the integration path can be chosen in any way provided only that the integrals (9.127), (9.151) exist. Moreover, the line integral representation of the polarisation field for a neutral collection of charges can be written so that any arbitrary origin is eliminated in favour of the particle coordinates (cf. Chapter 2). Note also that the integration constant that arises in the case of a charged system is actually irrelevant here since it is orthogonal to the Coulomb gauge vector potential. We thus arrive at a generalised PZW transformation by taking $\mathbf{g}(\mathbf{x}, \mathbf{x}')$ as any solution of the defining equation, (2.64),

$$\nabla \cdot \mathbf{g}(\mathbf{x}, \mathbf{x}') = -\delta^3(\mathbf{x} - \mathbf{x}'), \tag{9.152}$$

and regarding the formally unitary operator Λ as a functional of the Green's function $\mathbf{g}$ [44], [45]. Thus every $\mathbf{g}$ leads to the QED Hamiltonian in the form (9.142) in a distinct 'PZW' representation. It is pertinent to recall the discussion in §3.3.1 of the last term in (9.142) which is precisely the energy $\mathcal{E}_\mathbf{P}$ with the particle position variables interpreted as position operators. The contribution of the charges and the transverse radiation to the total energy of the electric field is quite arbitrarily divided between them, and only in the singular case $\mathbf{P}^\perp = 0$, that is, the Coulomb gauge, is there an explicit Coulomb interaction between two charges [46].

We must keep in mind that a definite physical situation (experimental setup) *cannot* be associated with any particular form for $\mathbf{g}$ and that a necessary condition for calculated quantities to be identified with physical observables is that they be independent of $\mathbf{g}(\mathbf{x}, \mathbf{x}')^\perp$. The vector potential operator $\mathbf{A}$ has only entered as a useful working variable that disappears from the final result. It is clear from the foregoing discussion, however, that, despite superficial appearances, the PZW Hamiltonian does not give a 'gauge-invariant' interaction potential; the arbitrary nature of the vector potential has simply been transferred to the arbitrariness of the polarisation densities (expressed through the freedom to choose paths at will). A particular choice of path has been made in (9.127), and so the task of establishing what quantities can be calculated that do not depend on such a choice is not thereby avoided.

The family of Hamiltonians H[$\mathbf{g}$], parameterised by the Green's function $\mathbf{g}(\mathbf{x}, \mathbf{x}')$, is very convenient for investigating the gauge invariance of calculations. We saw in Chapter 3 that the classical Hamiltonian $H[\mathbf{g}]$ yields the correct equations of motion independently of $\mathbf{g}^\perp$. Since Maxwell's equations and the Lorentz force law are linear in the dynamical variables, the quantum operators obey the same gauge-invariant equations of motion; this, however, is not the usual approach to calculation in non-relativistic quantum electrodynamics.

The main methods of quantum mechanical calculation in QED are based on the splitting of the full Hamiltonian shown in (9.103):

$$H[\mathbf{g}] = H_0 + V[\mathbf{g}]. \tag{9.153}$$

The reference Hamiltonian is customarily chosen to be independent of the arbitrary Green's function $\mathbf{g}(\mathbf{x}, \mathbf{x}')$. The interaction Hamiltonian, V[$\mathbf{g}$], contains terms that are linear and quadratic in the charge, and it is convenient to separate it into these two parts,

$$V[\mathbf{g}] = V^1[\mathbf{g}] + V^2[\mathbf{g}], \tag{9.154}$$

in an obvious notation. Using the expansion (9.115) and the fact that F, (9.109), is linear in the charge, we can write explicit relationships between the general interaction $V[\mathbf{g}]$ and the Coulomb gauge interaction $V[0]$, (9.107), namely

$$V^1[\mathbf{g}] = V^1[0] + \frac{i}{\hbar}\,[H_0,\ F]] \tag{9.155}$$

$$V^2[\mathbf{g}] = V^2[0] + \frac{i}{\hbar}\,[V^1[0],\ F] + \frac{1}{2!}\left(\frac{i}{\hbar}\right)^2 [\,[H_0,\ F]\,,F]\,. \tag{9.156}$$

These relations provide us with a tool for investigating in a general way the gauge invariance (or equivalently, path independence) of the S-matrix, which we shall take up in §10.2.1.

9.4 QED and Symmetry

The gauge invariance of the QED Hamiltonian H is an important dynamical symmetry that we have described in detail in this chapter. It leads to the fundamental principle that only gauge-invariant quantities can be candidates for physical observables. The other symmetries of the QED Hamiltonian are easily summarised. A simple additive construction of particle and field quantities suffices to construct the total linear momentum[5] ($\mathbf{P}$) and total angular momentum ($\mathbf{J}$) operators of the combined system of charges and field, for example,

$$\mathbf{P} = \sum_i^n \mathbf{p}_i + \sum_{\mathbf{k},\lambda} \hbar\mathbf{k}\ n_{\mathbf{k},\lambda}\,, \tag{9.157}$$

in terms of the individual particle momenta and the photon number operator. $\mathbf{J}$ can be constructed in a similar fashion. The operators $\mathbf{P}$ and $\mathbf{J}$ so formed, together with H, satisfy the Lie bracket relations for the relativity groups described in Chapters 3 and 5. This is true classically as well as in the quantum mechanical account. Thus the total linear momentum and the total angular momentum are conserved quantities. However, it is not possible to define a boost operator $\mathbf{K}$ with the requisite properties to complete the algebras of either the Galilean or Poincaré groups. The theory is conventionally described as 'non-relativistic'.

An evident limitation of a presentation of quantum electrodynamics based on canonical quantisation of classical electrodynamics is that the charges are necessarily spin 0; there are familiar examples in particle physics (π- and K-mesons), but they are not of interest here. It is well known that the overall gross properties of atomic matter can be described in terms of non-relativistic quantum mechanics without an explicit reference to the concept of particle spin, provided the Pauli exclusion principle (or more generally, the permutation group symmetry of the particle Hamiltonian) is recognised. Electrons are fermions with spin $\frac{1}{2}$, while nuclei can have either integer

---

[5] We use the conventional notation; $\mathbf{P}$ must not be confused with the electric polarisation field. The context is sufficient to keep this clear.

or half-integer spin ≥ 0. Interactions involving the particle spin operators occur naturally in relativistic (that is, Lorentz-invariant) formulations of quantum mechanics. The standard Lorentz-invariant formulation of quantum electrodynamics is a quantum theory of interacting electron and electromagnetic fields ('electrons and photons') in which particle number is not a conserved quantity. Nuclei sit rather uncomfortably in this framework other than as fixed classical sources of external fields [47], not least because their anomalous magnetic moments may make such a theory unrenormalisable. As discussed in Chapter 5, there is no known Lorentz-invariant quantum theory of an N-particle system involving electromagnetic interactions with fixed N, and so there is no such theory of atoms and molecules.

The Lorentz-invariant quantum theory of one electron is based on the Dirac equation [48]; to get some idea of 'relativistic effects' in atoms and molecules one can proceed in various ways, all of which involve some degree of 'approximation'. The characteristic parameters that divide 'non-relativistic' and 'relativistic' phenomena are firstly, the ratio of the particle speed to the speed of light, $\beta = v/c$, and secondly the particle's reduced Compton wavelength, $\lambda = \hbar/mc$; both can be used as expansion parameters to develop a series of terms that provide 'relativistic corrections' to the ordinary Schrödinger equation.[6] The result generally is a series of singular potentials involving r^{-3} terms, Dirac delta functions ('contact potentials') and so on which collectively should only be used in first-order perturbation theory since they do not translate into self-adjoint operators. They may be regularised as we have discussed earlier; physically this may be interpreted as recognising finite size effects for nuclei and nuclear structure.

Among the 'relativistic corrections' there is one term that can rationally be appended to the spin-free non-relativistic Hamiltonian to give an operator that is still self-adjoint and well behaved; this is the low-energy limit of the Pauli term,[7] which for a Dirac electron with g-factor precisely 2 is

$$V_{\text{Pauli}} = -\frac{e\hbar}{2m_e}\boldsymbol{\sigma}\cdot\mathbf{B}. \tag{9.158}$$

Here $\mathbf{B}$ is the quantised magnetic field operator and the $\{\sigma_i\}$ are the usual Pauli matrices. This interaction term can be extended to a many-particle system by linearity. It is important that the Pauli term is a gauge-invariant addition to the Hamiltonian, and this is not accidental. A free non-relativistic electron is a spin-half particle; in order to make the spin explicit we modify (9.33) to read

$$H_e = \frac{(\boldsymbol{\sigma}\cdot\mathbf{p})^2}{2m_e} \tag{9.159}$$

and write the Schrödinger equation (9.34) with a two-component wave function. If we introduce the components of the covariant derivative in place of the ordinary derivatives through minimal coupling, the Pauli term (9.158) appears automatically.

---

[6] The substantial and by now rather old literature describing these approaches: the Foldy–Wouthuysen transformation, Feynman Gell–Mann projection, Löwdin partitioning, the Breit equation and so on is reviewed in the context of atomic/molecular theory in [49].

[7] In Lorentz-invariant form this is proportional to $\alpha_4[\gamma^\mu,\gamma^\nu]F_{\mu\nu}$ in terms of the usual Dirac matrices [47]; for a slowly moving particle the electric field contribution in $F_{\mu\nu}$ can be dropped.

Optionally, the contribution of V_{Pauli} can be absorbed into the magnetisation term in the interaction in Eq. (9.142). Nuclear spin effects are, of course, extremely important in atoms and molecules (hyperfine interactions, NMR etc.); there is no fundamental equation for any nuclear particle analogous to the Dirac equation, and so the proportionality constant in the Pauli term for nuclei must be determined empirically. Another approach to 'relativistic corrections' is to modify (9.159) to the form appropriate for a Dirac electron by writing, for each particle of mass m,

$$H_e \to \boldsymbol{\alpha} \cdot \mathbf{p} + \beta m, \tag{9.160}$$

where $\{\boldsymbol{\alpha}, \beta\}$ are the usual Dirac matrices, and again this can be extended to electrodynamics by introducing the covariant derivative operator $\bar{\mathbf{p}}$.

9.5 The Semiclassical Radiation Model

A widely used approach to the theoretical description of the interactions between atoms or molecules and electromagnetic radiation is based on the notion that the field can be treated as a classical electromagnetic field described by Maxwell's theory (cf. Chapter 2) and that the quantum properties of the atomic system are given by an appropriate time-dependent Schrödinger equation. This is the 'semiclassical radiation model'. A static electric or magnetic field is always classical, and its interaction with charged particles can be described by the inclusion of additional terms in the time-independent Schrödinger equation that modify the spectrum of the atomic system; such perturbations may lead to shifts in eigenvalues (Zeeman effect) or the conversion of eigenvalues into (metastable) resonances (Stark effect). These topics are discussed thoroughly in numerous standard quantum mechanics texts.

On the other hand, the treatment of the interaction of atomic/molecular matter with an optical field using classical electromagnetism is not a trivial matter; its relationship to quantum electrodynamics does not seem to be well described in the literature. Given the extensive evidence that the electromagnetic field is a quantum mechanical system, one may enquire how an approach that eschews that information can possibly succeed, in an admittedly limited set of experimental situations. In the following we attempt to answer that question. The main limitation of such an approach is that the atom can only respond to a non-zero classical field; thus stimulated absorption and emission, and light scattering can be considered, but phenomena that derive from spontaneous emission, for example fluorescence, luminescence and, phosphorescence, or involve virtual photons, for example intermolecular interactions, resonant energy transfer processes and the problems of self-interaction are outside the scope of the semiclassical model.

Given a quantum Hamiltonian constructed by canonical quantisation of the corresponding classical theory in Hamiltonian form (P.B.s $\to$ quantum commutators, $x \to \mathsf{x}$ etc.), we know that the classical equations of motion for the classical variables are replaced by operator equations of motion for the corresponding quantum mechanical operators. Furthermore, *linear* equations of motion such as the Maxwell equations for the electromagnetic field have the same form in both cases with a suitable operator interpretation of the particle and field variables in the quantum case. The classical Hamiltonian equations of motion yield the wave equation for the vector potential[8]

$$\square\, \mathbf{a}(\mathbf{x},t) \;=\; \mu_0 \mathbf{j}(\mathbf{x},t), \tag{9.161}$$

which may be solved in the usual way by the Green's function technique,

$$\mathbf{a}(\mathbf{x},t) = \mu_0 \int \mathbf{G}(\mathbf{x},t;\mathbf{x}',t') \cdot \mathbf{j}(\mathbf{x}',t')\, \mathrm{d}^3\mathbf{x}'\, \mathrm{d}t', \tag{9.162}$$

where $\mathbf{G}$ satisfies the equation[9]

$$\square\, \mathbf{G}(\mathbf{x},t;\mathbf{x}',t') = \delta^3(\mathbf{x}-\mathbf{x}')\delta(t-t'). \tag{9.163}$$

The form of the charge-current density operator j_μ to use in the quantum mechanical equivalent of (9.162) is obtained by comparison of the Maxwell equations (Chapter 2) with the field equations of motion obtained from the full QED Hamiltonian for charged particles interacting with electromagnetic radiation. Its components are the charge density operator ($\mu = 0$),

$$\rho(\mathbf{x},t) = \sum_n e_n\, \delta^3(\mathbf{x}(t)_n - \mathbf{x}), \tag{9.164}$$

and the gauge-invariant current density operator ($\mu = 1,2,3$),

$$\mathbf{j}(\mathbf{x},t) = \sum_n \frac{e_n}{m_n}\big(\mathbf{p}(t)_n - e_n\, \mathbf{a}(\mathbf{x},t)\big)\, \delta^3(\mathbf{x}(t)_n - \mathbf{x}), \tag{9.165}$$

where the sums are taken over all the charges in the atomic system. There are also the equations of motion for the position and momentum operators of the particles, which we will not require here. The resulting coupled equations define QED in a Heisenberg representation (see Chapter 11).

The semiclassical radiation model of light scattering is based on the ansatz that the equations of motion generated by the fully quantised form of H are simplified by a mean-field approximation in which the charge-current density operator $j_\mu(\mathbf{x},t)$ is replaced in (9.162) by an expectation value in a basis of time-dependent atomic states $\{|\Psi(t)_u\rangle\}$,

$$j_\mu(\mathbf{x},t) \;\rightarrow\; \langle\Psi(t)_u|j_\mu(\mathbf{x},t)|\Psi(t)_u\rangle = j_\mu(\mathbf{x},t)_u. \tag{9.166}$$

This requires that the operator $\mathbf{a}$ be replaced by the classical vector potential $\mathbf{a}$ so that one has c-numbers on both sides of the $=$ signs in (9.162) and (9.165). The $\{|\Psi(t)_u\rangle\}$ are taken as solutions of the time-dependent Schrödinger equation for an atom in the presence of a classical time-dependent electromagnetic field (see what follows). If (9.164) and (9.165) are inserted in (9.166), the components of $(j_\mu)_u$ so defined satisfy the classical equation of continuity, as first shown by Schrödinger. This leads to the

---

[8] $\square$ is the four-dimensional d'Alembertian operator.
[9] The explicit form for the Green's function depends on both the gauge of the vector potential and the boundary conditions imposed on the solution. The Coulomb gauge Green's function with retarded boundary conditions is given in Appendix C and is remarkable [14] for containing a static term varying as x^{-3}.

interpretation that the atom/molecule under the influence of an external classical electromagnetic field acts as an oscillating source of radiation that can be treated by the
standard techniques of classical electromagnetism [6], [7].

An atom in a state $|\Psi_u\rangle$ under the influence of an incident monochromatic light wave
of frequency ω has a current density which has a purely harmonic time dependence,

$$\mathbf{j}(\mathbf{x},t) = e^{-i\omega t}\mathbf{j}(\mathbf{x}), \tag{9.167}$$

where $\mathbf{j}(\mathbf{x})$ is understood to be $\langle\Psi_u|\mathbf{j}(\mathbf{x})|\Psi_u\rangle$. Then, dropping terms that decrease more
rapidly than $1/|\mathbf{x}|$ and carrying out the t' integration in (9.162), with the Green's func-
tion satisfying retarded boundary conditions, one finds the classical vector potential
in the far-field is determined purely by the transverse current density:

$$\mathbf{A}(\mathbf{x},t) = \mathbf{A}(\mathbf{x},t)_0 - \left(\frac{1}{\varepsilon_0 c^2}\right)\int \frac{e^{ik|\mathbf{x}-\mathbf{x}'|}}{4\pi|\mathbf{x}-\mathbf{x}'|}\mathbf{j}(\mathbf{x}')^{\perp}\,d^3\mathbf{x}'. \tag{9.168}$$

Here $\mathbf{A}(\mathbf{x},t)_0$ is a transverse, that is, Coulomb gauge, solution of the homogeneous
wave equation. The coefficient of $\mathbf{j}(\mathbf{x}')^{\perp}$ will be recognised as the outgoing wave Green's
function for the scalar wave equation, $(\omega = kc)$, [50]

$$G(\mathbf{x},\mathbf{x}';k)^{\text{out}} = \frac{e^{ik|\mathbf{x}-\mathbf{x}'|}}{4\pi|\mathbf{x}-\mathbf{x}'|}. \tag{9.169}$$

If we introduce a unit vector $\hat{\mathbf{s}}$ that points in the direction of $\mathbf{x}$ and evaluate (9.168)
for $|\mathbf{x}| \to \infty$, we have

$$\mathbf{A}(\mathbf{x},t)^{\text{scat}} \approx \left(\frac{1}{\varepsilon_0 c^2}\right)\frac{e^{i(kx-\omega t)}}{4\pi x}\int \mathbf{j}(\mathbf{x}')^{\perp}\,e^{-i\mathbf{k}_s\cdot\mathbf{x}'}\,d^3\mathbf{x}', \quad \text{for large } x, \tag{9.170}$$

where $\mathbf{k}_s = k\hat{\mathbf{s}}$, and only terms in $1/x$ are retained. Thus the vector potential for the
scattered radiation in the far-field (the wave-zone) has the form of an outgoing spheri-
cal wave modulated by an angular factor (the scattering amplitude) determined by the
final wave vector $\mathbf{k}_s$,

$$\mathbf{f}(-\mathbf{k}_s) = \frac{1}{4\pi}\int \mathbf{j}(\mathbf{x}')^{\perp}\,e^{-i\mathbf{k}_s\cdot\mathbf{x}'}\,d^3\mathbf{x}'. \tag{9.171}$$

Introducing Fourier transform notation,

$$\mathbf{j}(\mathbf{p}) = \int \mathbf{j}(\mathbf{x})\,e^{i\mathbf{p}\cdot\mathbf{x}}\,d^3\mathbf{x}, \tag{9.172}$$

this angular factor is seen to be essentially the $-\mathbf{k}_s$ component of the Fourier transform of the transverse current density, $\mathbf{j}(-\mathbf{k}_s)^{\perp}$. Retaining only $1/x$ terms, the classical
electric field in the far-field region is

$$\mathbf{E}(\mathbf{x},t)^{\infty} = \frac{ik}{c}\frac{e^{i(kx-\omega t)}}{4\pi\varepsilon_0 x}\,\mathbf{j}(-\mathbf{k}_s)^{\perp}, \tag{9.173}$$

where

$$\mathbf{j}(-\mathbf{k}_s)^{\perp} = \hat{\mathbf{\varepsilon}}_{sx}\,j(-\mathbf{k}_s)_x + \hat{\mathbf{\varepsilon}}_{sy}\,j(-\mathbf{k}_s)_y, \tag{9.174}$$

taking $\hat{\mathbf{s}}$ along the z-axis, that is, the field is purely transverse with respect to the propagation direction of the scattered radiation, $\hat{\mathbf{s}}$. The asymptotic field (9.173) can be used to determine the intensity and polarisation characteristics of the radiation scattered by an atomic/molecular system through the calculation of the classical Stokes parameters described in §4.3 in terms of quadratic combinations of the electric field strength. Such calculations are described fully in the monograph [7].

Now we must consider how the semiclassical approach just summarised might be related to quantum electrodynamics. Clearly one cannot simply take the classical Hamiltonian system (Chapter 3: (3.254)–(3.257)) and reinterpret only the particle variables as quantum operators leaving the classical field variables and their Poisson brackets unchanged, not least because of (3.257) which contains both particle and field variables. Likewise, if the classical current is interpreted as a matrix element of a quantum mechanical operator as in (9.166), then for consistency the LHS of (9.162) would also be a matrix element, and there is no obvious reason for restricting such terms to just diagonal elements, $u = v$. In the semiclassical radiation model the electromagnetic field appears as an external field with its own time-dependent dynamics, and there is no Hamiltonian for the radiation field. The notion of an 'external field' in the Hamiltonian is generally taken as self-evident, but is it? The fact that an electromagnetic field may be produced such that it exhibits some 'classical properties' is really a statement about the *state* of the field, rather than its Hamiltonian. The Hamiltonian is surely fixed for a given physical system, here the electromagnetic field, which, of course, has quantum features that require a quantum mechanical Hamiltonian (cf. Chapter 7).

Quantum electrodynamics can be cast into a similar form to that used in the semiclassical model by making the transition to an interaction representation. We start with the Schrödinger equation for the full QED Hamiltonian $H[\mathbf{g}]$, (9.86),

$$i\hbar \frac{\partial \Psi_S}{\partial t} = H[\mathbf{g}]\Psi_S, \tag{9.175}$$

where Ψ_S is the Schrödinger representation wave function, and pass to an interaction representation generated by H_{rad}, so that the field operators have their free-field time dependence. Then, in the usual way, the interaction representation wave function, Φ_I, satisfies the equation

$$i\hbar \frac{\partial \Phi_I}{\partial t} = k([\mathbf{g}], t)\Phi_I, \tag{9.176}$$

where

$$k([\mathbf{g}], t) = H_{\mathrm{charges}} + V(t, [\mathbf{g}]), \tag{9.177}$$

with

$$V([\mathbf{g}], t) = U(t)_I\, V[\mathbf{g}]\, U(t)_I^{-1}, \qquad U(t)_I = \exp\left(i H_{\mathrm{rad}} t / \hbar\right). \tag{9.178}$$

The particle operators have no explicit time dependence, and H_{charges} is independent of $\mathbf{g}$. These relations are valid for any choice of $\mathbf{g}^\perp$ including $\mathbf{g}^\perp = 0$ (the Coulomb gauge). Since a unitary transformation does not disturb the commutation relations, the PZW transformation between the Coulomb gauge Hamiltonian and the Hamiltonian operator for arbitrary $\mathbf{g}^{\perp}$ can be carried out in any representation; $\mathsf{F}(t)$ is the generator in the interaction representation.

In view of the classical features that may be associated with coherent states $\{|\alpha\rangle\}$ of the electromagnetic field with large α parameters (see Chapter 7), it is interesting to consider an expansion of the interaction representation ket $|\Phi_I\rangle$, (9.176), in a set of coherent states, $\{|\alpha\rangle\}$,

$$|\Phi(t)_I\rangle = \sum_{\alpha} \Phi(t)_{\alpha} |\alpha\rangle. \tag{9.179}$$

The coefficients $\{\Phi(t)_{\alpha}\}$ depend on only the particle variables; in a sense they are the atomic wave functions in this representation, although the $\{\Phi(t)_{\alpha}\}$ have no straightforward interpretation as probability amplitudes since the coherent states are over-complete and non-orthogonal. We insert the expansion (9.179) into (9.176), with k taken as the QED Hamiltonian in the interaction representation, and form the scalar product with an arbitrary coherent state $|\beta\rangle$; provided $\mathsf{V}(t)$ is first written in normal ordered form, the result can be written in matrix notation as

$$\mathbf{S}\left(i\hbar\frac{\mathrm{d}\boldsymbol{\Phi}(t)}{\mathrm{d}t} - \mathcal{K}_{\text{charges}}\boldsymbol{\Phi}(t) - \mathbf{V}^N\boldsymbol{\Phi}(t)\right) = 0, \tag{9.180}$$

where $\boldsymbol{\Phi}(t)$ is a column vector of expansion coefficients $\{\Phi(t)_{\alpha}\}$,

$$\mathbf{S}_{\beta\alpha} = \langle\beta|\alpha\rangle, \tag{9.181}$$

and

$$\left(\mathbf{V}^N\right)_{\beta\alpha} = V(\beta^*, \alpha)_N. \tag{9.182}$$

The additional terms arising from $\frac{1}{2}\varepsilon_0^{-1}\int|\mathbf{P}(\mathbf{x})^{\perp}|^2\mathrm{d}^3\mathbf{x}$ and normal ordering, if required, have been added on to $\mathsf{H}_{\text{charges}}$ to give an operator we denote as $\mathcal{K}_{\text{charges}}$ that contains only particle operators. In writing (9.180) we have made use of the fundamental property of coherent states, (7.72). However, the matrix $\mathbf{S}$ is singular because the coherent states are over-complete; since the inverse matrix $\mathbf{S}^{-1}$ does not exist, the expression in parentheses in (9.180) involving the coefficient vector $\boldsymbol{\Phi}(t)$ does not have to vanish, although it may do so.

In view of (7.139) and (7.145), we take the classical limit of the electromagnetic field as being characterised by a particular coherent state with large $|\alpha|$ for which all couplings and overlap integrals with states of other α can be neglected, and assume that

$$i\hbar\frac{\mathrm{d}\Phi(t)_{\alpha}}{\mathrm{d}t} = \mathcal{K}_{\text{charges}}\Phi(t)_{\alpha} + \mathsf{V}(\alpha^*, \alpha; t)_N \Phi(t)_{\alpha}. \tag{9.183}$$

This is the time-dependent Schrödinger equation for the 'atomic' state $\Phi(t)_{\alpha}$ under the action of $\mathcal{K}_{\text{charges}}$ in the presence of the time-dependent classical electromagnetic field characterised by α (cf. (7.143) for the amplitude of the electric field vector).

Obviously processes involving the photon vacuum ($\alpha = 0$) such as radiative energy transfer, spontaneous emission, and resonance and damping effects are eliminated by such an ansatz, which reduces the superposition (9.179) to a simple Hartree product,

$$\Phi(t)_I \sim |\alpha\rangle \, \Phi(t)_\alpha. \tag{9.184}$$

We recall that the Hartree product is consistent with the statement that the classical field has its own dynamics that is not affected by its interaction with the atom; quantum mechanically one would expect the two parts of the system to be 'entangled' as implied by the superposition (9.179).

In the non-relativistic interaction (9.103) there are terms linear in the field variables, and a quadratic term. The linear terms require no comment since they are not altered by the normal ordering; if the quadratic term is brought to normal ordered form, the interaction Hamiltonian acquires additional terms which may involve the particle variables (but not the photon operators) and these have been put into $\mathcal{K}$ in order to arrive at the hoped-for atom-classical field interaction; their precise form depends on the gauge of the field potential.

The usual choice in the semiclassical radiation model is to specify the vector potential in the Coulomb gauge ($\mathbf{g}^\perp = \mathbf{0}$); in this gauge $\frac{1}{2}\varepsilon_0^{-1} \int |\mathbf{P}(\mathbf{x})^\perp|^2 \, d^3\mathbf{x}$ vanishes and the extra term from the normal ordering is an infinite constant. Thus $\mathcal{K}$ in this gauge reduces to $\mathrm{H_{charges}}$ (plus an infinite constant). However, this is the only gauge in which this is true. This statement is obvious since $\frac{1}{2}\varepsilon_0^{-1} \int |\mathbf{P}(\mathbf{x})^\perp|^2 \, d^3\mathbf{x}$ is an integral over a squared expression and so only vanishes if the integrand is zero; as regards the normal ordering of the field operators, the Coulomb gauge is a special case because the quadratic term in the vector potential is a local operator – for each charge both field operators are to be evaluated at the position of that charge.[10] This locality does not hold in any other gauge.

It should be noted especially that the extra terms from the normal ordering do not cancel the arbitrary operator $\frac{1}{2}\varepsilon_0^{-1} \int |\mathbf{P}(\mathbf{x})^\perp|^2 \, d^3\mathbf{x}$; to see what is involved in the normal ordering calculation for non-zero $\mathbf{g}^\perp$ we may take the generalised diamagnetic susceptibility interaction (9.136) that arises in the PZW transformation theory. The normal ordering of this interaction involves ordering $\mathbf{B}(\mathbf{x})\mathbf{B}(\mathbf{x}')$, which can be done using the results in Appendix D, followed by integration over the tensor $\mathcal{X}$ and this is nothing like $\frac{1}{2}\varepsilon_0^{-1} \int \mathbf{P}(\mathbf{x})^\perp \cdot \mathbf{P}(\mathbf{x})^\perp \, d^3\mathbf{x}$. In QED this arbitrary term involving the square of the electric polarisation field operator seems to have been first discussed by Power and Zienau [14]; they interpreted it as an atomic 'self-energy', hence something involving the photon vacuum (and actually all Fock space matrix elements diagonal in the photon occupation numbers). Its occurrence in a non-relativistic classical formulation of Hamiltonian electrodynamics seems to have been first noted in [15]. As we saw in earlier chapters, its appearance in the Hamiltonian does not require the quantisation of the field; apparently the semiclassical radiation model can say nothing about it beyond (tacitly) choosing the unique gauge that makes it vanish (the Coulomb gauge).

[10] The normal ordering calculation is the same as the result in (F.1.10) – see also Appendix D where the relevant calculation requires the ordering of $\mathbf{A}(\mathbf{x})\mathbf{A}(\mathbf{x}')$ with $\mathbf{x} = \mathbf{x}'$.

The imposition of a gauge condition carries with it the requirement that the gauge invariance of the results thereby obtained can be guaranteed. A general approach to this question is to determine what quantities can be obtained independently of the choice of $\mathbf{g}$ using the full interaction $V[\mathbf{g}]$. This is what is done in QED and, as we shall see in Chapter 10 (§10.2.1), there are quantities like the S-matrix that are independent of $\mathbf{g}$, that is, gauge invariant; in particular the contributions of $\mathbf{P}^\perp$ are cancelled in all orders of perturbation theory by other terms that arise from virtual photon processes involving the quantised field. This cancellation mechanism is not available for the classical electromagnetic field coupled to a quantised system. The adoption of the Coulomb gauge condition eliminates the problem, but offers only a calculation in a fixed gauge and no means of checking whether fixing the gauge affects the results.

Having made that choice, we may consider a semiclassical version of the PZW transformation in which the quantised transverse vector potential in the PZW generator is replaced by its time-dependent classical analogue. The Coulomb gauge Hamiltonian can be put in the form

$$H([0],t)_{\text{sc}} = \sum_n \frac{1}{2m_n}\left(\mathbf{p}_n - e_n\,\mathbf{A}(\mathbf{x}_n,t)\right)^2 + \tfrac{1}{2}\sum_{n,m}\frac{e_n e_m}{4\pi\varepsilon_0|\mathbf{x}_n - \mathbf{x}_m|},$$

$$= H_{\text{charges}} + V([0],t)_{\text{sc}}, \tag{9.185}$$

where

$$V([0],t)_{\text{sc}} = -\sum_n \frac{e_n}{m_n}\mathbf{p}_n\cdot\mathbf{A}(\mathbf{x}_n,t) + \sum_n \frac{e_n^2}{2m_n}\mathbf{A}(\mathbf{x}_n,t)\cdot\mathbf{A}(\mathbf{x}_n,t) \tag{9.186}$$

is the Coulomb gauge interaction Hamiltonian for charges interacting with an external classical electromagnetic field; the $[0]$ notation is as in §9.3.2. The atomic states evolve in time according to the time-dependent Schrödinger equation

$$i\hbar\frac{\partial\Psi}{\partial t} = H([0],t)_{\text{sc}}\Psi. \tag{9.187}$$

The semiclassical version of the PZW transformation is generated by the time-dependent unitary operator

$$\Lambda(t)_{\text{sc}} = \exp\left(-\frac{i}{\hbar}\int \mathbf{P}(\mathbf{x})\cdot\mathbf{A}(\mathbf{x},t)\,\mathrm{d}^3\,\mathbf{x}\right), \tag{9.188}$$

that acts on the atomic states. $\mathbf{A}(\mathbf{x},t)$ is the classical Coulomb gauge vector potential for the electromagnetic field, and $\mathbf{P}(\mathbf{x})$ is the quantised electric polarisation field. In the new representation generated by Λ_{sc} the time-dependent Schrödinger equation (9.187) is transformed to

$$i\hbar\frac{\partial\Psi_\Lambda}{\partial t} = H([\mathbf{g}],t)_{\text{sc}}\Psi_\Lambda, \tag{9.189}$$

where

$$\Psi_\Lambda = \Lambda_{\text{sc}}\Psi, \tag{9.190}$$

and

$$H([\mathbf{g}],t)_{sc} = \Lambda_{sc} H([0],t)_{sc} \Lambda_{sc}^{-1} + \frac{\partial}{\partial t} \int \mathbf{P}(\mathbf{x}) \cdot \mathbf{A}(\mathbf{x},t) d^3 \mathbf{x}. \qquad (9.191)$$

The time derivative of the Coulomb gauge vector potential is just the transverse electric field with a minus sign, so that the last term in (9.191) is

$$-\int \mathbf{P}(\mathbf{x}) \cdot \mathbf{E}(\mathbf{x},t)^\perp d^3 \mathbf{x}. \qquad (9.192)$$

Since the operator Λ_{sc} involves only the particle position operators, it commutes with the Coulomb interaction operator for the charges, and the unitary transform of $H([0],t)_{sc}$ involves only a modified kinetic energy term obtained from the replacement,

$$\Lambda_{sc} \mathbf{p}_n \Lambda_{sc}^{-1} \to \mathbf{p}_n + \boldsymbol{\nabla}_n \int \mathbf{P}(\mathbf{x}) \cdot \mathbf{A}(\mathbf{x},t) d^3 \mathbf{x}, \qquad (9.193)$$

in exact correspondence with the PZW calculation in §9.3.3.

If we introduce (9.150) into (9.193) and integrate out the delta function in the charge density operator, we can define a modified vector potential $\mathbf{a}$ exactly as in (9.104) and so write the transformed Hamiltonian $H([\mathbf{g}],t)_{sc}$ as

$$H([\mathbf{g}],t)_{sc} = = \sum_n \frac{1}{2m_n} \left(\mathbf{p}_n - e_n \mathbf{a}(\mathbf{x}_n,t) \right)^2 + \frac{1}{2} \sum_{n,m} \frac{e_n e_m}{4\pi\varepsilon_0 |\mathbf{x}_n - \mathbf{x}_m|}$$

$$- \int \mathbf{P}(\mathbf{x}) \cdot \mathbf{E}(\mathbf{x},t)^\perp d^3 \mathbf{x},$$

$$= H_{charges} + V([\mathbf{g}],t)_{sc}, \qquad (9.194)$$

where the semiclassical interaction potential in an arbitrary gauge,

$$V([\mathbf{g}],t)_{sc} = -\sum_n \frac{e_n}{2m_n} \left(\mathbf{p}_n \cdot \mathbf{a}(\mathbf{x}_n,t) + \mathbf{a}(\mathbf{x}_n,t) \cdot \mathbf{p}_n \right)$$

$$+ \sum_n \frac{e_n^2}{2m_n} \mathbf{a}(\mathbf{x}_n,t) \cdot \mathbf{a}(\mathbf{x}_n,t) - \int \mathbf{P}(\mathbf{x}) \cdot \mathbf{E}(\mathbf{x},t)^\perp d^3 \mathbf{x}, \qquad (9.195)$$

does not contain the term $\int (\mathbf{P}^\perp)^2 d^3 \mathbf{x}$ found in both the corresponding classical transformation (cf. Chapter 3), and the QED PZW transformation.

As discussed above, the classical vector potential $\mathbf{A}$ can be characterised by a coherent state parameter α using the relation

$$\langle \alpha | \mathbf{A}(\mathbf{x},t) | \alpha \rangle = \mathbf{A}(\mathbf{x},t,\alpha), \qquad (9.196)$$

and one can think of the generator of the semiclassical transformation, (9.188), as arising in this way. However this straightforward correspondence is only valid for non-linear functions of the annihilation and creation operators if they are normal-ordered.

Bringing Λ_{PZW} to normal order requires essentially the calculation leading to (11.20), and one has

$$\langle \alpha | \Lambda_{\text{PZW}}(t) | \alpha \rangle = \langle \Psi_0 | \Lambda_{\text{PZW}}(t) | \Psi_0 \rangle \Lambda(t)_{\text{sc}}. \tag{9.197}$$

For point particles the RHS vanishes unless Λ_{PZW} is the identity operator which leaves the Coulomb gauge formalism unchanged.

The perturbed state of the atom, $|\Psi_u\rangle$, required for the evaluation of (9.166) is obtained using the perturbation theory described in §6.4. Let us initially work in the Coulomb gauge and consider other gauges later on (see below); in this gauge the vector potential may be taken as a transverse plane wave with wave vector $\mathbf{k}_i$

$$\mathbf{A}(\mathbf{r},t) = -\frac{iE^{(0)}\hat{\boldsymbol{\varepsilon}}_\lambda}{2\omega} \left(e^{i(\mathbf{k}_i\cdot\mathbf{r}-\omega t)} - e^{-i(\mathbf{k}_i\cdot\mathbf{r}-\omega t)} \right), \tag{9.198}$$

where E^0 is an amplitude function. We may assume that the atom-radiation interaction is sharply concentrated around time $t = 0$ and that the coupling vanishes in the remote past $(t = -\infty)$ and in the remote future $(t = \infty)$. For a more realistic description of a laser pulse centred at $t = 0$ with characteristic time τ_0 for the pulse duration, the vector potential $\mathbf{A}(\mathbf{r},t)$ can be modulated by a pulse shape function $z(t,\tau_0)$ which is rapidly decreasing as $|t|$ increases, for example the Gaussian shape function $z = \exp(-t^2/\tau_0^2)$. The time-dependent perturbation is then, from (9.165),

$$V(t) = -\int \mathbf{j}(\mathbf{x},t) \cdot \mathbf{A}(\mathbf{x},t) \, \mathrm{d}^3\mathbf{x}. \tag{9.199}$$

We choose $\mathsf{H}_0 = \mathsf{H}_{\text{atom}}$ and for illustrative purposes consider the effect of the interaction $V(t)$ only to first-order in the charge e. The $|\mathbf{A}|^2$ term is $O(e^2)$ and so may be dropped from the interaction potential. The perturbation operator in this approximation is therefore

$$V(t) = \mathsf{v}e^{-i\omega t} + \mathsf{v}^+ e^{+i\omega t}, \tag{9.200}$$

with

$$\mathsf{v} = \frac{iE^{(0)}}{2\omega}\hat{\boldsymbol{\varepsilon}}_\lambda \cdot \sum_n \frac{e_n}{m_n}\mathsf{p}_n e^{i\mathbf{k}_i\cdot\mathbf{x}_n}. \tag{9.201}$$

A matrix element of $V(t)$ between two time-dependent unperturbed states is

$$V(t)_{uv} = e^{i\omega_{uv}t}\langle \phi_u^{(0)} | V(t) | \phi_v^{(0)} \rangle,$$

$$= \frac{iE^{(0)}}{2\omega}\hat{\boldsymbol{\varepsilon}}_\lambda \cdot \left(\mathbf{J}^0(\mathbf{k}_i)_{uv}e^{-i\omega t} + \mathbf{J}^0(\mathbf{k}_i)_{vu}^* e^{i\omega t} \right) e^{i\omega_{uv}t}, \tag{9.202}$$

where

$$\mathbf{J}^{(0)}(\mathbf{k}_i)_{uv} = \langle \phi_u^{(0)} | \sum_n \frac{e_n}{m_n}\mathsf{p}_n e^{i\mathbf{k}_i\cdot\mathbf{x}_n} | \phi_v^{(0)} \rangle. \tag{9.203}$$

Then the expectation value of an operator Λ in the first-order perturbed state is

$$\langle \Psi(t) | \Lambda | \Psi(t) \rangle = \Lambda_{vv}^0 - \frac{iE^{(0)}}{2\omega}\hat{\boldsymbol{\varepsilon}}_\lambda \cdot \sum_u \left(\frac{\Lambda_{vu}^0 \mathbf{J}^0(\mathbf{k}_i)_{uv}}{\hbar(\omega_{uv} - \omega)} + \frac{\Lambda_{uv}^0 \mathbf{J}^0(\mathbf{k}_i)_{vu}}{\hbar(\omega_{uv} + \omega)} \right) e^{-i\omega t}$$

$$- \frac{iE^{(0)}}{2\omega} \hat{\boldsymbol{\varepsilon}}_\lambda \cdot \sum_u \left(\frac{\Lambda^0_{vu} \mathbf{J}^0(\mathbf{k}_i)^*_{vu}}{\hbar(\omega_{uv} + \omega)} + \frac{\Lambda^0_{uv} \mathbf{J}^0(\mathbf{k}_i)^*_{uv}}{\hbar(\omega_{uv} - \omega)} \right) e^{i\omega t}. \qquad (9.204)$$

This formula is only valid for frequencies ω such that

$$E_{\min} - E_v > \hbar\omega, \, E_u - E_v \neq \hbar\omega, \qquad (9.205)$$

where $E_{\min}$ is the energy of the start of the continuum (essential spectrum) [51]; this requires that the particle centre of mass variables be removed from the Hamiltonian H_{atom}, so that $|\phi_v^0\rangle$ belongs to the discrete spectrum of internal states, and that the operators are similarly separated, as will be discussed in detail in Chapter 10.

The vector potential contribution to the gauge-invariant current density yields an harmonic contribution linear in $E^{(0)}$ in the first term in (9.204), but may be dropped from the sum-over-states terms in (9.204) which are already linear in $E^{(0)}$. Then after Fourier transformation Λ^0_{vv} yields a contribution to the scattering amplitude of

$$f(-\mathbf{k}_s)^{(1)}_\mu = \frac{iE^{(0)}}{4\pi\omega} \hat{\boldsymbol{\varepsilon}}_\lambda \cdot \hat{\boldsymbol{\varepsilon}}_\mu \langle \phi_v^{(0)} | \sum_n \frac{e_n^2}{2m_n} e^{i(\mathbf{k}_i - \mathbf{k}_s) \cdot \mathbf{x}_n} | \phi_v^{(0)} \rangle$$

$$= \frac{iE^{(0)}}{4\pi\omega} \hat{\boldsymbol{\varepsilon}}_\lambda \cdot \hat{\boldsymbol{\varepsilon}}_\mu \Upsilon(\mathbf{q})_{vv}, \qquad (9.206)$$

where we have anticipated the notation of Appendix F for $\Upsilon(\mathbf{q})_{vv}$ (cf. F.2.14), $\hbar\mathbf{q} = \hbar(\mathbf{k}_i - \mathbf{k}_s)$ is the momentum transfer vector, and the $\{\hat{\boldsymbol{\varepsilon}}_\mu\}$ are the polarisation vectors for the scattered radiation. With Λ chosen as the current operator, the sum term yields

$$f(-\mathbf{k}_s)^{(2)}_\mu = \frac{iE^{(0)}}{8\pi\omega} \sum_u \left(\frac{\mathbf{J}^0(-\mathbf{k}_s)_{vu} \cdot \hat{\boldsymbol{\varepsilon}}_\mu \mathbf{J}^0(\mathbf{k}_i)_{uv} \cdot \hat{\boldsymbol{\varepsilon}}_\lambda}{\hbar(\omega_{uv} - \omega)} \right.$$

$$\left. + \frac{\mathbf{J}^0(\mathbf{k}_i)_{vu} \cdot \hat{\boldsymbol{\varepsilon}}_\lambda \mathbf{J}^0(-\mathbf{k}_s)_{uv} \cdot \hat{\boldsymbol{\varepsilon}}_\mu}{\hbar(\omega_{uv} + \omega)} \right). \qquad (9.207)$$

Adding these two pieces together, we obtain the overall scattering amplitude in the form

$$f(-\mathbf{k}_s)_\mu = \frac{iE^{(0)}}{4\pi\omega} \Upsilon(-\mathbf{q})_{vv} \delta_{\alpha\beta} \hat{\varepsilon}_\mu^\alpha \hat{\varepsilon}_\lambda^\beta + \frac{iE^{(0)}}{8\pi\omega} \sum_u \left(\frac{J^0(-\mathbf{k}_s)^\alpha_{vu} J^0(\mathbf{k}_i)^\beta_{uv}}{\hbar(\omega_{uv} - \omega)} \right.$$

$$\left. + \frac{J^0(\mathbf{k}_i)^\beta_{vu} J^0(-\mathbf{k}_s)^\alpha_{uv}}{\hbar(\omega_{uv} + \omega)} \right) \hat{\varepsilon}_\mu^\alpha \hat{\varepsilon}_\lambda^\beta. \qquad (9.208)$$

In view of (9.203), (9.208) leads to the full non-relativistic form of the Kramers–Heisenberg (KH) dispersion formula written here in terms of the complete current operator $\mathbf{j}$ rather than just the leading term of its multipole expansion. We shall see later that this calculation reproduces the result obtained from quantum electrodynamics (Chapter 10); note that the semiclassical calculation is subject to the restriction (9.205).

It is now of interest to reconsider the time-dependent perturbation theory when $\mathbf{P}^\perp \neq 0$ is allowed; this may be done by using the unitary transformation generated by Eq. (9.188). There is nothing to be said about the calculation of the Kramers–Heisenberg formula since expectation values are invariant against unitary transformation. However, it is of interest to see how the development coefficients $\{b_n^1(t)\}$ behave under arbitrary gauge transformation. For this calculation it is convenient to write Eq. (9.191) in a slightly different way using the usual expansion of an exponential operator,

$$\exp(-i\chi/\hbar) = 1 - \frac{i}{\hbar}\chi + \frac{1}{2!}\left(\frac{i}{\hbar}\right)^2\chi^2 + \dots, \tag{9.209}$$

for the evaluation of the unitary transform of the Coulomb gauge Hamiltonian. If we define for ease of notation

$$\mathsf{F}([\mathbf{g}],t)_{sc} = \int \mathbf{P}(\mathbf{x})\cdot\mathbf{A}(\mathbf{x},t)\,d^3x, \quad \mathsf{F}([\mathbf{g}],\pm\infty)_{sc} = 0, \tag{9.210}$$

we can write the transformed Hamiltonian in the form

$$\mathsf{H}([\mathbf{g}],t)_{sc} = \mathsf{H}([0],t)_{sc} + \frac{i}{\hbar}[\mathsf{H}([0],t)_{sc},\mathsf{F}([\mathbf{g}],t)_{sc}] + \frac{\partial\mathsf{F}([\mathbf{g}],t)_{sc}}{\partial t} + \dots. \tag{9.211}$$

The development coefficients $\{\tilde{b}(t)_n\}$ at time t for the state $|\Psi(t)\rangle$ in an arbitrary gauge are, to first order,

$$\tilde{b}(t)_{nv} \approx -\frac{i}{\hbar}\int_{-\infty}^{t}\tilde{V}(t')_{nv}e^{i\omega_{nv}t'}\,dt', n \neq v, \tag{9.212}$$

$$\tilde{b}(t)_{vv} \approx 1 - \frac{i}{\hbar}\int_{-\infty}^{t}\tilde{V}(t')_{vv}\,dt', \tag{9.213}$$

$$\tilde{V}(t')_{nv} = \langle\phi_n^{(v)}|\mathsf{V}^{(1)}([\mathbf{g}],t')_{sc}|\phi_v^0\rangle, \tag{9.214}$$

and we take only the terms of $0(e)$ from (9.211). This matrix element may therefore be written as

$$\tilde{V}(t')_{nv} = \langle\phi_n^{(0)}|\mathsf{V}^{(1)}([0],t')_{sc}|\phi_v^{(0)}\rangle + i\omega_{nv}\langle\phi_n^{(0)}|\mathsf{F}([\mathbf{g}],t')_{sc}|\phi_v^{(0)}\rangle$$
$$+ \langle\phi_n^{(0)}|\frac{\partial\mathsf{F}([\mathbf{g}],t')_{sc}}{\partial t'}|\phi_v^{(0)}\rangle, \tag{9.215}$$

which is to be used in (9.212), (9.213).

An integration by parts shows that

$$\int_{-\infty}^{t}\left(\frac{\partial\mathsf{F}([\mathbf{g}],t')_{sc}}{\partial t'}\right)_{nv}e^{i\omega_{nv}t'}\,dt = \left(\mathsf{F}([\mathbf{g}],t)_{sc}\right)_{nv}e^{i\omega_{nv}t}$$
$$-i\omega_{nv}\int_{-\infty}^{t}\left(\mathsf{F}([\mathbf{g}],t')_{sc}\right)_{nv}e^{i\omega_{nv}t'}\,dt', \tag{9.216}$$

and so we may replace (9.212), (9.213) by a single equation valid for all n,

$$\tilde{b}(t)_{nv} = b(t)_{nv} - \frac{i}{\hbar}\left(\mathsf{F}([\mathbf{g}],t)_{sc}\right)_{nv}e^{i\omega_{nv}t}, \tag{9.217}$$

where $b(t)_{nv}$ is the development coefficient in the Coulomb gauge, and $\tilde{b}$ is the same quantity in an arbitrary gauge specified by some non-zero value for $\mathbf{g}^{\perp}$.

Since the vector potential for the laser pulse vanishes in the far future, the asymptotic development coefficients ($t = +\infty$) are strictly gauge invariant,

$$\tilde{b}(+\infty)_{nv} = b(+\infty)_{nv}, \quad \forall \mathbf{g}(\mathbf{x}, \mathbf{x}')^{\perp}. \tag{9.218}$$

For finite times t, however, the situation is entirely problematic. The probability $P(t)_n$ that the system will be found in the eigenstate $\phi_n^{(0)}$ at time t is, by the usual rules of quantum mechanics,

$$P(t)_n = |\tilde{b}(t)_{nv}|^2, \tag{9.219}$$

and is gauge dependent. The probability $P(t)_W$ that the initial wave packet formed at time $t = 0$ has survived at time t is, in an arbitrary gauge,

$$\begin{aligned} P(t)_W &= |\langle \tilde{\Psi}(0) | \tilde{\Psi}(t) \rangle|^2, \\ &= \sum_n \tilde{b}(0)_{nv}^* \tilde{b}(t)_{nv}, \end{aligned} \tag{9.220}$$

which is also gauge dependent. Consequently nothing physical can be said about the temporal evolution of an atomic wave packet 'prepared' by laser pulse excitation. In this sense the non-stationary state is not a physical state with its own dynamics. Physically the notion of a pulse duration is clear, but its implementation in the perturbation theory formalism is not well defined because, although a pulse shape function $z(t, \tau_0)$ can be taken to be strongly decreasing for $|t| >> \tau_0$, the gauge function $\mathbf{g}(\mathbf{x}, \mathbf{x}')$ can be arbitrarily large, so what value of t is large enough for the gauge-dependent term in (9.217) to be negligible is not defined. Obviously the magnitude of t that can be considered is limited by the timescales of the decay processes of the excited states $k \neq 0$, but that is a different matter requiring a better treatment than first-order perturbation theory; the conclusion about gauge dependence is not thereby nullified.

References

[1] Schrödinger, E. (1926), Ann. der Physik **81**, 109.

[2] Dirac, P. A. M. (1927), Proc. Roy. Soc. (London) **A114**, 243.

[3] Einstein, A. (1917), Physikalische Z. **18**, 121.

[4] Waller, I. (1930), Z. Physik **62**, 673.

[5] Oppenheimer, J. R. (1930), Phys. Rev. **35**, 461.

[6] James, R. W. (1965), *The Optical Principles of the Diffraction of X-Rays*, Cornell University Press.

[7] Barron, L. D. (1982), *Molecular Light Scattering and Optical Activity*, Cambridge University Press.

[8] Glauber, R. J. (1963), Phys. Rev. **131**, 2766.

[9] Loudon, R. (1983), *The Quantum Theory of Light*, 2nd ed., Clarendon Press.

[10] Fock, V. (1926), Z. Physik **38**, 242; ibid. **39**, 226.

[11] Weyl, H. (1929), Z. Physik **56**, 330.

[12] Dirac, P. A. M. (1931), Proc. Roy. Soc. (London) A**133**, 60.

[13] Bohm, D. and Aharonov, Y. (1959), Phys. Rev. **115**, 485.

[14] Power, E. A. and Zienau, S. (1959), Phil. Trans. Roy. Soc. (London) A**251**, 427.

[15] Atkins, P. W. and Woolley, R. G. (1970), Proc. Roy. Soc. (London) A**319**, 549.

[16] Woolley, R. G. (1971), Proc. Roy. Soc. (London) A**321**, 557.

[17] Babiker, M., Power, E. A. and Thirunamachandran, T. (1973), Proc. Roy. Soc. (London) A**332,** 187; ibid. A**338**, 235.

[18] Woolley, R. G. (1975), Adv. Chem. Phys. **33**, 153.

[19] Johnson, B. R., Hirschfelder, J. O. and Yang, K. H. (1983), Rev. Mod. Phys. **55**, 109.

[20] Andrews, D. L., Jones, G., Salam, A. and Woolley, R. G. (2018), J. Chem. Phys. **148**, 040901.

[21] Dirac, P. A. M. (1955), Can. J. Phys. **33**, 650.

[22] Mandelstam, S. (1962), Ann. Phys. (N.Y.) **19**, 1.

[23] Bialynicki-Birula, I. and Bialynicki-Birula, Z. (1975), *Quantum Electrodynamics*, Pergamon Press.

[24] Ryder, L. H. (1985), *Quantum Field Theory*, Ch. 3, Cambridge University Press.

[25] Feynman, R. P., Leighton, R. B. and Sands, M. (1964), *The Feynman Lectures on Physics*, vol. 2, 15-5, Addison-Wesley Publishing Co.

[26] Caprez, A., Barwick, B. and Batelaan, H. (2007), Phys. Rev. Letters **99**, 210401.

[27] Tonomura, A. and Non, F. (2008), Nature **452/20**, 298.

[28] Cabibbo, N. and Ferrari, E. (1962), Nuovo Cimento **23**, 1147.

[29] Schiff, L. I. (1967), Phys. Rev. **160**, 1257.

[30] O'Raifeartaigh, L. (1997), *The Dawning of Gauge Theory*, Princeton University Press.

[31] Wilson, K. G. (1974), Phys. Rev. D**10**, 2445.

[32] Dirac, P. A. M. (1934), Proc. Camb. Phil. Soc. **30**, 150.

[33] Peierls, R. E. (1934), Proc. Roy. Soc. (London) A**146**, 420.

[34] Fock, V. (1934), Phys. Zeits. Sowjetunion **12**, 404.

[35] Schwinger, J. (1951), Phys. Rev. **82**, 664.

[36] Valatin, J. G. (1954), Proc. Roy. Soc. (London) A**222**, 93.

[37] Chrètien, M. and Peierls, R. E. (1954), Proc. Roy. Soc. (London) A**223**, 468.

[38] Woolley, R. G. (1980), J. Phys. A: Math. Gen. **13**, 2795.

[39] Haagensen, P. E. and Johnson, K. (1997), arXiv:hep-th/9702204v1.

[40] Woolley, R. G. (1971), Mol. Phys. **22**, 1013.

[41] Power, E. A. (1964), *Introductory Quantum Electrodynamics*, Longmans.

[42] Goeppert-Mayer, M. (1931), Ann. der Physik **9**, 273.

[43] Fiutak, J. (1963), Can. J. Phys. **41**, 12.

[44] Woolley, R. G. (2000), Proc. Roy. Soc. (London) A**456**, 1803.

[45] Woolley, R. G. (2003), in *Handbook of Molecular Physics and Quantum Chemistry*, ed. S. Wilson, vol. 1, part 7, chapters 38–41, John Wiley and Sons Ltd.

[46] Woolley, R. G. (2020), Phys. Rev. Res. **2**, 013206 (11 pages).

[47] Weinberg, S. (1995), *The Quantum Theory of Fields*, vol. 1, *Foundations*, Cambridge University Press.

[48] Dirac, P. A. M. (1967), *The Principles of Quantum Mechanics*, 4th ed. revised, Oxford University Press.

[49] Woolley, R. G. (1975), Mol. Phys. **30**, 649.

[50] Roman, P. (1965), *Advanced Quantum Theory*, Addison-Wesley.

[51] Landau, L. D. and Lifshitz, E. M. (1977), *Quantum Mechanics, Non-relativistic Theory*, 3rd ed., Pergamon Press.

10 Quantum Electrodynamics and Perturbation Theory

10.1 Introduction

The previous chapter was devoted to an account of the general setting for non-relativistic quantum electrodynamics when the field variables are interpreted properly as quantum mechanical operators. The general QED Hamiltonian is obtained by canonical quantisation of the corresponding classical model of charged particles and electromagnetic fields; it is necessarily a quantum theory of spin zero charges. Its application to atomic and molecular phenomena rests largely on the introduction of the permutation symmetry required for identical particles in a quantum theory together with the spin-dependent Pauli interaction (9.158). An analysis of the properties of such a quantum mechanical model by direct calculation for the full Hamiltonian is a formidable task which properly lies in the purview of mathematical physics; Chapter 11 will give an outline of what is involved. Practical calculations in chemical physics and allied fields are based on a perturbation theory approach which makes the fundamental assumption that one can define a reference ('unperturbed') Hamiltonian, H_0, in the *same* Hilbert space as the full Hamiltonian H. Such an assumption is hard to justify in any straightforward way,[1] so the perturbation theory described in this chapter has a formal character. Nevertheless, it has proved immensely fruitful across many areas and perhaps should be thought of as more of an 'engineering solution' than a rigorous theory.

Non-relativistic quantum electrodynamics is applicable to light scattering phenomena which involve real photons and to intermolecular processes such as energy transfer and van der Waals interactions mediated by virtual photons, and to processes that involve both. In this chapter we concentrate mainly on spectroscopic processes. It should be recalled, however, that energy shifts are determined by diagonal matrix elements of the resolvent operator, (§6.3). They have essentially the same perturbation expansion as the scattering theory transition operator provided

---

[1] The immediate problem is that H has absolutely continuous spectrum above a simple eigenvalue, the vacuum state, so every state of the atomic/molecular system becomes a 'resonance'. Moreover, extensive use is made of 'complete sets of states' supposed to describe the atomic/molecular bound and continuum states. The mathematical properties of such 'continuum states', let alone even their existence, is unknown, so estimates of matrix elements and other quantities are impossible. This difficulty is independent of the description (classical or quantum) of the electromagnetic field.

certain restrictions on the allowed unperturbed states are respected, so similar methods of calculation may be expected.

In a spectroscopic experiment, a probe beam of particles ($\mathcal{P}$) (equivalently, an incident wave) interacts with a target ($\mathcal{T}$), and the scattered particles (waves) are detected in a region remote from the target; we consider both the particle beam (incident wave) and the target to be parts of a closed quantised system and that the conservation laws are fully applicable. The key idea of the theoretical description is that the interaction is localised in space and time, so that originally non-interacting parts come together to interact and then separate again, after which detection occurs, so that the non-interacting system of probe beam and target can be used as a reference system to describe the outcome of the experiment. The difference between an absorption experiment and a scattering experiment is then just that, in the former, the detector monitors particles moving in the same direction as the incident particle beam, whereas in scattering, particles that have suffered a change in direction are monitored, that is, the distinction is whether or not the particle beam and the target have exchanged momentum, which overall is conserved.

More precisely, if the direction of the detected radiation is different from that of the incident beam, the process is called 'non-forward' scattering; if there is no change in direction and the experiment is more than a simple absorption measurement, the process is 'forward' scattering. Forward scattering is responsible for optical birefringence phenomena (changes in the polarisation state of the beam such as optical rotation, the Kerr effect and so on), and involve the interference of the forward scattered component with the incident beam. Because the scattering into the forward direction by different molecules is a coherent process, the observed birefringence depends on the molecular density. By contrast, non-forward scattering (for example, Rayleigh scattering, the Raman effect) is an incoherent process, so that the total detected intensity is the sum of the intensities of individual scattered waves (particles); then the polarisation is independent of the density [1].

The quantum mechanical description will be based on the scattering theory outlined in Chapter 6; in order to implement this formalism for the interaction of electromagnetic radiation with atoms and molecules, we require that the full Hamiltonian H be divided into a reference part, denoted by H_0, and a perturbation part, V,

$$H = H_0 + V, \quad H\Psi_n = E_n\Psi_n. \tag{10.1}$$

It is usually the case that both H_0 and the full Hamiltonian H commute with the operators that generate time translations and space translations, and as will be discussed more fully in what follows, this leads to transition amplitudes that contain delta functions expressing the law of conservation of energy and the law of conservation of momentum. Our main interest will be the calculation of cross sections for processes in which an initial state of the probe beam with momentum $\hbar\mathbf{k}$ and polarisation λ is scattered to a final state with momentum $\hbar\mathbf{k}'$ and polarisation λ', with overall conservation of energy and momentum. There are other important symmetry operations which have analogous commutation properties, for example space rotations, space and time inversions, that can be exploited to characterise the transition amplitudes in terms of appropriate symmetry labels.

An operator that does not fit this pattern is the generator of generalised PZW transformations, and so the proof of the gauge invariance of the scattering theory for QED is more difficult than for the kinematical symmetries. The important point for now is that since the perturbation operators depend on the arbitrary Green's function $\mathbf{g}$ (cf. §9.3.3), so too will the transition operator. Let $\mathsf{T}[0]$ be the Coulomb gauge transition operator, and $\mathsf{T}[\mathbf{g}]$ be the same quantity for some non-zero $\mathbf{g}^\perp$, for example the PZW choice,[2] and let $\mathsf{W}[\mathbf{g}]$ be their difference:

$$\mathsf{T}[\mathbf{g}] = \mathsf{T}[0] + \mathsf{W}[\mathbf{g}]. \tag{10.2}$$

Since $\mathsf{W}[\mathbf{g}]$ can be expressed[3] as a commutator involving the reference Hamiltonian H_0, the matrix element $W[\mathbf{g}]_{kn}$ for initial and final states $|n\rangle$ and $|k\rangle$ with the *same energy* is identically zero, and cross sections calculated for such pairs of stationary states of equal energy are gauge invariant; thus, the choice of $\mathbf{g}$ is reduced to one of convenience in this case. A gauge-invariant description of transient processes over finite time intervals is much more problematic.

The Schrödinger equation for the probe beam and the target provides a basis of eigenstates for the reference system,

$$\mathsf{H}_{\mathcal{P}}|\varphi\rangle = E_\varphi|\varphi\rangle, \quad \mathsf{H}_{\mathcal{T}}|\psi_s\rangle = E_s|\psi_s\rangle, \quad \mathsf{H}_0 = \mathsf{H}_{\mathcal{P}} + \mathsf{H}_{\mathcal{T}}, \tag{10.3}$$

so that a typical basis element of the Hilbert space of the reference system is

$$|\Phi_n^0\rangle = |\varphi\rangle|\psi_s\rangle, \tag{10.4}$$

with energy

$$E_n^0 = E_\varphi + E_s. \tag{10.5}$$

We assume that the states Φ_n^0 and Ψ_n (into which Φ_n^0 evolves due to the interaction) have the same energy, $E_n^0 = E_n$. A free one-particle state with momentum $\mathbf{P}$ has a position representation wave function,

$$\varphi_{\mathbf{P}}(\mathbf{x}) = \langle \mathbf{x}|\varphi_{\mathbf{P}}\rangle = \frac{1}{(2\pi\hbar)^{3/2}} e^{i\mathbf{P}\cdot\mathbf{x}/\hbar}, \tag{10.6}$$

such that (cf. §5.2.1)

$$\langle \varphi_{\mathbf{P}'}|\varphi_{\mathbf{P}}\rangle = \delta^3(\mathbf{P}' - \mathbf{P}). \tag{10.7}$$

In an idealised experiment, we suppose that the detector picks out a particular state $|\Phi_k^0\rangle$ of the non-interacting system $(\mathcal{P} + \mathcal{T})$. Quantum mechanical scattering theory provides formal expressions for the transition rate to this state. The S-matrix element

---

[2] Here the [0] refers to the condition $\mathbf{g}^\perp = \mathbf{0}$ which is equivalent to the Coulomb gauge condition, and the arbitrary $\mathbf{g}$ has an arbitrary (non-zero) transverse part. Hereafter, we drop the $\perp$ symbol to simplify the notation.

[3] This is shown in §10.2.2.

S_{kn} is the probability amplitude for a transition from an initial state Φ_n^0 to a final state Φ_k^0; it can be written as (cf. §6.5)

$$S_{kn} = \delta_{kn} - 2\pi i \delta(E_k^0 - E_n^0)T_{kn}, \qquad (10.8)$$

and this leads [2]–[4] to the usual 'golden rule' type of formula for the transition rate to the state Φ_k^0,

$$\tau(k)_{n \to k}^{-1} = \frac{2\pi}{\hbar}|T_{kn}|^2\,\delta(E_k^0 - E_n^0). \qquad (10.9)$$

Note that the rate $\tau(k)^{-1}$ depends on matrix elements of the full transition operator T, rather than simply V. The first-order perturbation theory approximation to the T-matrix element based on its replacement by V is usually known as 'the Born approximation'.

For an isolated system, that is, translation invariant, the total momentum is a conserved quantity; accordingly, the transition matrix element T_{kn} in such a case has the structure

$$-2\pi i T_{kn} = \delta^3(\mathbf{P}_k^0 - \mathbf{P}_n^0)M_{kn}, \qquad (10.10)$$

where the matrix element M_{kn} has a smoother dependence on the momentum variables,[4] and $\mathbf{P}_n^0, \mathbf{P}_k^0$ are the initial and final total momenta, respectively. If the scattering is from a fixed centre of force, the momentum Dirac delta function does not appear. Since the probability for the transition $n \to k$ is proportional to $|S_{kn}|^2$, there is evidently the technical problem of making sense of both the square of the energy conservation delta function and that for momentum conservation. The solution is to imagine that the system is confined to a box with volume Ω, and that the interaction is turned on for only a finite time T; at the end of the calculations limits to infinity are taken [5].

The Dirac delta functions are replaced with the following well-behaved functions:[5]

$$\delta_\Omega^3(\mathbf{P} - \mathbf{P}') = \frac{1}{(2\pi\hbar)^3}\int_\Omega e^{i(\mathbf{P}-\mathbf{P}')\cdot\mathbf{x}/\hbar}\,d^3\mathbf{x}, \qquad (10.11)$$

$$\delta_T(E - E') = \frac{1}{2\pi\hbar}\int_T e^{i(E-E')t/\hbar}\,dt. \qquad (10.12)$$

When the limits $\Omega, T \to \infty$ are taken, the energy and momentum become continuous variables, and sums over states are replaced by integrals. Thus, we use

$$\left(\delta_\Omega^3(\mathbf{P}_k^0 - \mathbf{P}_n^0)\right)^2 = \frac{\Omega}{(2\pi\hbar)^3}\,\delta_\Omega^3(\mathbf{P}_k^0 - \mathbf{P}_n^0) \qquad (10.13)$$

and

$$\left(\delta_\Omega(E_k^0 - E_n^0)\right)^2 = \frac{T}{(2\pi\hbar)}\,\delta_\Omega(E_k^0 - E_n^0). \qquad (10.14)$$

---

[4] In the sense that it is a function rather than a distribution.
[5] They are actually the familiar discrete Kronecker delta function.

At the same time, free-particle wave functions are given 'box normalisation', for example

$$\langle \mathbf{x} | \mathbf{P} \rangle = \frac{1}{\sqrt{\Omega}} e^{i\mathbf{P} \cdot \mathbf{x}/\hbar}, \tag{10.15}$$

in place of Dirac's continuum normalisation. This change implies a renormalisation of the S-matrix element S_{kn} in Eq. (10.8) that depends on the number of particles involved; if N_n and N_k are the numbers of particles in the initial and final states, respectively, the modified S-matrix is

$$S_{kn}^{\text{box}} = \left(\frac{(2\pi\hbar)^3}{\Omega} \right)^{(N_n + N_k)/2} S_{kn}. \tag{10.16}$$

The transition rate is the probability of the transition occurring during the time T that the interaction acts, and this is

$$\tau_{n \to k}^{-1} = \frac{1}{2\pi\hbar} \left(\frac{(2\pi\hbar)^3}{\Omega} \right)^{(N_n + N_k - 1)} \delta_T (E_n^0 - E_k^0) \delta_\Omega^3 (\mathbf{P}_n^0 - \mathbf{P}_k^0) |M_{kn}|^2. \tag{10.17}$$

This is the rate for a transition to a particular final state Φ_k^0; in a large box, the final states are closely spaced with a number density per particle of

$$\rho_k = \frac{\Omega}{(2\pi\hbar)^3} \, \mathrm{d}^3 \mathbf{P}_k \tag{10.18}$$

in the infinitesimal volume element $\mathrm{d}^3 \mathbf{P}_k$. The measurable transition rate from the state Φ_n^0 is then

$$\tau_n^{-1} = \left(\frac{\Omega}{(2\pi\hbar)^3} \right)^{N_k} \tau_{n \to k}^{-1} \mathrm{d}\xi, \tag{10.19}$$

where $\mathrm{d}\xi$ is the product

$$\mathrm{d}\xi = \mathrm{d}^3 \mathbf{P}_k^1 \dots \mathrm{d}^3 \mathbf{P}_k^{N_k}. \tag{10.20}$$

Finally, combining (10.17) with (10.19) leads to the transition rate

$$\tau_n^{-1} = \frac{1}{2\pi\hbar} \left(\frac{(2\pi\hbar)^3}{\Omega} \right)^{N_n - 1} |M_{kn}|^2 \delta(E_n^0 - E_k^0) \delta^3(\mathbf{P}_n^0 - \mathbf{P}_k^0) \mathrm{d}\xi. \tag{10.21}$$

As an example consider the collision of two non-relativistic particles with masses m_1, m_2, and the flux is the relative velocity u divided by the volume Ω; if we work in the centre-of-mass reference frame, the particles have initial momenta $\mathbf{P}_n^1 = \mathbf{p}$, $\mathbf{P}_n^2 = -\mathbf{p}$ so that $\mathbf{P}_n^0 = \mathbf{0}$, and the relative velocity is $|\mathbf{p}|/\mu$ where μ is the usual reduced mass of the pair. Thus, we get the factor

$$\delta^3(\mathbf{P}_k^1 + \mathbf{P}_k^2) \delta(E_k^1 + E_k^2 - E_n^0) \, \mathrm{d}^3 \mathbf{P}_k^1 \mathrm{d}^3 \, \mathbf{P}_k^2 \to \mu P_k^1 \, \mathrm{d}\Omega(\mathbf{P}_k^1), \tag{10.22}$$

leading to a transition rate [5],

$$\tau_n^{-1} = \frac{(2\pi\hbar)^2}{\Omega} |M_{kn}|^2 \mu P_k^1 \, \mathrm{d}\Omega(\mathbf{P}_k^1). \tag{10.23}$$

The differential cross section for the scattering is then

$$d\sigma = \frac{\text{transition rate}}{\text{incident flux}}$$

$$= (2\pi\hbar)^2 |M_{kn}|^2 \mu^2 \frac{P_k^1}{P_n^1} d\Omega(\mathbf{P}_k^1). \tag{10.24}$$

As mentioned earlier, in case one of the pair is fixed, the momentum conservation delta function is dropped from the S-matrix and the reduced mass becomes the invariant mass of the moving particle. For photons in a mode $(\mathbf{k}, \lambda)$ scattered into the mode $(\mathbf{k}', \lambda')$, the flux[6] and density of states are

$$J_{\mathbf{k}} = \frac{n_{\mathbf{k}} c}{\Omega}; \quad \rho_{\mathbf{k}'} = \frac{\Omega}{(2\pi)^3} \left(\frac{k'^2}{\hbar c} \right) d\Omega_{\mathbf{k}'}. \tag{10.25}$$

For scattering of any kind of particle (electron, photon,…) from a free composite system (atom, molecule, nucleus…) the only role for the centre-of-mass coordinate of the composite system is its involvement in the expression of the law of conservation of linear momentum, while the details of the scattering cross section are determined purely by the internal variables. For this to be achieved, it is required to separate the overall motion of the composite system from the internal motions of the particles. The separation of the QED Hamiltonian into parts required for the conventional perturbation theory approach was discussed in §9.3.2; it is predicated on the explicit appearance of the static Coulomb interaction between pairs of charges in the unperturbed Hamiltonian, H_0.

The classical Coulombic potential energy of the particles is purely a function of interparticle distance vectors and is independent of the overall motion. The problem therefore comes down to separating the total kinetic energy of all the particles into kinetic energy of free translation, plus kinetic energy of internal motion, with no cross-coupling; and as described in Chapter 8, this can be done quite straightforwardly. The transformation of variables into centre-of-mass and internal coordinates and momenta is also applied to the operators involved in the perturbation part of the Hamiltonian. This, however, does not lead to a decomposition of the interaction operators analogous to that achieved for the reference Hamiltonian. Instead, the conservation of linear momentum is manifested as a momentum-conserving delta function in the matrix elements of the transition operator.

The atomic/molecular Hamiltonian is (9.102) while that for a charged particle beam at non-relativistic energies is[7]

$$H_e = \frac{p^2}{2m}. \tag{10.26}$$

---

[6] The irradiance of the light beam is then $I = J_{\mathbf{k}} \mathcal{E}_{\mathbf{k}}$ where $\mathcal{E}_{\mathbf{k}} = \hbar k c$ is the energy of a photon in the mode $(\mathbf{k}, \lambda)$.

[7] This is an approximation to simplify the discussion; it implies the neglect of the mutual Coulombic repulsions between the particles in the beam which are controlled experimentally by focusing external field configurations, so that a wave-packet description would be more realistic.

The Hamiltonian for the photons is Eq. (7.37); the photon vacuum $|0\rangle$ is annihilated by H_{rad} if the zero-point energy constant is omitted. In order to make the scattering kinematics apparent, we separate out explicitly the molecular centre-of-mass motion from the internal dynamics using (8.49) and (8.52) to write the Hamiltonian for the non-interacting system as

$$H_0 = H_e + H_{cm} + H_{int} + H_{rad}, \tag{10.27}$$

where H_{int} is the internal molecular Hamiltonian, and H_{cm} describes the motion of the molecular centre of mass,

$$H_e|\mathbf{p}\rangle = E_{\mathbf{p}}|\mathbf{p}\rangle, \quad E_{\mathbf{p}} = \frac{p^2}{2m}, \tag{10.28}$$

$$H_{cm}|\mathbf{P}\rangle = E_{\mathbf{P}}|\mathbf{P}\rangle, \quad E_{\mathbf{P}} = \frac{P^2}{2M}, \tag{10.29}$$

$$H_{int}|\phi_s\rangle = E_s|\phi_s\rangle. \tag{10.30}$$

The stationary state wave functions of the operators that make up H_0 are used as the set of reference states for the free system. In a coordinate representation we take

$$\langle \mathbf{x}|\mathbf{p}\rangle \sim e^{i\mathbf{p}\cdot\mathbf{x}/\hbar}, \quad \langle \mathbf{R}|\mathbf{P}\rangle \sim e^{i\mathbf{P}\cdot\mathbf{R}/\hbar}, \tag{10.31}$$

in accordance with (10.6), and specify orthonormality for the internal states

$$\langle \phi_s|\phi_{s'}\rangle = \delta_{s,s'}. \tag{10.32}$$

The eigenstates of H_{rad} are the Fock space states described in §7.2.1. A Fock space state describing n photons in the mode with wave vector $\mathbf{k}$ and polarisation λ will be denoted as usual by $|\lambda[n_{\mathbf{k}}]\rangle$; their orthonormality is expressed by

$$\langle \lambda'[m_{\mathbf{k}'}]|\lambda[n_{\mathbf{k}}]\rangle = \delta_{\lambda',\lambda}\,\delta_{\mathbf{k}',\mathbf{k}}\,\delta_{m,n}. \tag{10.33}$$

Practical calculations are based on a perturbation expansion of the formally exact equation [3]

$$T_{kn} = V_{kn} + \sum_m \frac{T_{km}T_{nm}^*}{E_n^0 - E_m^0 + i\varepsilon}. \tag{10.34}$$

In non-relativistic QED, the Hamiltonian contains terms linear in the charge e and terms proportional to e^2; in general, both types of term contribute to the perturbation operator, so it is convenient to denote explicitly this dependence with a superscript and set

$$T_{kn}^{(1)} = V_{kn}^{(1)},$$

$$T_{kn}^{(2)} = V_{kn}^{(2)} + \sum_m \frac{V_{km}^{(1)}V_{mn}^{(1)}}{E_n^0 - E_m^0 + i\varepsilon}. \tag{10.35}$$

Then a T-matrix element, T_{kn}^p, is properly proportional to e^p. With this notation, the perturbation expansion can be given a concise representation as a recurrence relation,

$$T_{kn}^{(p)} = \sum_m \frac{V_{km}^{(1)} T_{mn}^{(p-1)}}{E_n^0 - E_m^0 + i\varepsilon} + \sum_m \frac{V_{km}^{(2)} T_{mn}^{(p-2)}}{E_n^0 - E_m^0 + i\varepsilon}, \quad p \geq 3, \tag{10.36}$$

since $V_{kn} = V_{nk}^*$. The summation is over the full set of eigenstates of the reference Hamiltonian, H_0.

10.2 Gauge-invariant Calculation in Electrodynamics

10.2.1 Gauge Invariance of the S-Matrix

The issue of gauge-invariant calculation in quantum electrodynamics can be formulated succinctly using the formalism presented in §6.5; the interaction representation operator $V(t)$ obtained from (6.85) is a functional of the arbitrary Green's function $\mathbf{g}$ because V, (9.154), is, and thus the time development operator is too:

$$U(t, t_0) = U(t, t_0, [\mathbf{g}]). \tag{10.37}$$

Consider now two arbitrary gauges specified by $\mathbf{g}^1$ and $\mathbf{g}^2$; then in general

$$U(t, t_0, [\mathbf{g}^1]) \neq U(t, t_0, [\mathbf{g}^2]) \quad \mathbf{g}^1 \neq \mathbf{g}^2, \tag{10.38}$$

since t, t_0 are arbitrary. Physical observables are obtained from squared matrix elements of U, so a transition $\Phi_n \to \Phi_k$ is a physical process if and only if

$$|\langle \Phi_k | U(t, t_0, [\mathbf{g}^1]) | \Phi_n \rangle|^2 = |\langle \Phi_k | U(t, t_0, [\mathbf{g}^2]) | \Phi_n \rangle|^2, \tag{10.39}$$

whatever $\mathbf{g}^1$ and $\mathbf{g}^2$ may be.

By way of an elementary example, consider the first-order perturbation solution for the time development operator; only $V^1[\mathbf{g}]$, (9.155), can contribute to this order, and so (6.88) reduces to just the first-order approximation for $U = U^{(1)}$,

$$U(t, t_0; [\mathbf{g}])^{(1)} = 1 - \frac{i}{\hbar} \int_{t_0}^t V(t', [\mathbf{g}]) \, dt',$$

$$= 1 - \frac{i}{\hbar} \int_{t_0}^t V(t', [0]) \, dt' - \left(\frac{i}{\hbar}\right)^2 \int_{t_0}^t \left[H_0, F(t') \right] dt',$$

$$= U(t, t_0; [0])^{(1)} - \left(\frac{i}{\hbar}\right)^2 \int_{t_0}^t \left[H_0, F(t') \right] dt'. \tag{10.40}$$

In the energy representation defined by the reference Hamiltonian

$$H_0 | \Phi_n \rangle = E_n | \Phi_n \rangle, \tag{10.41}$$

this becomes the matrix equation,

$$\langle \Phi_k | \mathsf{U}(t, t_0; [\mathbf{g}])^{(1)} | \Phi_n \rangle = \langle \Phi_k | \mathsf{U}(t, t_0; [0])^{(1)} | \Phi_n \rangle$$
$$- \left(\frac{i}{\hbar} \right)^2 (E_k - E_n)(\mathsf{F})_{kn} \int_{t_0}^{t} e^{i(E_k - E_n)t'/\hbar} \, \mathrm{d}t'. \tag{10.42}$$

This result implies definite restrictions on the kinds of questions that can be asked about the time evolution of an atom in the presence of electromagnetic radiation. For example, suppose the state Φ_n describes an atom initially (t_0) in its ground-state $|\psi_0\rangle$, with the radiation field in a specified state $|i\rangle$; the probability that the atom is in a state $|\psi_p\rangle$ while the field is in a state $|j\rangle$ at a later time t is determined by

$$|\langle \psi_p, j | \mathsf{U}(t, t_0; [\mathbf{g}])^{(1)} | i, \psi_0 \rangle|^2. \tag{10.43}$$

This will only be gauge invariant if either $E_k = E_n$ or $\langle \Phi_k | \mathsf{F} | \Phi_n \rangle = 0$ since the integral never vanishes for $t \neq t_0$. Any state in the Hilbert space is a possible final state, however, and the difficulty for time-dependent perturbation theory is that it does not generally restrict the final states to those that give gauge-invariant amplitudes in (10.42). The customary appeal in time-dependent perturbation theory to the time-energy uncertainty relation as the guarantee of approximate energy conservation is not sufficient to eliminate the gauge-dependent contribution.

Although a question about the probability (10.43) may seem very natural, it evidently may have no physically meaningful answer. Crucially there is an exceptional case. Gauge invariance may be ensured, irrespective of the matrix elements of F, through the matrix $(\mathsf{F})_{kn}$ having a zero coefficient; this occurs in the asymptotic limit $t_0 \to -\infty, t \to +\infty$ because

$$\lim_{\substack{t \to +\infty \\ t_0 \to -\infty}} E_{kn} \int_{t_0}^{t} e^{iE_{kn}\tau/\hbar} \, \mathrm{d}\tau \propto (E_k - E_n)\delta(E_k - E_n) = 0. \tag{10.44}$$

The operator $\mathsf{U}(+\infty, -\infty)$ is precisely the S-matrix we met in §6.5. It is of special importance for the calculation of physical quantities. The first-order perturbation calculation just discussed will be substantially generalised in what follows; the main conclusion is the same, however. Probability amplitudes for transitions induced by electromagnetic radiation are gauge invariant provided that the initial and final states are stable, that is, stationary. A time-independent framework will therefore be used to investigate the gauge invariance of the S-matrix; time-dependent approaches will be discussed later in §10.2.5.

The gauge invariance of the S-matrix for QED that we shall demonstrate in this section is the expected result. Since $\mathsf{H}[\mathbf{g}]$ is the unitary transform of $\mathsf{H}[0]$ by Λ, (9.108), the S-matrix defined by (6.67) is obviously invariant. Moreover, the gauge invariance of the S-matrix in the full Lorentz invariant QED is a well-established and fundamental result [6]. However, we do not generally use the Lorentz invariant theory to calculate S-matrix amplitudes for processes involving atoms and molecules, and then take the non-relativistic limit; instead, we adopt the Hamiltonian (9.103) and start all over again with a non-covariant perturbation theory based on (10.36). In any case, the non-relativistic limit $c \to \infty$ is singular and the preservation of gauge invariance needs to be checked, so the study of the non-relativistic theory is of interest in its own right. We shall therefore try to use the same methods as the ones actually used in practice for light scattering calculations. The quantisation of both particles and radiation is essential for the logical consistency of the theory since the S-matrix, Eq. (6.67), is a quantum mechanical probability amplitude for the combined system of charged particles and the electromagnetic field described by the Schrödinger equation for H, Eq. (9.91).

The literature of specific perturbation theory calculations is very extensive; it includes absorption/emission of radiation, spectral lineshapes, the Kramers–Heisenberg formula for light scattering, nonlinear optical processes, intermolecular forces and atomic self-energies [7]–[16]. The majority of such calculations use only the leading terms of a multipole expansion of the interaction, usually just the electric dipole contribution (the long-wavelength approximation), and are based on either the Coulomb gauge interaction, (9.107), or the multipolar Hamiltonian that arises from the PZW transformation with Λ constructed from (9.121). It is obviously of interest to establish the circumstances in which these two formulations yield the same observables, and also when different answers might be expected. A familiar example of the first case is afforded by the calculation of light scattering cross sections. For example, the identity of the generalised Kramers–Heisenberg dispersion formula obtained from either of these gauge choices can be demonstrated by a direct transformation of the matrix elements of the two different forms of V without any multipole approximation [17]; this result can also be obtained by a more formal method involving the PZW transformation theory [18]. Such calculations have been limited to low-order perturbation theory and just two specific gauges. On the other hand, the two gauges give different results in, for example, line shape calculations based on time-dependent perturbation theory [19] – hardly surprising in view of the preceding discussion.

From the theoretical point of view it is very desirable to investigate the gauge invariance of the perturbation theory in a general way that goes beyond the low-order theory that is sufficient for most experiments, and is not tied to any particular gauge. One reason is to exclude possible gauge-dependent contributions in higher orders of perturbation theory; these cannot be ruled out just on the grounds of the smallness of the coupling constant (the dimensionless fine structure constant $\alpha \approx 1/137$) since terms involving **g** in perturbation theory have a polynomial dependence on it and so can be arbitrarily large (**g** is unbounded).

The idea of the following argument is to show that the difference between the T-matrices in two different gauges can be written as a commutator involving the reference Hamiltonian H_0; the (kn)-matrix elements of the difference term are then always proportional to $(E_k - E_n)$ and so will be annihilated by the energy conservation Dirac delta function (cf. (10.44)) provided the matrix element containing **g** does not have a pole at $E = E_n$. A proof by induction shows that this is true in every order of perturbation theory. To begin with, we will assume the validity of the perturbation expansion without further ado; this is certainly highly questionable when resonance and damping effects occur, and we shall have to consider the modifications required for such a case.

10.2.2 Non-resonant Scattering and Perturbation Theory

The interaction Hamiltonian (9.154) has the structure

$$V = \lambda V^{(1)} + \lambda^2 V^{(2)}, \tag{10.45}$$

with λ standing for the coupling constant. The idea in [20] is to write the perturbation series (10.36) for the transition operator, T, in an algebraic form as a power series in λ,

$$T = \sum_{p=1} \lambda^p T^{(p)}. \tag{10.46}$$

We then have a three-term recurrence relation for the components $\{T^{(p)}\}$,

$$T^{(p)} = V^{(1)} G_0 T^{(p-1)} + V^{(2)} G_0 T^{(p-2)}, \qquad p \geq 3, \tag{10.47}$$

with the starting values

$$T^{(1)} = V^{(1)}, \tag{10.48}$$

$$T^{(2)} = V^{(2)} + V^{(1)} G_0 T^{(1)}. \tag{10.49}$$

Suppose we perform such a calculation for two different transverse Green's functions, say $\mathbf{g}^\perp = \mathbf{0}$ (the Coulomb gauge), and some other (arbitrary) non-zero $\mathbf{g}^\perp$. The T-matrix elements $T[\mathbf{g}]_{kn}^{(p)}$ and $T[0]_{kn}^{(p)}$ obtained from (10.47) will be different, since the perturbation operators $V^{(1)}$ and $V^{(2)}$ are functionals of $\mathbf{g}$, so we need to investigate their difference.

Let

$$W[\mathbf{g}] = T[\mathbf{g}] - T[0], \tag{10.50}$$

where

$$T[\mathbf{g}] = V[\mathbf{g}] + V[\mathbf{g}] G_0 T[\mathbf{g}], \tag{10.51}$$

$$T[0] = V[0] + V[0] G_0 T[0], \tag{10.52}$$

and define

$$\Delta[\,\mathbf{g}] = V[\,\mathbf{g}] - V[0]. \tag{10.53}$$

Then using (10.51)–(10.53), we obtain an implicit equation for $W[\mathbf{g}]$,

$$W[\mathbf{g}] = \Delta[\mathbf{g}] \left(1 + G_0 T[0] + G_0 W[\mathbf{g}] \right) + V[0] G_0 W[\mathbf{g}]. \tag{10.54}$$

As with the transition operator, we can try to solve this equation using perturbation theory; we write

$$W[\mathbf{g}] = \sum_{p=1} \lambda^p W^{(p)}[\mathbf{g}], \tag{10.55}$$

$$\Delta[\mathbf{g}] = \lambda \Delta^{(1)}[\mathbf{g}] + \lambda^2 \Delta^{(2)}[\mathbf{g}], \tag{10.56}$$

and using (10.45), (10.46), (10.55), (10.56) we obtain an inhomogeneous three-term recurrence relation for the coefficients $\{W^{(p)}[\mathbf{g}]\}$. In order to simplify the notation from now on we suppress the explicit dependence on $\mathbf{g}$; note that all the dependence on $\mathbf{g}$ has been put into $\{W^{(p)}, \Delta^{(1)}$ and $\Delta^{(2)}\}$. Using this notation the recurrence relation may be written as

$$W^{(p)} = \left(\Delta^{(1)} + \Delta^{(2)}\right) G_0 T^{(p-1)} + \left(\Delta^{(1)} + V^{(1)}\right) G_0 W^{(p-1)}$$
$$+ \left(\Delta^{(2)} + V^{(2)}\right) G_0 W^{(p-2)}, \qquad (10.57)$$

with the starting values

$$W^{(1)} = \Delta^{(1)}, \qquad (10.58)$$

$$W^{(2)} = \Delta^{(2)} + \Delta^{(1)} G_0 T^{(1)} + \left(\Delta^{(1)} + V^{(1)}\right) G_0 W^{(1)}. \qquad (10.59)$$

The terms in (10.57) can be systematically regrouped so that the coefficients $W^{(p)}$ can be written with all the gauge dependence residing in the required form of a commutator,

$$W[\mathbf{g}]^{(p)} = \left[H_0, \Omega[\mathbf{g}]^{(p)}\right], \qquad (10.60)$$

where

$$\Omega[\mathbf{g}]^{(p)} = \sum_{m=1}^{p} c_m F^m G_0 T^{(p-m)}, \qquad (10.61)$$

with $F = F[\mathbf{g}]$ given by (9.109),

$$c_m = \frac{(-1)^{(m+1)}}{m!} \left(\frac{i}{\hbar}\right)^m, \qquad (10.62)$$

and we use the convention that $G_0 T^{(0)} = 1$. The result (10.60) is true for $p = 1$ since (9.155), (10.53) and (10.56) yield directly

$$\Delta^{(1)} = \frac{i}{\hbar} [H_0, F] = \left[H_0, \Omega[\mathbf{g}]^{(1)}\right] = W^{(1)}. \qquad (10.63)$$

On the other hand, if the general formula (10.60) is used to construct $W^{(p-2)}$ and $W^{(p-1)}$ for the RHS of (10.57), lengthy but elementary calculation shows that the result is precisely $W^{(p)}$, so by the principle of induction it is true for all p [20]. The difference between T-matrix elements (10.50) in the basis defined by (10.3) in any order, p, of perturbation theory thus depends on

$$W[\mathbf{g}]^{(p)}_{kn} = (E_k - E_n)\Omega[\mathbf{g}]^{(p)}_{kn}, \qquad (10.64)$$

and this is annihilated by the energy conservation delta function in (10.8), provided that Ω_{kn} does not have a pole at energy E_n; the S-matrix is then gauge invariant (independent of $\mathbf{g}^\perp$) in all orders of perturbation theory [20].

There are several ingredients that are essential for the success of the argument. The energy conservation condition for the process $n \to k$, which makes (10.64) vanish, is crucial for the argument to work; this can be interpreted as the requirement that the initial and final states in physical processes must be stable, that is, time-independent. Energy conservation is only required for real processes, and virtual processes which appear in perturbation theories have to be treated carefully if gauge dependence is to be avoided. The reduction of the gauge-dependent terms to the commutator (10.60), which involves extensive cancellation, is a purely algebraic calculation; since it involves non-commuting variables the commutation properties of $H_0, V^{(1)}$ and $V^{(2)}$ with the transformation operator $F[\mathbf{g}]$ play a crucial role. It is worth noting that in principle there must be an equivalent calculation based on the matrices that are produced by inserting a complete set of intermediate states in (10.57). This suggests that other forms of perturbation theory in which specified state(s) must be omitted from the matrix multiplications may be liable to gauge dependence because of incomplete cancellation of terms involving $\mathbf{g}$ (see §10.2.3, §10.2.5). On the other hand, the assumption that the perturbation series is meaningful is at best of only limited validity. The presence of other reference states that are degenerate with the initial state Φ_n makes $G_0(E_n)$ singular, and requires particular attention.

10.2.3 S-Matrix Theory and Linewidths

We now consider the modifications of the S-matrix theory required for the description of spectral linewidths. In the conventional presentation of the quantum theory of atoms and molecules, it is customary to regard the atomic system as a *closed* entity; it is recognised, of course, that an 'isolated molecule' is a fiction and that in real practical situations atoms and molecules are always coupled to their environment, to varying degrees, in what may be called 'persistent interactions' [21], [22]. As a consequence, infinitely long-lived discrete states with perfectly sharp energies in the sense of Bohr's 'stationary states', other than ground states, do not exist; excited energy levels gain widths (equivalently, finite life times) in various ways which are revealed in spectroscopy as lineshapes. The description of this situation in quantum theory is based on coupling between the idealised discrete states and one or more continua that characterise the 'environment', which typically is taken as macroscopic. Macroscopic systems in quantum theory have purely continuous spectra and are describable by quantum field theories; examples include the electromagnetic field itself, a crystalline host lattice which supports phonons, solutions and liquids characterised by Brownian motion and so on, which may be described as 'heat baths' or 'reservoirs'.

A formal solution to this problem can be achieved by the refinement of the scattering theory summarised at the end of §6.5 [23], [24]. The reaction matrix $\mathbf{D}$ is closely related to the matrix $\boldsymbol{\Sigma}$ in (6.32); only its on-energy-shell elements $\{D_{kn}\}$ are required for the T-matrix, and these can be seen to be gauge invariant from the analysis of the gauge dependence of $\boldsymbol{\Sigma}$ presented in what follows. In practice the Heitler equation (6.91) has not been used in atomic and molecular quantum electrodynamics. Instead the results of the non-resonant perturbation theory described in §10.2.2 (or other formulations that are equivalent to it) are modified by the incorporation of phenomenological damping factors that account for energy level shifts and lifetimes due to resonant interactions. In condensed media there are many other factors that give rise to linewidths that can be incorporated in this way; such factors are constructed as gauge-invariant quantities.

The natural lineshape was already considered by Low [25] within the covariant QED formalism, although that method has never been extended to a general account of resonant molecular light scattering and the general problem of line broadening. At least part of the difficulty identified by Power and Zienau [19] in non-relativistic lineshape theory can be avoided if the non-resonant part of the cross section is also retained [26], [27]. In fact, as we shall see, the suggested division of the T-matrix into resonant and non-resonant contributions is a gauge-dependent separation which leads to the difficulties encountered in practical applications.

Recall from the discussion of the S-matrix in Chapter 6 that for a transition $\Phi_n \to \Phi_k$ there is a formally exact closed form for the T-matrix,

$$T(E_n)_{kn} = V_{kn} + \sum_{m,j} V_{km} G^+(E_n)_{mj} V_{jn}, \tag{10.65}$$

where $G^+(E_n)_{mj}$ is a matrix element of the exact Green's function evaluated at the energy E_n,

$$G^+(E_n)_{mj} = \frac{\delta_{mj}}{E_n - E_m + i\varepsilon} + \frac{1}{E_n - E_m + i\varepsilon} \sum_r V_{mr} G^+(E_n)_{rj}. \tag{10.66}$$

Iteration of (10.66) leads to the usual perturbation series for $T(E_n)_{kn}$.

Suppose now that for some $m = p$ we have $E_p = E_n$, so that the iteration of (10.66) cannot be expected to converge. The customary procedure is to isolate this term and write (10.65) as a sum of resonant and non-resonant contributions,

$$T_{kn} = T_{kn}^{\text{res}} + T_{kn}^{\text{non-res}}, \tag{10.67}$$

with

$$\begin{aligned} T_{kn}^{\text{res}} &= V_{kp} G^+(E_n)_{pp} V_{pn}, \\ T_{kn}^{\text{non-res}} &= V_{kn} + \sum_{m,j} (1 - \delta_{m,j} \delta_{jp}) V_{km} G^+(E_n)_{mj} V_{jn}. \end{aligned} \tag{10.68}$$

Assuming there are no other resonances, the usual perturbation series is taken to be sufficient for the non-resonant contribution,

$$T_{kn}^{\text{non-res}} \approx V_{kn} + \sum_{m \neq p} \frac{V_{km} V_{mn}}{E_n - E_m + i\varepsilon} + \dots . \tag{10.69}$$

On the other hand, a more exact treatment is required for the resonant contribution; because of level shifts and radiation damping, the singularity at $E_n = E_p$ in $G_0(E_n)$ is shifted to a complex pole in $G^+(E_n)$ which gives rise to the characteristic resonance lineshape in spectra.

Let us first consider the non-resonant contribution in two different gauges, that is, evaluate

$$T[\mathbf{g}]_{kn}^{\text{non-res}} - T[0]_{kn}^{\text{non-res}}. \tag{10.70}$$

This calculation is obviously closely related to the proof of gauge invariance of the S-matrix outlined in §10.2.2, but there is now a difference due to the omission of the state Φ_p from the matrix multiplications. It is sufficient to write out (10.70) to $0(e^2)$, using

(10.69) to see the nature of the problem; the difference between the perturbation in the two gauges yields for $\Delta[\mathbf{g}]$, (10.53),

$$\Delta[\mathbf{g}] = (\frac{i}{\hbar})[\mathsf{H}_0, \mathsf{F}] + (\frac{i}{\hbar})[\mathsf{V}^{(1)}[0], \mathsf{F}] + \frac{1}{2!}(\frac{i}{\hbar})^2 [[\mathsf{H}_0, \mathsf{F}], \mathsf{F}]. \tag{10.71}$$

The (kn) matrix element of the first term in (10.71) vanishes on the energy shell, $E_k = E_n$, but the elimination of the remaining two terms requires cancellation by compensating terms that must originate from the second term in (10.69). Unlike the situation in §10.2.2, this cancellation is not complete, however, because of the deletion of the state Φ_p from the matrix multiplication in the second term. A straightforward calculation shows that to $0(e^2)$ (10.70) reduces to

$$\frac{i}{\hbar} \left(V_{kp} F_{pn} - F_{kp} V_{pn} \right), \tag{10.72}$$

which does not generally vanish and can therefore have any value. Obviously the missing term required to cancel (10.72) would have been found in (10.68) if we could have assumed $E_p \neq E_n$ and replaced $\mathsf{G}^+(E_n)$ by $\mathsf{G}_0(E_n)$; however, the more complete treatment of the Green's function $\mathsf{G}^+(E_n)$ spoils this relationship.

Standard formal manipulations in resolvent theory (cf. Chapter 6) show that

$$\mathsf{G}^+(E_n)_{pp} = \frac{1}{E_n - E_p - \Sigma(E_n)_{pp}}, \tag{10.73}$$

where

$$\Sigma(E_n)_{pp} = (\mathsf{V} + \mathsf{V}\mathsf{G}_0(E_n)\Sigma(E_n))_{pp}, \tag{10.74}$$

with the state Φ_p excluded from the matrix multiplications in (10.74). Clearly, the dependence of Σ on $\mathbf{g}$ can be investigated using the iteration method described in §10.2.2. This will be done in more detail in §10.2.5 where both diagonal and off-diagonal matrix elements are required; here it is sufficient to note from that discussion that $\Sigma(E_n)_{pp}$ is gauge invariant for $E_n = E_p$. This, however, is not sufficient to ensure that the separation of the T-matrix in (10.67) leads to a gauge-invariant result, as can be seen by looking at the gauge dependence of the numerator in the resonant contribution in two different gauges,

$$T[\mathbf{g}]^{\text{res}}_{kn} - T[0]^{\text{res}}_{kn} = \frac{V[\mathbf{g}]_{kp}V[\mathbf{g}]_{pn} - V[0]_{kp}V[0]_{pn}}{E_n - E_p - \Sigma(E_n)_{pp}}. \tag{10.75}$$

Introducing $\Delta[\mathbf{g}]$ from (10.71), the numerator in (10.75) reduces to

$$V[0]_{kp}\Delta[\mathbf{g}]_{pn} + \Delta[\mathbf{g}]_{kp}V[0]_{pn} + \Delta[\mathbf{g}]_{kp}\Delta[\mathbf{g}]_{pn}. \tag{10.76}$$

The lowest-order contribution, proportional to e^2, vanishes because the resonance condition $E_k = E_n = E_p$ removes the (kp) and (pn) matrix elements of the H_0 commutator term in $\Delta[\mathbf{g}]$, assuming that $\Sigma(E_n)_{pp} \neq 0$ in the denominator of (10.75). This means that the contribution required to cancel (10.72) in the non-resonant part of the T-matrix is lost. On the other hand there are gauge-dependent and generally non-zero contributions of order e^3 and e^4 in (10.76) which do not cancel out with compensating terms of the same order in $T_{kn}^{\text{non-res}}$ because they are multiplied with $G^+(E_n)_{pp}$ rather than $G_0(E_i)_{pp}$. It is very difficult to see how one can make a consistent gauge-invariant calculation of the T-matrix, T_{kn}, if one starts by isolating the pole associated with a resonance.

10.2.4 Friedrichs–Fano Models

An early treatment of linewidths in continuous absorption spectra, and one that is highly instructive, is the theory of atomic autoionisation due to Fano [28], [29]. Asymmetric lineshapes associated with so-called *Fano resonances* have been observed in diverse areas of physics [30]–[32]. In the simplest case considered, a single discrete state above the first ionisation threshold is coupled to a continuum.[8] These states correspond to specified atomic 'configurations' constructed in the independent electron approximation, so that they do not diagonalise the atomic Hamiltonian H, and the true states of the atom arise from 'configuration interaction'.

Let the normalised discrete state be denoted by φ_k, and the continuum states by $\psi(E')$ with $E_1 \leq E' \leq E_2$; the continuum states are normalised according to Dirac's delta function prescription, §5.2.1 and are orthogonal to the discrete state. Then we restrict attention to the sub-matrix block of the Hamiltonian matrix with elements

$$\langle \varphi_k | H | \varphi_k \rangle = E_k \tag{10.77}$$

$$\langle \psi(E') | H | \varphi_k \rangle = V_k(E') \tag{10.78}$$

$$\langle \psi(E'') | H | \psi(E') \rangle = E' \delta(E'' - E'), \tag{10.79}$$

with E_k lying in the energy range of the considered continuum.

The solution of the Schrödinger equation

$$H\Psi(E) = E\Psi(E) \tag{10.80}$$

restricted to this subspace may be expressed as the superposition

$$\Psi(E) = a_k(E)\varphi_k + \int_{E_1}^{E_2} b(E:E')\psi(E')\,\mathrm{d}E' \tag{10.81}$$

with coefficients a, b that satisfy

$$E_k a_k(E) + \int_{E_1}^{E_2} V_k(E')^* b(E:E')\,\mathrm{d}E' = E\,a_k(E) \tag{10.82}$$

$$V_k(E')a_k(E) + E'b(E:E') = E\,b(E:E') \tag{10.83}$$

according to (10.77)–(10.79).

---

[8] The discussion can be generalised to multiple discrete states and continua.

Equation (10.83) may be solved by generalising Dirac's prescription [33] for dealing with the singularity at $E = E'$,

$$b(E : E') = \left[\frac{1}{E - E'} + z(E)\delta(E - E')\right] V_k(E')a_k(E), \qquad (10.84)$$

where $z(E)$ is to be determined. Substituting in (10.82) and eliminating $a_k(E)$ yields

$$E_k + F_k(E) + z(E)|V_k(E)|^2 = E, \qquad (10.85)$$

with

$$F_k(E) = \int_{E_1}^{E_2} \frac{|V_k(E')|^2}{E - E'} \, dE', \qquad (10.86)$$

and the integral is interpreted as a principal part integral. One thus obtains

$$\frac{1}{z(E)} = \frac{|V_k(E')|^2}{E - E_k - F_k(E)}. \qquad (10.87)$$

Fano's analysis is completed by the careful demonstration that the coefficients a, b required for (10.81) can be obtained by requiring normalisation of $\Psi(E)$; there results

$$a_k(E) = \frac{\sin(\chi)}{\pi V_k(E)} \qquad (10.88)$$

$$b(E : E') = \left(\frac{V_k(E')}{\pi V_k(E)}\right)\left(\frac{\sin(\chi)}{E - E'}\right) - \cos(\chi)\delta(E - E'), \qquad (10.89)$$

where the *phase angle* χ is given by

$$\chi = -\arctan\left[\frac{\pi|V_k(E)|^2}{E - E_k - F_k(E)}\right]. \qquad (10.90)$$

The physical interpretation of this calculation is that the configuration interaction transforms the initial discrete state φ_k into a band of stationary states with a half-width of $\pi|V_k(E)|^2$; the lifetime, τ, of the state φ_k is $\hbar/(2\pi|V_k(E)|^2)$. It should be noted that this is a purely stationary state description in which overall energy is conserved.

The amplitude for the photo-excitation of the state $\Psi(E)$ from some lower discrete state, typically the ground state, φ_0, is given by the T-matrix element, $\langle\Psi(E)|T|\varphi_0\rangle$, according to §10.1; here T is the transition operator for the coupling with the quantised electromagnetic field. In terms of a modified atomic state,

$$\Phi_k = \varphi_k + \int_{E_1}^{E_2} \frac{V_k(E')}{E - E'} \, dE', \qquad (10.91)$$

this T-matrix element may be cast into the difference of two terms,

$$\langle\Psi(E)|T|\varphi_0\rangle = \frac{1}{\pi V_k(E)^*}\langle\Phi_k|T|\varphi_0\rangle\sin(\chi) - \langle\psi(E)|T|\varphi_0\rangle\cos\chi. \qquad (10.92)$$

Since $\sin(\chi)$ and $\cos(\chi)$ are even and odd functions, respectively, of the energy variable $(E - E_k - F_k(E))$, these two terms interfere with *opposite* phases on the two sides of the resonance, and since (10.92) determines the cross section, this explains the character-istic asymmetric lineshapes observed in atomic photoabsorption experiments.

This calculation can readily be interpreted in terms of the partitioning technique and resolvent theory sketched in §6.3. The quantity $F_k(E)$ in (10.86) is sometimes referred to as the 'self-energy' and can be identified with $\Sigma_i(E)$ in (6.31). The self-energy $F_k(E)$ is analytic in the whole complex plane apart from a branch cut on the real axis between E_1 and E_2. The evaluation of the integral (6.39) yields the survival probability amplitude as a sum determined by the residues $\{R_{mk}\}$ at the poles $\{E_{mk}\}$ of $\langle \varphi_k | R(E) | \varphi_k \rangle$ plus the contribution from the cut. One thus has

$$c_k(t) = \sum_{E_m} R_{mk} \exp(-iE_m t) + A(t). \tag{10.93}$$

In this case, there is typically a single pole with $\Im E_{mk} < 0$, and so the survival probability is given by (6.42).

Equations (10.77)–(10.79) provide an exact matrix representation of a *model* Ham-iltonian,

$$\begin{aligned} H = E_k |\varphi_k\rangle\langle\varphi_k| &+ \int_{E_1}^{E_2} E |\psi(E)\rangle\langle\psi(E)| \, dE \\ &+ \int_{E_1}^{E_2} \Big(V_k(E) |\varphi_k\rangle\langle\psi(E)| + V_k^*(E) |\psi(E)\rangle\langle\varphi_k| \Big) \, dE, \end{aligned} \tag{10.94}$$

which is capable of considerable generalisation. Fano already considered [29] the case of many discrete levels ($k = 1, \ldots N$). The integrations can be extended to an infinite domain,

$$\int dE \longrightarrow \int d^3\mathbf{k}, \tag{10.95}$$

usually interpreted as summation over the modes of a collection of harmonic oscil-lators in the infinite volume limit, in versions of a model originally introduced by Friedrichs [21], [36]. The oscillators can be interpreted in terms of a scalar boson quan-tum field (e.g. phonons in a crystalline lattice), or indeed the quantised electromagnetic field [37]. In the latter case, the coupling between the 'atom' (modelled as an N-level set of discrete states) and the field is restricted to the electric dipole approximation, $V = -\mathbf{d} \cdot \mathbf{E}^\perp$, and the magnitude of the wave vector, $|\mathbf{k}|$, in the interaction terms is cut off at some large value to maintain non-relativistic energies. These matters will be pursued further in the next chapter.

10.2.5 Time-dependent Perturbation Theory and Gauge Invariance

Historically the first non-perturbative attack on the problem, due to Heitler [38], was based on a time-dependent approach to electrodynamics; it can be formulated as follows. We seek a solution of the time-dependent Schrödinger equation

$$i\hbar \frac{\partial |\Psi\rangle}{\partial t} = H |\Psi\rangle, \tag{10.96}$$

subject to an initial condition that specifies $\Psi(t = 0)$ as an eigenstate Φ_p of H_0, (9.103), and then examine the solution for later times t; the method of solution involves transforming to an interaction representation by setting

$$|\Psi(t)\rangle = e^{-iH_0 t/\hbar}|\Phi(t)\rangle. \tag{10.97}$$

The projections of the eigenstates of the reference system $\{\Phi_i\}$ on $|\Phi(t)\rangle$ are the development coefficients $\{b(t)_{ip}\}$ of the traditional method of variation of constants discussed in §6.4.

A non-perturbative formal solution of the problem can be obtained by introducing a Fourier integral representation with

$$b(t)_k = -\frac{1}{2\pi i} \int\limits_{-\infty}^{+\infty} A(E)_k e^{i(E_k-E)t/\hbar} \, dE. \tag{10.98}$$

$A(E)_k$ is a meromorphic function of E with its poles lying in the lower-half complex E-plane so as to ensure that $b(t)_k = 0$ for $t < 0$. The coefficients $\{A(E)_k\}$ can be expressed in terms of the matrix $\Sigma(E)$ defined as the solution of the implicit equation

$$[\Sigma(E)]_{kp} = [V]_{kp} + [VG_0^+(E)\Sigma(E)]_{kp}, \tag{10.99}$$

on the understanding that the initial state p is omitted from the matrix multiplication in (10.99),

$$A(E)_p = \frac{1}{E - E_p - \Sigma(E)_{pp}}, \tag{10.100}$$

$$A(E)_k = \Sigma(E)_{kp} \frac{1}{E - E_k + i\varepsilon} A(E)_p, k \neq p. \tag{10.101}$$

In modern terminology these coefficients are matrix elements of the resolvent of H [3]. In practice, this solution can usually only be applied using an expansion of $\Sigma(E)$ in powers of the interaction V, and it is this expansion we discuss in what follows using the general Hamiltonian (9.86).

The computation of the Fourier integral (10.98) is in general highly complicated because $\Sigma(E)$ does not have a simple dependence on the variable E. Two main approximations have been used; firstly the probability for a transition $\Phi_p \to \Phi_k$, given by

$$P_k = |b(+\infty)_k|^2, \tag{10.102}$$

requires the limiting value for $t \to +\infty$. A straightforward calculation yields the asymptotic values [38],

$$b(+\infty)_{kp} = \frac{\Sigma(E_k)_{kp}}{E_k - E_p - \Sigma(E_k)_{pp}}, \quad k \neq p, \tag{10.103}$$

$$b(+\infty)_{pp} = 0. \tag{10.104}$$

For studying the time evolution of the system at finite times, the resonant state on-the-energy-shell approximation has been invoked. This amounts to the claims that the energy correction term $\Sigma(E)_{pp}$ in (10.100) is a slowly varying function of E close to E_p and that the real part of this term will be much smaller than E_p. There is then a pole in $A(E)_p$ close to E_p, and to sufficient accuracy $\Sigma(E)_{pp}$ may be approximated by $\Sigma(E_p + i\varepsilon)_{pp}$, with the usual limit $\varepsilon \to 0^+$ understood [39].

Comparison of the matrix $\boldsymbol{\Sigma}(E)$, (10.99), with the T-matrix (6.45) shows that they have similar structures, and we make a corresponding analysis of $\boldsymbol{\Sigma}(E)$. For example, instead of (10.48), (10.49), we have

$$\Sigma^{(1)} = \mathsf{V}^{(1)}, \tag{10.105}$$

$$\Sigma^{(2)} = \mathsf{V}^{(2)} + \mathsf{V}^{(1)}\mathsf{G}_0^+\Sigma^{(1)}. \tag{10.106}$$

There are, however, two essential differences; firstly $\boldsymbol{\Sigma}(E)$ is not required at $E = E_p$, the initial state energy, unless the resonant on-the-energy-shell approximation is made, and secondly there is a restriction on the sum over states in the matrix multiplication in (10.99) that does not apply to (6.45).

As in §10.2.2, (10.105) and (10.106) can be compared for two different gauges; a non-zero matrix element for this difference, which we call $\mathsf{K}(E, [\mathbf{g}])$, signals gauge dependence of the corresponding matrix element of $\boldsymbol{\Sigma}(E)$. Equation (10.105) with (9.155) leads at once to the first-order result,

$$\mathsf{K}(E, [\mathbf{g}])^{(1)} = \Sigma(E, [\mathbf{g}])^{(1)} - \Sigma(E[0])^{(1)},$$
$$= i/\hbar[\mathsf{H}_0, \mathsf{F}], \tag{10.107}$$

where from now on we do not show explicitly the dependence of F on $\mathbf{g}(\mathbf{x}, \mathbf{x}')$. Similarly, taking the difference between the terms in (10.106) for two different gauges gives

$$\mathsf{K}(E, [\mathbf{g}])^{(2)} = \Sigma(E, [\mathbf{g}])^{(2)} - \Sigma(E, [0])^{(2)},$$
$$= \mathsf{V}^{(2)}[\mathbf{g}] - \mathsf{V}^{(2)}[0] + \mathsf{V}^{(1)}[\mathbf{g}]\mathsf{G}_0^+\Sigma(E, [\mathbf{g}])^{(1)},$$
$$- \mathsf{V}^{(1)}[0]\mathsf{G}_0^+\Sigma(E, [0])^{(1)}, \tag{10.108}$$

which with the aid of (9.155) and (9.156) reduces to

$$-i\hbar\mathsf{K}(E, [\mathbf{g}])^{(2)} = [\mathsf{V}^{(1)}[0], \mathsf{F}] + \mathsf{V}^{(1)}[0]\mathsf{G}_0^+[\mathsf{H}_0, \mathsf{F}] + [\mathsf{H}_0, \mathsf{F}]\mathsf{G}_0^+\mathsf{V}^{(1)}[0]$$
$$+ \frac{i}{\hbar}[\mathsf{H}_0, \mathsf{F}]\mathsf{G}_0^+[\mathsf{H}_0, \mathsf{F}] - \frac{1}{2}\frac{i}{\hbar}[\mathsf{F}, [\mathsf{H}_0, \mathsf{F}]]. \tag{10.109}$$

Physical applications based on Eq. (10.102) have their gauge dependence contained in the (k, p) and (p, p) matrix elements of $\mathsf{K}(E_k, [\mathbf{g}])$. In first order we have

$$K(E_k, [\mathbf{g}])_{kp}^{(1)} = \frac{i}{\hbar}(E_k - E_p)F_{kp}. \tag{10.110}$$

The transformation operator F is linear in the vector potential; the Coulomb gauge vector potential (7.122) has non-zero matrix elements between states differing by one photon, so both factors in (10.110) vanish in the diagonal element, $k = p$, but matrix elements $K(E_k, [\mathbf{g}])_{kp}$ between states k, p differing by one photon are in general gauge-dependent unless they are degenerate. The matrix elements of the second-order term are easily found; they are

$$- i\hbar K(E_k, [\mathbf{g}])_{kp}^{(2)} = (E_k - E_p) \left(\left(\mathsf{V}^{(1)}[0] \mathsf{G}_0^+(E_k) \mathsf{F} \right)_{kp} + \frac{1}{2} \frac{i}{\hbar} (\mathsf{FF})_{kp} \right) k \neq p, \quad (10.111)$$

and

$$K(E_k, [\mathbf{g}])_{pp}^{(2)} = -\frac{2}{\hbar} (E_k - E_p) \mathrm{Im} \left(\mathsf{V}^{(1)}[0] \mathsf{G}_0^+(E_k) \mathsf{F} \right)_{pp}$$
$$- \frac{(E_k - E_p)}{\hbar^2} \sum_m \frac{E_{pm} f_{pm} f_{mp}}{E_{km}}, \quad (10.112)$$

which is purely real. One can continue in this fashion for higher terms, but the pattern is already clear; gauge-invariant probabilities (10.102) are only obtained if a condition of energy conservation ($E_k = E_p$) is imposed from outside the theory. Otherwise, the gauge-dependent terms in the numerator ($\Sigma(E_k)_{kp}$) and denominator ($\Sigma(E_k)_{pp}$) of (10.103) are non-zero (and arbitrary).

These calculations can be applied directly to the diagonal matrix element of the resolvent, $A(E)_p$, (10.100), at arbitrary energy E. The first non-zero gauge-dependent term in a perturbation series for $\Sigma(E)_{pp}$ is (10.112) with E_k replaced by E. Assuming that the coefficients of $(E - E_p)$ in (10.112) are regular, this vanishes at $E = E_p$, so that the final result of the resonant state on-the-energy-shell approximation [39] is gauge invariant. However, the arbitrariness of these coefficients spoils the argument that $\Sigma(E, [\mathbf{g}])_{pp}$ must be small compared to E_p for E near E_p, irrespective of $\mathbf{g}$, and the pole need not be close to E_p. The true poles of $A(E)_p$, (10.100), are the eigenvalues of the full Hamiltonian H [3] which are gauge invariant, so the difficulty lies in the use of this perturbation series.

An account of the interactions of molecules with quantised radiation including one- and two-photon absorption and emission, scattering of photons, and spectral lineshape theory including resonance fluorescence can be based on (10.98)–(10.104) [40]. For the calculation of transition probabilities per unit time, the matrix $\Sigma(E)$ is required on the energy shell and gives gauge-invariant results. For other applications such as the scattering of photons, however, approximate energy conservation was invoked only through the claim that the real and imaginary parts of the diagonal element Σ_{pp} are small relative to typical molecular excitation energies. As we have seen, this argument is invalidated by the gauge dependence of Σ when energy conservation is not strictly valid; the gauge-dependent terms in Σ are unbounded, and so the magnitude of Σ is completely arbitrary.

Another application of the time-dependent theory described here is the conventional account of the shapes of spectral lines, particularly for emission lines and resonance fluorescence [19], [38], [40], [41]. This formalism is taken to be a theoretical improvement over the original treatment of Weisskopf and Wigner [42] (see also [43], [44]) which starts from the assumption that the initial state decays exponentially in time and does not allow for any frequency variation of the matrix elements involved. A typical calculation is the theory of the natural lineshape which can be obtained by taking the initial state at $t = 0, \Phi_p$, to describe an excited atomic level with energy E_n and no photons present, while Φ_k refers to the atom in its ground state with energy E_0, and a photon of frequency ω is present. The required probability, dP, for emission of a photon in the frequency interval $[\omega, \omega + d\omega]$ is then

$$dP = |b(+\infty)_{kp}|^2 d\omega, \tag{10.113}$$

where $b(+\infty)_{kp}$ is given by (10.103) for this situation. In terms of the energy separation of the atomic levels, $\omega_{n0} = E_n - E_0$, and the decay constant Γ defined by $\frac{1}{2}\hbar\Gamma = \Im\Sigma(E_k)_{pp}$, this has the form [41]

$$dP = \frac{|\Sigma(E_k)_{kp}|^2 d\omega}{(\omega - \omega_{n0})^2 + \frac{1}{4}\Gamma^2}. \tag{10.114}$$

It is usually argued that $\Gamma << \omega$ and we can put

$$E_n - E_0 = \hbar\omega, \tag{10.115}$$

that is, take $E_k = E_p$, which would ensure gauge invariance. This argument is spoilt by the gauge dependence of the denominator in (10.103) that is implied by (10.112).

With narrow bandwidth radiation sources, moreover, the frequency variation of the lineshape can be measured accurately; it is not completely determined by the denominator in (10.114). This was investigated in detail by Power and Zienau [19], who considered a variety of lineshape problems including the experiment for the accurate determination of the Lamb shift in atomic hydrogen [45]. They showed that the Coulomb gauge theory (in electric dipole approximation) and the multipole Hamiltonian, restricted to its first, electric dipole term, gave different frequency dependencies as ω moved away from the resonance condition (10.115) because of the frequency variation of the numerator of (10.114). From the present perspective, this difference is the expected result of the gauge dependence of $\Sigma(E_k)_{kp}$ when $E_k \neq E_p$, and it suggests that an entirely different method of calculation that maintains gauge invariance throughout must be sought for the lineshape observable. This is so even though it appears that the multipole Hamiltonian leads to a predicted lineshape in good agreement with Lamb's experiment [19], [45] which is sufficiently precise to rule out definitely the Coulomb gauge calculation; this result remains to be explained in a QED framework.

In terms of the usual unitary time development operator $\mathsf{U}(t, t_0)$ and the given initial conditions, the development coefficients are

$$b(t)_k = \langle \Phi_k | \mathsf{U}(t, 0) | \Phi_p \rangle. \tag{10.116}$$

It is known from general formal considerations [3] that only in the doubly infinite limit $t_0 \to -\infty, t \to +\infty$ do the matrix elements of $\mathsf{U}(t_0, t)$ have the property

$$\langle \Phi_k | \mathsf{U}(+\infty, -\infty) | \Phi_p \rangle \propto \delta(E_k - E_p); \tag{10.117}$$

otherwise, as found in the preceding, matrix elements of $\mathsf{U}(t, t_0)$ do not enforce energy conservation, which appears to be intimately involved in the maintenance of gauge invariance. These calculations cast serious doubt as to whether physically meaningful (gauge-invariant) time-dependent probabilities can actually be obtained for processes involving electromagnetic radiation. Evident weaknesses are the unphysical initial condition which specifies that the atomic system is in a sharp energy eigenstate of H_0 at the initial instant $t = 0$, and, in the lineshape calculations, the involvement of states liable to decay by spontaneous emission. These limitations, though not the gauge dependence they lead to, were recognised long ago and may be remedied by explicit incorporation of the details of the excitation mechanism [38].

10.3 Diagrammatic Perturbation Theory

The evaluation of the terms in the expansion (10.36) for the transition matrix for a given physical process often involves cumbersome algebraic computations which may be simplified using graphical or diagrammatic techniques [46]. This is especially so for multiphoton processes if only the lowest-order multipolar interactions are retained in the PZW-transformed Hamiltonian, since in this approximation the perturbation operators are proportional to the charge parameter e. If the full Hamiltonian is used, V also contains terms proportional to e^2, and so there are more types of diagram to include. The application of such methods in non-relativistic QED became well known in the 1960s [7]–[11], [47]–[49].

Every term in the perturbation series corresponds to a particular diagram, and to recover the complete perturbation formulae of a given order in the interaction, all topologically distinct diagrams must be considered. There are three basic components to the diagrams, *open lines, vertices* and *propagators*. At a vertex a photon is absorbed or emitted by a 'particle', which may be a single charge or a collection of charges, for example, an atom or molecule, or distinct parts of molecules (e.g. chromophores) if they can be assumed to be electronically distinct, and so is translated into a matrix element of the interaction V. Open lines correspond to real particles, and in the absence of external fields the conservation laws for energy and momentum apply at the vertices where open lines terminate or start. The diagrams are to be read with time increasing from left to right. The basic absorption and emission vertex diagrams are shown in Figure 10.1. Thus, diagram (i) in Figure 10.1 represents absorption of a photon in the mode $\mathbf{Q}, \mu$ with the particle gaining energy $\hbar Q c$ and momentum $\hbar \mathbf{Q}$; similarly, diagram (ii) represents emission of a photon and loss of the same energy and momentum by the particle to the field.[9]

Diagrams are built up by glueing lines together; a closed line between two vertices represents the propagation of a virtual particle, and energy conservation does *not* apply to such vertices. A closed line is translated into a matrix element of a Green's function or propagator, for example,

$$\langle n|G^0(E_i)|m\rangle = \frac{1}{E_n^0 - E_m^0 + i\varepsilon}, \tag{10.118}$$

---

[9] These diagrams are evaluated in §10.4 where we will see that a single charge cannot satisfy both conservation laws in a real process, though atoms and molecules can.

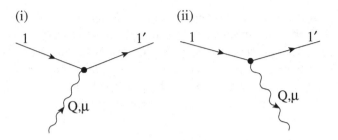

Figure 10.1   Primitive Feynman diagrams: the absorption, (i), and emission, (ii), vertices.

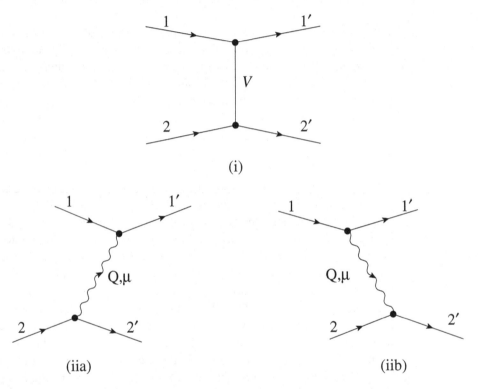

Figure 10.2   Diagrams for electrodynamical interaction between charge distributions to order α: in (i) V is the static contribution of $\int |\mathbf{P}|^2 \, d\tau$; (ii) the two possible virtual photon exchange processes.

and so closed lines give the characteristic energy denominators in the perturbation expansion. For a free system conservation of momentum is maintained at vertices between closed lines. Static fields act instantaneously, and so external static electric and magnetic fields may be included using diagrams that have a vertical line ending at a vertex (with incoming and outgoing lines for the particles). The static Coulomb interaction between charges[10] may be represented by a vertical line joining two vertices (see Figure 10.2). Conservation of momentum does not apply at external field vertices.

[10] More generally the contribution of $\int |\mathbf{P}|^2 \, d\tau$.

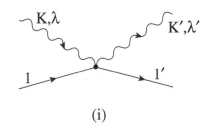

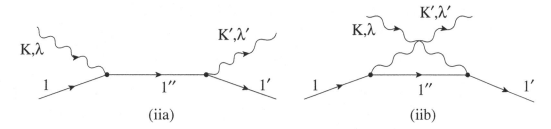

Figure 10.3 Diagrams for linear light scattering: (i) the $|\mathbf{A}|^2$ contact interaction in the Coulomb gauge; (ii) the two possible processes involving virtual excitation of the particle.

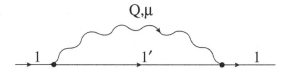

Figure 10.4 Self-interaction to order α.

An alternative procedure is to use the stationary states of the particle in the presence of the static field, and just include vertices involving photons. The overall transition amplitude to a given order for any physical process [7], [48], is equal to the sum of all topologically distinct diagrams having the same initial (n) and final (k) states satisfying energy conservation, $E_n^0 = E_k^0$.

To order e^2 we can picture three basic ways to combine the two diagrams in Figure 10.1, being careful to keep the temporal sequence of events clear:

1. Simply push the diagrams together and join the lines labelled $1, 1'$; this can be done with two time sequences corresponding to absorption of the photon before and after emission of the photon; the perturbation operator $\propto e^2$ also contributes to the amplitude of course. Such diagrams (Figure 10.3) describe scattering of a (real) photon by a charge distribution.

2. One can join the photon lines; again absorption may precede or follow emission of the (virtual) photon. Such diagrams (Figure 10.2) describe the interaction between two charge distributions mediated by the exchange of a virtual photon.

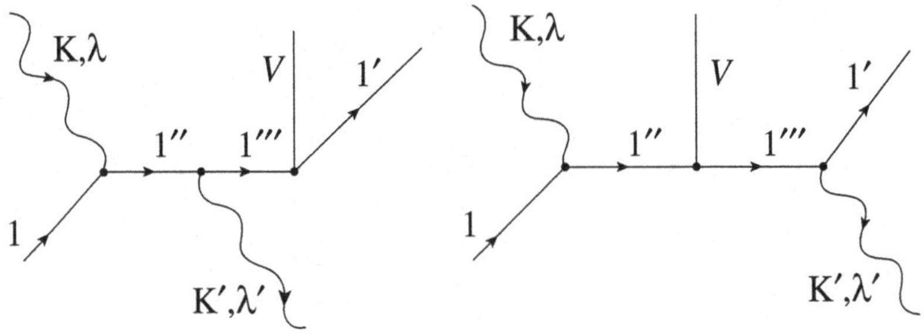

Figure 10.5 Third-order diagrams involving interaction with a static, external field.

3. There is just one way to take the diagrams created in event 1 and modify them by also joining the photon lines so describing the emission of a virtual photon which is later reabsorbed by the charge distribution that has propagated in a virtual state. Such diagrams (Figure 10.4) give rise to the infinities due to 'self-interaction' which will be discussed in §10.6.

The order n amplitude in the perturbation expansion for interaction operators linear in the charge e is represented by diagrams containing n vertices and $n-1$ propagators; one must not forget to include the relevant diagrams when interactions proportional to e^2 are also considered. The number of diagrams increases rapidly with the order of the amplitude; for example, there are in all six diagrams of the third-order type shown in Figure 10.5 if specialised to the case of the Kerr effect [8] with a static external field interaction[11] of $-\mathbf{d}\cdot\mathbf{E}$, and $\mathbf{k}=\mathbf{k}'$. For the non-linear light scattering process depicted in Figure 10.6, there are 24 topologically distinct diagrams. There are 48 of each type of diagram in Figure 10.7 which arise in a model of optical activity involving pairs of chromophores [50]. The three different types of diagram just identified can occur in combination in higher-order processes; for example, the self-interaction diagram can be inserted between adjacent vertices. As we will see in §10.6, such diagrams can be interpreted as (divergent) contributions to the masses of the charges. They are not explicitly dealt with, and instead one assumes the experimental values of the mass and charge parameters of the charges are to be used. The matrix elements of the photon operators may be evaluated completely in the Fock space basis in terms of fundamental constants and properties of the beam; for virtual photons the intermediate state summations amount to summation over the two polarisation states and integration over the photon momentum. There then result amplitudes expressed entirely in terms of molecular matrix elements and energy denominators, among which are the usual susceptibilities of the molecular system.

The rules for the construction of the T-matrix element of order n for a scattering process involving photons and molecules are as follows [7], [11], [48]:

1. An nth-order diagram contains n vertices.

---

[11] A static electric field applied across an isotropic fluid causes birefringence.

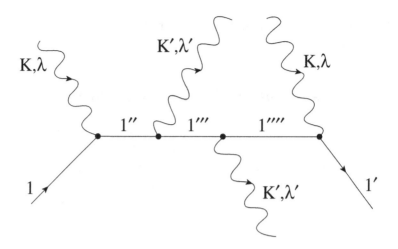

Figure 10.6   Non-linear light scattering to order α^2.

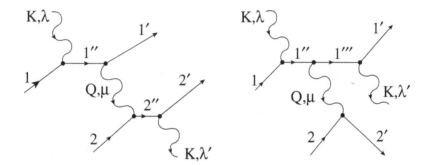

Figure 10.7   Two fourth-order diagrams involving a virtual photon interaction between two chromophores that leads to optical activity in a dissymmetric molecular structure [50].

2. Each vertex is associated with a matrix element of the interaction potential V; the numerator for the diagram is the product of these matrix elements listed in order, starting with the final vertex.

3. The denominator of a nth-order diagram, D_n, requires $n-1$ factors corresponding to the first $n-1$ vertices, that is,

$$D_n = D_1 \times D_2 \times \ldots D_{n-1}. \tag{10.119}$$

4. The pth factor, D_p in a denominator is given by

$$
\begin{aligned}
D_p \;=\;& \text{energy of the initial molecular state } E_n^0 \\
-\;& \text{energy of } p\text{th intermediate state } E_m^0 \\
-\;& \hbar \,(\text{sum of emitted photon frequencies in the first } p \text{ vertices}) \\
+\;& \hbar \,(\text{sum of absorbed photon frequencies in the first } p \text{ vertices}) \\
+\;& i\varepsilon.
\end{aligned}
$$

$$\tag{10.120}$$

The use of the diagrams and these rules will be illustrated through a few examples in the next section.

10.4 Absorption, Emission and Scattering – the Basic Processes

This section is devoted to an outline of the evaluation of the basic diagrams, Figures 10.1–10.3 using the Coulomb gauge Hamiltonian for a charged particle interacting with the quantised electromagnetic field; after that, the extension to the physically interesting cases involving many charges (atoms, molecules, condensed matter, plasmas etc.) will be seen to be quite straightforward. The matrix elements required are given in Appendix F.

Figure 10.1 shows the primitive absorption and emission vertices that correspond to first-order perturbation theory; there is no denominator to evaluate. Consider the absorption vertex; according to Eq. (F.1.5), the perturbation operator is

$$K_a^1 = -\sum_{\mathbf{q},\sigma} \mathbf{f}_e(\mathbf{q}) \cdot \hat{\boldsymbol{\varepsilon}}(\mathbf{q})_\sigma \, c_{\mathbf{q},\sigma}. \tag{10.121}$$

If the initial and final states for the absorption of a photon by a free charge are $\Phi_n^0 = |\varphi_{\mathbf{P}}, \mu[n_{\mathbf{Q}}]\rangle$ and $\Phi_k^0 = |\varphi_{\mathbf{P}'}, \mu[n_{\mathbf{Q}} - 1]\rangle$, respectively, the matrix element is (cf. (F.1.8))

$$\langle \Phi_k^0 | K_a^1 | \Phi_n^0 \rangle = -\frac{e}{m} \sqrt{\frac{\hbar^2}{2\varepsilon_0 \Omega \mathcal{E}_{\mathbf{Q}}}} \mathbf{P} \cdot \hat{\boldsymbol{\varepsilon}}(\mathbf{Q})_\mu \delta^3(\mathbf{P} + \hbar\mathbf{Q} - \mathbf{P}'). \tag{10.122}$$

The emission vertex has the same form, with $\mathbf{Q} \to -\mathbf{Q}$.

We have to recognise that a free charge cannot absorb or emit a (real) photon because energy cannot be conserved in such a transition. To see this, consider a charge initially at rest and an incident photon with wave vector $\mathbf{Q}$. After absorbing the photon, the particle must have momentum $\hbar\mathbf{Q}$. Thus, we have

$$E_n^0 = \hbar Q c, \qquad \mathbf{P}_n^0 = \hbar\mathbf{Q}, \tag{10.123}$$

$$E_k^0 = \frac{\hbar^2 Q^2}{2m}, \qquad \mathbf{P}_k^0 = \hbar\mathbf{Q}, \tag{10.124}$$

for the initial and final energy and momentum of the (charge + photon) system. But since the final speed of the particle is $v_k = \hbar Q/m$, conservation of energy would require $v_k = 2c$ which is impossible.[12] Photons are absorbed and emitted by free charges in *virtual* transitions to which energy conservation does not apply.

It is easy to see that collections of bound charges have a different behaviour as regards absorption and emission of photons because the requirement for momentum conservation can be met by the centre-of-mass motion of the composite ('recoil'), and photon absorption/emission can be associated with transitions involving the internal

---

[12] The same conclusion is obtained if one uses the relativistic relation between energy and momentum, $(E_k^0)^2 = |\mathbf{P}|_k^2 c^2 + m^2 c^4$, and adds the invariant energy mc^2 to E_n^0; evidently, $(E_k^0)^2 < (E_n^0)^2$.

states. It is then possible to satisfy the requirements for energy and momentum conservation simultaneously; thus, atoms and molecules undergo real energy-conserving
transitions involving the absorption and emission of photons. The first-order amplitude $V_{fi}^{(1)}$ in (10.35) is non-zero and yields the Einstein coefficients for absorption and
emission of radiation [41].

To demonstrate this, consider the interaction between a pair of molecular states
(N, N') and a photon in the mode of the radiation field $(\mathbf{k}, \lambda)$, described by the diagrams
in Figure 10.1. The relevant reference states are

$$|1\rangle = |N, \lambda [n_{\mathbf{k}} + 1]\rangle \quad |1'\rangle = |N', \lambda [n_{\mathbf{k}}]\rangle, \tag{10.125}$$

with energies

$$E_1^0 = E_N + (n_{\mathbf{k}} + 1)\hbar kc, \quad E_{1'}^0 = E_{N'} + n_{\mathbf{k}}\hbar kc. \tag{10.126}$$

The energy conservation condition for a real transition requires

$$\hbar kc = E_{N'} - E_N, \tag{10.127}$$

which is the Bohr frequency rule.

The one-photon absorption is described by the matrix element $\left(\mathsf{M}_a^{(1)}\right)_{n_{\mathbf{k}}(n_{\mathbf{k}}+1)}^{N',N}$, and
the usual Fermi golden rule construction yields the absorption rate as

$$\tau_{\mathbf{k},\lambda}^{-1} = \frac{2\pi}{\hbar} \left| \left(\mathsf{M}_e^{(1)}\right)_{n_{\mathbf{k}}(n_{\mathbf{k}}+1)}^{N',N} \right|^2 \rho_{\mathbf{k}}, \tag{10.128}$$

with

$$\rho_{\mathbf{k}} = \frac{\Omega}{(2\pi)^3} \left(\frac{k^2}{\hbar c}\right) d\Omega_{\mathbf{k}}. \tag{10.129}$$

The accompanying transfer of momentum to the atomic system is the microscopic
account of radiation pressure. The force exerted on the particle is equal to the rate
of absorption (10.128) multiplied by the momentum absorbed. Since each photon in
the mode $(\mathbf{k}, \lambda)$ carries linear momentum $\hbar \mathbf{k}$, the momentum transferred is $n_{\mathbf{k}}\hbar \mathbf{k}$ for $n_{\mathbf{k}}$
absorbed photons. Although the effect was confirmed experimentally at the start of the
twentieth century [51], [52], it was generally of little experimental significance before
the advent of the laser. Ashkin[13] was the first to show that a laser beam could be used
to manipulate spatially small dielectric particles with the discovery of optical trapping
[53].

The matrix element for one-photon emission is $\left(\mathsf{M}_e^{(1)}\right)_{(n_{\mathbf{k}}+1)n_{\mathbf{k}}}^{N',N}$, and this is associated
with the loss of momentum by the atomic system. The rates of absorption and emission
for the pair of molecular states in the one-photon approximation are related by

$$\frac{R_e}{R_a} = \frac{|M_e^{(1)}|^2}{|M_a^{(1)}|^2} = \frac{n_{\mathbf{k}} + 1}{n_{\mathbf{k}}}, \tag{10.130}$$

[13] Arthur Ashkin (1922–2020) shared the 2018 Nobel Prize for Physics.

as first shown by Einstein [54]. The first term in the numerator of (10.130) describes stimulated emission, while the 1 accounts for the spontaneous emission rate.

The reduction of the internal state matrix element in (10.128) takes various forms depending on the magnitude of the photon wave vector and whether the molecular internal states are bound. For bound states and long wavelengths a multipole expansion,

$$e^{i\mathbf{k}\cdot\mathbf{r}_n} \approx 1 + i\mathbf{k}\cdot\mathbf{r}_n + \dots, \tag{10.131}$$

is appropriate; the expansion of the exponential must be made after the centre-of-mass coordinate has been separated out. The $\mathbf{k} \to 0$ limit is usually called the 'long-wavelength' or 'dipole' approximation, and the required matrix element reduces to $i\omega\mathbf{d}_{N'N}$, where $\mathbf{d}$ is the electric dipole operator, and ω is the circular frequency of the radiation. The approximation amounts to ignoring the photon momentum in comparison with the average momentum of the bound particles. For X-ray photons, the complete exponential may need to be retained.

In the electric dipole approximation, (10.128) yields

$$\tau_{\mathbf{k},\lambda}^{-1} = \frac{(2\pi)^2 v^3}{\hbar c^3} \frac{|\mathbf{d}_{N'N} \cdot \hat{\boldsymbol{\varepsilon}}_\lambda|^2}{4\pi\varepsilon_0} \, d\Omega_{\mathbf{k}}. \tag{10.132}$$

Suppose the angle between the atomic dipole vector and the photon wave vector is θ; then the overall rate for all polarisation directions, $\tau_{\mathbf{k}}^{-1}$, is obtained by summing (10.132) over λ (using (2.26) with $i = j$):

$$\tau_{\mathbf{k}}^{-1} = \frac{(2\pi)^2 v^3}{\hbar c^3} \frac{|\mathbf{d}_{N'N}|^2}{4\pi\varepsilon_0} \sin^2(\theta) \, d\Omega_{\mathbf{k}}. \tag{10.133}$$

The total rate for emission of photons regardless of solid angle is obtained by integration over $d\Omega_{\mathbf{k}}$. The integration is elementary, and the result is that

$$\tau^{-1}(\text{total}) = \frac{32\pi^3}{3} \frac{v^3}{\hbar c^3} \frac{|\mathbf{d}_{N'N}|^2}{4\pi\varepsilon_0}, \tag{10.134}$$

which is the Einstein A-coefficient for electric dipole radiation.

A free particle can exchange energy and momentum with any other particle (atom, electron, photon etc.) in a scattering event. Before making any detailed calculations, it may be helpful to review briefly the kinematics for the collision of two freely moving particles with velocities $\mathbf{v}_1, \mathbf{v}_2$ at non-relativistic energies. In the centre-of-mass frame before the collision, we have

$$\mathbf{P}_n^1 = \mathbf{p}, \quad \mathbf{P}_n^2 = -\mathbf{p}, \quad \mathbf{P}_n^0 = \mathbf{0} \tag{10.135}$$

$$E_n^0 = E_n^1 + E_n^2 = \frac{p^2}{2m_1} + \frac{p^2}{2m_2} \equiv \frac{p^2}{2\mu}, \tag{10.136}$$

where μ is the reduced mass of the pair,

$$\mu = \frac{m_1 m_2}{m_1 + m_2}. \tag{10.137}$$

After the collision we have, in order to satisfy the conservation laws,

$$\mathbf{P}_k^1 = \mathbf{p}', \quad \mathbf{P}_k^2 = \mathbf{P}', \tag{10.138}$$

$$E_k^0 = E_k^1 + E_k^2 = \frac{p'^2}{2m_1} + \frac{P'^2}{2m_2} = E_n^0 = \frac{p^2}{2\mu}, \tag{10.139}$$

which require

$$\mathbf{p}' = -\mathbf{P}' \longrightarrow |\mathbf{p}'| = |\mathbf{P}'| = p. \tag{10.140}$$

The magnitude of the momentum transfer vector $\mathbf{q}$ is then simply expressed in terms of the scattering angle θ,

$$q^2 = |\mathbf{p} - \mathbf{p}'|^2 = 2p^2(1 - \cos(\theta)) \equiv 4p^2 \sin^2(\theta/2). \tag{10.141}$$

The quantum mechanical description is based on the golden rule formula (10.9) for the transition rate. In first approximation this requires the evaluation of the off-diagonal T-matrix element $T_{kn}^{(2)}$ defined in Eq. (10.35) for pairs of states Φ_n^0, Φ_k^0 compatible with the conservation laws. It corresponds to the diagrams in Figure 10.2. A wholly conventional account of this process for two charges, $\{e_1, e_2\}$, is based on the *choice* of the Coulomb gauge condition so that (cf. Chapter 9) $V^{(2)}$ in Eq. (10.35) is the familiar static Coulomb interaction for the pair, and the virtual photon exchange terms involving $V^{(1)}$ are dropped. In the initial and final states there are then no photons present, so these are

$$|n\rangle = |\varphi_n, [0]\rangle, \quad |k\rangle = |\varphi_k, [0]\rangle \tag{10.142}$$

with energies and momenta given by (10.135)–(10.139). It should be noted that in electrodynamics it is the T-matrix element that is gauge invariant so that in an arbitrary gauge all three diagrams will contribute. In view of the kinematics just described, the matrix element of this Coulomb interaction for a transition with momentum transfer $\mathbf{q}$ yields its Fourier transform evaluated at $\mathbf{q}$ and the momentum conservation delta function so that M_{kn} in (10.10) is

$$M_{kn} = -2i\left(\frac{1}{2\pi\hbar}\right)\left(\frac{e_1 e_2}{4\pi\varepsilon_0}\right)\left(\frac{1}{q^2}\right), \tag{10.143}$$

and (10.24) reduces to the familiar form of Rutherford's classical formula for Coulomb scattering of an α-particle ($e_1 = +2e$) from an atomic nucleus ($e_2 = +Ze$),

$$d\sigma = \left(\frac{e_1 e_2}{4\pi\varepsilon_0}\right)^2 \frac{1}{(4E_n^0)^2 \sin^4(\theta/2)} d\Omega. \tag{10.144}$$

It is notable that it is independent of Planck's constant; the quantum corrections in the exact scattering amplitude obtained from solving the two-body Schrödinger equation with a Coulomb potential appear only in its phase,[14] and so they do not modify the cross section (10.144).

Next we must evaluate the contributions from Figures 10.2iia,b. The initial and final states for these diagrams are as before, but now the photon vacuum, $|[0]\rangle$, has a role. The intermediate state in Figure 10.2iia is

$$|a\rangle = |\varphi_a^0 \mu[1_\mathbf{Q}]\rangle, \tag{10.145}$$

---

[14] This result is only true for the $1/r$ potential!

where $|a\rangle$ has e_1 in its initial state, e_2 in its final state and a virtual photon. According to the rules for diagrams we therefore write

$$\text{Numerator} = \langle \varphi_k^0[0]|V^1|\mu[1\,\mathbf{Q}]\varphi_a^0\rangle\langle\varphi_a^0\mu[1\,\mathbf{Q}]|V^1|\varphi_n^0[0]\rangle$$
$$\text{Denominator} = E_n^2 - E_k^2 - \mathcal{E}_{\mathbf{Q}}, \tag{10.146}$$

where $\mathcal{E}_{\mathbf{Q}}$ is the energy of the virtual photon ($\hbar Q c$).

The perturbation operator V^1 in the Coulomb gauge is given in Appendix F: equations G.0.1–G.0.4. Integrating out the photon variables leads directly to the diagram value,

$$\text{Diagram} 10.2\text{iia} \;\rightarrow\; \sum_{\mathbf{Q}} \frac{\mathbf{f}_{ka}^{(1)}(\mathbf{Q})\mathbf{f}_{an}^{(2)}(-\mathbf{Q}):\mathcal{B}(\mathbf{Q})}{E_n^2 - E_k^2 - \mathcal{E}_{\mathbf{Q}}}, \tag{10.147}$$

where we have used Eq. (2.26) to carry out the sum over the virtual photon's polarisation vectors, and

$$\mathbf{f}^{(\xi)}(\mathbf{Q}) = \frac{e_\xi}{m_\xi}\sqrt{\frac{\hbar^2}{2\varepsilon_0\Omega\mathcal{E}_{\mathbf{Q}}}}\mathbf{p}_\xi e^{i\mathbf{Q}\cdot\mathbf{x}_\xi}, \qquad \xi = 1,2 \tag{10.148}$$

is essentially a Fourier component of the current density operator for the charge ξ.

Diagram 10.2iib describes the same pair of interactions with the roles of 1 and 2 interchanged; this has the effect of simply putting $\mathbf{Q} \rightarrow -\mathbf{Q}$ in Eq. (10.147), and so the two diagrams have the same numerator since they are to be summed over all $\mathbf{Q}$. Adding the two contributions together yields

$$\sum_{\mathbf{Q}} \mathbf{f}_{ka}^{(1)}(\mathbf{Q})\mathbf{f}_{an}^{(2)}(-\mathbf{Q}):\mathcal{B}(\mathbf{Q})\left[\frac{1}{E_n^2 - E_k^2 - \mathcal{E}_{\mathbf{Q}}} + \frac{1}{E_n^1 - E_k^1 - \mathcal{E}_{\mathbf{Q}}}\right]. \tag{10.149}$$

Conservation of energy enforced by the energy conservation Dirac delta function in the transition rate implies that $E_n^1 + E_n^2 = E_k^1 + E_k^2$ so that $E_n^2 - E_k^2 = -(E_n^1 - E_k^1)$ and hence

$$\frac{1}{E_n^2 - E_k^2 - \mathcal{E}_{\mathbf{Q}}} + \frac{1}{E_n^1 - E_k^1 + \mathcal{E}_{\mathbf{Q}}} = \frac{2\mathcal{E}_{\mathbf{Q}}}{(E_n^1 - E_k^1)^2 - \mathcal{E}_{\mathbf{Q}}^2}. \tag{10.150}$$

Thus, the electrodynamical contribution to the T-matrix element from these two diagrams is

$$2\sum_{\mathbf{Q}} \frac{\mathcal{E}_{\mathbf{Q}}\mathbf{f}_{ka}^{(1)}(\mathbf{Q})\mathbf{f}_{an}^{(2)}(-\mathbf{Q}):\mathcal{B}(\mathbf{Q})}{(E_n^1 - E_k^1)^2 - \mathcal{E}_{\mathbf{Q}}^2}. \tag{10.151}$$

Written out fully using the one-particle wave functions from (10.6), this is

$$\frac{e_1 e_2}{\varepsilon_0}\frac{\hbar^2}{m_1 m_2}\frac{1}{\Omega}\sum_{\mathbf{Q}}\langle\varphi_{\mathbf{P}_k^1}(\mathbf{x}_1)|e^{i\hbar\mathbf{Q}\cdot\mathbf{x}_1/\hbar}\mathbf{p}_1|\varphi_{\mathbf{P}_n^1}(\mathbf{x}_1)\rangle$$
$$\times\;\langle\varphi_{\mathbf{P}_k^2}(\mathbf{x}_2)|e^{-i\hbar\mathbf{Q}\cdot\mathbf{x}_2/\hbar}\mathbf{p}_2|\varphi_{\mathbf{P}_n^2}(\mathbf{x}_2)\rangle:\mathcal{B}(\mathbf{Q})\frac{1}{(E_n^1 - E_k^1)^2 - \mathcal{E}_{\mathbf{Q}}^2}. \tag{10.152}$$

Explicit evaluation using the rule

$$\sum_{\mathbf{Q}} \rightarrow \frac{\Omega}{(2\pi)^3} \int d^3\mathbf{Q} \tag{10.153}$$

puts $\mathbf{Q} \rightarrow \mathbf{q} = \mathbf{p} - \mathbf{p}'$ and produces an overall factor of $\delta^3(\mathbf{P}_n^0 - \mathbf{P}_k^0)$, making explicit the conservation of momentum in the scattering process. This leads to

$$M_{kn} = -2i\left(\frac{1}{2\pi\hbar}\right)\left(\frac{e_1 e_2}{4\pi\varepsilon_0}\right)\frac{\mathbf{P}_n^1}{m_1}\frac{\mathbf{P}_n^2}{m_2} : \mathcal{B}(\mathbf{q})\frac{1}{(E_n^1 - E_k^1)^2 - \mathcal{E}_{\mathbf{q}}^2}, \tag{10.154}$$

which must be added to (10.143).

It is of interest to compare this calculation with the contribution from the static Coulomb potential. The two-particle matrix element of the Coulomb potential can be put in a similar form by using the relation

$$\frac{1}{4\pi|\mathbf{x}_1 - \mathbf{x}_2|} = \frac{1}{(2\pi)^3}\int \frac{e^{i\mathbf{Q}\cdot(\mathbf{x}_1 - \mathbf{x}_2)}}{Q^2} d^3\mathbf{Q}. \tag{10.155}$$

Then $\langle\Phi_k^0|V^{(2)}|\Phi_n^0\rangle$ factorises into a product of one-particle matrix elements and, mindful of (10.153), a sum over a 'virtual' wave vector $\mathbf{Q}$, explicitly,

$$\langle\Phi_k^0|V^{(2)}|\Phi_n^0\rangle = \frac{e_1 e_2}{\varepsilon_0}\frac{1}{\Omega}\sum_{\mathbf{Q}}\langle\varphi_{\mathbf{P}_k^1}(\mathbf{x}_1)|e^{i\hbar\mathbf{Q}\cdot\mathbf{x}_1/\hbar}|\varphi_{\mathbf{P}_n^1}(\mathbf{x}_1)\rangle$$

$$\times \langle\varphi_{\mathbf{P}_k^2}(\mathbf{x}_2)|e^{-i\hbar\mathbf{Q}\cdot\mathbf{x}_2/\hbar}|\varphi_{\mathbf{P}_n^2}(\mathbf{x}_2)\rangle\frac{1}{Q^2}. \tag{10.156}$$

The integrals over the coordinates in the matrix elements in (10.152) and (10.156) are the same and lead to the same delta functions, and the sum over $\mathbf{Q}$ yields the momentum conservation delta function as a factor of $T_{kn}^{(2)}$. Assuming $|E_n^1 - E_k^1|$ is 'small' compared to $\mathcal{E}_{\mathbf{q}}$ (it vanishes for elastic scattering), we see that (10.152) has the same dependence on $|\mathbf{q}|$ as (10.156), $(1/q^2)$, but is $O(v_1 v_2/c^2)$ compared to the static contribution.

The diagrams in Figure 10.2 can be used for a perturbation theory calculation of a wide variety of physical phenomena. Their original application was to the scattering interaction between two charged particles, for example, e-e scattering, or the scattering of an α-particle by an atomic nucleus as previously. However, there is no restriction to charged species, as overall neutral charge distributions can still be involved; thus, an important application is to electron diffraction by atoms and molecules in both rarefied and condensed phases of matter. A perturbation theory description of electron scattering from a molecular system is at best only appropriate for 'fast' electrons such as are used in electron diffraction experiments; it requires the neglect of the consequences of the Pauli exclusion principle (Chapter 5) for fast collisions [55]. The conventional account is based on the first-order perturbation treatment of the instantaneous Coulombic interaction. At lower energies the antisymmetry requirement for all of the electrons must be respected; such cases are dealt with as scattering theory problems for some approximation to the Coulomb Hamiltonian, and electrodynamical effects are not considered.

The T-matrix elements for electron-molecule scattering can be obtained directly by an obvious modification of (10.152) and (10.156); we identify 'particle 1' as the fast electron, and 'particle 2' as the molecule. Then we make the substitutions

$$\frac{e_2}{m_2}\langle\varphi_{\mathbf{P}_k^2}(\mathbf{x}_2)|e^{-i\hbar\mathbf{Q}\cdot\mathbf{x}_2/\hbar}\mathbf{p}_2|\varphi_{\mathbf{P}_n^2}(\mathbf{x}_2)\rangle \rightarrow \langle\varphi_k|\sum_\alpha\frac{e_\alpha}{m_\alpha}\mathbf{p}_\alpha e^{-i\hbar\,\mathbf{Q}\cdot\mathbf{x}_\alpha/\hbar}|\varphi_n\rangle$$

$$= \mathbf{J}(-\mathbf{Q})_{kn} \tag{10.157}$$

$$e_2\langle\varphi_{\mathbf{P}_k^2}(\mathbf{x}_2)|e^{-i\hbar\mathbf{Q}\cdot\mathbf{x}_2/\hbar}|\varphi_{\mathbf{P}_n^2}(\mathbf{x}_2)\rangle \rightarrow \langle\varphi_k|\sum_\alpha e_\alpha e^{-i\hbar\mathbf{Q}\cdot\mathbf{x}_\alpha/\hbar}|\varphi_n\rangle \tag{10.158}$$

in (10.152) and (10.156), respectively. The coordinate and momentum operators in (10.157) and (10.158) can be separated into the centre-of-mass and internal variables, as discussed in Chapter 8.

$$\mathbf{x}_\alpha = \mathbf{R}_{cm} + \mathbf{r}_\alpha \qquad \mathbf{p}_\alpha = \frac{m_\alpha}{M}\mathbf{P}_{cm} + \boldsymbol{\sigma}_\alpha. \tag{10.159}$$

The $\{\mathbf{r}_\alpha, \boldsymbol{\sigma}_\alpha\}$ are linearly dependent but can be transformed to independent translationally invariant variables. At the same time the molecular states are factorised as $\varphi_n = \varphi_{cm}^n\phi_s(\{\mathbf{r}_\alpha\})$. Conservation of momentum can be satisfied formally by choosing the centre-of-mass wave functions as momentum eigenstates,

$$\varphi_{cm}^n \sim e^{i\mathbf{P}_{cm}^n\cdot\mathbf{R}_{cm}/\hbar}. \tag{10.160}$$

However, in many applications recoil is neglected and the molecule is treated as fixed in space:

$$\varphi_{cm}^n \sim \delta^3(\mathbf{R}_{cm} - \mathbf{X}). \tag{10.161}$$

The matrix element for the direct Coulomb interaction may then be obtained by the replacement,

$$\frac{e_1e_2}{q^2} \rightarrow e^2\frac{\Gamma(-\mathbf{q})_{s's}}{q^2}, \tag{10.162}$$

where we have put $e_\alpha = Z_\alpha e$ for each charge and defined the electric form-factor as a matrix element involving only internal molecular variables,

$$\Gamma(\mathbf{k})_{s's} = \langle\phi_{s'}|\sum_\alpha Z_\alpha e^{i\hbar\mathbf{k}\cdot\mathbf{r}_\alpha/\hbar}|\phi_s\rangle. \tag{10.163}$$

Similarly the electrodynamical contribution is obtained by the substitution

$$e_2\frac{\mathbf{P}_n^2}{m_2} \rightarrow \mathbf{J}(-\mathbf{q})_{s's} \tag{10.164}$$

after integrating out the centre-of-mass variables using (10.160).

A quantum formula for the scattering of electromagnetic radiation by matter was first obtained from a correspondence principle argument applied to classical light dispersion theory [56]; shortly afterwards it was derived by Dirac using quantum mechanics [57] with the electric dipole approximation. It is known as the Kramers–Heisenberg dispersion formula. Intensity-independent light scattering again requires simply the T-matrix elements in (10.35) to second order which can be obtained by evaluating the diagrams in Figure 10.3. As was the case for Figure 10.2, these diagrams can be interpreted to correspond to a diverse range of light scattering phenomena. First, however, consider the process in which a photon in the mode $(\mathbf{k}, \lambda)$ is scattered into the mode $(\mathbf{k}', \lambda')$ by a single charge e. The extension to scattering by many-body systems is quite straightforward.

The initial and final states can be taken to be

$$\Phi_n^0 = |\varphi_n, \lambda[n_{\mathbf{k}}]\rangle, \quad E_n^0 = E_n + n_{\mathbf{k}}\mathcal{E}_{\mathbf{k}}$$
$$\Phi_k^0 = |\varphi_k, \lambda[n_{\mathbf{k}} - 1], \lambda'[1_{\mathbf{k}}]\rangle, \quad E_k^0 = E_k + (n_{\mathbf{k}} - 1)\mathcal{E}_{\mathbf{k}} + \mathcal{E}_{\mathbf{k}'}, \quad (10.165)$$

where the $\{\varphi\}$ are wave functions for the charge. The contribution of diagram 10.3i can be written down directly from the results in Appendix F,

$$\langle \Phi_k^0 | V^2 | \Phi_n^0 \rangle = \frac{e^2}{2m} \frac{\hbar}{2\varepsilon_0 \Omega c} \sqrt{\frac{n_{\mathbf{k}}}{kk'}} \hat{\mathbf{e}}(\mathbf{k})_\lambda \cdot \hat{\mathbf{e}}(\mathbf{k}')_{\lambda'} \langle \varphi_k | e^{i(\mathbf{k}-\mathbf{k}')\cdot\mathbf{x}} | \varphi_n \rangle. \quad (10.166)$$

If the $\{\varphi\}$ are momentum eigenstates (e.g. for a free charge), the remaining matrix element reduces simply to the delta function, expressing overall conservation of momentum.

For diagrams 10.3iia,b, we need to consider two intermediate states:

$$\Phi_a^0 = |\varphi_\sigma, \lambda[n_{\mathbf{k}} - 1]\rangle, \quad E_a^0 = E_\sigma + (n_{\mathbf{k}} - 1)\mathcal{E}_{\mathbf{k}}$$
$$\Phi_b^0 = |\varphi_\sigma, \lambda[n_{\mathbf{k}}], \lambda'[1_{\mathbf{k}'}]\rangle, \quad E_b^0 = E_\sigma + n_{\mathbf{k}}\mathcal{E}_{\mathbf{k}} + \mathcal{E}_{\mathbf{k}'}. \quad (10.167)$$

Diagram 10.3iib then yields for the numerator and denominator:

$$\text{Num.} : \langle \varphi_k | \mathbf{f}_a(\mathbf{k}) \cdot \hat{\mathbf{e}}(\mathbf{k})_\lambda | \varphi_\sigma \rangle \langle \varphi_\sigma | \mathbf{f}_e(-\mathbf{k}') \cdot \hat{\mathbf{e}}(\mathbf{k}')_{\lambda'} | \varphi_n \rangle$$
$$= \frac{e^2}{m^2} \frac{\hbar^2}{2\varepsilon_0 \Omega c} \sqrt{\frac{n_{\mathbf{k}}}{kk'}} \langle \varphi_k | \mathbf{P} \cdot \hat{\mathbf{e}}(\mathbf{k})_\lambda e^{i\mathbf{k}\cdot\mathbf{x}} | \varphi_\sigma \rangle \langle \varphi_\sigma | \mathbf{P} \cdot \hat{\mathbf{e}}(\mathbf{k}')_{\lambda'} e^{-i\mathbf{k}'\cdot\mathbf{x}} | \varphi_n \rangle \quad (10.168)$$
$$\text{Den.} : E_n^0 - E_b^0 + i\varepsilon \equiv E_n - E_\sigma - \mathcal{E}_{\mathbf{k}'} + i\varepsilon. \quad (10.169)$$

The numerator for diagram 10.3iia is obtained from (10.168) by simply exchanging the positions of $\mathbf{f}_a$ and $\mathbf{f}_e$ in the two matrix elements, while its denominator has $-\mathcal{E}_{\mathbf{k}}$ replaced by $\mathcal{E}_{\mathbf{k}}$. As before, if the $\{\varphi\}$ are momentum eigenstates, the operator $\mathbf{P}$ is replaced by the appropriate eigenvalues and the matrix elements can then be evaluated to yield the familiar delta functions with momentum transfer arguments. The sum over states in the perturbation formula (10.35) finally yields the overall conservation of momentum delta function. The T-matrix element T_{kn}^2 is then the sum of these three terms.

For the case of high-energy photons, a widely made approximation is to retain only the contribution of the $|\mathbf{A}|^2$ term which gives a differential cross section for scattering by a free charge,

$$d\sigma = r_q^2 |\hat{\boldsymbol{e}}_\lambda \cdot \hat{\boldsymbol{e}}_{\lambda'}|^2 \, d\Omega, \tag{10.170}$$

where r_q is the classical parameter, Eq. (1.6).[15] In a typical case, the observable cross section is a Boltzmann weighted average; moreover, if rotational invariance can be invoked, as, for example, scattering from an isotropic distribution of electrons in a plasma, the whole cross section is proportional to (10.170). If we define the scattering angle, θ, to be the angle between the incident ($\mathbf{k}$) and scattered ($\mathbf{k}'$) wave vectors, and the azimuthal angle, ϕ, to be the angle between the scattering plane and the initial plane of polarisation, then [58]

$$|\hat{\boldsymbol{e}}_\lambda \cdot \hat{\boldsymbol{e}}_{\lambda'}|^2 \to 1 - \cos^2(\theta) \sin^2(\phi). \tag{10.171}$$

For unpolarised radiation and no measurement of the polarisation of the scattered radiation, the cross section (10.170) reduces to the classical Thomson formula for scattering by an electron,

$$d\sigma^T(\theta) = \tfrac{1}{2} r_e^2 (1 + \cos^2(\theta)) \, d\Omega. \tag{10.172}$$

The extension to photon scattering from a many-body system is immediate; we simply have to add an index to the variables for the charges and sum over all charges. With the notation of (10.158) and setting

$$\Upsilon(\mathbf{k})_{kn} = \langle \varphi_k | \sum_\alpha \frac{e_\alpha^2}{2m_\alpha} e^{i\mathbf{k} \cdot \mathbf{x}_\alpha} | \varphi_n \rangle, \tag{10.173}$$

the result is

$$T_{kn}^{(2)} = \frac{\hbar}{2\varepsilon_0 \Omega c} \sqrt{\frac{n_\mathbf{k}}{kk'}} \, \hat{\boldsymbol{e}}(\mathbf{k})_\lambda^\mu \hat{\boldsymbol{e}}(\mathbf{k}')_{\lambda'}^\nu$$

$$\times \left[\Upsilon(\mathbf{q})_{kn} \delta_{\mu\nu} + \sum_p \frac{J(\mathbf{k})_{kp}^\mu J(-\mathbf{k}')_{pn}^\nu}{E_n - E_p + \hbar kc + i\varepsilon} + \frac{J(-\mathbf{k}')_{kp}^\nu J(\mathbf{k})_{pn}^\mu}{E_n - E_p - \hbar kc + i\varepsilon} \right]. \tag{10.174}$$

The calculation can be used to describe Thomson and Compton scattering of a high-energy photon by an electron in matter. The former case is associated with non-relativistic or near-relativistic energies (X-rays) and is an example of coherent scattering. Polarisation studies are important now that polarised X-rays are readily available from synchotron light sources. X-ray Thomson scattering is widely used for the study of plasmas where there are free electrons; the electron density can be obtained from the intensity of scattering, and plasma temperature is accessible through Doppler broadening of transitions. Compton scattering is associated with photons of much higher energy such that significant momentum is exchanged in the collision.[16] In this case, the electron is effectively free and the scattering is incoherent [58]. Its potential use to give a quantum mechanical description of X-ray diffraction will be considered in Chapter 12.

---

[15] Evaluated with the appropriate charge parameter, q.

[16] In terms of the *Compton wavelength*, $\lambda_C = \hbar/mc$, Thomson scattering corresponds to $\lambda^{\mathrm{ph}} > \lambda_C$ and Compton scattering to $\lambda^{\mathrm{ph}} \ll \lambda_C$. Since the Compton effect is only significant for high-energy photons, a realistic calculation requires the full apparatus of covariant QED; the result is the Klein–Nishina formula [6].

The diagrams are equally applicable to familiar atomic/molecular scattering processes involving bound electrons and radiation of much longer wavelengths than X-rays. Suppose that the initial state has a molecule in its ground state $|\phi_N\rangle$ and a radiation field state containing $n_{\mathbf{k}}$ photons polarised along $\hat{\mathbf{e}}(\mathbf{k})_\lambda$ with wave vector $\mathbf{k}$, thus

$$|n\rangle = |\phi_N, \lambda[n_{\mathbf{k}}]\rangle, \quad E_n^0 = E_N + n_{\mathbf{k}}\mathcal{E}_{\mathbf{k}}, \tag{10.175}$$

where $\mathcal{E}_{\mathbf{k}} = \hbar k c$ is the energy of a photon. The final state of the radiation field monitored at the detector far from the interaction zone must correspond to a state of the (molecule + field)-system having the same overall energy as the initial state, and this can be achieved in several ways corresponding to different physical processes. If the photons monitored have the same energy as the initial beam, one infers that the final state of the molecule is its ground state; the possible changes in the radiation field can involve the photon momentum and its polarisation state. If there is no change in momentum, the process is forward scattering and there arises a description of birefringence manifested through changes in the photon polarisation. Measurement of non-forward scattering involves placing the detector at some angle θ to the initial beam direction and corresponds to a change in the photon momentum $\mathbf{k}$, and possibly the polarisation state. If $k = |\mathbf{k}|$ is unchanged, one has Rayleigh scattering, and one infers that the final state of the molecule is again the ground state (since $\hbar k c$ is unchanged); finally, if the photon energy is changed, the final state of the molecule must be an excited state, and one has Raman Stokes scattering.[17] Not surprisingly, the angular dependence of the Thomson scattering cross section for an electron is a factor of the general Kramers–Heisenberg dispersion formula for atomic/molecular light scattering if polarisation characteristics are not measured.

By way of introduction, we assume the electric dipole approximation so that the interaction operator in the lowest-order approximation is

$$\mathsf{V}^0 = -\mathbf{d}\cdot\mathbf{E}^\perp; \tag{10.176}$$

this is the limiting case of the PZW-transformed Hamiltonian with the magnetic field interaction terms dropped. The mode expansion of the electric field operator contains exponential factors which are approximated by setting

$$e^{\pm i\mathbf{k}\cdot\mathbf{x}_n} = e^{\pm i\mathbf{k}\cdot(\mathbf{R}_{cm}+\mathbf{r}_\alpha)}$$
$$\approx e^{\pm i\mathbf{k}\cdot\mathbf{R}_{cm}}. \tag{10.177}$$

According to the preceding discussion, there are just two terms to evaluate that may be thought of as the molecule absorbing a (real) photon either before or after emitting a (real) photon. The required diagrams are those in Figure 10.3.ii(a,b) which must be evaluated with the interaction (10.176). With the restrictions just listed, Figure 10.3.i makes no contribution since in the PZW formalism it corresponds to an interaction

---

[17] Raman anti-Stokes scattering takes place from excited vibration-rotation states of the molecule, with the final state of the molecule having lower energy; this situation is best described by a density matrix formulation.

mediated by the generalised diamagnetic susceptibility contracted with the square of the magnetic field. For definiteness take the final state to be described by

$$|k\rangle = |\varphi_s, \lambda[n_{\mathbf{k}} - 1], \lambda'[1_{\mathbf{k}'}]\rangle, \quad E_k^0 = E_{s'} + \mathcal{E}_{\mathbf{k}'} + (n_{\mathbf{k}} - 1)\mathcal{E}_{\mathbf{k}}, \tag{10.178}$$

which describes Rayleigh scattering.[18] The two possible intermediate states are

$$|a\rangle = |\varphi_s'', \lambda[n_{\mathbf{k}} - 1]\rangle, \quad E_a^0 = E_{s''} + (n_{\mathbf{k}} - 1)\mathcal{E}_{\mathbf{k}} \tag{10.179}$$

$$|b\rangle = |\varphi_{s''}, \lambda[n_{\mathbf{k}}], \lambda'[1_{\mathbf{k}'}]\rangle, \quad E_b^0 = E_{s''} + n_{\mathbf{k}}\mathcal{E}_{\mathbf{k}} + \mathcal{E}_{\mathbf{k}'}. \tag{10.180}$$

Figure 10.3.ii(a) is evaluated according to the diagram rules as

$$B\sum_{s''} \frac{\langle s'|\mathbf{d}\cdot\hat{\boldsymbol{\varepsilon}}_\lambda|s''\rangle\langle s''|\mathbf{d}\cdot\hat{\boldsymbol{\varepsilon}}_{\lambda'}|s\rangle}{E_{ss''} + \hbar kc + i\varepsilon}, \tag{10.181}$$

while Figure 10.3.ii(b) gives

$$B\sum_{s''} \frac{\langle s'|\mathbf{d}\cdot\hat{\boldsymbol{\varepsilon}}_{\lambda'}|s''\rangle\langle s''|\mathbf{d}\cdot\hat{\boldsymbol{\varepsilon}}_\lambda|s\rangle}{E_{ss''} - \hbar kc + i\varepsilon}, \tag{10.182}$$

where

$$B = \frac{\hbar kc\sqrt{n_\lambda}}{2\Omega\varepsilon_0}. \tag{10.183}$$

The T-matrix element is then the sum of these two terms. We define

$$\mathcal{R}(k)_{s's}^{\alpha\beta} = \sum_{s''} \left[\frac{d_{s's''}^\alpha d_{s''s}^\beta}{k_{ss''} + k + i\varepsilon} + \frac{d_{s's''}^\beta d_{s''s}^\alpha}{k_{ss''} - k + i\varepsilon} \right], \quad k_{ss''} = \frac{E_s - E_{s''}}{\hbar c} \tag{10.184}$$

so that

$$T_{s's}^{(2)} = \frac{k\sqrt{n_{\mathbf{k}}}}{2\Omega\varepsilon_0} R_{s's}, \tag{10.185}$$

where

$$R_{s's} = \mathcal{R}_{s's}^{\alpha\beta} \hat{\boldsymbol{\varepsilon}}(\mathbf{k}')_{\lambda'}^\alpha \hat{\boldsymbol{\varepsilon}}(\mathbf{k})_\lambda^\beta. \tag{10.186}$$

With the usual construction of the scattering cross-section this leads to the familiar Rayleigh law depending on $1/\lambda^4$,

$$\mathrm{d}\sigma = \alpha^2 k^4 |R_{s's}|^2 \, \mathrm{d}\Omega(\mathbf{k}'), \tag{10.187}$$

where α is the fine structure constant, as usual, and $k = 2\pi/\lambda$.

We can isolate the polarisation vectors in the preceding cross section by defining

$$\Xi(\mathbf{k}', \mathbf{k})_{\alpha,\beta,\gamma,\delta} = \hat{\varepsilon}_{\lambda'}(\mathbf{k}')^\alpha \hat{\varepsilon}_\lambda(\mathbf{k})^\beta \hat{\varepsilon}_{\lambda'}(\mathbf{k}')^\gamma \hat{\varepsilon}_\lambda(\mathbf{k})^\delta. \tag{10.188}$$

An important special case occurs when the detector monitors photons of momentum $\mathbf{k}'$ without regard to polarisation; then we must sum the cross section over λ'. If the incident beam is also unpolarised, one gets

$$\Xi(\mathbf{k}', \mathbf{k})_{\alpha,\beta,\gamma,\delta} \to \frac{1}{2}[\mathcal{B}(\mathbf{k}')]_{\alpha,\gamma}\,[\mathcal{B}(\mathbf{k})]_{\beta,\delta}. \tag{10.189}$$

---

[18] We keep in mind here that $|\mathbf{k}| = |\mathbf{k}'| = k$ is the condition for Rayleigh scattering.

Its scalar part reduces Ξ to

$$\Xi_{\alpha,\beta,\gamma,\delta}\delta_{\alpha,\beta}\delta_{\gamma,\delta} = \frac{1}{2}(1+\cos^2(\theta)), \tag{10.190}$$

where θ is the usual scattering angle, and we recognise the classical Thomson angular dependence of the scattering as a factor in the cross-section formula.

Such calculations can be carried through with higher-order multipoles in the expansion, though usually the PZW perturbation operator is most useful when only the first few multipoles are retained. Equivalent calculations can be carried out using the Coulomb gauge Hamiltonian, and on the energy shell the same final results are obtained [17], [18], as expected from the general discussion in §10.2.1. Once again, the only role for the molecular centre of mass is to make conservation of momentum explicit; however, in non-forward scattering there is a change of momentum and the scattering is a further contribution to the radiation pressure. Forward scattering is the special case when the initial and final states are the same in an expression like (10.186). There are no transitions, and so a rate cannot be defined; instead it is interpreted as a change in the energy of the molecule, and this too can produce a so-called gradient force which is important in the spatial manipulation of 'particles' using a very precise design of the laser radiation source [59].

10.5 Birefringence

The light scattered by a molecule in a uniform static field (either electric or magnetic), when this field makes an angle to the propagation direction of the light beam, is in general elliptically polarised. The Kerr and Cotton–Mouton effects correspond to the electric and magnetic field cases, respectively [60]. The rotation of the plane of polarisation in the absence of any external fields ('optical activity') is the characteristic property of chiral substances; the same effect can be induced in any fluid substance by an external magnetic field applied along the direction of the light (the Faraday effect). Analogous birefringence phenomena in the absence of applied external fields may be induced by the intense optical fields of powerful lasers.

Certain kinds of processes cannot be described using the simple interaction (10.176); for example, it cannot describe chirality (a change in the polarisation state of the beam) in isotropic media since the optical rotation angle far from resonance depends purely on the imaginary part of the T-matrix. Since $\mathbf{d}$ is a real operator, (10.176) leads to a real T-matrix element; the generalised diamagnetic susceptibility (9.136) is also purely real and so cannot contribute to optical activity. For such a case one must introduce the magnetic dipole interaction involving the magnetic induction vector $\mathbf{B}$; the magnetic dipole operator is pure imaginary. This means giving up the assumption that the electric field is approximately uniform, and for consistency one must also include the electric quadrupole term that couples to the electric field gradient; thus, in the next multipolar approximation one has

$$V^1 = -\mathbf{d} \cdot \mathbf{E}^{\perp} - \mathbf{m} \cdot \mathbf{B} - \mathbf{Q} : \nabla\mathbf{E}^{\perp}. \qquad (10.191)$$

It should also be mentioned that recent work has shown that (laser) light can be engineered to possess a twisting or helical phase structure that can be characterised by assigning orbital angular momentum to photons. The plane wave description of the field variables $(\mathbf{A}, \mathbf{B}, \mathbf{E}^{\perp})$ cannot describe such properties, and a different formulation is required. Given the requisite field variable expansions (e.g. (7.246)), the perturbation theory of light scattering summarised here, based on the Kramers–Heisenberg dispersion formula, can be reworked and novel phenomena identified. A detailed study can be found in [61]; a striking prediction is of novel chiroptical birefringence effects in which the molecular quadrupole operator plays an essential role.

The quantum mechanical approach to the optical birefringence of a rarefied medium considers a beam of photons being scattered by a molecule. For such a system, the initial state can be represented by a molecule in a given initial state and photons linearly polarised along one direction of polarisation and in a single specified mode $\mathbf{k}$ of the field. In the distant future, the final state of the system has the molecule in its original state but recognises that there is a non-zero probability that photons have transferred from one polarisation direction λ to the other λ', without a change of momentum. Thus, although this is a case of forward scattering, a transition (the 'polarisation flip') has occurred and the scattering theory based on the T-matrix is still appropriate for the calculation of this probability. The observations that one makes on the incident and emergent light beams in a birefringence experiment are their intensities and the characteristics of the polarisation ellipse of each expressed through the azimuth and ellipticity angles; these observables are summarised elegantly by the Stokes parameter formalism (Chapter 7).

When polarisation changes are of most interest, it is convenient to work directly with either the Stokes operators or the polarisation density matrix, in conjunction with the quantum mechanical scattering theory. One may either compute expectation values of the Stokes operators in the final scattered state or express the density matrix of the final state in terms of the Stokes operators and evaluate traces. This approach seems to have been considered first by Stephen [60] and was developed in a series of papers by Atkins and his students [1], [7]–[10], [50], [62], [63]. Non-linear (intensity-dependent) optical effects are described within the scattering theory framework through higher-order terms in the perturbation expansion of the T-matrix.

The birefringence parameters are obtained by considering scattering from a laminar of unit area and infinitesimal thickness transverse to the beam, and integrating up the contributions of all the laminae in the sample length L, on the assumption that the fluid is dilute and there is negligible absorption of radiation in the specified mode $(\mathbf{k}, \lambda)$. The birefringence is produced by interference[19] between the probability amplitudes for a photon to be transmitted undisturbed and to be scattered into the forward direction with a polarisation flip [62], [63]. The result is that the ellipticity η and the azimuth θ of the emergent light beam may be characterised by a complex angle Φ, in terms of

---

[19] This is the quantum mechanical expression of the classical statement that birefringence is due to interference between the transmitted wave and waves scattered into the forward direction [12].

which

$$\eta = \Re\Phi \quad \text{and} \quad \theta = \Im\Phi, \tag{10.192}$$

$$\Phi \propto \sigma\kappa T_{\lambda'\lambda}^{(N)}, \tag{10.193}$$

where σ is the number of molecules per unit volume, $\kappa = L/\hbar c$ for a sample length L, and the complex quantity $T_{\lambda'\lambda}^{(N)}$ is an abbreviation for the T-matrix element $\langle N; \mathbf{k}, \hat{\boldsymbol{\varepsilon}}_{\lambda'} | \mathsf{T} | N; \mathbf{k}, \hat{\boldsymbol{\varepsilon}}_\lambda \rangle$.

Returning to the electric dipole approximation, one can easily demonstrate the connection between the scattering theory and the linear response formalism in the case that the final molecular state is the same as the initial state (so $\phi_{s'} = \phi_s$ in (10.181) and (10.182)) as in birefringence. The sum over the intermediate states can be removed by transformation to time-dependent operators using the molecular Hamiltonian H', since [64]

$$\langle \phi_s | \mathsf{d}_u | \phi_{s''} \rangle e^{i\omega_{ss''}\tau} = \langle \phi_s | \mathsf{d}(\tau)_u | \phi_{s''} \rangle, \tag{10.194}$$

where

$$\mathsf{d}(\tau)_u = e^{i\mathsf{H}'\tau/\hbar} \mathsf{d}_u e^{-i\mathsf{H}'\tau/\hbar}. \tag{10.195}$$

Then using the rule

$$\frac{1}{\omega \pm \omega_{ss'} + i0} = -i \int_0^\infty e^{i(\omega \pm \omega_{ss'})\tau} \, \mathrm{d}\tau, \tag{10.196}$$

the T-matrix to order e^2 is proportional to

$$\chi(\omega; s)_{uv} = -i \int_0^\infty e^{i\omega\tau} \langle \phi_s | [\mathsf{d}(\tau)_u, \mathsf{d}(0)_v] | \phi_s \rangle \, \mathrm{d}\tau. \tag{10.197}$$

Noting that $|\phi_s\rangle\langle\phi_s| = \rho_s$ is the density matrix for state s, this is also

$$\chi(\omega; s)_{uv} = -i \int_0^\infty e^{i\omega\tau} \mathrm{Tr}\big(\rho_s[\mathsf{d}(\tau)_u, \mathsf{d}(0)_v]\big) \, \mathrm{d}\tau, \tag{10.198}$$

which is essentially the susceptibility found in linear response theory (cf. §6.7.2).

More generally, if there is a change in the polarisation of the light beam due to the scattering, the contribution from the A^2 term to the Kramers–Heisenberg dispersion formula (10.174) is simply proportional to $\hat{\boldsymbol{\varepsilon}}_\lambda \cdot \hat{\boldsymbol{\varepsilon}}_{\lambda'}$, and so vanishes. Transformation to time-dependent operators shows that the susceptibility tensor $\chi(\mathbf{k}, \omega, N)_{\alpha,\beta}$ is

$$\chi(\mathbf{k}, \omega, N)_{\alpha\beta} = \frac{i}{\hbar} \int_0^\infty e^{i\omega\tau} \mathrm{Tr}\{\rho_N[\mathsf{j}(\mathbf{k}, \tau)^\alpha, \mathsf{j}(-\mathbf{k}, 0)^\beta]\} \, \mathrm{d}\tau, \tag{10.199}$$

when the full form for the time-dependent current operator is retained:

$$\mathsf{j}(\mathbf{k}, \tau) = \sum_\alpha \frac{e_\alpha}{m_\alpha} \mathbf{p}(\tau)_\alpha e^{i\mathbf{k}\cdot\mathbf{x}(\tau)_\alpha}. \tag{10.200}$$

Here ρ_N is the density matrix for the target, and we have deferred the separation into centre-of-mass and internal variables.

For a molecule in a gas in thermal equilibrium at temperature T with Hamiltonian H, we define the Gibbs state ρ_G as ($\beta = 1/k_B T$)

$$\rho_G = \frac{e^{-\beta H}}{Z}, \quad \text{Tr}\{\rho_G\} = 1, \tag{10.201}$$

and for such a system the susceptibility tensor is

$$\chi(\mathbf{k}, \omega, T)_{\alpha\beta} = \frac{i}{\hbar} \int_0^\infty e^{i\omega\tau} \text{Tr}\left\{ \rho_G \left[\mathbf{j}(\mathbf{k}, \tau)^\alpha, \mathbf{j}(-\mathbf{k}, 0)^\beta \right] \right\} d\tau. \tag{10.202}$$

A general quantum theory of birefringence induced in molecular gases by external perturbations at finite temperatures can be based on using the Dyson series with (10.202). For a molecular fluid in thermal equilibrium at temperature T in the presence of an external static field, the Gibbs state may be written formally as in (10.201), where $H = H_0 + V$ with H_0 the Hamiltonian for unperturbed molecules, and V describing their interaction with the external field. Then, as in §6.8.2, the exponential operator can be expanded in powers of V using the Dyson series; introducing the unperturbed states in the trace, we have directly

$$\text{Tr}(e^{-\beta H} O) = \sum_{m,n} e^{-\beta E_n^0} U(-\beta)_{nm} O_{mn}$$
$$= \sum_n e^{-\beta E_n^0} U(-\beta)_{nn} O_{nn} + \sum_{m,n}' e^{-\beta E_n^0} U(-\beta)_{nm} O_{mn}, \tag{10.203}$$

where O is an abbreviation for the two-point current fluctuation function in (10.202). The diagonal elements of the matrix $\mathbf{U}(-\beta)$ give terms in ascending powers of β, while the off-diagonal elements introduce the matrices $\mathbf{x}(-\beta)$ and $\mathbf{y}(-\beta)$ (Eqs. (6.104) and (6.105) in Chapter 6), whose variation with β depends on the energy level spacing relative to β. Then (10.199) can be evaluated in the unperturbed basis as a series in ascending powers of V,

$$\chi(\omega, T)_{\alpha,\beta} = \chi(\omega, T)^0_{\alpha,\beta} + \chi(\omega, T)^1_{\alpha,\beta} + \chi(\omega, T)^2_{\alpha,\beta} + \cdots, \tag{10.204}$$

where $\chi(\omega, T)^0_{\alpha,\beta}$ describes light scattering in the absence of the field. A notable feature of the approach is the elimination of the question of origin invariance in the traditional sense, since no arbitrary molecular origin is ever invoked; this follows from the explicit treatment of the kinematics of the experiment in which the centre-of-mass variables are *dynamical* variables. It was illustrated by the application of the theory to the example of birefringence induced by static electric quadrupole fields, **E**. The effect of the field gradient is contained in $\chi(\omega, T)^1_{\alpha,\beta}$, while $\chi(\omega, T)^2_{\alpha,\beta}$ represents the usual Kerr effect which is quadratic in $|\mathbf{E}|$ [65]–[67].

The results of the quantum mechanical scattering theory generally parallel those obtained from the traditional semi-quantal approach to birefringence [12], [68]–[71]. There are differences in some of the formulae for susceptibilities [35], but these are only apparent; as shown in Chapter 6, the Rayleigh–Schrödinger perturbation theory used in the semi-quantal accounts treats certain diagonal terms in a different way from the Dyson series. The Dyson expansion, which is related directly to the diagrammatic perturbation theory, utilises sums over complete sets of intermediate states without any restrictions. However, either formulation can be transformed into the other which shows that all that has happened is that the terms in the infinite sums in the perturbation formulae have simply been rearranged (§6.5). Some features of the semi-quantal approach survive in the scattering theory formulations referred to previously. For example, finite temperature effects are handled by classical statistical mechanics; for molecules in the presence of an external static electric field, the classical distinction between non-polar and polar substances due to Debye in terms of the absence or presence of permanent electric dipole moments is invoked. This will be discussed further in Chapter 12. It should also be noted that electric and magnetic multipoles are given their traditional origin-dependent definitions, and the kinematics of scattering events are not discussed.

10.6 Level Shifts and Self-Interaction

The diagram expansion is equally applicable to the perturbation theory approach to *energy level shifts*; recall from Chapter 6 that a discrete energy level of the reference system can be related formally to an energy level of the full problem by solving for the roots of the equation,

$$
\begin{aligned}
E_n &= E_n^0 + \Delta_n(E) \\
&\approx E_n^0 + \Delta_n(E_n^0),
\end{aligned}
\tag{10.205}
$$

where

$$
\Delta_n(E_n^0) = \langle \Phi_n^0 | \left(\mathsf{V} + \mathsf{V}\mathsf{G}^0(E_n^0)\mathsf{V} + \ldots \right)' | \Phi_n^0 \rangle.
\tag{10.206}
$$

This is essentially the same expansion as for the T-matrix with the supplementary condition that Φ_n^0 must be excluded from sums over complete sets of states.

A diagram in Figure 10.2 can also be repurposed as the building block of the diagrammatic expansion for the energy shift ΔE due to Van der Waals interactions of pairs of neutral atoms/molecules/chromophores, supposed electronically distinct, via the exchange of virtual photons. One can imagine making a copy of diagram (10.2iia) and, taking the two copies together, joining the external lines 1 to $1'$, 2 to $2'$; the composite diagram is then relabelled so that the initial and final states are the same, (1,2), and the virtual intermediate states are $(1',2')$. There are six topologically distinct diagrams that can be formed in this way when all time orderings of the vertices are allowed for (excluding interchange of the two particles); the diagrams have four vertices and so, if Φ_n^0 in (10.206) is taken to be the tensor product of the ground state of the atomic system and the photon vacuum, the diagrams describe the van der Waals interaction to order e^4 (i.e. α^2); within the electric dipole approximation they lead to [11], [72],

$$
\Delta E \approx \frac{C}{R^7}, \quad R \gg \lambda; \quad \Delta E \approx \frac{C'}{R^6}, \quad R \ll \lambda,
\tag{10.207}
$$

where C, C' are related to the polarisabilities of the two atomic groups that are separated by R. λ is a typical optical excitation wavelength. The weakening of the force law for the potential energy as R increases was the property that points to the involvement of the electromagnetic field, and thus quantum electrodynamics, because the interaction cannot be instantaneous [72].

Another very important case of considerable current interest is *resonance energy transfer – RET* in which an excited molecule transfers energy to a neighbouring unexcited molecule, a 'donor-acceptor' situation. Here the term 'molecule' can be given a wide interpretation including nano-structures [73]–[75]. If the excitation/de-excitation interactions are included, the overall process conserves energy and the resulting diagrams (cf. Figure 10.7) can be interpreted in terms of the scattering theory with energy transfer as a feature. An old example is a model of optical activity arising from the energy exchanges through virtual photon transfer between pairs of achiral chromophores that are arranged in a dissymmetric molecular structure [50].

In an instructive calculation, Power [76], [77] has shown how the vacuum state wave functional (7.167) can be used in a perturbation theory of an energy shift, using the resolvent expansion (10.206) to calculate intermolecular forces. Such forces can be thought of as arising from finite shifts in the zero point energy of the electromagnetic field which occur when static polarisable bodies are placed in it. Working directly with the field strength operators is obviously a gauge-invariant method of calculation. Perhaps the most striking illustration of such an effect is the Casimir force which is a nanoscale, room-temperature phenomenon of a purely quantum nature [78]. In the prototype calculation, which can be generalised in many ways (various geometries, finite temperatures, real materials, surface roughness effects etc.; see [79]), two perfectly conducting metal plates are brought together to a separation of a at zero temperature. The plates have area $A \gg a$.

There are two ways to proceed. The first method focuses on the difference between the zero-point energy of the field in vacuum and in the presence of the plates. The plates can be modelled by introducing the boundary condition

$$\mathbf{E}^{\perp} = 0, \quad B_z = 0, \tag{10.208}$$

because the field is excluded from a perfect conductor, so the calculation can be based on the mode frequencies for a box as in Chapter 4. The zero point energies are infinite, but their difference is finite; this difference can be obtained if the divergent quantities are regularised before the subtraction. This can be done in various ways, and importantly, the finite result is independent of the method chosen. The derivative of this energy with respect to the separation a (with a minus sign) is the force.

Alternatively, we can aim to calculate directly the force on one of the plates which can be obtained from the vacuum expectation value of the normal-normal component of the Maxwell stress-tensor, T_{zz}, for the electromagnetic field in the presence of the plates. According to (2.100), this is

$$T_{zz} = \frac{1}{2}\varepsilon_0 \left[\mathbf{E}\cdot\mathbf{E} + c^2\mathbf{B}\cdot\mathbf{B} \right] - \varepsilon_0 E_z^2 - \varepsilon_0 c^2 B_z^2 \qquad (10.209)$$

$$= \frac{1}{2}\varepsilon_0 \left[\mathbf{E}^\perp\cdot\mathbf{E}^\perp + c^2\mathbf{B}^\perp\cdot\mathbf{B}^\perp - E_z^2 - c^2 B_z^2 \right]. \qquad (10.210)$$

The physical force can be calculated as the difference between the stress with the plates present and the stress in free space; once again these quantities are infinite, but their difference is finite and is the Casimir force [80]. Explicit calculation in either way shows that for this geometrical configuration

$$F = -\frac{\pi^2}{240}\frac{\hbar c A}{a^4}. \qquad (10.211)$$

If the plates have $A = 1$ cm^2 and the separation distance $a \sim 1$ micron, the force obtained from (10.211) is $\sim 10^{-7}$ N which is accessible to modern experimental techniques using, for example, atomic force microscopy. Notice the force is proportional to the reduced Planck constant $\hbar$, a signature of its quantum mechanical origin. The absence of the charge e is a consequence of the idealisation of the plates by the boundary condition (10.208); real materials may be described by a frequency-dependent dielectric constant $\varepsilon(v)$, and the Casimir force formula in general involves a function $g(\varepsilon(v))$ which takes the value 1 for an idealised perfect conductor. It is now known that this purely quantum mechanical effect can lead to both attractive and repulsive forces, depending on the materials used and their geometrical arrangement; chirality is also a relevant feature [81].

We conclude this chapter with an elementary discussion of the self-interaction of charged particles based on the quantum mechanical perturbation theory. The treatment is of only qualitative value since it is carried out in the non-relativistic formalism. Consider a system consisting of a single charged particle moving freely with no external electromagnetic field. A related 'elementary' problem is a consideration of the self-interaction of a charge in a bound state; the important cases are a free electron, which was first discussed by Waller [83], and the electron in the hydrogen atom, first discussed by Oppenheimer [84]. In both cases, the first non-vanishing contribution to the energy shift (10.206) was found to be infinite, and until the early post-WWII years it was believed that quantum electrodynamics was fundamentally wrong, and something else was needed.

In QED the $p = 1$ term in the perturbation expansion of (10.206) vanishes since $V^{(1)}$ is an off-diagonal operator in every gauge; the lowest-order term in the expansion is therefore the second-order term

$$\Delta_n^{(2)}(E_n^0) = V_{nn}^{(2)} + \sum_k \frac{|V_{nk}^{(1)}|^2}{E_n^0 - E_k^0}, \qquad (10.212)$$

which we shall evaluate in the following, assuming the photon reference state is the vacuum state $|[0]\rangle$. The second-order self-interaction can be visualised by the Feynman diagram shown in Figure 10.4 in which the wavy line represents the virtual photon, and the straight line the electron, both with the wave vectors shown.

For now we work with the Coulomb gauge theory.[20] As in §10.4, in second order the initial and final states Φ_n^0 describe a charge with momentum $\mathbf{P}$ and mass m moving in the photon vacuum, while the intermediate states $\{\Phi_k^0\}$ describe a charge with momentum $\mathbf{q}$ and a virtual photon with momentum $\hbar\mathbf{Q}$. The only diagonal contribution from the $\mathbf{A}\cdot\mathbf{A}$ part of the second-order interaction to the matrix element (10.212) is the formally infinite last term in (F.1.11); in terms of a cut-off $\hbar\Lambda$ on the photon momentum this yields

$$\frac{\alpha}{\pi}\left(\frac{\hbar^2\Lambda^2}{2m}\right), \tag{10.213}$$

where α is the fine structure constant (5.183). Equation (10.213) is independent of the dynamical variables and so may be interpreted as a shift in the origin for measuring energies; like the infinite 'zero-point energy' of the free radiation field, it is usually ignored. The perturbation operator linear in e ($\propto \sqrt{\alpha}$) has only off-diagonal matrix elements and contributes to the sum-over-states term in (10.212).

As suggested previously (§10.3), the contribution of the diagram in Figure 10.4 can be obtained directly from that of Figure 10.2iib with the obvious replacements in (10.168) and (10.169) of $\mathbf{k}, \mathbf{k}' \rightarrow \mathbf{Q}$. Conservation of energy does not hold for the intermediate states, and we have energies

$$E_n^0 = \frac{P^2}{2m}, \quad E_k^0 = \frac{q^2}{2m} + \hbar Qc. \tag{10.214}$$

In summing over the virtual states of the charge and the photon, we recognise that there is no momentum transfer in this process which leads to a factor of $\delta^3(0)$. Thus, we use

$$\delta^3(0) = \frac{\Omega}{(2\pi)^3}$$

$$\mathbf{P} = \mathbf{q} + \hbar\mathbf{Q}. \tag{10.215}$$

The energy shift is then

$$\Delta E_n = \frac{1}{(2\pi)^3}\frac{e^2}{m^2}\frac{\hbar}{2\varepsilon_0 c}\int \frac{P_\alpha q_\beta (1-\hat{\mathbf{Q}}\hat{\mathbf{Q}})_{\alpha\beta}}{Q\left[\frac{P^2}{2m} - \frac{q^2}{2m} - \hbar Qc\right]}\, d^3\mathbf{Q}, \tag{10.216}$$

where the integrand is to be evaluated at $\mathbf{q} = \mathbf{P} - \hbar\mathbf{Q}$ because of overall conservation of momentum. This means we can replace q_α by P_α in the numerator since $\mathbf{Q}$ is orthogonal to the polarisation vectors, and put $q^2 = |\mathbf{P} - \hbar\mathbf{Q}|^2$ in the denominator which reduces to

$$E_n^0 - E_k^0 = -\left(\frac{\hbar^2 Q^2}{2m} + \hbar Qc - \frac{\hbar\mathbf{P}\cdot\mathbf{Q}}{m}\right). \tag{10.217}$$

Choosing the polar axis along $\mathbf{p}$ gives

$$(1-\hat{\mathbf{Q}}\hat{\mathbf{Q}})_{\alpha\beta}\, P_\alpha P_\beta = P^2(1-\cos^2(\theta)), \quad \mathbf{P}\cdot\mathbf{Q} = PQ\cos(\theta), \tag{10.218}$$

---

[20] The generalised PZW Hamiltonian will be considered afterwards.

and so the energy shift is

$$\Delta E_n = -\frac{1}{(2\pi)^2}\frac{e^2}{m^2}\frac{(\hbar p)^2}{2\varepsilon_0\hbar c}\int_0^\infty Q\int_{-1}^{+1}\frac{(1-x^2)}{\left[\frac{Q^2}{2m}+\frac{Qc}{\hbar}-\frac{pQx}{m}\right]}\,dx\,dQ. \tag{10.219}$$

The integral over Q shows an ultraviolet divergence and must be regulated in some way; if a cut-off on the upper limit of the integral, $Q_{max} = \Lambda$, is introduced, the energy shift may be brought to the form

$$\Delta E_n = -\left(\frac{2}{\pi}\right)I(\sigma,\xi)\,\alpha\left(\frac{p^2}{2m}\right), \tag{10.220}$$

to first order in the fine structure constant. The integral $I(\sigma,\xi)$ in dimensionless form is

$$I(\sigma,\xi) = \int_0^\xi\int_{-1}^{+1}\frac{(1-x^2)}{(1+t-\sigma x)}\,dx\,dt. \tag{10.221}$$

Here $\xi = \hbar\Lambda/2mc \le 1$ and the dimensionless parameter $\sigma = \hbar p/mc \ll 1$ for non-relativistic particles.

The electric dipole approximation is the assumption that the momentum of the virtual photon may be regarded as negligible with respect to the charged particle's momentum, while its energy must be retained. It imposes the following simplifications:

1. The neglect of the exponential factor in the photon operators $K_{a,e}^{(1)}$ (cf. Appendix F); that is, we take $\mathbf{Q} = 0$.
2. At the same time we must take $P^2 = q^2$ in the energy denominator of (10.216), while hQc is retained.

In the dipole approximation for (10.219), the angular integration yields simply $8\pi/3$, so

$$\Delta E_n^{(2)}(E_n^{(0)}) = -C(\alpha)\left(\frac{p^2}{2m}\right), \quad C(\alpha) = \left(\frac{8}{3\pi}\right)\xi\alpha. \tag{10.222}$$

The dipole approximation can be obtained directly from (10.221) by putting $\sigma = 0$ and using $\log_e(1+\xi)\approx\xi$ for ξ small enough.

The energy of the particle in the reference state Φ_n^0 is $P^2/2m$; the negative sign in (10.220) implies that the particle's kinetic energy is decreased by its self-interaction, or equivalently, its mass is increased. We write

$$E \equiv \frac{P^2}{2m_{obs}} = \frac{P^2}{2m} - \mu(\alpha)\frac{P^2}{2m}, \tag{10.223}$$

where to order α

$$\mu(\alpha) = \frac{2}{\pi}I(\sigma,\xi)\,\alpha. \tag{10.224}$$

Hence we should use

$$m_{obs} = \frac{m}{(1-\mu(\alpha))} \tag{10.225}$$

in the reference Hamiltonian, so that in the mass renormalised theory there are only the effects of the field due to other charges. A relativistic calculation equivalent to the preceding one for a free electron was first given by Waller [83].

The diagram in Figure 10.4 also describes the self-interaction of a molecule to order α if we make suitable relabelling. The required diagonal matrix element can be obtained directly from the generalised Kramers–Heisenberg dispersion formula by joining up the photon lines in Figure 10.3. The quantum state of the molecule before and after the virtual interaction, say $|N\rangle$, is unchanged; it propagates in a virtual state $|N'\rangle$, while a virtual photon is emitted and then reabsorbed. Thus, we put $\mathbf{K}' = \mathbf{K}, s' = s, \mathbf{k}' = \mathbf{k}, \lambda' = \lambda, n_{\mathbf{k}} = 1$, and sum over the photon momentum and polarisation variables and the virtual states $\{|N\rangle\}$ of the molecule. Figure 10.3.i becomes a loop attached to a single vertex, while Figures 10.3.iia,b collapse into a single diagram like Figure 10.4 with emission of the virtual photon preceding its absorption.

The energy shift $\Delta_N^{(2)}$ may then be obtained directly from (10.174); since $\mathbf{K}' = \mathbf{K}$, we put $\mathbf{q} = 0$ and then

$$\Delta_N^{(2)} = \sum_{\mathbf{Q},\mu} \frac{\hbar}{2\varepsilon_0 \Omega Q c} \left(\Upsilon(0)_{NN} \delta_{\alpha,\beta} + \mathcal{R}(\mathbf{Q},N)_{\alpha,\beta} \right) \hat{\varepsilon}_\mu(\mathbf{Q})^\alpha \hat{\varepsilon}_\mu(\mathbf{Q})^\beta, \qquad (10.226)$$

with

$$\mathcal{R}(\mathbf{Q},N)_{\alpha,\beta} = \sum_{N'} \frac{J(\mathbf{Q})_{NN'}^\alpha J(-\mathbf{Q})_{N'N}^\beta}{E_{NN'} - \hbar Q c}, \qquad (10.227)$$

and

$$\mathbf{J}(\mathbf{Q})_{NN'} = e\langle N| \sum_n \frac{Z_n}{m_n} \mathbf{p}_n e^{i\,\mathbf{Q}\cdot\mathbf{x}_n} |N'\rangle. \qquad (10.228)$$

The sum over the photon polarisation vectors yields $\mathcal{B}(\mathbf{Q})$, (7.156), and the sum over the photon momentum becomes

$$\sum_{\mathbf{k}} \rightarrow \frac{\Omega}{(2\pi)^3} \int d^3\mathbf{Q}, \qquad (10.229)$$

as usual. From (F.2.15) we have

$$\Upsilon(0)_{NN'} = e^2 \sum_n \frac{Z_n^2}{m_n} \equiv \frac{e^2}{\overline{m}}, \qquad (10.230)$$

so the first term in $\Delta_n^{(2)}(E_n^{(0)})$ is

$$\frac{\hbar e^2}{2\varepsilon_0 c \overline{m}} \frac{1}{(2\pi)^3} \int \frac{d^3\mathbf{Q}}{Q} = \frac{\alpha}{\pi} \left(\frac{\hbar^2 \Lambda^2}{2\overline{m}} \right), \qquad (10.231)$$

where the divergent Q-integral has again been cut off at $Q_{max} = \Lambda$. This term is obviously the generalisation of (10.213) to the many-particle case. The second term gives

$$\alpha \left(\frac{\hbar}{2\pi} \right)^2 \int \sum_{N'} \frac{\mathbf{J}(\mathbf{Q})_{NN'}\mathbf{J}(-\mathbf{Q})_{N'N} : (\mathbf{1} - \hat{\mathbf{Q}}\hat{\mathbf{Q}})}{Q(E_{NN'} - \hbar Q c)} d^3\mathbf{Q}, \qquad (10.232)$$

which is obviously divergent and must also be regulated.

If the 'molecule' is specialised to the simplest case, namely the hydrogen atom, this is essentially Oppenheimer's calculation for a bound electron [84]. For a bound electron what is of physical significance is the shift in energy for pairs of initial states that can be connected by transitions observable in spectroscopy, so we now consider the real part of (10.232). The hydrogen atom with its single electron moving in the purely Coulombic field of the proton has an accidental symmetry which causes states with the same principal quantum number but different angular momenta to have the same energy; this is so for both the Schrödinger and Dirac equation descriptions. Thus, the $2s_{\frac{1}{2}}$ and $2p_{\frac{1}{2}}$ levels are predicted to be degenerate; experimentally they have an energy separation known as the Lamb shift that is observable in a fully allowed electric dipole microwave transition of about $\nu = 1040$ Hz [85]. This result is readily derived from (10.232); in the electric dipole approximation we ignore the photon momentum and put $\mathbf{k} = 0$, so that, for the electron in the hydrogen atom,

$$\mathbf{J}(0)_{NN'} = \frac{\mathbf{p}_{NN'}}{m}. \tag{10.233}$$

The shift in energy for the level N is then, to order α,

$$\Delta E_N = \alpha \left(\frac{2}{3\pi} \right) \frac{1}{(mc)^2} \sum_{N'} |\mathbf{p}_{NN'}|^2 \!\!\int_0^\infty \frac{E\, \mathrm{d}E}{E_{NN'} - E}. \tag{10.234}$$

Now as we saw previously, the self-interaction for an electron makes a contribution to its mass; if we use m^{obs} in the wave equation, this contribution is already accounted for in the energy levels and so must be subtracted from the shift (10.234). With a cut-off energy of $E_{\max} = mc^2$ imposed on the integration and further assumptions specific to the hydrogen atom, Bethe was able to make a reasonably accurate estimate of the Lamb shift which is due solely to a shift in energy of the $j = 0$ state [86], [87]. While the most accurate estimations of the Lamb shift rely on the full Lorentz-invariant QED (see, for example, [6]), it is notable that quite different accounts of its physical basis based on the perturbation of the zero-point energy of the field due to the presence of the atom agree with Bethe's calculation [88], [89]. In this respect they resemble the preceding account of the Casimir effect. So the finite energy shift can either be attributed to 'radiation reaction' in the sense of classical electrodynamics or be regarded as a consequence of an effect involving the vacuum state of the quantised electromagnetic field; both interpretations are sustainable. However, what the preceding calculation might mean for an arbitrary molecule where the limitation to s-states ($j = 0$) for a single electron is not defined is unknown.

Another approach to the problem of infinities is a reorganisation of the QED Hamiltonian such that the energy shifts due to self-interaction are absorbed into the basic parameters of the charged particles to leave only interaction terms corresponding to external electromagnetic fields (that is, fields produced by other charges). In the non-relativistic theory only the mass parameter can be adjusted in this way.[21] This is the

---

[21] At energies above the electron–positron pair production threshold the basic diagram in Figure 10.4 can be modified by the insertion of one or more 'loops', each loop corresponding to the virtual photon breaking up into a virtual electron–positron pair that later recombines into a photon which is finally

approach to classical electrodynamics that was described in Chapter 3 with the difference that only point particles are considered. The required reformulation of the QED Hamiltonian can be achieved through a unitary transformation first discussed by Schwinger [90]. It is closely related to the transformation of the classical Lagrangian discussed in §3.8.6 with the difference that it is intrinsically a perturbation theory–based approach using the *free* fields. Such transformations are known as 'dressing transformations'.

We introduce a new Hamiltonian through the relation

$$H_N = \Lambda_N^{-1} H[0] \Lambda_N, \tag{10.235}$$

where $H[0]$ is the full Coulomb gauge Hamiltonian. Schwinger proposed to write the unitary operator as

$$\Lambda_N = e^{iN/\hbar} \qquad N = \frac{e}{m_0 c^2} \mathbf{p} \cdot \mathbf{Z}(\mathbf{x}), \tag{10.236}$$

where

$$\mathbf{Z}(\mathbf{x}) = -\int \frac{\boldsymbol{\pi}(\mathbf{x}')^\perp}{4\pi\varepsilon_0 |\mathbf{x} - \mathbf{x}'|} \, d^3\mathbf{x}'. \tag{10.237}$$

We write m_0 explicitly in (10.236) to indicate that this is the 'bare' mass which is to be renormalised; the same identification is made in $H[0]$.

Evidently Λ does not commute with the vector potential $\mathbf{A}$ but leaves the conjugate momentum $\boldsymbol{\pi}^\perp$ unchanged. The conjugate operators for the charge, $\mathbf{x}$ and $\mathbf{p}$, are also modified; one easily finds

$$\mathbf{x} \to \mathbf{x}_N \quad = \mathbf{x} - \frac{e}{m_0 c^2} \mathbf{Z}(\mathbf{x})$$

$$\mathbf{p} \to \mathbf{p}_N \quad = \mathbf{p} + \mathbf{J}(\mathbf{Z}(\mathbf{x}), \mathbf{p})$$

$$\boldsymbol{\pi}(\mathbf{r})^\perp \to \boldsymbol{\pi}(\mathbf{r})_N^\perp = \boldsymbol{\pi}(\mathbf{r})^\perp$$

$$A(\mathbf{r})^\alpha \to A(\mathbf{r})_N^\alpha = A(\mathbf{r})^\alpha + \left(\frac{e}{m_0 c^2}\right) p^\beta \int \frac{\delta_{\alpha\beta}^\perp (\mathbf{r} - \mathbf{x}')}{4\pi\varepsilon_0 |\mathbf{x} - \mathbf{x}'|} \, d^3\mathbf{x}'. \tag{10.238}$$

$\mathbf{J}$ vanishes in the electric dipole approximation (spatially constant $\mathbf{Z}$) but otherwise is a complicated expression involving derivatives of $\mathbf{Z}$. Notice that the particle position operator is translated by a distance proportional to $\mathbf{Z}$ so that the Coulomb interaction between two particles with *different* charges and/or masses is modified. This fact was crucial in Schwinger's calculation of the non-relativistic part of the Lamb shift in the hydrogen atom [90]. As expected, the new vector potential is singular at the position of the particle, $\mathbf{r} = \mathbf{x}$; using Fourier integrals, one finds

$$\mathbf{A}(\mathbf{x})_N = \mathbf{A}(\mathbf{x}) + \left(\frac{4}{3\pi}\right)\left(\frac{e\Lambda}{4\pi\varepsilon_0 m_0 c^2}\right)\mathbf{p}, \tag{10.239}$$

where a wave vector cut-off Λ has been introduced as before.

absorbed. This 'loop' process is interpreted as an electrical polarisation of the vacuum that modifies the electric charge parameter e; in the Lorentz-invariant theory vacuum polarisation is accommodated through charge renormalisation.

Using the power series expansion of the exponential operator, the transformed Hamiltonian is

$$H_N = H[0] + \frac{ie}{\hbar}[H[0], N] + \frac{1}{2!}\left(\frac{ie}{\hbar}\right)^2 [[H[0], N], N] + \dots. \tag{10.240}$$

This may be written as

$$H_N = H_0 + V_N, \tag{10.241}$$

where H_0 is the reference Hamiltonian and, as in §9.3.2, we have

$$V_N = V^{(1)}[N] + V^{(2)}[N] + \dots, \tag{10.242}$$

with

$$V^{(1)}[N] = V^{(1)}[0] + \frac{ie}{\hbar}[H_0, N], \tag{10.243}$$

$$V^{(2)}[N] = V^{(2)}[0] + \frac{ie}{\hbar}\left[V^{(1)}[0], N\right] + \frac{1}{2!}\left(\frac{ie}{\hbar}\right)^2 [[H_0, N], N], \tag{10.244}$$

in terms of the (free) reference Hamiltonian and the Coulomb gauge perturbation operators.

Both terms in the reference Hamiltonian,

$$H_0 = \frac{p^2}{2m_0} + H_{rad}, \tag{10.245}$$

contribute to the commutator with N required for (10.243):

$$\begin{aligned}
[H_0, N] &= \frac{p^\alpha}{2m_0^2 c^2}[p^2, Z(\mathbf{x})^\alpha] + \frac{p^\alpha}{m_0^2 c^2}[H_{rad}, Z(\mathbf{x})^\alpha] \\
&= -\frac{\hbar^2}{2m_0^2 c^2}p^\alpha\left(\nabla^2 Z(\mathbf{x})^\alpha\right) \\
&\quad + \frac{i}{m_0}p^\alpha[H_{rad}, \sum_{\mathbf{k},\lambda}\frac{1}{\omega}\left(A(\mathbf{x})^\alpha_{\mathbf{k},\lambda}c_{\mathbf{k},\lambda} - A(\mathbf{x})^{*\alpha}_{\mathbf{k},\lambda}c^+_{\mathbf{k},\lambda}\right)] \\
&= \left(\frac{\hbar^2}{2m_0^2 c^2 \varepsilon_0}\right)\mathbf{p}\cdot\mathbf{E}(\mathbf{x})^\perp - \left(\frac{i\hbar}{m_0}\right)\mathbf{p}\cdot\mathbf{A}(\mathbf{x}).
\end{aligned} \tag{10.246}$$

Combining (10.243) and (10.246), we see that the $\mathbf{p}\cdot\mathbf{A}$ terms cancel, and so

$$V^{(1)}[N] = \frac{i\hbar e}{2m_0^2 c^2 \varepsilon_0}\mathbf{p}\cdot\mathbf{E}(\mathbf{x})^\perp. \tag{10.247}$$

In the electric dipole approximation this term would be identically zero, since the approximation amounts to taking the field operators as spatially constant.

Using the explicit forms for $V^{(1)}[0], N$ and $[H_0, N]$, we see that for the evaluation of the second-order terms we require two commutators of the same general type that we can write as

$$[\mathbf{p}\cdot\mathbf{F}(\mathbf{x}), \mathbf{p}\cdot\mathbf{Z}(\mathbf{x})] \tag{10.248}$$

where $\mathbf{F}$ is either $\mathbf{A}$ or $\mathbf{E}^\perp$. Now

$$[\mathbf{p} \cdot \mathbf{F(x)}, \mathbf{p} \cdot \mathbf{Z(x)}] = p^\alpha p^\beta [F(x)^\alpha, Z(x)^\beta]$$
$$+ p^\alpha \left([p^\beta, Z(x)^\alpha] F(x)^\beta - [p^\beta, F(x)^\alpha] Z(x)^\beta \right) \qquad (10.249)$$

If $\mathbf{F} = \mathbf{E}^\perp$, the first term vanishes since both parts of the commutator are proportional to $\pi^\perp$; for $\mathbf{F} = \mathbf{A}$ the Fourier expansions of the field operators at the same position yield

$$[A(x)^\alpha, Z(x)^\beta] = -\frac{i\hbar}{2\varepsilon_0 \Omega} \sum_{k,\lambda} \frac{\varepsilon_\lambda(\mathbf{k})^\alpha \varepsilon_\lambda(\mathbf{k})^\beta}{k^2} \qquad (10.250)$$

$$= -\frac{i\hbar}{2\varepsilon_0} \frac{\delta_{\alpha,\beta}}{(2\pi)^3} \frac{8\pi}{3} \int_0^\infty dk, \qquad (10.251)$$

so there is a term proportional to p^2 with a divergent coefficient. Putting in all the constants and collecting terms shows that its contribution to $V^{(2)}[N]$ is

$$\frac{4}{3\pi} \frac{\hbar\Lambda}{m_0 c} \alpha \left(\frac{p^2}{2m_0} \right), \qquad (10.252)$$

where a cut-off $k_{\max} = \Lambda$ has been used. It is of the same form as (10.222), and as before it can be removed by renormalising the bare mass m_0 in H_0.

It remains to discuss the bracketed term in (10.249). If we use the Fourier expansion of the field operators again, we see that a commutator like $[p^\beta, F(x)^\alpha]$ can be simply replaced by the field operator expansion with a factor of $\hbar k^\beta$ inside the $\mathbf{k}$ summation. For self-interaction effects to order α we only require the diagonal matrix element of $V^{(2)}[N]$ in the photon vacuum Ψ_0, and this is zero since the vacuum matrix elements reduce to constants $\times k^\beta (1 - \hat{\mathbf{k}}\hat{\mathbf{k}})_{\alpha,\beta} = 0$. In the electric dipole approximation these terms are identically zero at the operator level. Collecting these results together, the renormalised non-relativistic Hamiltonian to order α takes the form

$$H_N = \frac{p^2}{2m_{\text{obs}}} + H_{\text{rad}} + \frac{e^2}{2m_{\text{obs}}} \mathbf{A(x)} \cdot \mathbf{A(x)}$$
$$- \frac{ie^2}{2\hbar m_{\text{obs}}^2 c^2} p^\alpha \left([p^\beta, Z(x)^\alpha] A(x)^\beta - [p^\beta, A(x)^\alpha] Z(x)^\beta \right). \qquad (10.253)$$

The many-particle case can be treated in a similar way; one must not forget to include the Coulomb interaction between pairs of charges. Under the Schwinger transformation, the separation of the two charges $|\mathbf{x}_i - \mathbf{x}_j|$ is modified according to (10.238).

The loop diagram, Figure 10.4, can be joined onto or inserted into any straight line in any other diagram, as often as one likes, provided we respect its particular feature of having the incoming particle state the same as the outgoing particle state, with no overall change for the field; the appearance of the virtual photon is transient. Recall that in the perturbation expansion for the T-matrix, *complete* sets of states are inserted to give matrix elements without any restrictions on their indices, so that such diagonal matrix elements must occur. Thus, in the perturbation expansion (10.206) these self-energy contributions are always of the form of (10.212) where Φ_n^0 is interpreted as some virtual state of the particle-field system. In higher orders of perturbation theory, it is also possible to have loops inside loops; we shall not consider any higher-order terms further.

To conclude this account, we describe briefly the calculation of self-interactions in an arbitrary gauge specified by some non-zero value for the Green's function $\mathbf{g}(\mathbf{x}, \mathbf{x}')$ (cf. §9.3.3). Everything so far in this discussion has used the Coulomb gauge formalism ($\mathbf{g}^{\perp} = 0$) in which the contribution of $V_{nn}^{(2)}$ is simply (10.213) with the Coulomb energies[22] of a many-particle system absorbed in H_0, and there is just the $\mathbf{p} \cdot \mathbf{A}$ contribution to $V_{nk}^{(1)}$. According to Eqs. (9.154)–(9.156) the general perturbation operators are related to the Coulomb gauge expressions by

$$V[\mathbf{g}] = V^1[\mathbf{g}] + V^2[\mathbf{g}], \tag{10.254}$$

$$V^1[\mathbf{g}] = V^1[0] + \frac{i}{\hbar} [H_0, F]] \tag{10.255}$$

$$V^2[\mathbf{g}] = V^2[0] + \frac{i}{\hbar} [V^1[0], F] + \frac{1}{2!} \left(\frac{i}{\hbar} \right)^2 [[H_0, F], F], \tag{10.256}$$

where F is the action integral for an arbitrary polarisation field specified by some $\mathbf{g}^{\perp}$:

$$F = \int \mathbf{P}(\mathbf{x}) \cdot \mathbf{A}(\mathbf{x}) \, d^3\mathbf{x}. \tag{10.257}$$

The commutator of H_{rad} in the last term in (10.256) gives rise to the contribution $\frac{1}{2\varepsilon_0} \int \mathbf{P}^{\perp} \cdot \mathbf{P}^{\perp} d^3\mathbf{x}$. As discussed in Chapter 3 it is firstly, highly singular and secondly, cancels identically the Coulomb energies when the two parts are put together to give the term $\frac{1}{2\varepsilon_0} \int \mathbf{P} \cdot \mathbf{P} \, d^3\mathbf{x}$ in the general Hamiltonian. There is no physical significance to be attached to any particular choice of $\mathbf{P}^{\perp}$ (including the value $\mathbf{g}^{\perp} = \mathbf{0}$, i.e. the Coulomb gauge value!) because the division of the energy between the transverse electric field and the electric polarisation field is quite arbitrary. In the perturbation theory context based on the use of the Fock space for non-interacting photons and atoms/molecules, it is obviously important to verify the gauge invariance of the self interaction calculation and so provide a secure justification for thinking in terms of 'Coulomb energies'. The perturbation expansion of the diagonal matrix element of the resolvent operator is closely related to that of the T-matrix discussed earlier in this chapter, and one can indeed use the same formal approach to verify gauge invariance using the method outlined in §10.2.2 [20].

References

[1] Atkins, P. W. and Barron, L. D. (1969), Mol. Phys. **16**, 453.

[2] Goldberger, M. L. and Watson, K. M. (1964), *Collision Theory*, J. Wiley.

---

[22] Recall that these come from the longitudinal part of the electric polarisation field through the energy integral $\frac{1}{2\varepsilon_0} \int \mathbf{P}^{\perp} \cdot \mathbf{P}^{\perp} d^3\mathbf{x}$ as described in Chapter 3.

[3] Roman, P. (1965), *Advanced Quantum Theory*, Addison-Wesley Publishing Co. Ltd.

[4] Gottfried, K. (1966), *Quantum Mechanics 1: Fundamentals*, 1st ed., W. A. Benjamin.

[5] Weinberg, S. (2013), *Lectures on Quantum Mechanics*, Cambridge University Press.

[6] Weinberg, S. (1995), *The Quantum Theory of Fields*, vol. 1, *Foundations*, Cambridge University Press.

[7] Atkins, P. W. and Barron, L. D. (1968), Proc. Roy. Soc. (London) A**304**, 303.

[8] Atkins, P. W. and Barron, L. D. (1968), Proc. Roy. Soc. (London) A**306**, 119.

[9] Atkins, P. W. and Miller, M. H. (1968), Mol. Phys. **15**, 491.

[10] Atkins, P. W. and Miller, M. H. (1968), Mol. Phys. **15**, 503.

[11] Power, E. A. (1964), *Introductory Quantum Electrodynamics*, Longmans.

[12] Barron, L. D. (1982), *Molecular Light Scattering and Optical Activity* Cambridge University Press.

[13] Loudon, R. (1983), *The Quantum Theory of Light*, 2nd ed., Clarendon Press.

[14] Craig, D. P. and Thirunamachandran, T. (1998), *Molecular Quantum Electrodynamics*, Dover.

[15] Andrews, D. L. and Allcock, P. (2002), *Optical Harmonics in Molecular Systems*, Wiley-VCH.

[16] Salam, A. (2010), *Molecular Quantum Electrodynamics*, John Wiley & Sons, Inc.

[17] Healy, W. P. (1977), Phys. Rev. A**16**, 1568.

[18] Healy, W. P. and Woolley, R. G. (1978), J. Phys. B**11**, 1131.

[19] Power, E. A. and Zienau, S. (1959), Phil. Trans. Roy. Soc. (London) A**251**, 427.

[20] Woolley, R. G. (1998), Mol. Phys. **94**, 409.

[21] Prigogine, I. and George, C. (1977), Int. J. Quant. Chem. **12(S1)**, 177.

[22] Woolley, R. G. (1988), in *Molecules in Physics, Chemistry and Biology*, **I**, edited by J. Maruani, Kluwer Academic Publishers, 45.

[23] Goldberger, M. L. (1951), Phys. Rev. **84**, 929.

[24] Bethe, H. and de Hoffmann, H. (1954), *Mesons and Fields*, vol. 2 *Mesons*, Row, Peterson and Co.

[25] Low, F. (1952), Phys. Rev. **88**, 53.

[26] Fried, Z. (1973), Phys. Rev. A**8**, 2835.

[27] Davidovich, L. and Nussenzveig, H. M., (1980), *Foundations of Radiation Theory and Quantum Electrodynamics*, edited by A.O. Barut, p. 83, Plenum Press.

[28] Fano, U. (1935), Il Nuovo Cimento **12**, 156.

[29] Fano, U. (1961), Phys. Rev. **124**, 1866.

[30] Faist, J., Capasso, F., Sirtori, C., Weiss, K. W. and Pfeiffer, L. N. (1997), Nature **390**, 589.

[31] Madhavan, V. V., Chen, W., Jamneala, T., Crommie, M. F. and Wingreen, N. S. (1998), Science **280**, 567.

[32] Miroshnichenko, A. E., Flach, S. and Kivshar, Y. S. (2010), Rev. Mod. Phys. **82**, 2257.

[33] Dirac, P. A. M. (1967), *The Principles of Quantum Mechanics*, 4th ed., Clarendon Press.

[34] Chu, W.-C. and Lin, C. D. (2010), Phys. Rev. A**82**, 053415.

[35] Mackrodt, W. C. (1971), Mol. Phys. **20**, 251.

[36] Friedrichs, K. O. (1948), Comm. Pur. Appl. Math. **1**, 361.

[37] Ballesteros, M., Faupin, J., Fröhlich, J. and Schubnel, B. (2015), Comm. Math. Phys. **337**, 633.

[38] Heitler, W. (1954), *The Quantum Theory of Radiation*, 3rd ed., Oxford University Press.

[39] Blake, N. P. (1990), J. Chem. Phys. **93**, 6165.

[40] Hameka, H. F. (1965), *Advanced Quantum Chemistry*, Addison-Wesley.

[41] Berestetskii, V. B., Lifshitz, E. M. and Pitaevskii, L. P. (1971), *Course of Theoretical Physics*, vol. 4, *Relativistic Quantum Theory*, Part 1. Translated from the Russian by J. B. Sykes and J. S. Bell, Pergamon Press.

[42] Weisskopf, V. and Wigner, E. P. (1930), Z. Physik **63**, 54; ibid. **65**, 18.

[43] Sakurai, J. J. (1967), *Advanced Quantum Mechanics*, Addison-Wesley.

[44] Sakurai, J. J. (1994), *Modern Quantum Mechanics*, Addison-Wesley.

[45] Lamb, W. E. (1952), Phys. Rev. **85**, 259.

[46] Feynman, R. P. (1962), *Quantum Electrodynamics*, W. A. Benjamin Inc.

[47] Ward, J. F. (1965), Rev. Mod. Phys. **37**, 1.

[48] Wallace, R. (1966), Mol. Phys. **11**, 457.

[49] Wallace, R. (1968), Mol. Phys. **15**, 249.

[50] Atkins, P. W. and Woolley, R. G. (1970), Proc. Roy. Soc. (London) A**314**, 251.

[51] Lebedev, P. N. (1901), Ann. Phys. **6**, 433.

[52] Nichols, E. F. and Hull, G. (1901), Phys. Rev. **13**, 307.

[53] Ashkin, A. (1970), Phys. Rev. Lett. **24**, 156.

[54] Einstein, A. (1917), Physikalische Z. **18**, 121.

[55] Mott, N. F. and Massey, H. S. W. (1965), *The Theory of Atomic Collisions*, 3rd ed., p. 477, Oxford University Press.

[56] Kramers, H. A. and Heisenberg, W. (1925), Z. Physik, **31**, 681.

[57] Dirac, P. A. M. (1927), Proc. Roy. Soc. (London), A**114**, 710.

[58] Crowley, B. J. B. and Gregori, G. (2014), High Energy Density Phys., **13**, 55.

[59] Bradshaw, D. S. and Andrews, D. L. (2017), Eur. J. Phys. **38**, 034008 (17 pages).

[60] Stephen, M. J. (1958), Proc. Camb. Phil. Soc. **54**, 81.

[61] Forbes, K. A. and Salam, A. (2019), Phys. Rev. A**100**, 053413 (13 pages).

[62] Atkins, P. W. and Barron, L. D. (1970), Mol. Phys. **18**, 721.

[63] Atkins, P. W. and Barron, L. D. (1970), Mol. Phys. **18**, 729.

[64] Landau, L. D. and Lifshitz, E. M. (1980), *Course of Theoretical Physics*, vol. 5, *Statistical Physics*, Part 1. Translated from the Russian by J. B. Sykes and M. J. Kearsley, Pergamon Press.

[65] Dumitru, A. G. and Woolley, R. G. (1998), Mol. Phys. **94**, 581.

[66] Dumitru, A. G. and Woolley, R. G. (1998), Mol. Phys. **94**, 595.

[67] Woolley, R. G. (2001), Mol. Phys. **99**, 547.

[68] Buckingham, A. D. (1956), Proc. Phys. Soc. B**69**, 344.

[69] Buckingham, A. D. (1959), J. Chem. Phys. **30**, 1580.

[70] Buckingham, A. D. and Longuet-Higgins, H. C. (1968), Mol. Phys. **14**, 63.

[71] Imrie, D. A. and Raab, R. E. (1991), Mol. Phys. **74**, 833.

[72] Power, E. A. and Zienau, S. (1957), Il Nuovo Cimento **6**, 7.

[73] Andrews, D. L. and Bradshaw, D. S. (2014), Annalen der Physik (Berlin), **526**, 173.

[74] Salam, A. (2018), Atoms, **6**, 56.

[75] Jones, G. A. and Bradshaw, D. S. (2019), Front. Phys., **7**, doi: 10.3389/fphys.2019.00100.

[76] Power, E. A. (1972), in *Magic without Magic*, edited by J. Klauder, p. 135, W. H. Freeman.

[77] Power, E. A. (1974), Phys. Rev. **A10**, 756.

[78] Strange, A., Campbell, D. and Bishop, D. (2021), Phys. Today, **74**, 43.

[79] Bordag, M., Mohideen, U. and Mostepanenko, V. M. (2001), Phys. Reports **353**, 1.

[80] Ambjørn, J. and Wolfram, S. (1983), Ann. Phys. (N.Y.) **147**, 1.

[81] Jiang, Q. D. and Wilczek, F. (2019), Phys. Rev. **B99**, 125403.

[82] Woolley, R. G. (1971), Mol. Phys. **22**, 1013.

[83] Waller, I. (1930), Z. Physik **62**, 673.

[84] Oppenheimer, J. R. (1930), Phys. Rev. **35**, 461.

[85] Lamb, W. E. Jr. and Retherford, R. C. (1947), Phys. Rev. **72**, 241.

[86] Bethe, H. A. (1947), Phys. Rev. **72**, 339.

[87] Milonni, P. W. (1980), in *Foundations of Radiation Theory and Quantum Electrodynamics*, edited by A. O. Barut, p. 1, Plenum Press.

[88] Welton, T. A. (1948), Phys. Rev. **74**, 1157.

[89] Power, E. A. (1966), Amer. J. Phys. **34**, 516.

[90] Schwinger, J. (1949), Phys. Rev. **75**, 651.

11 Quantum Electrodynamics beyond Perturbation Theory

11.1 Introduction

Perturbation theory as used in non-relativistic quantum electrodynamics in the conventional fashion was described in Chapter 10. The S-matrix description is predicated on the assumption that the reference Hamiltonian, H_0, and the full Hamiltonian, H, are related by a unitary transformation, that is, they are operators on the same Hilbert space $\mathcal{H}$. In the time-dependent view of scattering this requires that H_0 and H coincide at $t = \pm\infty$. As shown in Chapter 6, in this framework the unitary transformation operator can be constructed approximately as a perturbation expansion in powers of the coupling constant. We start here with some general remarks about how and why problems can arise when a quantum field is involved. Let us first work in a Schrödinger representation in which H is time-independent. We can gain an insight into the construction in the following way. Given any vector in $\mathcal{H}$ chosen as an initial state $|u_0\rangle$, one may develop an orthonormal sequence of states from a three-term recurrence relation generated by the full Hamiltonian,

$$H|u_n\rangle = a_n|u_n\rangle + b_{n+1}|u_{n+1}\rangle + b_{n-1}|u_{n-1}\rangle, \tag{11.1}$$

with starting coefficients [1]

$$a_0 = \langle u_0|H|u_0\rangle, \tag{11.2}$$

$$b_1 = \langle u_1|H|u_0\rangle, \quad b_{-1} = 0, \ b_0 = 1, \tag{11.3}$$

$$|u_1\rangle = b_1^{-1}(H - a_0)|u_0\rangle. \tag{11.4}$$

Requiring normalisation of $|u_1\rangle$ shows that the first off-diagonal coefficient is determined by the variance of the Hamiltonian in the starting state $|u_0\rangle$,

$$b_1^2 = \langle u_0|H^2|u_0\rangle - a_0^2. \tag{11.5}$$

The quantities $\{a_n, b_n\}$ give a tri-diagonal (Jacobi) representation of the Hamiltonian which may be brought to diagonal form by a further unitary transformation. The normalisation requires that the states $\{u_n\}$ are square integrable.[1]

---

[1] Since H contains ∇^2 operators, one requires the additional condition that the derivatives up to order 2 are square-integrable.

If we apply this construction to the QED Hamiltonian, we immediately encounter difficulties, as may be seen in what follows. The Hamiltonian for the free electromagnetic field, H_{rad}, has a Fock representation with eigenstates that are simple photon occupation number states. For simplicity, we consider a single charge e, m with canonical operators $(\mathbf{x}, \mathbf{p})$. The full Hamiltonian in the Coulomb gauge is then

$$H = \frac{|\mathbf{p}|^2}{2m} + \sum_{\mathbf{k},\lambda} \hbar\omega c^+_{\mathbf{k},\lambda} c_{\mathbf{k},\lambda} - \frac{e}{m}\mathbf{p}\cdot\mathbf{A}(\mathbf{x}) + \frac{e^2}{2m}\mathbf{A}(\mathbf{x})\cdot\mathbf{A}(\mathbf{x}), \qquad (11.6)$$

where we have dropped the zero-point energy of the field. The matrix elements required for the Coulomb gauge theory are described in detail in Appendix F.

Consider using a state Φ_i constructed as a product of a state of the particle, $|N\rangle$, and the photon Fock space vacuum, $|\Psi_0\rangle$,

$$\Phi_i = |N, \Psi_0\rangle, \quad \Phi_i \in \mathcal{H} \qquad (11.7)$$

as the initial state in the recurrence, $|u_0\rangle$. The first step, (11.2), yields

$$\begin{aligned}
a_0 &= \langle \Phi_i | H | \Phi_i \rangle \\
&= \langle \frac{\mathbf{p}^2}{2m} \rangle_N + \frac{e^2}{2m} \langle \mathbf{A}(\mathbf{x})\cdot\mathbf{A}(\mathbf{x}) \rangle_{\Psi_0} \\
&= \langle \frac{\mathbf{p}^2}{2m} \rangle_N + \frac{\alpha}{\pi}\left(\frac{\hbar^2}{m}\right)\int_0^\infty k\,dk,
\end{aligned} \qquad (11.8)$$

which is quadratically divergent,[2] because it involves the Fock space average of the product of the vector potential operator with itself evaluated at the point $\mathbf{x}$, exactly as in (10.213). This formally infinite term is the first diagonal element in the Jacobi matrix for H and must be subtracted from H for the determination of $|u_1\rangle$, according to (11.4). In the spirit of perturbation theory let us make explicit the dependence on α and write

$$a_0 = a_{00} + \alpha\, a_{01}, \qquad (11.9)$$

where a_{00}, a_{01} can be identified from (11.8). The normalisation of the resulting state requires the evaluation of (11.5). Then $(a_0)^2$ will contribute $2a_{00}a_{01}$ to $O(\alpha)$ which cancels a term of the same order arising from the matrix element of H^2, leaving

$$\frac{e^2}{m^2}\langle \Phi_i | \mathbf{p}^\alpha \mathbf{p}^\beta A(\mathbf{x})^\alpha A(\mathbf{x})^\beta | \Phi_i \rangle = \frac{e^2}{m^2}\langle N | \mathbf{p}^\alpha \mathbf{p}^\beta | N \rangle \langle A(\mathbf{x})^\alpha A(\mathbf{x})^\beta \rangle_{\Psi_0} \qquad (11.10)$$

from the squaring of the first-order perturbation operator $(\mathbf{p}\cdot\mathbf{A}(\mathbf{x}))$. This term is also singular in the continuum limit, and after summing over polarisation states, we get

$$\begin{aligned}
b_1^2(\alpha, m) &= \frac{e^2}{m^2}\langle N | \mathbf{p}^\alpha \mathbf{p}^\beta | N \rangle \frac{\hbar}{2\varepsilon_0 \Omega c}\sum_{\mathbf{k}}\frac{(1 - k^\alpha k^\beta)}{k} \\
&= \frac{8\alpha}{3\pi}\left(\frac{\hbar^2}{m}\right)\langle \frac{\mathbf{p}^2}{2m}\rangle_N \int_0^\infty k\,dk.
\end{aligned} \qquad (11.11)$$

---

[2] α is the fine structure constant, as usual.

This quadratic divergence means that the perturbed state $|u_1\rangle$ does not lie in the Hilbert space $\mathcal{H}$ for any non-zero value of the coupling constant α or any finite value of the mass m, and so perturbation theory cannot be discussed in terms of unitary transformation as in quantum mechanics. Clearly, making $\alpha \to 0$ and/or $m \to \infty$ so that $b_1^2 \to 0$ is of no interest since it makes the theory trivial – the particles and the field have no interaction. The problem with (11.11) is that it includes photons with arbitrarily large momentum (and hence high energy) in the calculation; in other words, energies far beyond what is reasonable for non-relativistic QED. A simple, if crude, method is to impose a finite upper limit, Λ, on the integral where $\hbar\Lambda c$ should be much less than the electron invariant mass energy (the threshold of electron–positron pair production). Then the recurrence relation is regularised.

After a perturbation theory calculation is finished the result will depend on the ultraviolet cut-off applied, and so one has to find a way of incorporating the cut-off(s) in the parameters for charge and mass. From the foregoing it is apparent that the divergence arises from a diagonal matrix element; if we confine attention to off-diagonal matrix elements in the photon Fock space when making the T-matrix expansion, terms like the infinity in (F.1.10) cannot contribute and a wide range of phenomena can be given a formal perturbation theory description using QED. For the calculation of all off-diagonal transition amplitudes, we simply assume that the parameters for the charges and masses of the particles take the experimentally observed values. This was the approach taken in the last chapter. The quantity in (11.11) is interpreted in terms of a renormalisation of the mass of the particle. It should be recognised that even if these self-interactions had turned out to be finite, they would still have to be absorbed in the formal charge and mass parameters introduced in setting up the Lagrangian, so that eventually everything is expressed in terms of the experimentally observed values. The final possibility, which seems not to have been explored, is to recognise that in a quantum field theory the Hilbert space has to be *chosen*, and the choice of the 'free charges $\oplus$ free field' tensor product which gives the infinities is simply not a good choice. Perhaps the charges should appear 'as they are' with their associated photon dressing, and presumably no self-interactions. We next give two simple examples of how such non-equivalent spaces can arise. We do not know how to use them.

11.1.1 A Coherent State Formulation of the PZW Representation

First, we consider the relationship of the Coulomb gauge and the Power–Zienau–Woolley representation of non-relativistic QED in terms of generalised coherent states [2]. We noted in §7.3 that unitary operators of the form of (9.109) can be presented in the form of a product of the familiar coherent state displacement operators or boson translators. Since the generator F is a product of charged particle and field variables and is proportional to the charge e, the resulting coherent state 'parameters' involve mixtures of particle and field variables and e, and only make sense for the interacting system.[3] Using the mode expansion of the Coulomb gauge vector potential, (7.122), and proceeding to the continuum limit, the transformation operator (9.109) may be cast as

$$
\Lambda_C = \exp\left[\sum_{\lambda=1,2}\int\left(\alpha(\mathbf{k}:\mathcal{C})_\lambda c(\mathbf{k})_\lambda^+ - \alpha(\mathbf{k}:\mathcal{C})_\lambda^* c(\mathbf{k})_\lambda\right)d^3\mathbf{k}\right],
\tag{11.12}
$$

where for each mode $\mathbf{k}, \lambda$ and path $\mathcal{C}$, the coherent state 'parameter' is

$$
\alpha(\mathbf{k}:\mathcal{C})_\lambda = -i\sqrt{\frac{1}{2\varepsilon_0\hbar ck(2\pi)^3}}\mathbf{P}(\mathbf{k}:\mathcal{C})\cdot\hat{\boldsymbol{\varepsilon}}(\mathbf{k})_\lambda.
\tag{11.13}
$$

Here $\mathbf{P}(\mathbf{k}:\mathcal{C})$ is the Fourier transform of the polarisation field for the specified path evaluated at the wave vector $\mathbf{k}$.

The transformed annihilation and creation operators for the mode $\mathbf{k}, \lambda$ are

$$
\begin{aligned}
C(\mathbf{k}:\mathcal{C})_\lambda &= c(\mathbf{k})_\lambda + \alpha(\mathbf{k}:\mathcal{C})_\lambda\\
C(\mathbf{k}:\mathcal{C})_\lambda^+ &= c(\mathbf{k})_\lambda^+ + \alpha(\mathbf{k}:\mathcal{C})_\lambda^*
\end{aligned}
\tag{11.14}
$$

and one still has

$$
[C(\mathbf{k}:\mathcal{C})_\lambda, C(\mathbf{k}':\mathcal{C})_{\lambda'}^+] = \delta_{\lambda,\lambda'}\delta^3(\mathbf{k}-\mathbf{k}').
\tag{11.15}
$$

We can define a new vacuum state, $\Psi_0(\mathcal{C})$, by setting

$$
C(\mathbf{k}:\mathcal{C})_\lambda|\Psi_0(\mathcal{C})\rangle = 0, \forall \mathbf{k}
\tag{11.16}
$$

to give a new representation of the Fock space for the field. A straightforward generalisation from the single-mode case (Chapter 7) to the continuum limit shows that $\Psi_0(\mathcal{C})$ is related to the free-field vacuum Ψ_0 by

$$
\begin{aligned}
|\Psi_0(\mathcal{C})\rangle = \exp\left(-\tfrac{1}{2}\int\sum_{\lambda=1,2}|\alpha(\mathbf{k}:\mathcal{C})_\lambda|^2 d^3\mathbf{k}\right)\\
\times \exp\left(\int\sum_{\lambda=1,2}\alpha(\mathbf{k}:\mathcal{C})_\lambda C(\mathbf{k}:\mathcal{C})_\lambda^+ d^3\mathbf{k}\right)|\Psi_0\rangle.
\end{aligned}
\tag{11.17}
$$

For a given polarisation field $\mathbf{P}(\mathbf{k}:\mathcal{C})$, the number of photons with wave vector $\mathbf{k}$ and polarisation λ in the coherent state (11.17) is (cf. (7.69))

$$
n[\mathbf{P}(\mathbf{k},\lambda:\mathcal{C})] = \langle\Psi_0(\mathcal{C})|c(\mathbf{k})_\lambda^+ c(\mathbf{k})_\lambda|\Psi_0(\mathcal{C})\rangle = |\alpha(\mathbf{k}:\mathcal{C})_\lambda|^2
\tag{11.18}
$$

and the total number of photons in the state is

$$
\overline{N} = \int\sum_{\lambda=1,2}|\alpha(\mathbf{k}:\mathcal{C})_\lambda|^2 d^3\mathbf{k}.
\tag{11.19}
$$

The overlap between the old and new vacuum states is also known as the vacuum expectation value (vev) for the transformation operator Λ_C; this is

---

[3] For $e = 0$, the transformation is the trivial identity.

$$\langle \Psi_0 | \Psi_0(\mathcal{C}) \rangle = \langle \Psi_0 | \Lambda_{\mathcal{C}} | \Psi_0 \rangle$$

$$= \exp\left(-\tfrac{1}{2} \int \sum_{\lambda=1,2} |\alpha(\mathbf{k}:\mathcal{C})_\lambda|^2 d^3k \right)$$

$$= \exp(-\tfrac{1}{2}\overline{N}). \tag{11.20}$$

The evaluation of the integral in (11.20) is closely related to the calculation of $\mathcal{E}_{\mathbf{P}}$ in Chapter 2; for the neutral two-particle case and the straight-line path, the Fourier transform required is just (2.173). The sum over the product of two photon polarisation vectors is given by (2.26),

$$\sum_{\lambda=1,2} \hat{e}(\mathbf{k})_{\lambda i} \hat{e}(\mathbf{k})_{\lambda j} = \delta_{ij} - \frac{k_i k_j}{k^2}, \tag{11.21}$$

and by choosing $\mathbf{r}$ as the polar axis we obtain, after carrying out the polarisation sum,[4]

$$\tfrac{1}{2} |\alpha(\mathbf{k}:\mathcal{C})|^2 = \frac{\alpha}{2\pi} \left(\frac{1 - \cos^2(\theta)}{\cos^2(\theta)} \right) \frac{\sin^2(kr\cos(\theta)/2)}{k^3} \chi_a^2(k), \tag{11.22}$$

where we have included a regulator. The angular integration over $d\Omega(\mathbf{k})$ is exactly as in the calculation of (3.99), and if, for example, we put $\chi_a(k) = \exp(-ak/\pi)$, there remains the dimensionless quantity

$$\frac{\alpha}{\pi} \int_0^\infty \frac{e^{-st}}{t} \left[t\sin(t) + \frac{\sin(t)}{t} + \cos(t) - 2 \right] dt = R(s), \quad s = \frac{2a}{\pi r} \tag{11.23}$$

for the exponent in (11.20). The integral differs from (3.100) only by the factor t^{-1} in the integrand. It is readily verified that

$$\frac{\pi}{\alpha} R(s) = \frac{1}{1+s^2} + 1 - s\,\text{arccot}(s) - \ln(-s^2) - \ln(-1-s^2), \tag{11.24}$$

and

$$\lim_{a\to 0} R(2a/\pi r) = \infty. \tag{11.25}$$

In the limit $a \to 0$, (11.20) vanishes, that is, there are an infinite number of photons in the transformed state, and the two vacuum states are orthogonal. This implies that the Hilbert spaces $\mathcal{H}[c]$ and $\mathcal{H}[C(\alpha)]$ are orthogonal. Thus, one has realisations[5] of the canonical commutation relations for the annihilation and creation operators $(c_{\mathbf{k},\lambda}, c_{\mathbf{k},\lambda}^+; \{C(\mathbf{k}:\mathcal{C})_\lambda, C(\mathbf{k}:\mathcal{C})_\lambda^+\})$ which are *not* related by unitary transformation; they are said to be *unitary inequivalent* [3], [4]. The practical consequence is that there is no longer a guarantee that a Coulomb gauge theory based on the free-field Fock space and its PZW transform will lead to identical expectation values. Quantities like $\mathcal{E}_{\mathbf{P}}$, (3.104), and R, (11.24), can be kept finite by insisting that the parameter a satisfies $a > 0$; this amounts to the introduction of a minimum length. Note, however, that this move, though it makes these quantities finite and enforces unitary equivalence, does

---

[4] The coefficient α on the RHS is the fine structure constant (in conventional notation), not to be confused with (11.13).

[5] A non-denumerable set since there is an uncountable infinity of possible paths $\mathcal{C}$.

not remove the cancellation[6] of the static Coulombic energy between pairs of charges revealed by (3.104).

11.1.2 The Heisenberg Equations of Motion

In the Heisenberg representation the state vector is constant in time, and the problems arise in the relationship between operators at different times, and so we need to look at the integration of the equations of motion. Let us revisit §3.8 and Appendix C from a quantum mechanical perspective in which the classical variables $\{\mathbf{q}, \mathbf{p}, \mathbf{A}, \boldsymbol{\pi}\}$ are reinterpreted as operators in the usual way. Equations (3.315), (3.316), with the Poisson brackets replaced by commutators, are the Heisenberg equations of motion for the charged particle; likewise the Fourier variables for the field are to be reinterpreted as the photon annihilation and creation operators.

Consider first the equation of motion for the annihilation operator $c_{\mathbf{k},\lambda}$; after quantisation, (C.0.5) is replaced by

$$\dot{c}_{\mathbf{k},\lambda}(t) = -i\omega c_{\mathbf{k},\lambda}(t) + \frac{ie\chi_a(k)}{\sqrt{2\Omega\hbar kc\varepsilon_0}}\dot{\mathbf{q}}\cdot\varepsilon(\mathbf{k})_\lambda\, e^{-i\mathbf{k}\cdot\mathbf{q}}, \qquad (11.26)$$

which in integrated form with retarded boundary conditions is

$$c_{\mathbf{k},\lambda}(t) = c_{\mathbf{k},\lambda}(t_0) + \frac{ie\chi_a(k)}{\sqrt{2\Omega\hbar kc\varepsilon_0}}\int_{-t_0}^{+\infty}\theta(t-t')\dot{\mathbf{q}}(t')\cdot\varepsilon(\mathbf{k})_\lambda e^{-i\left(\mathbf{k}\cdot\mathbf{q}(t')+\omega(t-t')\right)}\,\mathrm{d}t'. \quad (11.27)$$

Here $\mathbf{q}$ and $\dot{\mathbf{q}}$ are now to be interpreted as non-commuting operators, and $\omega = kc$. In the Heisenberg picture a scattering process is described in terms of in- and out-operators corresponding to the physical situations at $t = -\infty$ and $t = +\infty$, respectively. The S-matrix provides the relationship between the in- and out-operators according to

$$\Gamma^{\text{out}} = \mathsf{S}^{-1}\,\Gamma^{\text{in}}\,\mathsf{S}. \qquad (11.28)$$

Thus, in (11.27) we replace $c_{\mathbf{k},\lambda}(t_0)$ by $c_{\mathbf{k},\lambda}^{\text{in}}$ and make the lower limit of the integral $-\infty$; $c_{\mathbf{k},\lambda}(t)$ and its adjoint determine the field operators at time t.

If we attempt to solve the coupled operator equations in the way described in §3.8, we must pay careful attention to the non-commutation of the operators involved. Recalling that $\dot{\mathbf{q}}$ is given by Hamilton's equation, (3.316), there are two points to mention:

1. The coordinate operators at different times do not commute; hence the product of their exponential factors is not simply equal to an exponential with the exponents added together. The required modification may be computed using the Baker–Campbell–Hausdorff formula.
2. While it is true that $\mathbf{q}(t')$ and $\mathbf{p}(t')$ satisfy the fundamental equal-time canonical commutation relation,

$$[\mathsf{q}(t')_\alpha, \mathsf{p}(t')_\beta] = i\hbar\,\delta_{\alpha\beta}, \qquad (11.29)$$

---

[6] The cancellation is independent of the functional form chosen for $\chi_a(k)$ which merely affects the explicit expression of $R(s)$.

we are only concerned with the transverse component of the field variable arising from the equal-time combination. Thus, we can replace $\dot{\mathsf{q}}$ with $\mathbf{p}/m$, and note that $(\delta_{\alpha\beta} - \hat{k}_\alpha\hat{k}_\beta)\,\mathsf{p}_\alpha\,e^{i\mathbf{k}\cdot\mathbf{q}}$ is identical to $(\delta_{\alpha\beta} - \hat{k}_\alpha\hat{k}_\beta)\,e^{i\mathbf{k}\cdot\mathbf{q}}\,\mathsf{p}_\alpha$, since the different orders of the operators give rise to a commutator that is proportional to k_α, and this is annihilated by the transverse projector $\delta_{\alpha\beta} - \hat{k}_\alpha\hat{k}_\beta$.

First, we need to calculate the time evolution of the coordinate using

$$\mathsf{q}(t')_\alpha = e^{-i\mathsf{H}\tau/\hbar}\,\mathsf{q}(t)_\alpha\,e^{+i\mathsf{H}\tau/\hbar}, \quad t' = t - \tau. \tag{11.30}$$

The field energy contribution to H commutes with $\mathbf{q}$, so using the expansion of an exponential operator (11.30) may be evaluated to yield

$$\mathsf{q}(t')_\alpha = \mathsf{q}(t)_\alpha - \frac{i\tau}{\hbar}[\mathsf{H},\mathsf{q}(t)_\alpha] + \left(\frac{i\tau}{\hbar}\right)^2\frac{1}{2!}[\mathsf{H},[\mathsf{H},\mathsf{q}(t)_\alpha]] + \dots \tag{11.31}$$

$$= \mathsf{q}(t)_\alpha - \frac{\tau}{m}\left(\mathbf{p}(t) - e\mathbf{A}_\xi(\mathbf{q}(t))\right)_\alpha, \tag{11.32}$$

and we infer that

$$[\mathsf{q}(t)_\alpha,\mathsf{q}(t')_\beta] = -\frac{i\hbar\tau}{m}\,\delta_{\alpha\beta}, \tag{11.33}$$

with $\tau = t - t'$.

It follows that with sufficient accuracy the transverse component of the velocity operator, $\dot{\mathsf{q}}(t')$, can be taken to commute with the exponential factors, $e^{i\mathbf{k}\cdot\mathbf{q}(\tau)}$, irrespective of the time τ. Hence in place of (C.0.27), we can write the quantum mechanical vector potential operator as the symmetrised expression

$$\mathsf{A}_\xi(\mathbf{q}(t))_\alpha = \frac{1}{2}\left(\frac{e}{8\pi^3\varepsilon_0 c}\right)\iint_{-\infty}^t\left(\frac{\chi_a^2(k)}{k}\right)\sin[kc\tau](\mathbf{1} - \mathbf{kk})_{\alpha\beta}$$
$$\times\left(\dot{\mathsf{q}}(t')_\beta e^{i\mathbf{k}\cdot\mathbf{q}(t')}e^{-i\mathbf{k}\cdot\mathbf{q}(t)} + e^{-i\mathbf{k}\cdot\mathbf{q}(t)}\dot{\mathsf{q}}(t')_\beta e^{i\,\mathbf{k}\cdot\mathbf{q}(t')}\right)\mathrm{d}^3\mathbf{k}\,\mathrm{d}t', \tag{11.34}$$

where the position of the velocity operator in the ordering is immaterial. Now considering the exponential factors in (11.34), we see that the first exponential correction in the Baker–Campbell–Hausdorff formula is a c-number, so we have

$$e^{i\mathbf{k}\cdot\mathsf{q}(t')}\,e^{-i\mathbf{k}\cdot\mathsf{q}(t)} = e^{i\,\mathbf{k}\cdot\left(\mathsf{q}(t')-\mathsf{q}(t)\right)}e^{\frac{1}{2}[\,\mathbf{k}\cdot\mathsf{q}(t'),\mathbf{k}\cdot\mathsf{q}(t)]}$$

$$= e^{i\mathbf{k}\cdot\left(\mathsf{q}(t')-\mathsf{q}(t)\right)}e^{-i\hbar\tau k^2/2m},$$

$$e^{-i\mathbf{k}\cdot\mathsf{q}(t)}e^{i\mathbf{k}\cdot\mathsf{q}(t')} = e^{i\,\mathbf{k}\cdot\left(\mathsf{q}(t')-\mathsf{q}(t)\right)}e^{-\frac{1}{2}[\,\mathbf{k}\cdot\mathsf{q}(t'),\mathbf{k}\cdot\mathsf{q}(t)]}$$

$$= e^{i\mathbf{k}\cdot\left(\mathsf{q}(t')-\mathsf{q}(t)\right)}e^{+i\hbar\tau k^2/2m}. \tag{11.35}$$

Thus, the quantum mechanical vector potential is given by essentially the same integral as in the classical case except that the integrand contains an additional cosine factor,

$$A_\xi(\mathbf{q}(t))_\alpha = C \int\int_{-\infty}^{t} \left(\frac{\chi_a^2(k)}{k}\right) \sin(kc\tau)$$

$$\times (\mathbf{1} - \hat{\mathbf{k}}\hat{\mathbf{k}}) \cdot \dot{\mathbf{q}}(t') e^{i\,\mathbf{k}\cdot(\mathbf{q}(t')-\mathbf{q}(t))} \cos\left(\frac{\hbar k^2 \tau}{2m}\right) d^3\mathbf{k}\, dt'. \tag{11.36}$$

The quantum mechanical generalisations of the complete solution for the vector potential and its separation into linear and non-linear parts are obvious.

The mass renormalisation procedure described in §3.8.1 can be repeated here with a surprising result; we make the same choice for 'u' and put

$$\mathrm{d}v = \sin[kc(t-t')]\cos\left[\left(\frac{\hbar k^2}{2m}\right)(t-t')\right], \tag{11.37}$$

so that

$$v = \left(\frac{1}{2}\right)\left[\frac{\cos\left[(kc+\frac{\hbar k^2}{2m})\tau\right]}{(kc+\frac{\hbar k^2}{2m})} + \frac{\cos\left[(kc-\frac{\hbar k^2}{2m})\tau\right]}{(kc-\frac{\hbar k^2}{2m})}\right], \tag{11.38}$$

with $\tau = (t - t')$. In the integrations the pole at $k = 2mc/\hbar$ is interpreted in terms of the Cauchy principal value. As before, the resulting boundary term vanishes in the far past, and at $t' = t$ it reduces to

$$uv\Big|^{t} = \left(\frac{\Delta m^{Q}}{e}\right)\dot{\mathbf{q}}(t), \tag{11.39}$$

with Δm^{Q} given by

$$\Delta m^{Q} = \Delta m^{Cl} \left(\frac{2a}{\pi}\right)\int_0^\infty \frac{\chi_a^2(k)}{1 - \frac{\hbar^2 k^2}{4m^2 c^2}}\, dk, \tag{11.40}$$

where the integral is a Cauchy principal value. Here Δm^{Cl} is the classical expression in Eq. (3.335). This obviously gives the expected classical result for $\hbar \to 0$ (or $m \to \infty$, a 'static charge') since the integral reduces to $\pi/2a$. Now, recalling that the point particle limit is given by

$$a \to 0 \Rightarrow \chi_a(k) \to 1, \tag{11.41}$$

we see that, for fixed $\hbar$ and m, (11.40) leads to

$$a \to 0 \Rightarrow \Delta m^{Q} \to 0, \tag{11.42}$$

since the principal value integral vanishes,

$$\int_0^\infty \frac{dx}{1-x^2} = 0. \tag{11.43}$$

Quantum mechanics leads to vanishing electromagnetic mass for the point particle. We recognise the appearance of the reduced Compton wavelength, $\lambdabar = \hbar/mc$, in (11.40). Of course, one is far from solving for the full quantum mechanical vector potential, (11.36). But one does learn that a classical point charge should not be thought of as the point limit of an extended classical charge distribution; rather, it should be considered to be the classical limit of a point quantum mechanical particle. Further discussion of these findings can be found in [7]–[9]; more recently, self-interaction has become an active topic in non-perturbative QED.

As in §11.1.1 we can view the relationship between $c_{\mathbf{k},\lambda}(t)$ and $c_{\mathbf{k},\lambda}^{in}$ displayed in (11.27) as a boson translation transformation with coherent state 'parameter'

$$\alpha(\mathbf{k})_\lambda = \frac{ie\chi_a(k)}{\sqrt{2\Omega\hbar kc\varepsilon_0}} \int_{-\infty}^{+\infty} \theta(t-t')\dot{\mathbf{q}}(t')\cdot\varepsilon(\mathbf{k})_\lambda e^{-i\left(\mathbf{k}\cdot\mathbf{q}(t')-\omega(t')\right)}\,dt'. \tag{11.44}$$

The important question then is: is $\alpha(\mathbf{k},\lambda)$ square integrable? While a full answer requires the solution of the coupled equations of motion for a specified physical situation which in general is very hard to achieve, simple model cases already suggest that this boson translation is problematic. Consider the special case described in §3.8.1 of a particle with constant velocity, $\dot{\mathbf{q}}(t) = \mathbf{v}$ with $|\mathbf{v}| > 0$. Evaluation of the integral (11.44) yields

$$\alpha(\mathbf{k})_\lambda = \frac{e\chi_a(k)}{\sqrt{2\Omega\hbar kc\varepsilon_0}}\mathbf{v}\cdot\varepsilon(\mathbf{k})_\lambda e^{-i\mathbf{k}\cdot\mathbf{q}_0}\frac{e^{i(\omega-\mathbf{k}\cdot\mathbf{v})}}{\omega - \mathbf{k}\cdot\mathbf{v}}. \tag{11.45}$$

For small $|\mathbf{k}|$ this behaves as $|\mathbf{k}|^{-3/2}$ and

$$\int |\alpha(\mathbf{k})_\lambda|^2 d^3\mathbf{k} \to \infty. \tag{11.46}$$

A closely related example [4] is a particle with initial constant velocity $\mathbf{v}$ that is given a kick at $t = 0$ and thereafter moves with velocity $\mathbf{v}'$. There are additional terms in $\alpha(\mathbf{k})_\lambda$ but its squared integral reduces to (11.45) for $\mathbf{v} = \mathbf{v}'$ and the additional terms do not alter the $|\mathbf{k}|^{-3/2}$ behaviour. Thus, the Fock representation for the in-operators cannot be used for operators at later times which have a unitarily inequivalent Fock representation. This is the origin of the infrared divergence problem which arises in a perturbation theory calculation (and does try to use a single Fock space) of charged particle scattering; the moving particle can emit an unlimited number of very-low-frequency ($\to 0$) photons. We note in passing that the Power–Zienau–Woolley transformation for an overall neutral system removes the infrared divergence problem entirely, at the expense of worse behaviour at large k;[7] however, as seen in (11.1.1) the transformed state does not stay in the initial free-field Fock space [2].

11.2 Non-perturbative Ideas

The conventional approach to quantum electrodynamics is based on perturbation theory for the S-matrix as described in Chapters 6, 9 and 10. Such calculations lead to

---

[7] Recall the EM field mode expansions vary as $\sqrt{k}$, as compared to the $1/\sqrt{k}$ dependence in (11.26) that originates from the mode expansion of the Coulomb gauge vector potential. The generator, N, of the Schwinger transformation (10.236), can be cast in the form of a boson translation with an $\alpha(\mathbf{k})_\lambda$-parameter that varies as $1/k^{3/2}$, and so shows an infrared divergence associated with 'soft photons', $k \to 0$.

cross sections that can be related to scattering experiments. The difficulties in perturbation theory are of two different sorts. Firstly, the 'loop' diagrams like Figure 10.4 allow the involvement of intermediate states with virtual photons of unrestricted momentum, and hence energies far beyond the regime of validity of the non-relativistic theory. These are the 'ultraviolet' divergences dealt with by, for example, a maximum momentum cut-off so as to suppress their contributions. The use of a cut-off is a crude realisations of the notion that high-momentum (high-energy) states must be eliminated in order to construct an 'effective' theory that is adequate for the low-energy physics of interest. This can be achieved with the systematic use of Feshbach projection (Löwdin's partitioning technique).

Secondly, charged particles in the field can be associated with an arbitrarily large number of virtual photons with energy close to zero; these require an infrared cut-off. With the full apparatus of covariant QED and an invariant method of calculation (for example, Feynman diagrams) one can extract finite values. When that is done for interacting electrons and photons, the agreement with experiment is remarkable, perhaps the most accurate quantities that can be calculated by quantum mechanics [6]. Nevertheless, the occurrence of infinities is an ugly feature which hints at underlying problems in the formalism of QED. Furthermore, there are important questions in QED which cannot be answered using a perturbation expansion, for example, the demonstration of the existence of a ground state for interacting charges and field required for the stability of bulk matter in the presence of the field, and the nature of the excitations. These require analytical techniques that are not based on perturbation methods. Over the past several decades, a mathematical approach to non-relativistic QED has been developed using the techniques of modern functional analysis; there is now a considerable research literature, and several monographs available too [10]–[12]. This chapter aims to give some introductory remarks about this programme and to indicate some connections with the ideas in the earlier chapters.

The use of the Coulomb gauge condition is the normal choice in the mathematical literature, though as we will see, the PZW transformation makes an appearance. We know from Chapter 9 that the full Hamiltonian for charged particles interacting with the quantised electromagnetic field can be written in the form

$$H_\lambda = H_0 + \lambda H_1 + \lambda^2 H_2, \tag{11.47}$$

where the coupling constant λ is identified with the fundamental charge e. Here H_0 is the same as the unperturbed Hamiltonian used in the perturbation theory approach, that is, the sum of (9.102) for the charges and (7.37) for the free electromagnetic field. The terms in (11.47) involving λ and λ^2 are, respectively, the familiar $\mathbf{p} \cdot \mathbf{A}$ and $|\mathbf{A}|^2$ terms in this gauge. The nuclei are treated as spin-zero particles, while the electrons are properly regarded as spin $\frac{1}{2}$ fermions with the 'semi-relativistic' Pauli interaction for the electrons (9.158) sometimes included in the interaction Hamiltonian. Importantly, if this is done, the operator $\mathbf{B}$ in it is the quantised field operator (7.125); it is of order λ. Thus, the terms in (11.47) are obtained as a straightforward generalisation of (11.6):

$$H_0 = \sum_{i,j} \frac{e_i e_j}{4\pi\varepsilon_0 |\mathbf{x}_i - \mathbf{x}_j|} + \sum_{\mathbf{k},\lambda} \hbar\omega \, c_{\mathbf{k},\lambda}^+ c_{\mathbf{k},\lambda}$$

$$H_1 = \sum_i \frac{e_i}{m_i} \mathbf{p}_i \cdot \mathbf{A}(\mathbf{x}_i) - \frac{e\hbar}{2m_e} \sum_\alpha \boldsymbol{\sigma}_\alpha \cdot \mathbf{B}(\mathbf{x}_\alpha)$$

$$H_2 = \sum_i \frac{e_i^2}{m_i} \mathbf{A}(\mathbf{x}_i) \cdot \mathbf{A}(\mathbf{x}_i), \tag{11.48}$$

where the sums over i, j are over electrons *and* nuclei, while those over α are restricted to the electrons; the $\boldsymbol{\sigma}_i$ are the usual Pauli matrices. We suppose there are N electrons (so $\alpha = 1, \ldots N$) and M nuclei with charges eZ_M; the total charge is then

$$Q = e\left(-N + \sum_{n=1}^{M} Z_n\right). \tag{11.49}$$

The total linear momentum of the system of charges and photons is

$$\mathbf{P}_\mathrm{T} = \sum_i \mathbf{p}_i + \hbar \sum_{\mathbf{k},\sigma} \mathbf{k} c_{\mathbf{k},\sigma}^+ c_{\mathbf{k},\sigma}. \tag{11.50}$$

It commutes with H_λ, which is an expression of the translation invariance of the whole system. If $H(\mathbf{P})$ is the Hamiltonian at fixed total momentum $\mathbf{P}$, the full Hamiltonian may be written as a direct integral,

$$H_\lambda = \int_{\Re^3}^\oplus H(\mathbf{P})\, d\mathbf{P}. \tag{11.51}$$

It is then sufficient to analyse the properties of $H(\mathbf{P})$ for some fixed $\mathbf{P}$, and in particular it is essential to establish whether $H(\mathbf{P})$ has an eigenvalue (i.e. a *bound state*) at the bottom of its spectrum. The obvious physical interpretation of such a state is a stable atom/molecule dressed with a cloud of photons in motion [13], [14].

The discussion in Chapter 8 for the case where electrons and nuclei interact through purely the Coulombic part of the electromagnetic field did not require specification of the momentum, nor was the overall charge particularly emphasised since it is physically reasonable that neutral and positively charged species are much more likely to be stable than ones with an excess of electrons. Here one may reasonably surmise that the charge Q will be a crucial parameter to be considered, not least because of the infrared singularity noted in §11.1.2 for a charged particle, and the observation that, if not up close, a charged molecule looks much like a charged 'particle', and the (spatial) far-field is related to the $k \to 0$ limit of the modes. If $Q = 0$, the photons see an electrically neutral charge distribution, and the resulting vector potential (which determines the fields) decays faster than $1/|x|$ which can be accommodated in the Fock space description. Thus, there is no infrared divergence, and a stable ground state is found for some range of values of $|\mathbf{P}| \geq 0$ [13]. The situation is more delicate if $Q \neq 0$; classically the radiation field is determined by (C.0.5) which reduces to the free field if the ion is at rest.[8] In the quantum mechanical account the equivalent condition is expressed in

---

[8] Here being at rest means $\dot{\mathbf{R}}_{cm} = 0$, where $\dot{\mathbf{R}}_{cm}$ is the velocity of the ion's centre of mass.

terms of an expectation value of the momentum being zero. Otherwise for $|\mathbf{P}| \neq 0$ and $Q \neq 0$ there is no ground state unless an infrared cut-off is applied.

11.3 The Spectrum of the QED Hamiltonian

In order to make the vector potential a well-defined operator in the Hilbert (Fock) space of the free field, its mode expansion (7.122) must be modified by the inclusion of an ultraviolet cut-off. This may be done in the way suggested in §2.4.2 in terms of a function $\chi(k)$; thus, we write

$$\mathbf{A}(\mathbf{x}) = \sqrt{\frac{\hbar}{2\varepsilon_0 \Omega c}} \sum_{\mathbf{k},\lambda} \chi(k) \frac{\hat{\boldsymbol{\varepsilon}}(\mathbf{k})_\lambda}{\sqrt{k}} \left(c_{\mathbf{k},\lambda} e^{i\mathbf{k}\cdot\mathbf{x}} + c_{\mathbf{k},\lambda}^+ e^{-i\mathbf{k}\cdot\mathbf{x}} \right). \tag{11.52}$$

The convergence factor is chosen so as to make the integral in (11.11) finite; its precise form is often unimportant, but typical examples are

$$\chi(k) \leq \begin{cases} 1 & \text{for } k \text{ near } 0 \\ (\frac{k}{\kappa})^{-3} & \text{for large } k \end{cases}$$

$$\chi(k) = \exp(-k^2/\kappa^2)$$

$$\chi(k) = \begin{cases} 0 & \text{if } k > \kappa \\ 1 & \text{if } k < \kappa. \end{cases} \tag{11.53}$$

where $\alpha^2 \ll \kappa\lambda_c \ll 1$. This defines the non-relativistic regime, where such effects as pair production and polarisation of the vacuum which result in charge renormalisation in standard QED cannot occur, while giving an energy, $\hbar c\kappa$, much greater than the typical ionisation energies of the atomic system [15].

The mode expansion of the free-field Hamiltonian does not require a cut-off in order that H_{rad} be a well-defined self-adjoint operator on the free-field Fock space. The electrons are treated in a fully quantum mechanical way (as fermions) using two-component wave functions; in early work the nuclei were regarded as fixed classical sources of a Coulomb field. This is unimportant in atoms, since one can reinterpret the origin (the nucleus) as the true centre of mass and bring in the reduced mass of the electron without losing any symmetries of the atomic states. For molecules, however, this would be a highly non-trivial assumption since nuclear permutation symmetry is a feature of the generic molecule if the nuclei are quantum mechanical particles. However, in more recent work, attention has changed to moving atoms and ions so that the nuclei are treated as quantum particles. This is a non-trivial matter since, as noted earlier, a distinction between neutral and charged species becomes apparent. The earliest investigations required much smaller values for the coupling constant λ than the actual physical values for electrons and nuclei determined by the fine structure constant [16], but many of these restrictions have been removed in later calculations, for example [17]. The systematic analysis of the consequences of the quantum mechanical Hamiltonian (11.47) can be traced back at least as far as a pioneering investigation published in 1938 [18], such that in the mathematical literature H_λ is commonly known as 'the Pauli–Fierz Hamiltonian'.

Even with the restriction of (11.52) to the non-relativistic regime there is still the problem of its behaviour as $|\mathbf{k}| \to 0$ which gives rise to the infrared divergence problem for a charged particle interacting with the quantised electromagnetic field. For the *neutral* atom/molecule, this may be ameliorated by making a unitary transformation of H_λ with a generator used by Pauli and Fierz [16], [17]; in the mathematical physics literature this transformation commonly bears their name. In atomic/molecular physics, it is known as the electric dipole approximation to the PZW transformation (§9.3.3), that is, exactly as in (9.123) but with the classical Coulomb gauge vector potential replaced by the corresponding operator form including the cut-off $\chi(k)$, and its spatial variation suppressed so there is no magnetic field,

$$\Lambda = \exp(-i\mathsf{F}/\hbar), \quad \mathsf{F} = \mathbf{d} \cdot \mathbf{A}(\mathbf{0}). \tag{11.54}$$

The cost of making such a transformation is an interaction term, $-\mathbf{d} \cdot \mathbf{E}^\perp(\mathbf{0})$, that increases as $|\mathbf{d}| \to \infty$. The Combes dilatation transformation [19] of both the particle coordinates and the photon momenta, described in what follows, acts sufficiently to control this growth. Alternatively, one can argue that since one is interested in bound states in which the charges are exponentially localised, this is sufficient to bound the dipole contribution. More recently, a 'generalised' Pauli–Fierz transformation has been described [15]; for charges $\{e_n\}$ with position operators $\{\mathbf{x}_n\}$ this involves the following quantity as the generator to be used in (11.54) in place of the dipole approximation

$$\overline{\mathsf{F}} = \sum_n e_n \sum_{\mathbf{k},\lambda} \frac{1}{\sqrt{k}} \left(f(\mathbf{x}_n)^*_{\mathbf{k},\lambda} \mathsf{c}_{\mathbf{k},\lambda} + f(\mathbf{x}_n)_{\mathbf{k},\lambda} \mathsf{c}^+_{\mathbf{k},\lambda} \right), \tag{11.55}$$

where[9]

$$f(\mathbf{x}_n)_{\mathbf{k},\lambda} = \sqrt{\frac{\hbar}{2\varepsilon_0 \Omega k c}} e^{-i\mathbf{k}\cdot\mathbf{x}_n} \chi_a(k) \varphi(\sqrt{k}\hat{\boldsymbol{\varepsilon}}(\mathbf{k})_\lambda \cdot \mathbf{x}_n), \quad \varphi'(0) = 1. \tag{11.56}$$

This reduces to (11.54) if φ is taken as a linear function, and the exponential factors are neglected as required for the dipole approximation. As usual we write

$$\overline{\mathsf{H}}_\lambda = e^{-i\overline{\mathsf{F}}} \mathsf{H}_\lambda e^{+i\overline{\mathsf{F}}}, \tag{11.57}$$

and this is evaluated as in Chapter 9 by expanding the exponentials; as with the PZW transformation, $\overline{\mathsf{F}}$ commutes with the field and particle 'position' variables, and produces new terms from the particle and field 'momenta' exactly as in (9.115).

In order to set the scene we begin with a qualitative description of the spectrum of the QED Hamiltonian presented in Chapter 9, beginning with the reference Hamiltonian H_0, Eq. (9.103). The spectrum of the particle Hamiltonian, H_{charges}, was described in §8.3; according to the HVZ theorem, there is a continuum corresponding to the half-axis $[\Sigma, \infty)$ for some $\Sigma \leq 0$, and isolated discrete energy levels $E_0, E_1, \ldots E_i$ below the

---

[9] A simple form for φ is $\varphi(r) = r$ if $|r| \leq 1/2$ and $|\varphi| = 1$ if $|r| \geq 1$; this controls the long-distance behaviour [21].

continuum, that is, $E_0 \leq E_1 \leq, \ldots, < \Sigma$; this spectrum is shown in Figure 8.1. The spectrum of the free electromagnetic field Hamiltonian[10] consists of a simple eigenvalue at 0, corresponding to the vacuum state, Ψ_0, and absolutely continuous spectrum on the half-axis $[0, \infty)$. The eigenstates of H_0 corresponding to the eigenvalues $\{E_i\}$ are of the form (11.7); what happens to such states in the presence of interactions is a significant question in the quantum theory of radiation [16]. These facts about the unperturbed spectra of the particles and field mean that the reference Hamiltonian, H_0, (9.103) has the same discrete spectrum as H_{charges}, that is, $\{E_i\}$, and a continuous spectrum covering the half-axis $[E_0, \infty)$ consisting of a union of branches $[E_i, \infty)$ starting at the energy levels E_i and the branch $[\Sigma, \infty)$. Thus, all the discrete energy levels of the atomic system including E_0 become thresholds of continuous spectra; they are said to be 'embedded' eigenvalues. This is the mathematical reason for the difficulties in perturbation theory; non-relativistic QED, which is focused on the behaviour of the discrete states of atoms and molecules in the presence of electromagnetic radiation, requires the perturbation theory of continuous spectra.

The spectrum of the full Hamiltonian, $\overline{H}_\lambda$, is most usefully defined in terms of matrix elements of its resolvent (cf. §6.3); the discrete and continuous spectra are the poles and cuts, respectively, of

$$\langle \phi | \frac{1}{\overline{H}_\lambda - z} | \psi \rangle \quad \text{for all } \phi, \psi \in \mathcal{H}, \tag{11.58}$$

where z is a complex variable. The structure of the resolvent can be exposed by using the idea of dilatation (or complex coordinate rotation [20]) transformations, described in Chapter 8 for the Coulomb Hamiltonian. Consider the family of transformed Hamiltonians defined by

$$\overline{H}_\lambda(\theta) = U(\theta) \, \overline{H}_\lambda \, U(\theta)^{-1}, \tag{11.59}$$

where θ is a real parameter and $U(\theta)$ is chosen to transform the particle positions and photon momenta as

$$x_n \to e^\theta x_n, \quad n = 1, \ldots N$$
$$k \to e^{-\theta} k. \tag{11.60}$$

The transformed Hamiltonian, $\overline{H}_\lambda(\theta)$, has an analytic continuation in the variable θ in a disc $D(0, \theta)$ about $\theta = 0$ in the complex θ plane. If $\Phi_\theta = U(\theta)\Phi$, then for $z \in C^+$ (i.e. $\Im z > 0$)

$$\langle \Phi | R(z) | \Phi \rangle = \langle \Phi | U(\theta)^{-1} R(\theta, z) U(\theta) | \Phi \rangle = \langle \Phi_\theta | R(\theta) | \Phi_\theta \rangle. \tag{11.61}$$

The quantity (cf. (6.31))

$$F(\theta, z) = \langle \Phi(\overline{\theta} | R(\theta, z) | \Phi(\theta) \rangle \tag{11.62}$$

has an analytic continuation into a neighbourhood of $\theta = 0$ [19], and subject to certain technical requirements the same is true for the untransformed resolvent (the LHS of

---

[10] The zero-point energy has been dropped for this account.

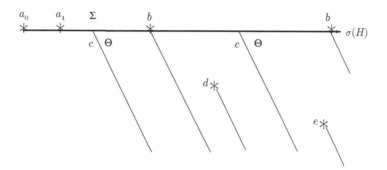

Figure 11.1 The spectrum of the rotated Hamiltonian, $H_\lambda(\theta)$; a_0, a_1: discrete eigenvalues of H_0; a_0 is the ground state, b: continuum embedded eigenvalues, c: thresholds, d: discrete, complex eigenvalue (resonance), e: complex threshold of $H_\lambda(\theta)$. Σ is the start of the essential spectrum of the atomic system in the absence of the field; the solid line is the spectrum of the electromagnetic field. $\Theta = \Im\theta$.

(11.61)). The real eigenvalues of $\overline{H}(\theta)$ give real poles of the RHS of (11.61), and so they are the real eigenvalues of $\overline{H}_\lambda$; the point of the complex coordinate rotation transformation, however, is that it reveals new structure in the RHS of (11.61), namely complex eigenvalues for $\Im\theta > 0$. These are the poles of the meromorphic continuation of the LHS of (11.61) across the essential spectrum of $\overline{H}_\lambda$ onto the second Riemann sheet, that is, into the lower complex half-plane. They are interpreted as the *resonances* of $\overline{H}_\lambda$, and since the transformation (11.57) is unitary these are also the resonances of the original Hamiltonian, H_λ. Every eigenvalue of H_0 apart from the ground state behaves in this way; thus, all excited stationary states of the free atomic/molecular system become metastable states because of their interaction with the quantised electromagnetic field.

Comparing Figure 11.1 with Figure 8.1, the significant difference is that for the transformed QED Hamiltonian every resonance is attached to a branch of the essential spectrum, whereas in the Coulomb Hamiltonian case the resonances were *isolated*. This occurs because the photon has zero mass and leads to the problem of infrared divergences.

An essential idea used in the characterisation of the spectrum of $\overline{H}_\lambda$ is to concentrate on a limited energy range. A similar idea is used in the mathematical analysis of the Born–Oppenheimer approximation (Chapter 8); it is based on the systematic reduction of the degrees of freedom of the Hilbert space to the energy range of interest using projection operator techniques. This is the precise formulation of the idea of using cut-offs to eliminate unwanted degrees of freedom. We sketch some of the ideas in what follows; the formidable technical details can be found in the original literature references. The formal setting, which is essentially that familiar from Löwdin's partitioning technique [22], is as follows [16]. Suppose we have a Hamiltonian H acting on a Hilbert space $\mathcal{H}$. As in Chapter 6 we define a pair of projection operators P and Q such that

$$P^2 = P, \quad P + Q = I \tag{11.63}$$

and construct the projected Hamiltonians

$$H_P = P\,H\,P, \quad H_Q = Q\,H\,Q, \tag{11.64}$$

which are operators on $P\mathcal{H}$ and $Q\mathcal{H}$, respectively. Now let $\rho(\Omega)$ denote the resolvent set of Ω, that is, the set of complex numbers z such that $\Omega - z\mathsf{I}$ has a bounded inverse. Then, provided 0 lies in $\rho(H_Q)$, the inverse operator H_Q^{-1} exists on $Q\mathcal{H}$ and is bounded.

A *Feshbach map*, $f_P(H)$, is defined on the reduced space $P\mathcal{H}$ by

$$f_P(H) = PHP - PHQ(H_Q)^{-1}QHP, \tag{11.65}$$

provided 0 belongs to $\rho(H_Q)$. Then we have, for example,

1. ε belonging to the resolvent set $\rho(H)$ implies 0 belongs to $\rho(f_P(H - \varepsilon))$.
2. For an eigenstate (E, Ψ) of H we have

$$H\Psi = E\Psi \iff f_P(H - E)\Phi, \quad \Phi = P\Psi. \tag{11.66}$$

The map is *isospectral* in the sense that it leads to an 'effective' operator, f_P, that in the energy range of interest has the *same* spectrum as the original operator.

We now focus on a particular discrete state of the unperturbed Hamiltonian for the charges with energy E_k obtained from the Schrödinger equation

$$H_{charges}|\phi_k\rangle = E_k|\phi_k\rangle \tag{11.67}$$

and enquire about its fate in the presence of quantised radiation using a Feshbach map constructed as follows. We define a new field operator by the relation

$$\xi(H_{rad} : \rho_0) = \sum_\lambda \int \left(c_{k,\lambda}^+ c_{k,\lambda}\right)\hbar\omega\, d^3k, \quad \hbar\omega < \rho_0, \tag{11.68}$$

which describes photons with energies $< \rho_0$. The projection operator required for the Feshbach map is then

$$P = \sum_{m=1}^{N} |\phi_k\rangle\langle\phi_k| \otimes \xi(H_{rad} : \rho_0), \tag{11.69}$$

which is combined with (11.65) and the Hamiltonian $\overline{H}_\lambda$; the maximum photon energy ρ_0 is related to the coupling constant λ. Iteration of Feshbach maps is like using a microscope to inspect tiny regions of the spectrum and gain ever finer information as the energy interval examined is reduced; in the limit of an infinite sequence of such maps one can in principle obtain the exact ground state of the Hamiltonian of interest and the precise location of its resonances [23], [24].

The Feshbach map construction just outlined has a straightforward physical interpretation but suffers from a technical disadvantage for computation because the sharp cut-off in the photon energies implies that it is not differentiable. A significant improvement is apparent with the introduction of so-called 'smooth' Feshbach maps, which, though lacking a straightforward interpretation in terms of a block diagonalisation of the Hamiltonian, have much nicer mathematical properties. In place of P and Q in (11.63) one introduces a pair of operators χ and $\overline{\chi}$ with

$$\chi^2 + \overline{\chi}^2 = \mathsf{I}. \tag{11.70}$$

The Hamiltonian may be decomposed in the usual perturbation theory form,

$$H = H_0 + V, \tag{11.71}$$

where H_0 is independent of the coupling between the field and charges. One requires that H_0 commutes with both χ and $\overline{\chi}$, and then the 'smooth' Feshbach map $F(H, H_0)$ can be defined in exactly the same way as in (11.65) with χ and $\overline{\chi}$ replacing P and Q, respectively [15], [23]–[26].

The main results of detailed mathematical analysis of the iterated Feshbach map construction are summarised in what follows. They refer to a *neutral* atomic or molecular system with linear momentum *less* than some critical value $|\mathbf{P}|_c$ [13]–[17]:

1. There is a ground state of H_λ derived from the ground state of H_0; it is exponentially localised in the coordinates of the charges. The existence of the ground state can be demonstrated for the physical coupling constant, α.
2. There are complex eigenvalues $\{E(\lambda)_{k,m}, m = 1, \ldots n_k\}$ associated with each eigenvalue E_k of H_0 with multiplicity n_k; they are independent of the angle θ used in the dilatation just described. The energy $E_{0,1}$ with the smallest real part is real and is the ground state energy of H_λ. Under certain further technical assumptions, all the $\{E(\lambda)_{k,m}, k \geq 1\}$ for non-zero λ can be shown to be complex quantities with *negative* imaginary parts, that is, they are the complex resonance energies of $\hat{H}_\lambda$.
3. The radiative corrections are of the form

$$E(\lambda)_{k,m} - E(0)_k \approx \lambda^2 \varepsilon_{k,m} + O(|\lambda|^2), \tag{11.72}$$

where $\Re \varepsilon_{k,m}$ is given by Bethe's formula (the Lamb shift in the case of the hydrogen atom; cf. Chapter 10) and $\Im \varepsilon_{k,m}$ is given by Fermi's golden rule for the decay rate. This identification is valid when allowance is made for the effects of the ultraviolet cut-off κ.

The demonstration that the non-relativistic QED Hamiltonian for a system of N electrons and M nuclei has a lowest energy eigenvalue $E_0 > -\infty$ confirms 'stability of the first kind'. As discussed in Chapter 8, the demonstration of 'stability' of the second kind' is also an important result since it is the guarantee that the energy of a n-body system is *extensive*, that is, proportional to the number of particles. What is required is the inequality

$$E_0 > C(N + K) \tag{11.73}$$

for some constant $C \leq 0$ that does not depend on N and M, but will generally involve the basic physical parameters of the system (mass, charge, Planck's constant etc.); $C < 0$ means that the binding energy per particle is positive. A comprehensive account of the stability of matter with or without quantised radiation can be found in [12], to which we refer; here we just note an extension of the argument given earlier in Chapter 8, (8.44)–(8.47). There is a similar lower bound in Chapter 11 of [12], namely the result of Lieb and Seiringer's Theorem 11.1 which involves some constant, C', that again does not depend on the nuclear positions. In this case, the *electronic* Hamiltonian H_0 is to be understood as the Pauli Hamiltonian for properly fermionic electrons interacting with quantised radiation (the vector potential is taken in the Coulomb gauge), so it includes the $\boldsymbol{\sigma} \cdot \mathbf{B}$ interaction; and the nuclei are fixed, so φ involves the electronic variables, including spin, and field variables in the usual Fock space description (denoted by $\mathcal{F}$). With this understanding we can recopy (8.44) as

$$\langle \varphi | \mathsf{H}_0 | \varphi \rangle_{el,\mathcal{F}} \geq - C' \langle \varphi | \varphi \rangle_{el,\mathcal{F}}; \qquad (11.74)$$

If the nuclei are taken as dynamical (quantum) particles, then we must make the following change,

$$\mathsf{T}_N \rightarrow \frac{1}{2M_N}(\mathbf{P}_N - e_N\mathbf{A}(\mathbf{R}_N))^2, \qquad (11.75)$$

to include the nuclear *kinetic* momentum; the symbols stand for all the nuclei, so a sum over them is understood. Here H is to be interpreted as the Pauli Hamiltonian for electrons and moving spin-zero nuclei interacting with the quantised EM field. The contribution of (11.75) to the expectation value is non-negative: T_N is, as before, the term linear in the vector potential, $\mathbf{P}_N \cdot \mathbf{A}$, contributes nothing to the expectation value since $\mathbf{A}$ is off-diagonal in the Fock space number basis, while the $|\mathbf{A}|^2$ term is divergent and must be cut off at some maximum momentum, but anyway is non-negative. One can then repeat more or less verbatim the argument at the end of §8.3 mutatis mutandis, to conclude that a lower bound on the energy satisfying (11.73) holds also in the case of spin-zero moving nuclei at $T = 0$.

References

[1] Haydock, R. (1980), Solid State Phys. **35**, 216.

[2] Woolley, R. G. (2020), Phys. Rev. Res. **2**, 013206 (11 pages).

[3] Haag, R. (1955), Mat.-fys. Meddr. **29**, 12.

[4] Strocchi, F. (1985), *Elements of Quantum Mechanics of Infinite Systems*, World Scientific.

[5] Kibble, T. W. (1968), J. Math. Phys. **9**, 315.

[6] Weinberg, S. (1995), *The Quantum Theory of Fields*, vol. 1, *Foundations*, Cambridge University Press.

[7] Moniz, E. J. and Sharp, D. H. (1977), Phys. Rev. D**15**, 2850.

[8] Grotch, H. and Kazes, E. (1977), Phys. Rev. D**16**, 3605.

[9] Rohrlich, F. (1997), Amer. J. Phys. **65**, 1051.

[10] Spohn, H. (2004), *Dynamics of Charged Particles and Their Radiation Field*, Cambridge University Press.

[11] Gustafson, S. and Sigal, I.M. (2006), *Mathematical Concepts of Quantum Mechanics*, 2nd ed., Springer.

[12] Lieb, E. H. and Seiringer, R. (2010), *The Stability of Matter in Quantum Mechanics*, Cambridge University Press.

[13] Loss, M., Miyao, T. and Spohn, H. (2007), J. Func. Anal. **243**, 353.

[14] Hasler, D. and Herbst, I. (2008), Comm. Math. Phys. **279**, 769.

[15] Sigal, I. M. (2009), J. Stat. Phys. **134**, 899.

[16] Bach, V., Fröhlich, J., and Sigal, I. M. (1998), Adv. Math. **137**, 299.

[17] Lieb, E. H. and Loss, M. (2003), Adv. Theor. Math. Phys. **7**, 667.

[18] Pauli, W. and Fierz, M. (1938), Il Nuovo Cimento **15**, 167.

[19] Balslev, E. and Combes, J.-M. (1971), Comm. Math. Phys. **22**, 280.

[20] Reinhardt, W. P. (1982), Ann. Rev. Phys. Chem. **33**, 223.

[21] Faupin, J. and Sigal, I. M. (2012), arXiv:1211.0268v1[math-ph].

[22] Löwdin, P. O. (1966), in *Perturbation Theory and Its Application in Quantum Mechanics*, Proceedings of Madison Symposium, edited by C. H. Wilcox, John Wiley and Sons, Inc.

[23] Fröhlich, J., Griesemer, M. and Sigal, I. M. (2011), Rev. Math. Phys. **23**, 179.

[24] Ballesteros, M., Faupin, J., Fröhlich, J. and Schubnel, B. (2015), Comm. Math. Phys. **337**, 633.

[25] Bach, V., Chen, T., Fröhlich and Sigal, I. M. (2003), J. Funct. Anal. **203**, 44.

[26] Griesemer, M. and Hasler, D. (2008), J. Func. Anal. **254**, 2329.

Quantum Electrodynamics and Molecular Structure

12.1 Introduction

Towards the end of his long scientific career, one of the pioneers of quantum chemistry, Per-Olov Löwdin,[1] promoted a research programme directed towards a 'mathematical' definition of a molecule. Löwdin noted that 'quantum chemistry is to a large extent still based on a great deal of chemical and physical insight of mostly experimental nature', and there is a very long road from the statement that a molecule is a collection of electrons and nuclei, subject to the laws of quantum mechanics, to an account of chemistry according to quantum mechanical principles [1]. Chapter 8 is an expression of that programme as envisaged by Löwdin. It does not lead very far, for as he remarked [1] (see also [2]):

> The Coulombic Hamiltonian H' does not provide much obvious information or guidance, since there is [*sic*] no specific assignments of the electrons occurring in the systems to the atomic nuclei involved – hence there are no atoms, isomers, conformations *etc*. In particular one sees no molecular symmetry, and one may even wonder where it comes from. Still it is evident that all this information must be contained somehow in the Coulombic Hamiltonian.

This quotation captures the essential problem for a fundamental quantum theory of chemistry; where should one start, for is it really evident that 'all this information' must emerge from the Coulomb Hamiltonian for a molecule? This chapter suggests that the answer must surely lie in another direction, although it cannot provide an answer.

A characteristic feature of practically all of organic chemistry, and much of inorganic chemistry too, is that molecules typically contain one or more sets of identical nuclei in sufficient numbers that the identical nuclei cannot be rendered distinguishable by isotopic substitution. We refer to such species as examples of the generic molecule; from the chemical point of view, the generic molecular formula always contains identical nuclei and supports isomerism. As an example consider the chemical formula C_3H_4; this is the stoichiometric formula of three classical organic compounds: cyclopropene, propa-1,2-diene (allene) and propyne (methylacetylene). At STP they

---

[1] P.-O. Löwdin, 1916–2000.

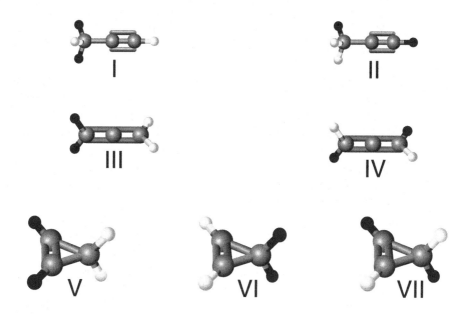

Figure 12.1 The isomers of $C_3H_2D_2$.

are stable, isolable gases that can be made by relatively straightforward synthetic procedures. Their classical molecular structures have two, three and four distinct H–H distances respectively.

The choice of the compound C_3H_4 is on the grounds of simplicity; the discussion should be seen to be really quite general once one gets beyond thinking purely about very small molecules which are not the typical case. Substitution of two hydrogen atoms with deuterium in C_3H_4 leads to the seven isomers shown in Figure 12.1; all of these isomers were prepared many years ago and their structures have been studied by physical techniques (microwave, infrared, Raman spectroscopies). Species I and II are the di-deuterated derivatives of propyne [3], [4]; III and IV are derived from propa-1,2-diene [5]–[9], and there are three distinct di-deuterated derivatives of cyclopropene (V–VII), [10]–[12]. It is generally the case that as the number of atoms in the molecular formula increases, so too does the number of possible structures, including chiral ones; however, provided a structure is drawn in accordance with the usual valency rules, one can expect there to be a rational plan for its preparation in the synthesis laboratory. Clearly chirality is beyond the scope of the Coulomb Hamiltonian, or indeed, more generally, the non-relativistic quantum electrodynamics of a closed system of electrons and nuclei.

The idea of a chemical bond between two different atoms suggests there will be an electrical imbalance, because the charges of their nuclei are different, and this leads to the very important idea of molecular dipole moments. The occurrence of permanent electric dipole moments in molecules has long been intimately connected with our ideas of classical molecular structure and the nature of the chemical bond. Polarity is an automatic feature of the shared electron theory of chemical bonding developed by Lewis and others in the early twentieth century [13]. Dipoles can be associated with charge displacements in chemical bonds between different atomic species, and the overall molecular dipole moment can be understood as the vector sum of the bond contributions.

Importantly, the special case of zero molecular electric dipole moment can occur; such species are said to be *non-polar*. This leads to a criterion for investigating molecular structures through a physical measurement, namely the temperature variation of the dielectric constant (or electric susceptibility) in fluid media. The *absence* of an overall moment requires the presence of a symmetry[2] in a proposed structure; for example, molecular structures in which the spatial arrangement of the atoms yields a centre of inversion are necessarily non-polar since the symmetry forces the bond dipole contributions to cancel completely. Likewise, if the molecular structure has multiple axes of symmetry, an axis perpendicular to a plane of symmetry, or three planes of symmetry, it too must be non-polar [14]. This feature of chemical substances, and what quantum mechanics might say about it, will be explored in the next section. Similar symmetry considerations circumscribe strongly the possibility of electrical polarisation in fluids and solids. In the absence of an applied electric field there is no net electric polarisation in an ordinary fluid medium, even if the fluid is composed of polar molecules, because of the rotational invariance and inversion symmetry of the fluid. In crystalline solids a net non-zero polarisation in the absence of an applied electric field is only permitted in certain space groups; so, for example, ferroelectricity, and pyroelectricity are quite limited phenomena.

For now it is sufficient to note that in terms of the classical bond dipole argument, replacement of H by D in any molecular structure would not be expected to give rise to polarity because H and D have the same nuclear charge. The conventional quantum chemistry account is in terms of a static dipole moment defined in a frame fixed in the molecule; a body fixed frame requires the nuclei to be treated as distinguishable particles, whether identical or not, because the nuclei must be identified for a frame to be defined. Within the usual Born–Oppenheimer approach based on clamped nuclei, §8.5, a potential energy surface is obviously independent of the masses of the fixed nuclei so that isotopic substitution should not affect the polarity of a bond. Nevertheless even the simple diatomic molecule HD has a small 'dipole moment' ($\mu \approx 10^{-2}$ Debye) according to its pure microwave spectrum [15] (and to electric deflection experiments), and the same is the case with the deuterated forms of C_3H_4 which have structures that can be characterised accurately through analysis of their pure microwave spectra. Furthermore, as noted above, once the nuclei are fixed there may even be loss of the space inversion symmetry possessed by a collection of electrons and nuclei according to quantum electrodynamics. And so it is here, since according to the classical rules of stereochemistry, isomers IV and VII are *chiral* and so might have isolable enantiomers; indeed the enantiomers[3] of IV have been prepared [9].

According to quantum theory the inter-proton distances in C_3H_4 are of the form $|\mathbf{x}_g^H - \mathbf{x}_h^H|$ in the notation of §8.5; since we are dealing with identical particles, the

---

[2] The symmetry here is described by some small finite group.

[3] They are the lightest known chiral species composed of stable elements.

corresponding self-adjoint quantum mechanical operator Q^{HH} must involve a symmetrised sum of such terms so that it is independent of the labels g, h; it only involves the translationally invariant coordinates $\{t^n\}$. Given a state Φ_n in the Hilbert space of the internal Hamiltonian operator for the system 'C_3H_4', the mean proton distance is the expectation value of Q^{HH} in this state:

$$\overline{HH}_n = \langle \Phi_n | Q^{HH} | \Phi_n \rangle. \tag{12.1}$$

What is essential for any kind of 'structural' account is the distribution of H−H distances; clearly the protons are correlated because this is a system with Coulomb interactions. Now as discussed in Chapter 5, every self-adjoint operator has a spectral family $E(\lambda)$, and we may write formally

$$Q^{HH} = \int_{-\infty}^{+\infty} \lambda \, dE(\lambda). \tag{12.2}$$

Combining (12.1) and (12.2), we have

$$\overline{HH}_n = \int_{-\infty}^{+\infty} \lambda \, d\langle \Phi_n | E(\lambda) | \Phi_n \rangle \equiv \int_{-\infty}^{+\infty} \lambda \rho(\lambda)_n \, d\lambda \tag{12.3}$$

in terms of a proton bond-length distribution function $\rho(\lambda)_n$ for the state Φ_n.

The distribution function should describe the correlations; however, it has never been calculated for any generic molecule, and in any case its relationship to the classical structural description is quite obscure. A sceptical reader might well ask what exactly is the "internal Hamiltonian for the compound labelled 'C_3H_4'" and hence what the state Φ_n is in explicit form?[4] From the foregoing discussion the answer is not provided by the Coulomb Hamiltonian theory described in Chapter 8. Of course, molecular bond lengths are not and never have been obtained through such a calculation, and this fact simply highlights the disconnection between quantum mechanics and what chemists actually do. As the late Hans Primas remarked, without the classical concept of molecular structure there is no chemistry, and "chemistry is not spectroscopy" [16]. There is a duality here, for there are both boson and fermion nuclei, and the consequences of the characteristic quantum mechanical property of nuclear spin statistics can be seen in various spectroscopies (e.g. rovibrational spectra, NMR etc.). On the other hand if one views molecular structure as being associated with some sort of classical limit of a quantum theory involving the nuclei, one has to accommodate the fact that the classical limit for boson particles is a classical *field* theory.[5]

The traditional picture of polarity in terms of molecular structure is essentially *static* insofar as the electron pairs in bonds are associated with *fixed* nuclei – there is no molecular motion. This is a powerful basis for accounting for chemical properties of substances, as Lewis demonstrated. But it is not suitable for an account of their physical properties. Suppose we tried to characterise a molecular dipole by sending in a slow

---

[4] The question can be understood as asking for the mathematical equations that have Φ_n as a possible solution. At finite temperature one must replace Φ_n by some density operator $\rho(T)$ and ask the same question.

[5] Cf. footnote 1 in Chapter 7 in connection with the electromagnetic field, and footnote 7 in Chapter 5 regarding the nonexistence of a classical limit for the 'spin' operators.

test charge and measuring how it is affected (i.e. a scattering experiment – this is a toy model of slow electron–molecule scattering). The interaction potential for a charge and a fixed-point dipole is given by electrostatics as

$$V = -\mathbf{d} \cdot \mathbf{E}(\mathbf{r}) \sim \frac{C \cos(\theta)}{r^2}, \tag{12.4}$$

where C is a constant, θ is the angle between the dipole vector and the wave vector of the charge, and r is their separation. An elementary calculation yields an infinite total elastic scattering cross section [17]. To remedy the situation we must allow the dipole to move freely and with rotational degrees of freedom; for even more physical realism we should consider finite temperature. Then, as we shall see, in quantum theory we can use space-fixed axes and account for 'polarity', even though the 'dipole moment' has disappeared.

In this chapter we outline quantum mechanical accounts of two topics that have long been interpreted in terms of classical molecular structure. As just described, dielectric constants are interpreted conventionally in terms of molecular polarity, that is, the presence or absence of 'permanent electric dipole moments'. The quantum mechanical derivation of the Langevin–Debye formula, originally due to van Vleck [18], is now largely forgotten, presumably because of the lack of chemical content. van Vleck showed that it can be understood in terms of energy levels, electric dipole transition matrix elements and the temperature T; importantly the expectation value of the electric dipole moment operator does not contribute. Nevertheless the distinction between 'polar' and 'non-polar' molecules is captured without reference to chemical bonds, electronegativity or molecular symmetry.

X-ray crystallography is essentially understood in terms of classical electromagnetism and the classical conception of the structure of matter in terms of collections of atoms (Chapter 1). One can describe the experiment formally using the quantum mechanical scattering theory (Chapter 10) and think of the differential cross section for elastic scattering of X-ray photons as a particular case of the Kramers–Heisenberg dispersion formula. The states involved in the formal matrix elements must describe the crystal and so derive from condensed matter physics; there, structure is assumed at the outset in the many-body theory.

12.2 Dielectric Phenomena

For most dielectric materials in laboratory electric fields the induced electric polarisation $\mathbf{P}$ is simply proportional to the external electric field $\mathbf{E}$,

$$P_i = \chi \varepsilon_0 E_i, \quad i = x, y, z, \tag{12.5}$$

where ε_0 is the permittivity of free space. This is the linear response regime (see §6.7). The proportionality constant χ is the scalar part of the electric susceptibility tensor and is directly related to the dielectric constant (or 'relative permittivity'), ε_r:

$$\varepsilon_r = \varepsilon_0(1+\chi). \tag{12.6}$$

The temperature dependence of the susceptibility provides an experimental criterion for distinguishing between polar and non-polar gases; polar gases show Curie-law
behaviour,

$$\chi \propto \frac{1}{T}, \tag{12.7}$$

whereas non-polar gases have susceptibilities that are, to good approximation, temperature independent. These facts are accounted for by the Langevin–Debye formula
which gives the electric susceptibility for an ensemble of molecules with number density $n = N/V$ in a static electric field in terms of their permanent moments $\boldsymbol{\mu}$, and mean
static polarisability α [19], [20],

$$\chi \sim \frac{n}{\varepsilon_0}\left(\frac{\mu^2}{3k_BT}+\alpha\right). \tag{12.8}$$

Suitable experimental apparatus for the measurement of the temperature variation
of the dielectric constants of gases and solutions became available from the mid 1920s
onwards and proved useful for structural studies of organic substances in the first
half of the twentieth century [14]. In his review Sutton noted that similar information could be obtained from Stark effect microwave spectroscopy and molecular beam
electric deflection experiments, although at the time they were of very limited applicability. While the experimental capability has been comprehensively advanced with
the developments in laser physics and molecular beam technologies, the framework for
interpretation has not changed much.

The classical theory of molecular electric moments and their relationship to the
experimentally observable dielectric constant was initiated by Debye in 1912, who identified two quite distinct mechanisms underlying the Langevin–Debye formula (12.8)
[19]. The polarisability α is a measure of the deformation of the electron distribution in an atom or molecule caused by the applied field; since the energy required
to redistribute the electrons over their energy levels $\gg k_BT$ for ambient temperatures, the temperature plays no role in this effect. Even so, there are many molecular
gases that show much greater susceptibilities than atomic gases, and this cannot be
attributed to a much greater mobility of the electrons. Langevin had explained the
Curie-law behaviour of paramagnetic materials in terms of the competition between
the orientation of permanent elementary magnets by an applied magnetic field, and
the randomising effects of thermal motion. Debye was thus led to conjecture [19], [20]
that molecules might have permanent electrical moments which would behave in the
same way in an applied electric field, and give rise to the μ^2 term in the susceptibility
(12.8).

At room temperature ($T \sim 300$ K) the average thermal energy, k_BT, has the value
4.12×10^{-21} J; a permanent dipole moment μ of 1 Debye[6] in an electric field, $|\mathbf{E}|$,
of 2×10^7 V m^{-1} has an energy $\mathcal{E}_F \sim \mu F$ of 6.67×10^{-23} J, only 1/60th of the thermal energy. On the other hand, as a spectroscopic energy, $\mathcal{E}_F$ is a few cm^{-1}, which

---

[6] 1 Debye $= 3.335 \times 10^{-30}$ C m.

is considerable in comparison with the typical spacing of molecular rotational energy levels. The polarisability determines the magnitude of the dipole moment induced by the applied field, $\mu_{\text{ind}} \sim \alpha|\mathbf{E}|$. Polarisability data are commonly quoted using a modified cgs unit $\mathring{A}^3 = 10^{-24}$ cm^3; these values represent $\alpha/4\pi\varepsilon_0$. We may take the water molecule as an example; the polarisability volume of water is $1.48\mathring{A}^3$ which in a reasonably strong electric field $|\mathbf{E}| = 2 \times 10^7$ V m^{-1} yields an induced dipole moment of only 10^{-3} Debye. However, large molecules, for example the peptides studied by electric beam deflection measurements [21], have polarisability volumes typically in the range $500 - 1500$ $\mathring{A}^3$, giving induced moments in the field $|\mathbf{E}|$ that are much more comparable with their permanent moments.

The quantum mechanical account of a molecule in an electric field is really quite straightforward. As in Chapter 8 (§8.4) we write the full Hamiltonian for a free molecule in the form

$$H = H_{\text{CM}} + H', \tag{12.9}$$

where H$'$ can be expressed in terms of a set of internal coordinates that are independent of the centre-of-mass variables. To do this we simply transform from the laboratory frame to a space-fixed frame by introducing the centre-of-mass coordinate operator explicitly and setting

$$\mathbf{x}_i = \mathbf{R} + \mathbf{r}_i. \tag{12.10}$$

According to (8.18) the solution to the Schrödinger equation

$$H\psi = E\psi \tag{12.11}$$

in a coordinate representation can be written in the product form

$$\psi = \Theta(\mathbf{R})_{\mathbf{K}}\Psi(\{\mathbf{r}_i\})^0_n, \tag{12.12}$$

where Ψ^0_n is an eigenfunction of H$'$ and, strictly speaking, Θ should be a square integrable wave packet (cf. 5.77). In the following for brevity we will use the device of normalisation in a box of volume Ω for the centre-of-mass motion and so write

$$H_{\text{CM}}\Theta(\mathbf{R})_{\mathbf{K}} = \frac{\hbar^2 K^2}{2M}\Theta(\mathbf{R})_{\mathbf{K}}, \quad \Theta(\mathbf{R})_{\mathbf{K}} = \frac{1}{\sqrt{\Omega}}e^{i\mathbf{K}\cdot\mathbf{R}}. \tag{12.13}$$

The interaction operator for a charge density $\rho(\mathbf{x})$ coupled to a static electric field with scalar potential ϕ is

$$V = \int \rho(\mathbf{x})\phi(\mathbf{x})\,d^3\mathbf{x}. \tag{12.14}$$

Using (12.13) for the centre-of-mass wave functions yields a general matrix element as

$$\langle\Theta_{\mathbf{K}}\Psi^0_n|V|\Psi^0_{n'}\Theta_{\mathbf{K}'}\rangle = \frac{1}{\Omega}\int e^{-i\mathbf{q}\cdot\mathbf{R}}\langle\Psi^0_n|\sum_i e_i\phi(\mathbf{R}+\mathbf{r}_i)|\Psi^0_{n'}\rangle\,d^3\mathbf{R}, \tag{12.15}$$

where $\hbar\mathbf{q} = \hbar(\mathbf{K}-\mathbf{K}')$ is the momentum transfer vector. For a neutral molecule we have then approximately

$$\langle\Theta_{\mathbf{K}}\Psi^0_n|V|\Psi^0_{n'}\Theta_{\mathbf{K}'}\rangle \sim \frac{1}{\Omega}\left(-\mathbf{d}_{nn'}\cdot\mathbf{E}(-\mathbf{q}) - \mathbf{Q}_{nn'}:\nabla\mathbf{E}(-\mathbf{q}) + \dots\right), \tag{12.16}$$

in terms of the Fourier components[7] of the field $\mathbf{E}$ and its gradient $\nabla\mathbf{E}$, and the molecule's electric dipole and quadrupole operators.

An obvious consequence of the space-reflection symmetry of the Coulomb Hamiltonian is that a molecule in one of its stationary states cannot have a permanent dipole moment; even when there are degeneracies due to other symmetries, all the eigenfunctions associated with a given energy eigenvalue can be assigned the same parity, and as the electric dipole operator $\mathbf{d}$ is odd under reflection, its expectation value must vanish. As a result, if one applies the RS perturbation theory to calculate perturbed energy eigenvalues in a uniform applied static electric field, one has

$$E_n = E_n^0 + \sum_{n' \neq n} \frac{\langle \Psi_n^0|\mathbf{d}|\Psi_{n'}^0\rangle\langle\Psi_{n'}^0|\mathbf{d}|\Psi_n^0\rangle}{E_n^0 - E_{n'}^0} : \mathbf{EE} + \dots$$

$$= E_n^0 + \boldsymbol{\alpha}(n) : \mathbf{EE} + \dots , \qquad (12.18)$$

with no first-order contribution; $\boldsymbol{\alpha}(n)$ is the polarisability tensor for the state Ψ_n^0. This expansion is at best an asymptotic series estimate of the 'perturbed eigenvalue' E_n since the Hamiltonian for the molecule in the presence of the electric field has purely continuous spectrum. The mathematical notion of *spectral concentration* shows that the perturbed (continuous) spectrum can be concentrated where the perturbed eigenvalues computed by the formal Rayleigh–Schrödinger calculus should lie [22], [23]. This is in agreement with the experimental observation that when a gas of hydrogen atoms is perturbed by a weak electric field, the gross spectrum of the atom persists, each line splitting into several reasonably sharp lines. Thus the heuristic guide to the 'validity' of the Rayleigh–Schrödinger perturbation theory, that is, the requirement

$$|\langle\Psi_n^0|\mathbf{d}\cdot\mathbf{E}|\Psi_{n'}^0\rangle| \ll |E_n^0 - E_{n'}^0|, \quad n' \neq n, \qquad (12.19)$$

may still be useful.

Exceptional Stark effect behaviour occurs when the condition (12.19) is not satisfied, the so-called strong-field case; this may be because the states Ψ_n^0, Ψ_n^0 are almost degenerate and have opposite parities and/or the electric field is 'large enough'. The textbook example of anomalous behaviour is the linear Stark effect in atomic hydrogen which occurs in states with the same principal quantum number n, but different angular momentum quantum numbers jm. The Dirac Hamiltonian for the hydrogen atom has the group O(4) as its symmetry group because of a special feature of the $1/r$ Coulomb potential; the group O(3) required by the relativity principle is a subgroup of O(4). As a consequence, states such as $|2\,^2S_{\frac{1}{2}}\rangle$ and $|2\,^2P_{\frac{1}{2}}\rangle$ have the same energy but opposite parities. This accidental degeneracy is removed when radiative corrections are included in the calculation (the Lamb shift), but even so, with 'ordinary' laboratory electric fields the Stark effect for atomic hydrogen is linear in the field. However, in extremely

---

[7] The Fourier integrals are defined by

$$\mathbf{A}(-\mathbf{q}) = \int \mathbf{A}(\mathbf{R})e^{-i\mathbf{q}\cdot\mathbf{R}}\,d^3\mathbf{R}, \quad \mathbf{A} = \mathbf{E}, \nabla\,\mathbf{E}. \qquad (12.17)$$

For a spatially uniform electric field one thus has $\mathbf{q} = 0$, that is, $\mathbf{K} = \mathbf{K}'$ and the centre-of-mass motion is not affected by the field.

weak fields one observes [24] the expected second-order Stark effect in agreement with (12.18). Examples of first-order Stark-effect behaviour in molecules have been known for many years, for example in the millimetre-wavelength region of the spectra [25] of CH_3F and CH_3CN; such species are conventionally referred to as 'rigid' molecules. By contrast, molecules like NH_3 show standard second-order Stark effects in accordance with (12.18) except in very strong electric fields; these molecules are classified as 'non-rigid' and the Stark-effect behaviour can be used as an indicator of molecular 'rigidity'.

From the quantum mechanical perspective, this is an obscure notion. We know from statistical physics that if the number of particles in the system is of the order of Avogadro's number we find that a new idea, namely temperature, is an essential aspect of physical description [26]. If the temperature is low enough, a large system can become *rigid*, and so the 'shape' of a macroscopic body has an obvious meaning. Rigid bodies are so familiar that it took a long time after the discovery of quantum theory before rigidity was fully recognised as an example of a *broken symmetry* – in this case, broken translational invariance – in which the observed (ground) state of the system has lower symmetry than its associated Hamiltonian. Many other examples are now known in condensed macroscopic matter [27]; broken symmetries are associated with quantum systems that have continuous spectra. They are not expected for the linear Schrödinger equation describing a system with a finite number of degrees of freedom. So at most the notion of molecular rigidity is an analogy; it is not something one expects to be associated with the Coulomb Hamiltonian for atoms and molecules.

The quantum mechanical account of the polarity of atoms and molecules was originally given by van Vleck [18]; a modern formulation can be based on the finite temperature perturbation theory described in §6.8, or equivalently by linear response theory. It is inevitably an essentially spectroscopic account with no reference to 'molecular structure', that is, the ingredients are energy levels, matrix elements and the temperature. The numerical values just presented suggest that finite temperature perturbation theory should converge quickly, so we work only to second order in β ($= 1/k_B T$). Provided parity is a symmetry of the system, the average value of the perturbation operator for the coupling to the electric field $\mathbf{E}$ vanishes,

$$\langle \Psi_n^0 | V | \Psi_n^0 \rangle = 0, \tag{12.20}$$

and $r(\beta)$ in (6.190) for this case reduces to

$$r(\beta)^{\mathbf{E}} = 1 + \beta^2 \overline{\alpha(\beta)_{nn}}, \tag{12.21}$$

where

$$\overline{\alpha(\beta)_{nn}} = \sum_{n,n'}{}' p_n^0 |\langle \Psi_n^0 | \mathbf{d} \cdot \mathbf{E} | \Psi_{n'}^0 \rangle|^2 f(x_{n'n}) \tag{12.22}$$

is an average over the unperturbed states of the system; $x_{n'n}$ is the dimensionless variable $\beta \varepsilon_{n'n}$.

For non-interacting molecules (an idealisation of a dilute fluid) with number density n, the average electric polarisation, P_i, of the fluid is then obtained from the perturbed partition function as

$$P_i = nk_BT \frac{1}{Z}\frac{\partial Z}{\partial E_i}, \quad i = x, y, z, \tag{12.23}$$

where now Z is understood to be evaluated with molecular states. At ordinary, ambient temperatures, and for realistic field strengths ($|\mathbf{E}| < \sim 10$ MV m^{-1}), $r(\beta) \sim 1$ so that $1/Z$ can be replaced by its unperturbed value, $1/Z_0$. Then the polarisation $\mathbf{P}$ is simply determined by the derivative of Z with respect to the electric field and is obviously linear in the field; the corresponding electric susceptibility tensor $\boldsymbol{\chi}$ is proportional to the second derivative of Z with respect to $\mathbf{E}$ and is given by

$$\chi_{ij} = \frac{n}{\varepsilon_0 k_B T}\sum_{n,n'}{}' p_n^0 \langle\Psi_n^0|\mathsf{d}_i|\Psi_{n'}^0\rangle\langle\Psi_{n'}^0|\mathsf{d}_j|\Psi_n^0\rangle f(x_{n'n}), \quad i, j = x, y, z. \tag{12.24}$$

The quantum mechanical distinction between non-polar and polar gases is based on the evaluation of (12.24) in the two limiting cases in (6.195) which involve a comparison of the thermal energy with the energy level spacing. Non-polar gases absorb i.r. and/or u.v. visible radiation but have no microwave spectrum at room temperature for which $k_B T \approx 200$ cm^{-1}. Accordingly they are characterised by the condition that non-zero dipole transition moments $\langle\Psi_n^0|\mathbf{d}|\Psi_{n'}^0\rangle$ connect states for which $|E_n^0 - E_{n'}^0| \gg k_B T$. Then $x \gg 1, f(x) \sim x^{-1}$, and (12.22) is

$$\overline{\alpha(\beta)_{nn}} \approx \frac{|\mathbf{E}|^2}{\beta}\sum_{n,n'}{}' p_n^0 \frac{|\mathsf{d}_{nn'}|^2}{E_{n'}^0 - E_n^0}, \tag{12.25}$$

where d is the component of $\mathbf{d}$ along the field direction. Since $x \gg 1, p_n^0 \sim \delta_{n0}$ and so, after averaging over the M sub-levels of the ground state $|0JM\rangle$,

$$\beta^2\overline{\alpha(\beta)_{nn}} \sim \frac{\alpha|\mathbf{E}|^2}{k_B T}, \tag{12.26}$$

where α is the scalar polarisability. Thus for the non-polar gas the electric susceptibility is temperature independent:

$$\chi = \frac{n}{\varepsilon_0}\alpha. \tag{12.27}$$

Polar gases have the property that they also have non-zero dipole transition moments $\mathsf{d}_{nn'}$ for which $|E_n^0 - E_{n'}^0| \ll k_B T$ for ambient temperatures, and so have microwave spectra. For these states $x \ll 1, f(x) \sim 1/2$ and equation (12.22) yields an additional contribution,

$$\delta\overline{\alpha(\beta)_{nn}} = \tfrac{1}{2}\sum_{n,n'} p_n^0 |\langle\Psi_n^0|\mathbf{d}\cdot\mathbf{E}|\Psi_{n'}^0\rangle|^2, \tag{12.28}$$

where the sum is restricted to the states that make $x \ll 1$. In practice this means the $\{|JM\rangle\}$ manifold of the ground electronic state $|0JM\rangle$. Following van Vleck [18], we write

$$\langle JM|\mu_q|J'M'\rangle = \langle 0JM|\mathsf{d}_q|0J'M'\rangle \quad (q = x, y, z) \tag{12.29}$$

for the matrix elements of $\mathbf{d}$ restricted to this manifold, and then taking the field along the z-direction, the sum in (12.28) can be expressed in terms of the reduced matrix elements of μ^2,

$$\sum_{J',M',M} |\langle 0JM|\mathsf{d}_z|0J'M'\rangle|^2 E^2 = \tfrac{1}{3}\langle J||\mu^2||J\rangle|\mathbf{E}|^2, \tag{12.30}$$

so that

$$\delta\overline{\alpha(\beta)_{nn}} \approx \tfrac{1}{6}E^2 \sum_J p^0_{0J}\langle J||\mu^2||J\rangle. \tag{12.31}$$

van Vleck's hypothesis of a permanent dipole moment requires a further assumption about the molecular internal states, namely that the reduced matrix element $\langle J||\mu^2||J\rangle$ is independent of J, so that the matrix representative of the operator μ^2 becomes a constant (c-number) multiple of the unit matrix [18]. With this further assumption, Eq. (12.26) has an extra term,

$$\beta^2\delta\overline{\alpha(\beta)_{nn}} \approx \frac{1}{6}\left(\frac{\mu|\mathbf{E}|}{k_BT}\right)^2, \tag{12.32}$$

where μ is the magnitude of the classical moment, and there is an additional contribution to the susceptibility of

$$\frac{n}{\varepsilon_0}\left(\frac{\mu^2}{3k_BT}\right), \tag{12.33}$$

which gives the Curie-law behaviour.

The form of the Langevin–Debye formula,

$$\chi = a + \frac{b}{T}, \tag{12.34}$$

survives without van Vleck's assumption, provided we interpret μ^2 as the average over the low-lying J levels on the RHS of (12.31). Moreover, molecular energy levels do not universally fall into one or other of the limiting cases (6.195), and in general the function $f(x)$ in Figure 6.1 can be expected to modulate the sum over the dipole matrix elements. This is the origin of the phenomenological description in terms of temperature-dependent dipole moments, expressed through the substitution in (12.34) of $b \to b(T)$.

This quantum mechanical account of the Langevin–Debye formula is of interest for several reasons; at finite temperatures and for ordinary laboratory fields, it is often an excellent approximation. It demonstrates explicitly the important difference in convergence between the finite temperature perturbation theory and the ordinary RS perturbation theory; for the polar gas, the perturbation by the external field is not small compared to the energy level spacings and RS perturbation theory does not converge. More importantly, it shows how the notion of an effective classical electric dipole moment emerges in the quantum theory through the interplay of matrix elements, small energy denominators and the temperature in the average (12.22), without any conflict with the obvious statement that the mean value of the electric dipole moment operator vanishes if the system's density matrix commutes with the parity operator.

It is also notable that at $T = 0$ van Vleck's distinction between non-polar and polar molecules breaks down, since a discrete spectrum can only give the 'high-frequency' contribution for T small enough. On the other hand, polarity is an essential feature of the electronic theory of the chemical bond, and this suggests that the traditional conception of chemical bonding is highly problematic in quantum mechanics (quantum theory at $T = 0$).

Electric deflection has been used in conjunction with spectroscopic techniques for many years [28], [29] since it is an effective means of state selection. A static inhomogeneous electric field applied transversely to the propagation direction of a molecular beam acts effectively as a filter, since the deflection depends on the molecular internal state; thus unwanted states can be rejected. This technique was essential in the development of the ammonia maser [30]. More complicated arrays of multipole electric fields are used in electrostatic lenses that produce focusing. Inhomogeneous electric fields directed along the beam direction cause acceleration; pulsed (time-dependent) longitudinal field configurations can be used effectively to slow down beams of polar molecules, and hence provide 'cooling' [31]. It is usual to interpret such observations in terms of permanent electrical dipole moments in the sense of Debye, and classical mechanics. A classical description in terms of forces and particle trajectories is equivalent to a wave description in the geometrical optics (ray) approximation, and so in quantum mechanical terms is most naturally related to a W.K.B. or eikonal approximation description of a scattering experiment. A general quantum mechanical description of these beam phenomena makes no reference whatsoever to the chemist's conception of molecular structure.

The electric field in electrostatics satisfies the Maxwell equations

$$\mathbf{\nabla} \cdot \mathbf{E} = 0, \quad \mathbf{\nabla} \wedge \mathbf{E} = 0, \tag{12.35}$$

outside the sources; taken together these equations imply that $\mathbf{E}$ has non-zero spatially varying components of comparable magnitudes in at least two directions. This does not imply that the forces in the two directions are comparable. The force exerted on the charge distribution is the gradient of the energy $\mathcal{E}_{\mathbf{E}}$ (cf. §2.4.3):

$$\mathbf{F} = -\mathbf{\nabla}\mathcal{E}_{\mathbf{E}}. \tag{12.36}$$

An external force changes the momentum of a molecule according to

$$\mathbf{F} = M\dot{\mathbf{v}}, \quad \mathbf{v} = \dot{\mathbf{R}}, \tag{12.37}$$

where $\mathbf{R} = X, Y, Z$ is the molecular centre-of-mass coordinate, and M is the molecular mass; $\mathbf{\nabla}$ in (12.36) is then $\mathbf{\nabla}_{\mathbf{R}}$. For a neutral system the force is non-zero only if the electric field is inhomogeneous.

As a simple example, suppose a molecular dipole is moving along the X-axis and enters a region of length L (along X), where there is a field gradient transverse to X, so that $\mathbf{F} = -\mathbf{\nabla}\mathcal{E}_{\mathbf{E}}$. If the initial speed is v, the force only acts for a time $\tau = L/v$ and the molecule is given a transverse momentum of magnitude

$$P_{\perp} = F\frac{L}{v}. \tag{12.38}$$

The angle of deflection of the molecule can then be expressed in terms of the initial momentum $P = Mv$ as

$$\Delta\theta = \frac{P_\perp}{P} \equiv \frac{FL}{Mv^2}, \tag{12.39}$$

where $F = -\boldsymbol{\mu} \cdot \boldsymbol{\nabla}_\perp \mathcal{E}_\mathbf{E}$. Suppose the transverse force F_Z can be represented as a series in powers of Z, thus

$$F_Z \sim F_0 + F_1 Z + \dots . \tag{12.40}$$

According to the standard classical theory of particle beams a constant force, F_0, leads to deflection ('steering') of the beam, while a linear term produces focusing if $F_1 < 0$, whereas $F_1 > 0$ leads to defocusing ('broadening'). An experimental configuration satisfying both (12.35) and (12.40) is the quadrupole electric field,

$$E_X = 0, \; E_Y = -fY, \; E_Z = E_0 + fZ. \tag{12.41}$$

The deflection effects are measurable downstream of the inhomogeneous field, provided the deflection is not dispersed too much by the velocity distribution of the beam and by variation of the field gradient across the beam width. Such experiments were performed successfully by Scheffers [32] and Scheffers and Stark [33], who made deflection experiments on alkali atom beams. Atoms have no permanent moment and the quantity $\boldsymbol{\mu}$ is the induced dipole moment $(\boldsymbol{\alpha} : \mathbf{E})$, so the deflection effect is proportional to $|\mathbf{E}|^2$ and is in the direction of increasing field gradient. However, when analogous experiments were performed on molecular beams formed from polar gases, a different behaviour was found; there was only a widening of the beam [20].

Modern deflection experiments benefit greatly from much improved collimation, velocity selection of the beam and improved field gradient stability. In recent years deflection measurements have been proposed for structural studies and the characterisation of much larger molecules including biomolecules; a typical example is the study of tryptophan and its complexes with glycine and alanine [34], [35]. The experiment is first run at a specified beam temperature with zero voltage and yields a beam described by a Gaussian profile function transverse to the beam direction with the maximum centred on the beam axis. This is the reference measurement against which the effects of the external field are compared. Let the beam axis be denoted by X; the beam profile is then a measurement of the number of molecules per unit time passing through area elements in the YZ plane located at a remote distance from the deflector along the X-axis. The temperature of the beam can be varied between 85K and 300K. The beam profiles for a given species in the presence of the inhomogeneous electric field are strongly temperature dependent; there are also marked differences between species. For some species the beam profiles are symmetrically broadened in the deflector with a reduction in the central maximum, whereas for others the beam is globally deflected towards the high field region. Intermediate behaviour where shifting and broadening occur together have also been observed [34].

The customary interpretation of these results is based on essentially the same ideas as that given previously for the Stark effect. For a 'rigid' molecule the dipole operator gives a non-zero first-order perturbation of the rotational levels. The force on the molecule varies with the rotational level, and so the final beam profile is calculated as a Boltzmann-weighted average over the rotational levels. Such a calculation leads to broadening, but not deflection [36]. A different approach is used for 'non-rigid' molecules for which the force is determined by the Langevin–Debye electric susceptibility; all the molecules are shifted by the same amount, so the beam is deflected, but not broadened [34]. In this classification W/85K is 'rigid' WG/300K and WG_2/300K are 'non-rigid' and the others are 'intermediate'. Such observations evidently require a more differentiated view of molecular polarity than a distinction between simple 'non-polar' and 'polar' gases. Tryptophan (W), for example, has the behaviour of a 'polar' species at 85K but undergoes deflection at 300K. The deflection is accurately proportional to the square of the applied field, as expected for a 'non-polar' species. The chemical explanation of these different behaviours is based on the conformational (potential) energy surface and the extent to which it can be explored under the conditions of the experiment.

Although a detailed quantum mechanical account of molecular electric deflection does not seem to have been given, one can readily imagine the lines along which it would go. These experiments are amenable to the scattering theory summarised in §10.1 which can be adapted to the present situation. It is sensible to formulate the problem by incorporating the specific features of a beam of neutral molecules passing through a force field that can be used for steering, focusing, or defocusing the beam. Intermolecular forces are neglected in comparison with the force due to the deflection apparatus so that a single molecule description should be adequate. Back scattering is negligible. We assume the optic axis of the force field is a straight line, denoted by X, and introduce a two-dimensional vector $\mathbf{R}_\perp = (Y, Z)$ to describe the planes transverse to the beam direction. These coordinates are identified with the centre of mass, $\mathbf{R}$, of the molecule. Similarly we divide the conjugate momentum into longitudinal, P_X and transverse $\mathbf{P}_\perp = (P_Y, P_Z)$ components. The particular features of a beam experiment to be exploited are that the beam propagates along the forward X-direction, and only a transverse force acts within the apparatus, which we take to be of length L (along the beam direction, X), so the interaction $V = V^\perp S(X)$ where $S(X) = 1$ for $0 \le X \le L$ and is otherwise 0, with $L \gg \lambda$, where λ is the average de Broglie wavelength of the molecules.

The Hamiltonian H is

$$H = \frac{P_X^2}{2M} + \frac{|\mathbf{P}_\perp|^2}{2M} + H' + V, \tag{12.42}$$

where the perturbation operator is

$$V = -\mathbf{d} \cdot \mathbf{E}. \tag{12.43}$$

Here $\mathbf{d}$ is the molecular electric dipole operator that depends on only the internal coordinates (12.10), and $\mathbf{E}$ is, for example, the quadrupole field (12.41) acting on the centre-of-mass coordinate. Deflection or other changes to the beam can be monitored through the expectation values of the transverse components of the centre-of-mass momentum operator, or through the expectation value and variance of its transverse coordinates evaluated at some position, X_f, along the beam propagation direction.[8] Thus one looks to the evaluation of expressions like

$$\langle \Lambda \rangle_{X_f} = \frac{1}{Z} \mathrm{Tr} \left(e^{-\beta H} \Lambda \, \delta(X - X_f) \right), \tag{12.44}$$

where Λ can be the position or momentum operators of the centre of mass. Using a basis of states for the unperturbed system and the Dyson series, this may be transformed to

$$\langle \Lambda \rangle_{X_f} = \frac{1}{Z} \sum_{n,n'} e^{-\beta E_n^0} U(\beta)_{nn'} \langle n' | \Lambda \, \delta(X - X_f) | n \rangle$$

$$\approx \frac{1}{Z} \sum_{n} e^{-\beta E_n^0} U(\beta)_{nn} \langle n | \Lambda \, \delta(X - X_f) | n \rangle + \dots \tag{12.45}$$

where … represents the off-diagonal contributions from $U(\beta)_{nn'}$. The distinction between 'polar' and 'non-polar' molecules in the earlier discussion on the electric susceptibility that is contained in U_{nn} appears again. Of course, nothing has been said about the off-diagonal matrix elements $\{ U(\beta)_{nn'} \}$. But perhaps nothing further needs to be said since the account can only be based on energy levels, transition matrix elements and the temperature, and so lacks contact with the chemists' notion of structure. The situation here is similar to the account of the birefringence of 'dipolar' molecules described in §10.5.

12.3 The Electron Density and Diffraction Experiments

12.3.1 The Diffraction Experiment

In the X-ray crystallography experiment a beam of X-rays impinges on a material target and the intensity and angular distribution of the scattered radiation is measured. For this discussion only the *elastic scattering*, that is, scattering without change in the energy of the radiation, is of interest.[9] Experiments essentially similar in concept can be carried out with particle beams (e.g. electrons, neutrons) at energies that correspond to de Broglie wavelengths ($\lambda = \hbar/\sqrt{2mE}$) comparable to the wavelengths of the X-rays.[10]

---

[8] Alternatively, one can aim to evaluate the differential cross section for the centre-of-mass scattering process $\mathbf{K} \to \mathbf{K}'$ averaged over the Gibbs state for H and the unobserved internal states. A plausible 'adiabatic' approximation would be to use the energy (12.18) as the potential for the scattering with $\mathbf{E}$ interpreted as $\mathbf{E}(\mathbf{R})$ in the inhomogeneous field.

[9] Absorption and inelastic scattering of X-rays are important spectroscopic techniques but are not the concern here.

[10] For example, copper and molybdenum produce K_α radiation with wavelengths of 1.54Å and 0.71Å respectively, corresponding to photon energies of 8 keV and 17 keV. A 1 keV electron has a de Broglie wavelength of about 0.4Å. The strength of the interaction of the 'probe' particle with the 'target' is, of course, very important; the interaction with X-rays is very much less than with electrons, for example.

The diffraction pattern for a single crystal consists of a number of separate diffraction 'spots' of varying intensity and positions; the scattered radiation is still commonly referred to as a set of 'reflections'.

The first systematic experiments of this type were carried out by W. H. Bragg and W. L. Bragg (father and son), who concentrated on simple inorganic crystals (e.g. diamond, rock salt, ZnS [37], [38]). The Braggs adopted the classical structural picture of materials devised by chemists, and proposed that crystals could be thought of as a *regular* array of atoms with a spacing comparable to the wavelength of the X-rays, and so would act like a three-dimensional diffraction grating. The planes of atoms give rise to specular reflection of the X-rays; in a regular *periodic* structure one expects substantial constructive and destructive interference of the scattered radiation. The observed scattering intensity pattern is quite different from that seen with ordinary fluid media. The Bragg method[11] uses monochromatic X-rays with a crystal and detector mounted on a device called a goniometer which rotates the crystal by an angle θ and simultaneously the detector by 2θ.

Let θ be the angle of incidence measured relative to the surface. For the single crystal we find that there is a high intensity of scattered X-rays only in certain directions given by the *Bragg diffraction condition*,

$$N\lambda = 2d_\alpha \sin(\theta), \quad N = 1, 2, 3 \ldots, \quad \alpha = h, k, l, \tag{12.46}$$

where d_α is a characteristic length labelled by three integers h, k, l (the Miller indices). Many planes contribute to the scattering, and in the Braggs' experimental arrangement a plot of intensity versus θ approximates to a line spectrum known as the *Bragg spectrum* of the crystal. The intensity peaks can be labelled by the Miller indices.

The classical wave theory of crystal diffraction requires the solution of the Maxwell–Lorentz equations (cf. Chapter 2) applied to the electromagnetic fields of the X-rays interacting with the charged particles in the crystal, so as to give the scattered radiation field received at a detector remote from the crystal [40], [41]. Quantitative measurements of the intensities in the diffraction pattern are interpreted in terms of the precise positions of the atoms, with the understanding that d_α is the separation between planes of atoms in the crystal. The classical theory of the crystal diffraction experiment shows that the Thomson differential cross section for the elastic scattering of X-rays by a single charge (10.172) is modified to the form

$$d\sigma = d\sigma^T(\theta)F(\mathbf{q})\, d\Omega$$
$$F(\mathbf{q}) = \langle |f(\mathbf{q})|^2 \rangle. \tag{12.47}$$

---

[11] Other classical techniques include Laue diffraction which uses 'white' (polychromatic) X-rays and a fixed single crystal, and the Debye–Scherrer powder method which uses monochromatic radiation and a polycrystalline sample, such that the individual crystallites have random orientations with respect to the incident radiation. Modern X-ray crystallography is an extremely sophisticated combination of crystal sample preparation, instrumentation and computing technologies; for a recent review see [39].

Here $\mathbf{q}$ is the momentum transfer vector, and $f(\mathbf{q})$, the 'structure factor', is the Fourier transform of a density distribution, $n(\mathbf{r})$, for the N charged particles in the crystal:

$$f(\mathbf{q}) = \int_V n(\mathbf{r})e^{i\mathbf{q}\cdot\mathbf{r}}\,d^3\mathbf{r}$$

$$n(\mathbf{r}) = \sum_i^N \lambda_i \delta^3(\mathbf{r}-\mathbf{x}_i). \tag{12.48}$$

The angular brackets in (12.47) imply that a statistical average is required since the experiment is run at finite temperature $T > 0$. For the scattering[12] of X-rays, $\lambda_i = e_i^2/m_i$, and the sum is restricted to just the electrons in the target, on the assumption that the nuclei can be neglected (or that the nuclei are fixed classical charges purely responsible for the attractive Coulombic interaction with the electrons). Obviously this is already a 'semi-classical' description since the stability of the system, guaranteed by quantum mechanics, is tacitly assumed.

The observation of well-defined diffraction spots (or Bragg peaks) at wave vectors $\{\mathbf{q}_i\}$ implies that the inferred density is periodic in real space,

$$n(\mathbf{x}) = n(\mathbf{x}+\mathbf{D}_i), \tag{12.49}$$

with translation vectors satisfying the *Laue condition*,

$$e^{i\mathbf{q}_i\cdot\mathbf{D}_i} = 1 \;\;\rightarrow\;\; \mathbf{q}_i\cdot\mathbf{D}_i = 2N\pi, \;\;\text{with integer } N. \tag{12.50}$$

Such a periodic structure is called a lattice, a basic unit that is repeated to build the lattice, which is called a unit cell, can be defined in terms of a parallelogram of four lattice points. It may contain one or more molecules. The integration in (12.48) is over the unit cell volume with the origin of the vector $\mathbf{r}$ chosen as one corner of the unit cell; the components of $\mathbf{q}$ are just the Miller indices (h,k,l) [42].

In the classical theory waves are described by complex-valued functions characterised by amplitude and phase. The modulus signs in (12.47) imply that the phase information is lost from the measured diffraction pattern; this is the *phase problem* in crystallography. If it can be overcome, the Fourier inversion required to obtain $n(\mathbf{r})$, (12.48), for the unit cell can be carried through with an accuracy determined by the extent of the diffraction pattern that can be satisfactorily measured. Various techniques have been developed to reconstruct the phase information (Patterson methods, direct methods, use of X-ray absorption edges etc.); we do not go into details (for a summary see, for example, [43]). Given the obvious possibilities for structure determination, it wasn't long after the Braggs' work on simple ionic crystals that the technique was applied successfully to molecular crystals; the first organic molecular structure to be 'solved' was that of hexamethylenetetramine reported by Dickinson and Raymond in 1923 [44]. Of course, the picture of DNA as a double helix is perhaps the most potent molecular structure image obtained in this way [45], [46].

The structure determination procedure is entirely classical in nature. It is important to recognise that the computational transformation of the measured diffraction

---

[12] For electron diffraction (ED), the structure factor can be written in a similar form with $\lambda_i = e_i$; also the ED cross section does *not* contain the factor r_0^2 contained in $d\sigma^T$.

intensities into a molecular structure is not guaranteed to yield a unique, chemically sensible structure, not least because of the finite amount of data that can be collected. Chemical knowledge is a required input so that a proposed structure can be checked as making good chemical sense, for example, that it has realistic bond angles and bond lengths, that all H atoms are found and located as could be expected from stereochemistry, and so on. Hence there is a procedure of 'structure refinement' involving an iterative comparison of the experimental data with simulated data for a proposed structure using 'atomic scattering factors'.[13]

12.3.2 Diffraction and Quantum Theory

In the quantum theory 'wave' and 'particle' versions of the theory are entirely equivalent, so one can just as well adopt a 'particle' viewpoint based on the scattering of X-ray photons. The change in direction of the radiation can then be viewed as a change in the momentum of the scattered photons, and the experiment can be understood as monitoring an exchange of momentum between the crystal and the X-rays. Suppose the incident and reflected photons have momenta $\mathbf{p}$ and $\mathbf{p}'$ respectively (for elastic scattering $|\mathbf{p}| = |\mathbf{p}'|$), and that reflections take place from infinitely extended parallel plane surfaces. The magnitude of the momentum transfer vector, $\mathbf{q} = \mathbf{p}' - \mathbf{p}$, satisfies

$$q = 2|\mathbf{p}|\sin(\theta), \tag{12.51}$$

where θ is the glancing angle. On the other hand combining (12.51) and (12.46) yields

$$q_N = N\frac{1}{d_\alpha}|\mathbf{p}|\lambda, \quad N = 1, 2, 3\ldots, \quad \alpha = h, k, l. \tag{12.52}$$

This odd-looking formula has a simple physical interpretation in terms of quantum theory which tells us that the magnitude of the momentum of a photon is simply $\hbar/\lambda$, and hence the momentum transfer (12.52) is quantised [47]:

$$q_N = N\frac{\hbar}{d_\alpha}, \quad N = 1, 2, 3\ldots, \quad \alpha = h, k, l. \tag{12.53}$$

The experiment can be described in terms of the 'probe' – 'target' – 'detector' schema described in Chapter 6. In the following we have in mind that the 'target' is an atom or molecule, and we make the assumption that similar ideas can be extended to a macroscopic quantum system – the crystal. An 'exact' calculation of the cross section for X-ray diffraction is out of the question since it would require a full treatment of a many-body problem; accordingly we seek approximation techniques guided by physical considerations. The essence of diffraction is that the probe beam particles have de

---

[13] Atomic scattering factors encapsulate the intensity variation of X-ray scattering as a function of scattering angle and wavelength; they are known for atoms and ions of all the elements and are standard reference data incorporated into many crystallographic software applications. They have been obtained from accurate quantum mechanical calculations made in atomic physics [42].

Broglie wavelengths much more nearly comparable with the length scale over which the target charge density varies. This is just the condition that requires the retention of the plane-wave factors, $e^{i\mathbf{k}\cdot\mathbf{x}}$, that describe the momentum state of the probe beam. The energies of the beams used in diffraction experiments are very much greater than chemical binding energies (typically a few eV) and greater than most electron ionisation energies; they are, however, still much less than the electron rest-energy and so a non-relativistic calculation should be adequate.

The X-ray diffraction experiment can be described in the terms presented in §10.4 for deriving the general form of the Kramers–Heisenberg dispersion formula for linear light scattering; for photons with energies in the X-ray region of the spectrum, there is no question of making a multipole expansion, so the complete expression for the cross section, derived from the T-matrix element, Eq. (10.174), could be the starting point. Only the cross section for elastic scattering is needed, and furthermore the use of an unpolarised incident beam and detection of all scattered photons without regard to their polarisation is an important special case; this is sufficient to obtain the factor of $d\sigma^T(\theta)$ in (12.47).

In elastic scattering $\Phi_k^0 = \Phi_n^0$ and $|\mathbf{k}| = |\mathbf{k}'|$; the first term in (10.174) yields a formula of the form of the classical result (12.47), (12.48) with the 'density' $n(\mathbf{r})$ appearing as the expectation value of a 'density' operator,[14]

$$n(\mathbf{r}) = \sum_i \frac{e_i^2}{m_i} \delta^3(\mathbf{r} - \mathbf{x}_i)$$

$$n(\mathbf{r}) = \langle \Phi_n^0 | n(\mathbf{r}) | \Phi_n^0 \rangle. \tag{12.54}$$

It is not uncommon to see the discussion stop at this point; for example, there are these two accounts 50 years apart: [41], [48]. In fact it is only by stopping at this point in the calculation that a cross section independent of Planck's constant, $\hbar$, and thus agreement with the classical account of crystallography, can be obtained. There is more to be said about (12.54); for example, we are only at the lowest order of perturbation theory and there is the second term in (10.174) that is required, formally at least, to maintain gauge invariance. It is not of the classical form and certainly involves Planck's constant through the particle momentum operators. Upon reflection it is evident that the perturbation theory approach must yield an intractable cross section because the T-matrix elements contain inter-particle correlations in every order of the expansion, apart from the contribution of (12.54), through the energy denominators (Green's functions), so that the resulting cross-section contains very complicated many-body terms.

Nothing has been said about what the wave function Φ_n^0 actually might be, so a mathematical analysis of the complicated perturbation theory terms (e.g. the demonstration of bounds on their contribution) is not possible. Physical arguments suggest that they may well be unimportant; this is the import of the so-called *impulse approximation* originally devised by Chew in nuclear physics which claims that for sufficiently

---

[14] In order to make the Thomson cross section, $d\sigma^T(\theta)$, explicit, we write the charge and mass parameters in terms of the electron's charge, e, and mass, m_e. Then e^2/m_e is a factor of every term in the sum over α and can be taken out, and inside the sum $e_\alpha^2/m_\alpha \to 1$ for electrons and Z_N^2/M_N for the nuclei.

high-energy beams the scattering takes place as though the particles in the target are 'free'; this requires the neglect of the Coulomb interactions between the charges, that is, the use of free-particle Green's functions, $G_0(E)$ in the T-matrix expansion (Chapter 10). Then the T-matrix elements simplify to a sum of simple two-body scattering amplitudes. The contribution T_n of particle n to this sum does, however, contain positional information such that

$$\langle \mathbf{k}'|T_n|\mathbf{k}\rangle = \langle \mathbf{k}'|t_n|\mathbf{k}\rangle e^{i\mathbf{q}\cdot\mathbf{x}_n}, \tag{12.55}$$

where t_n is the scattering amplitude for a probe beam particle (X-ray photon) and a free target particle (electron or nucleus) at position $\mathbf{x}_n$, and $\hbar\mathbf{q} = \hbar(\mathbf{k}' - \mathbf{k})$ defines the momentum transfer as usual. The approximation eliminates the awkward inter-particle correlations and so can lead to a structure factor in which (12.54) is a factor [49], [50].

One might suppose that this generalised cross section interpreted in terms of a 'measurement' of molecular structure or 'electron density' of molecular crystals could be calculated theoretically, at least in principle, in accordance with Dirac's dictum (Chapter 1) that 'all of chemistry' is explained by quantum mechanics, so that the theory could be checked against the experiment. Dirac seemed to think that the only problem was the practical difficulty of making the appropriate calculations. In this writer's opinion the difficulty is not knowing what concrete calculation one could do rather than not having a big enough computer. The Coulomb Hamiltonian describes a collection of charged particles identified through experimental charge and mass parameters as electrons and nuclei, with free space boundary conditions;[15] spin statistics have to be adjoined to this description ad hoc. As we have seen, it is a special case of non-relativistic QED associated with a particular gauge condition.

While realistic total molecular wave functions $\{\Phi_n^0\}$ for the possible initial states at temperature T are available for very small molecules, such a calculation could never give (12.49). Its discrete translation symmetry is lower than the symmetry of the QED Hamiltonian, which is indicative of a *broken symmetry state*; the system it is associated with is the crystal, and this requires consideration of the quantum mechanics of macroscopic matter provided by condensed matter physics where 'structure' is simply assumed [27], [53]. For a quantum mechanical system with a finite number of degrees of freedom described by a linear Schrödinger equation there is no known mechanism for producing states (wave functions/density matrices) with lower symmetry than the Hamiltonian. The phenomenon of broken symmetry familiar in condensed matter physics and high-energy physics is associated with quantum theories involving purely continuous spectra that can be described by quantum field theories (thus an infinite number of degrees of freedom).

This is not the place to go into details, but the general approach towards a many-body system of interacting electrons and nuclei is not based on diagonalising the Hamiltonian and the Rayleigh–Ritz variational principle as in quantum chemistry, but rather aims at transforming to a self-consistent description involving 'quasi-particles'. It encompasses the whole spectrum from 'adiabatic' (band theory) to

---

[15] This defines the 'isolated molecule' model, an essentially ideal conception [51], [52].

strongly 'vibronic' (polarons, superconductivity etc.) interactions between the electrons and nuclei. Nevertheless, even with nuclei described by orbitals, each nucleus is taken to be confined to its *own* 'potential well' so that the many-body wave function as far as the nuclei are concerned should be thought of as being a Hartree product of localised nuclear orbitals. A sketch of a 'quantum solid' based on the Hartree–Fock approximation (or better) in which exchange between sites is included is given in [27]. Such a solid is not the object of chemical crystallography.

The molecular quantum electrodynamics described in this book is usually thought of as the low-energy approximation to the Lorentz-invariant QED. The latter is very much a quantum theory of interacting electrons and photons; nuclei have a very limited role in practice being viewed mainly as fixed sources of classical external fields (the 'external field approximation'). Each nucleus contributes a term,

$$\mathcal{L}(x)_{\text{ext}} = \mathcal{A}_\mu(x) J_e^\mu(x), \tag{12.56}$$

to the QED Lagrangian, where $\mathcal{A}_\mu(x)$ is a classical 4-vector potential for the external field associated with it, and J_e^μ is the relativistic electric current operator for the electrons [54]; of course, the notion of 'structure' simply enters through the specification of the external field.

Quantum chemistry also conceives its role largely as an *electronic theory* of molecules and in practice asserts that nuclei are to be treated initially as fixed, classical, distinguishable particles, even if identical. In other words the chemists' notion of structure is simply put into the quantum mechanical formalism by hand at the outset as a response to a powerful 'felt need' to make contact with the classical idea of molecular structure [52]. The traditional quantum chemical approach to the energy levels of a diatomic molecule is based on the fixed-nuclei modification of the Coulomb Hamiltonian and potential energy curves. This framework, which originated in the Old Quantum Theory, is either seen as an empirical scheme in which spectroscopic data is translated into 'molecular constants' or is formulated within the Born–Oppenheimer approach (Chapter 8). It is, however, *optional* since the quantum states of a diatomic molecule labelled by all requisite quantum numbers can be formulated using a non-adiabatic formalism based on the Coulomb Hamiltonian with no reference to adiabatic or diabatic states derived from a prior Born–Oppenheimer approximation approach. There is, for example, the Generator Coordinate Method described in §8.5.1 which inherits some notions about structure to define intrinsic states [55]; there are also various computational approaches that simply assign 'orbitals' to all the particles and rely on the Rayleigh–Ritz variational principle. There are very few such non-adiabatic calculations of this kind beyond the simplest diatomic molecules, however; this is not just because of the computational burden. In the view presented here the quantum theories of the atom and the diatomic molecule (and their ions), based on the Coulomb Hamiltonian and treating all particles as quantum entities, are exceptional cases that cannot provide a reliable basis for a quantum theory of chemistry.

Difficulties arise already when we have three or more nuclei. The smallest increase in complexity would seem to be the formation of a chemical bond between some diatomic molecule and a proton. As an example consider adding a proton to dinitrogen, N_2, to give the diazenylium cation, N_2H^+. This species is well known in astronomical spectroscopy;[16] it is readily identified in the spectrum of the interstellar medium where the density of molecules is much less than the best laboratory high-vacuum instruments and temperatures are close to absolute zero; however, there is also a permanent background of electromagnetic radiation and cosmic rays so the 'isolated molecule' model cannot really be considered appropriate. Astronomical spectroscopy is also familiar with the isoelectronic homologues of N_2H^+, the pairs of *isomeric* species (HCN, HNC), and (HCO^+, HOC^+) [56], [57].

There lies the problem; any straightforward appeal to the Coulomb Hamiltonian taken at face value as the quantum mechanical description of these molecules must fail for, as noted by Löwdin, there is nothing in it to distinguish such isomers, and nothing to justify recognising all of these species as 'linear molecules' which is their accepted description [58]. One cannot doubt the universally accepted account of chemistry in terms of the classical molecular structures of *individual* molecules as capturing the physical content of ordinary chemical situations where finite temperatures and persistent interactions in the sense of Prigogine [59] due to, for example, the quantised electromagnetic field, other molecules in bulk media and so on, are the norm. This is a very long way from the quantum mechanics ($T = 0$) of the 'isolated molecule' that the Coulomb Hamiltonian claims to represent. It seems clear that the difficulty lies with the electrodynamical description of the nucleus but what to do is unknown. Thus it is very difficult to see how one can get much further without the semi-empirical input that so troubled Löwdin [1].

References

[1] Löwdin, P.-O. (1989), Pure Appl. Chem. **61**, 2065.

[2] Löwdin, P.-O. (1994), Int. J. Quant. Chem. **51**, 473.

[3] Thomas, L. F., Sherrard, E. I. and Sheridan, J. (1955), Trans. Faraday. Soc. **51**, 619.

[4] Duncan, J. L., McKean, D. C., Mallinson, P. D. and McCulloch, R. D. (1973), J. Mol. Spect. **46**, 232.

[5] Morse, A. T. and Lietch, L. C. (1958), J. Org. Chem. **23**, 990.

[6] Vogelsanger, B., Oldani, M. and Bauder, A. (1986), J. Mol. Spect. **119**, 214.

[7] Meyer, V., Sutter, D. H. and Vogelsanger, B. (1991), J. Mol. Spect. **148**, 436.

[8] Meyer, V. and Sutter, D. H. (1993), Zeits. f. Naturforschung, A: Physical Sciences, **48**, 725.

[9] Keserü, Gy. M., Nógrádi, M., Rétey, J. and Robinson, J. (1997), Tetrahedron, **53**, 2049

[10] Wiberg, K. B. and Bartley, W. J. (1960), J. Amer. Chem. Soc. **82**, 6375.

[11] Mitchell, R. W., Dorko, E. A. and Merritt, J. A. (1968), J. Mol. Spect. **26**, 197.

---

[16] It has also been characterised in terrestial laboratory spectroscopy.

[12] Yum, T. Y. and Eggers, D. F. Jr. (1979), J. Phys. Chem. **83**, 501.

[13] Lewis, G. N. (1923), *Valence and the Structure of Atoms and Molecules*, Chemical Catalog Co.

[14] Sutton, L. E. (1955), ch. 9 in *Determination of Organic Structures by Physical Methods*, edited by E. A. Braude and F. C. Nachod, Academic Press.

[15] Nelson, J. B. and Tabisz, G. C. (1983), Phys. Rev. A**28**, 2157.

[16] Primas, H. (1980), in *Quantum Dynamics of Molecules: The New Experimental Challenge to Theorists*, ed. R. G. Woolley, NATO ASI B**57**, Plenum Press.

[17] Burke, P. G. (1980), in *Quantum Dynamics of Molecules: The New Experimental Challenge to Theorists*, edited by R. G. Woolley, NATO ASI B57, p. 483, Plenum Press.

[18] van Vleck, J. H. (1932), *The Theory of Electric and Magnetic Susceptibilities*, §46, Clarendon Press.

[19] Debye, P. (1912), Z. Physik **13**, 97.

[20] Debye, P. (1929), *Polar Molecules*, Dover. See also Debye's Nobel lecture (1936) accessed at www.nobel.se/chemistry/laureates/1936/debye-lecture.pdf.

[21] Antoine, R., Compagnon, I., Rayane, D. et al. (2003), Anal. Chem. **75**, 5512.

[22] Kato, T. (1966), *Perturbation Theory for Linear Operators*, Springer-Verlag.

[23] Riddell, R. C. (1967), Pacific J. Math. **23**, 377.

[24] Buckingham, A. D. (1972), *M.T.P. International Review of Science, Physical Chemistry* Series One, **3**, 73.

[25] Steiner, A. and Gordy, W. (1966), J. Mol. Spect. **21**, 291.

[26] Rosenfeld, L. (1961), Nature, London **190**, 384.

[27] Anderson, P. W. (1984), *Basic Notions in Condensed Matter Physics*, Benjamin/Cummings/Addison-Wesley.

[28] Townes, C. H. and Schawlow, A. L. (1955), *Microwave Spectroscopy*, McGraw-Hill Book Co.

[29] Wharton, L., Gold, L. P. and Klemperer, W. (1962), J. Chem. Phys. **37**, 2149.

[30] Landshoff, P. V., Metherell, A. and Rees, W. G. (1997), *Essential Quantum Physics*, Cambridge University Press.

[31] Bethlem, H. J. and Meijer, G. (2003), Int. Rev. Phys. Chem. **22**, 73.

[32] Scheffers, H. (1934), Z. Physik, **35**, 425.

[33] Scheffers, H. and Stark, J. (1934), Z. Physik, **34**, 625.

[34] Antoine, R., Compagnon, I., Rayane, D. et al. (2002), Eur. Phys. J. D**20**, 583.

[35] Antoine, R., Broyer, M., Dugourd, Ph. et al. (2003), J. Amer. Chem. Soc. **125**, 8996.

[36] Dugourd, P., Compagnon, I., Lepine, F. et al. (2001), Chem. Phys. Letters, **336**, 511.

[37] Bragg, W. L. (1912), Proc. Camb. Phil. Soc. **17**, 43.

[38] Bragg, W. L. (1913), Proc. Roy. Soc. (London) A**89**, 90.

[39] Howard, J. A. K. and Probert, M. R. (2014), Science, **343**, 1098.

[40] Landau, L. D., Lifshitz, E. M. and Pitaevskii, L. P. (1984), *Electrodynamics of Continuous Media*, 2nd ed., Pergamon Press.

[41] Ciccariello, S. (2005), Progr. Colloid Polymer Sci. **130**, 20.

[42] Clegg, W. (2015), *X-Ray Crystallography*, 2nd ed., Oxford University Press.

[43] Cowtan, K. (2001), *Encyclopedia of Life Sciences*, Macmillan Publishers Ltd.

[44] Dickinson, R. G. and Raymond, A. L. (1923), J. Amer. Chem. Soc. **45**, 23.

[45] Franklin, R. E. and Gosling, R. G. (1953), Acta Cryst. **6**, 673.

[46] Watson, J. D. and Crick, F. C. (1953), Nature **17**, 964.

[47] Compton, A. H. (1923), Proc. Natl. Acad. Sci. USA **9**, 350.

[48] James, R. W. (1965), *The Optical Principles of the Diffraction of X-Rays*, ch. 3, Cornell University Press.

[49] Fowler, T. K. (1958), Phys. Rev. **112**, 1325.

[50] Goldberger, M. L. and Watson, K. M. (1964), *Collision Theory*, John Wiley & Sons, Inc.

[51] Woolley, R. G. (1976), Adv. Phys. **25**, 27.

[52] Woolley, R. G. (1978), J. Amer. Chem. Soc. **100**, 1073.

[53] Woolley, R. G. (1988), in *Molecules in Physics, Chemistry and Biology*, **2**, edited by J. Maruani, p. 651, Kluwer Academic Publishers.

[54] Weinberg, S. (1995), *The Quantum Theory of Fields*, vol. 1, *Foundations*, Cambridge University Press.

[55] Lathouwers, L. and van Leuven, P. (1982), Adv. Chem. Phys. **49**, 115.

[56] Saykally, R. J., Dixon, T. A., Anderson, T. G., Szabo, P. G. and Woods, R. C. (1976), Astrophys. Journ. **205**, L101.

[57] Cazzoli, G., Cludi, L., Buffa, G. and Puzzarini, C. (2012), Astrophys. Journ. Suppl. Series **203:11**, 1, doi: 10.1088/0067-0049/203/1/11.

[58] Koput, J. (2019), J. Chem. Phys. **150**, 154307, doi: 10.1063/1.5089718.

[59] Prigogine, I. and George, C. (1983), Proc. Natl. Acad. Sci. U.S.A. **80**, 4590.

Appendix A The Functional Calculus

Informally, a *function* f is a relation that takes an input (the 'argument'), commonly denoted by x, and outputs a value $f(x)$, expressed symbolically by equations like

$$y = f(x) \quad \text{or} \quad x \mapsto f(x) \tag{A.0.1}$$

so that the pair (x, y) (or $(x, f(x))$) belongs to the set of pairs defining f. In much broader mathematical settings, the input x may itself be a function; in such cases, we speak of f as a *functional*, and instead write

$$y = f[x] \tag{A.0.2}$$

with square brackets, where now x is understood as a function.

The definite integral of a function f over some region of values where f is defined,

$$F[f] = \int_\Omega f(\sigma) \, d\sigma, \tag{A.0.3}$$

is a typical functional relation. In this book many functionals have been met, though it has seldom been necessary to characterise them. In Chapter 2, we met the Green's function for the divergence equation, (2.69), essentially

$$\mathbf{g}(\mathbf{x}, \mathbf{x}'; \mathbf{O}, \mathcal{C}) = \int_{\mathcal{C}}^{\mathbf{x}'} \delta^3(\mathbf{z} - \mathbf{x}) \, d\mathbf{z}, \tag{A.0.4}$$

which is a function of $\mathbf{x}, \mathbf{x}'$ but a *functional* of the path $\mathcal{C}$. The electric polarisation field, $\mathbf{P}$, is a functional of this Green's function, Eq. (2.142), and thus also a functional of the path $\mathcal{C}$. In Chapter 3, we met for the first time a quantity F with dimensions of the mechanical quantity action which is a functional of both $\mathbf{P}$ and the vector potential $\mathbf{a}$ defined by the *functional scalar product*:

$$F = \int \mathbf{P} \cdot \mathbf{a} \, d^3\mathbf{x}. \tag{A.0.5}$$

Also in that chapter we met the action integral for a mechanical system which depends on the trajectories of the particles and is the basis for the principle of least action; S is a functional of the trajectories.

The functional calculus considers the same problems as the ordinary calculus of functions, that is, differentiation and integration with the difference that there are an infinite number of variables. One can consider how F is altered by small changes in f; this gives rise to the idea of the *functional derivative*. The equation

$$\frac{\delta F}{\delta f} = 0 \tag{A.0.6}$$

states that F is stationary to any first-order change in the function f. This is the form of the calculus of variations required for the principle of least action (Chapter 3), which we will now reconsider. Equation (A.0.6) can also be written with the usual notation of the multi-variable calculus of functions,

$$\frac{\partial F}{\partial f(\sigma)} = 0, \tag{A.0.7}$$

which is to be understood as holding for every value of σ. The $f(\sigma)$ are treated as an infinite set of independent variables.

The formal rule for functional differentiation is

$$\left\langle \frac{\delta F[\phi(\mathbf{x})]}{\delta \phi(\mathbf{x})}, s(\mathbf{x}) \right\rangle = \int \frac{\delta F[\phi(\mathbf{x})]}{\delta \phi(\mathbf{x}')} s(\mathbf{x}') \, \mathrm{d}^3 \mathbf{x}' \tag{A.0.8}$$

$$= \lim_{\varepsilon \to 0} \frac{F[\phi(\mathbf{x}) + \varepsilon s(\mathbf{x})] - F[\phi(\mathbf{x})]}{\varepsilon} \tag{A.0.9}$$

$$= \frac{\mathrm{d}}{\mathrm{d}\varepsilon} F[\phi + \varepsilon s] \Big|_{\varepsilon = 0}, \tag{A.0.10}$$

where s is a suitable test function. In terms of (A.0.7), this leads to the fundamental relation

$$\frac{\partial y(x')}{\partial y(x)} = \delta(x - x'), \tag{A.0.11}$$

where x, x' can be multidimensional. Thus, if we take the elementary example

$$y = \int_0^a z(x)^2, \tag{A.0.12}$$

we have

$$\frac{\delta y}{\delta z} = \frac{\partial}{\partial z(x)} \int_0^a z(x')^2 \, \mathrm{d}x'$$

$$= \int_0^a 2z(x') \delta(x - x') \, \mathrm{d}x' = 2z(x), \ \ 0 \le x \le a, \text{ zero otherwise.} \tag{A.0.13}$$

The practical rule then is to differentiate in the usual way under the integral sign and introduce (A.0.11).

The principle of least action (Chapter 3) can be formulated using the functional calculus. In terms of a Lagrangian $L = L(x(t), \dot{x}(t), t)$, where $x(t)$ can have any dimensionality, we require the action integral

$$S[x(t)] = \int_{t_1}^{t_2} L(x(t), \dot{x}(t), t) \, \mathrm{d}t \tag{A.0.14}$$

to attain its *minimum value* when $x(t)$ is the physically realised trajectory. This is expressed by the condition

$$\frac{\delta S[x(t)]}{\delta x(t)} = 0. \tag{A.0.15}$$

Thus, using (3.2) as a simple example, this is

$$0 = \frac{\delta}{\delta x(t')} \int_{t_1}^{t_2} \left[\tfrac{1}{2}m \left(\frac{dx}{dt'} \right)^2 - V[x(t')] \right] dt'$$

$$= \int_{t_1}^{t_2} \left[m \left(\frac{dx}{dt'} \right) \frac{d\delta(t-t')}{dt'} - \frac{dV[x(t')]}{dx} \delta(t-t') \right] dt', \qquad \text{(A.0.16)}$$

which when evaluated yields Newton's Law:

$$m\frac{d^2 x(t)}{dt^2} = -\frac{dV[x(t)]}{dx}. \qquad \text{(A.0.17)}$$

Given that functional derivatives can be given a definite meaning, one can go on to consider functional differential equations. We met an example in Chapter 9 when the quantum mechanical version of Gauss's Law is implemented as a condition determining the physical states of the system (Eq. 9.62). Likewise, one can formulate functional integrals with a domain of integration that is a function space; an extremely important example is Feynman's path integral formulation of quantum mechanics [1] which is based on the idea that the temporal propagator $\mathcal{K}(x',t':x,t)$ for a wave function,

$$\psi(x',t') = \int \mathcal{K}(x',t':x,t)\psi(x,t)\,dx, \quad t < t', \qquad \text{(A.0.18)}$$

can be written as a *functional integral* of the *classical* action,

$$\mathcal{K}(x',t':x,t) = \mathcal{N} \int e^{i/\hbar \int S[x(\tau)]} \mathcal{D}x. \qquad \text{(A.0.19)}$$

The integration is over all continuous paths $x(\tau), t \leq \tau \leq t'$ with $x(t'), x(t)$ *fixed* end points for all paths and $\mathcal{D}x$ is the generalisation of $d^N(x)$ for $N \to \infty$. This idea has proved extremely valuable in both condensed matter physics and high-energy physics, and has required powerful tools to make sure that an expression like (A.0.19) both makes mathematical sense and leads to a 'correct' value; the questions concern the 'normalisation constant' $\mathcal{N}$ and the 'integration measure' $\mathcal{D}x$. We do not go into details – see for example [2] – and simply note that the functional integral generalisation of the standard Gaussian integral has wide applicability in physical problems.

Appendix B Longitudinal and Transverse Fields

Let $v(x)$ be a function on a d-dimensional space. Its Fourier transform is defined by the integral

$$u(k) = \int v(x)e^{i(kx)}\,dx, \tag{B.0.1}$$

where (kx) is the d-dimensional scalar product, and the integral is taken over the whole space for which dx is the volume element. The inverse Fourier transform is then

$$v(x) = \frac{1}{(2\pi)^d} \int u(k)e^{-i(kx)}\,dk. \tag{B.0.2}$$

In physical applications x is often referred to as 'real' space and its dual k as 'reciprocal' or 'wave-vector' space. If we combine (B.0.1) with (B.0.2), we see that

$$v(x) = \frac{1}{(2\pi)^d} \iint v(x')e^{ik(\overline{x-x'})}\,dx'\,dk, \tag{B.0.3}$$

and this leads to the identification of the Fourier integral representation of Dirac's delta distribution,[1]

$$\frac{1}{(2\pi)^d} \int e^{ik(\overline{x-x'})}\,dk = \delta(x-x'). \tag{B.0.4}$$

For the case of $d = 3$, the usual vector notation is widely used:

$$(kx) \to \mathbf{k} \cdot \mathbf{x}, \quad dx \to d^3\mathbf{x}, \quad \delta(x-x') \to \delta^3(\mathbf{x}-\mathbf{x}'). \tag{B.0.5}$$

In the following, we derive some special representations of the delta distribution applicable to vector fields.

The potential associated with the Coulomb field plays an important role throughout electrodynamics. An elementary calculation shows that it has the Fourier integral representation

$$\frac{1}{(2\pi)^3} \int \left(\frac{1}{k^2}\right) e^{i\mathbf{k}\cdot\mathbf{x}}\,d^3\mathbf{k} = \frac{1}{4\pi x}, \tag{B.0.6}$$

---

[1] See Chapter 5.

where $x = |\mathbf{x}|$. A technique that leads to some useful relations is to differentiate the integral (B.0.6) with respect to $\mathbf{x}$. Assuming that $\nabla_{\mathbf{x}}$ can be brought under the integral sign, this increases the powers of $\mathbf{k}$ in the integrand. Thus, for example,

$$-\nabla_{\mathbf{x}}^2 \frac{1}{(2\pi)^3} \int \left(\frac{1}{k^2}\right) e^{i\mathbf{k}\cdot\mathbf{x}} d^3\mathbf{k} = \frac{1}{(2\pi)^3} \int \left(\frac{k^2}{k^2}\right) e^{i\mathbf{k}\cdot\mathbf{x}} d^3\mathbf{k}, \qquad (B.0.7)$$

so that, using (B.0.4), we have the useful identity as a distribution

$$-\nabla_{\mathbf{x}}^2 \left(\frac{1}{4\pi x}\right) = \delta^3(\mathbf{x}). \qquad (B.0.8)$$

This relation is to be understood as an equality when either side appears under an integral sign with integration over $\mathbf{x}$, so that, for example,

$$\mathbf{v}(\mathbf{x}) = \int \mathbf{v}(\mathbf{x}') \, \delta(\mathbf{x} - \mathbf{x}') d^3 \mathbf{x}'$$
$$= -\int \mathbf{v}(\mathbf{x}') \nabla_{\mathbf{x}'}^2 \left(\frac{1}{4\pi|\mathbf{x} - \mathbf{x}'|}\right) d^3\mathbf{x}'. \qquad (B.0.9)$$

Now integrate the second form in (B.0.9) by parts twice; since the integral is over all space, the boundary terms are to be evaluated at infinity and so vanish. The effect therefore is to move $\nabla_{\mathbf{x}'}^2$ simply from the $1/r$ term to act on $\mathbf{v}$ with no change in sign. Hence, an equivalent form for (B.0.9) is

$$\mathbf{v}(\mathbf{x}) = -\int \frac{[\nabla_{\mathbf{x}'}^2 \mathbf{v}(\mathbf{x}')]}{4\pi|\mathbf{x} - \mathbf{x}'|} d^3 \mathbf{x}'. \qquad (B.0.10)$$

For a vector field, we can define an orthogonal decomposition according to

$$\mathbf{v}(\mathbf{x}) = \mathbf{v}(\mathbf{x})^{\|} + \mathbf{v}(\mathbf{x})^{\perp},$$
$$\mathbf{v}(\mathbf{x})^{\|} \cdot \mathbf{v}(\mathbf{x})^{\perp} = 0,$$
$$\nabla \cdot \mathbf{v}(\mathbf{x})^{\perp} = \nabla \wedge \mathbf{v}(\mathbf{x})^{\|} = 0. \qquad (B.0.11)$$

With $\mathbf{v}(\mathbf{x})$ specified to be a transverse vector field, there is clearly no difference between

$$\int \nabla_{\mathbf{x}'} \left[\frac{\nabla_{\mathbf{x}'} \cdot \mathbf{v}(\mathbf{x}')}{4\pi|\mathbf{x} - \mathbf{x}'|}\right] d^3\mathbf{x}' \quad \text{and} \quad \int \frac{1}{4\pi|\mathbf{x} - \mathbf{x}'|} \nabla_{\mathbf{x}'} [\nabla_{\mathbf{x}'} \cdot \mathbf{v}(\mathbf{x}')] \, d^3\mathbf{x}', \qquad (B.0.12)$$

since both are zero anyway. Without this restriction, the former integral is 0 since it is a boundary term at infinity, but the second integral will not vanish in general if $\mathbf{v}(\mathbf{x})$ has a longitudinal component. With this restriction, we can add the second integral to (B.0.9) and invoke the vector identity

$$\nabla \wedge \nabla \wedge \mathbf{v}(\mathbf{x}) = \nabla(\nabla \cdot \mathbf{v}(\mathbf{x})) - \nabla^2 \mathbf{v}(\mathbf{x}) \qquad (B.0.13)$$

to write

$$\mathbf{v}(\mathbf{x}) = \int \frac{[\nabla_{\mathbf{x}'} \wedge \nabla_{\mathbf{x}'} \wedge \mathbf{v}(\mathbf{x}')]}{4\pi|\mathbf{x} - \mathbf{x}'|} d^3\mathbf{x}', \qquad (B.0.14)$$

which is Belinfante's formula (2.49) in Chapter 2 relating the magnetic field to the Coulomb gauge vector potential. This relationship was also used in the same chapter to display the dependence of the magnetisation $\mathbf{M}$ on the Green's function for the divergence operator, $\mathbf{g}(\mathbf{x}, \mathbf{x}')$.

From the point of view of the theory of Green's functions, we see that

$$\mathcal{Z}(\mathbf{x} - \mathbf{x}') = \frac{1}{4\pi|\mathbf{x} - \mathbf{x}'|}, \tag{B.0.15}$$

is a Green's function for the Laplacian,

$$\nabla_{\mathbf{x}}^2 \mathcal{Z}(\mathbf{x} - \mathbf{x}') = -\delta^3(\mathbf{x} - \mathbf{x}'). \tag{B.0.16}$$

This idea can be extended to define longitudinal and transverse delta dyadics which are used to project out the longitudinal and transverse components of vector fields; the dyadics satisfy

$$\delta_{nm}\delta^3(\mathbf{x}) = \delta(\mathbf{x})_{nm}^{\parallel} + \delta(\mathbf{x})_{nm}^{\perp}. \tag{B.0.17}$$

The longitudinal delta dyadic is defined by

$$\delta(\mathbf{x})_{nm}^{\parallel} = -\frac{1}{(2\pi)^3}\nabla_n\nabla_m \int \left(\frac{1}{k^2}\right) e^{i\mathbf{k}\cdot\mathbf{x}}\,d^3\mathbf{k},$$

$$= \frac{1}{(2\pi)^3}\int \hat{k}_n\hat{k}_m e^{i\mathbf{k}\cdot\mathbf{x}}\,d^3\mathbf{k}, \tag{B.0.18}$$

and is such that

$$v(\mathbf{x})_n^{\parallel} = \int \delta(\mathbf{x})_{nm}^{\parallel} v(\mathbf{x})_m\,d^3\mathbf{x}. \tag{B.0.19}$$

The dyadic field of a dipole is

$$W(\mathbf{x})_{nm} = -\nabla_n\nabla_m\left(\frac{1}{4\pi x}\right) = \frac{(\mathbf{1} - 3\hat{\mathbf{x}}\hat{\mathbf{x}})_{nm}}{4\pi x^3}, \tag{B.0.20}$$

in terms of which the longitudinal delta dyadic may be expressed as

$$\delta(\mathbf{x})_{nm}^{\parallel} = \tfrac{1}{3}\delta_{nm}\delta^3(\mathbf{x}) + W(\mathbf{x})_{nm}. \tag{B.0.21}$$

The Fourier integral representation of the transverse delta dyadic is evidently

$$\delta(\mathbf{x})_{nm}^{\perp} = \frac{1}{(2\pi)^3}\int (\mathbf{1} - \hat{\mathbf{k}}\hat{\mathbf{k}})_{nm} e^{i\mathbf{k}\cdot\mathbf{x}}\,d^3\mathbf{k}. \tag{B.0.22}$$

The integrand will be recognised as a factor in many calculations involving the electromagnetic field and its Green's functions. It follows at once from (B.0.21) that the transverse delta dyadic in real space is

$$\delta(\mathbf{x})_{nm}^{\perp} = \tfrac{2}{3}\delta_{nm}\delta^3(\mathbf{x}) - W(\mathbf{x})_{nm}. \tag{B.0.23}$$

If we have to integrate over a vector field $\mathbf{v}$ that has spherical symmetry, we may use

$$\int \mathbf{W}(\mathbf{x})\,d\Omega(\mathbf{x}) = 0, \tag{B.0.24}$$

and so (B.0.19) reduces to

$$\mathbf{v}(\mathbf{x})^{\parallel} \to \tfrac{1}{3}\mathbf{v}(\mathbf{x}), \tag{B.0.25}$$

with a similar expression for the transverse component (replace $\tfrac{1}{3}$ by $\tfrac{2}{3}$).

Appendix C Hamilton's Equations for the Electromagnetic Field of a Charge

The classical Hamiltonian in the Coulomb gauge for a single extended charge is

$$H[0] = \frac{1}{2m}\left(\mathbf{p} - e\mathbf{A}_\xi(\mathbf{q})\right)^2 + \tfrac{1}{2}\varepsilon_0 \int \left(\varepsilon_0^{-2}\boldsymbol{\pi}(\mathbf{x})\cdot\boldsymbol{\pi}(\mathbf{x}) + c^2\mathbf{B}(\mathbf{x})\cdot\mathbf{B}(\mathbf{x})\right)d^3\mathbf{x}, \quad (C.0.1)$$

where we use canonical variables $\mathbf{p}$, $\mathbf{q}$ for the particle, and mass, m and charge, e, parameters. The longitudinal component of $\boldsymbol{\pi}$ contributes the Coulomb self-energy, which is independent of the particle variables and so may be dropped. Thus, the field variables are purely transverse and describe radiation. The notation is

$$\mathbf{A}(\mathbf{q}(t))_\xi = \int \xi(\mathbf{x} - \mathbf{q}(t))\mathbf{A}(\mathbf{x})\,d^3\mathbf{x}, \quad (C.0.2)$$

and the Fourier transform of ξ is

$$\chi(\mathbf{k}) = \int e^{i\mathbf{k}\cdot\mathbf{x}}\,\xi(\mathbf{x})\,d^3\mathbf{x}, \quad (C.0.3)$$

as in Chapter 2.

The electromagnetic field in the interacting system can be obtained by integrating the Hamiltonian equations of motion for the variables $a_{\mathbf{k},i}(t)$ and $a^*_{\mathbf{k},i}(t)$ we introduced in Chapter 4 for the mode expansions of the vector potential and its conjugate momentum in a box of volume Ω; the P.B. relation

$$\dot{a}_{\mathbf{k},i}(t) = \{a_{\mathbf{k},i}, H[0]\} \quad (C.0.4)$$

that results from (C.0.1) is easily evaluated to give

$$\dot{a}_{\mathbf{k},i}(t) = -i\omega a_{\mathbf{k},i}(t) + \frac{ie\chi_a(k)}{\sqrt{2\Omega kc\varepsilon_0}}\dot{\mathbf{q}}\cdot\hat{\boldsymbol{\varepsilon}}(\mathbf{k})_i\, e^{-i\mathbf{k}\cdot\mathbf{q}}, \quad (C.0.5)$$

where we used (3.316) for $\dot{\mathbf{q}}$. This equation can be integrated formally to yield

$$a_{\mathbf{k},i}(t) - a_{\mathbf{k},i}(t_0) = \frac{ie\chi_a(k)}{\sqrt{2\Omega kc\varepsilon_0}}\int_{t_0}^{+\infty}\Theta(t - t')\dot{\mathbf{q}}(t')\cdot\hat{\boldsymbol{\varepsilon}}(\mathbf{k})_i e^{-i\left(\mathbf{k}\cdot\mathbf{q}(t')+\omega(t-t')\right)}\,dt', \quad (C.0.6)$$

where $a_{\mathbf{k},i}$ is an integration constant, and we have incorporated retarded boundary conditions with the Heaviside step function:

$$\Theta(\tau) = \begin{cases} 1 & \text{if } \tau \geq 0 \\ 0 & \text{if } \tau < 0. \end{cases} \quad (C.0.7)$$

For a system of N charges, (C.0.6) is modified in an obvious way with a sum over the contributions of the individual charges. $a_{\mathbf{k},i}^*(t)$ is the complex conjugate of (C.0.6). In the following we put $t_0 = -\infty$.

The mode variables are now substituted in equations (4.32) and (4.33) for the canonically conjugate field variables $\mathbf{A}$ and $\boldsymbol{\pi}$; as always we have

$$\mathbf{E} = -\frac{\boldsymbol{\pi}}{\varepsilon_0}, \quad \mathbf{B} = \boldsymbol{\nabla} \wedge \mathbf{A}. \tag{C.0.8}$$

For now we drop the integration constants which define solutions of the homogeneous wave equations for $\mathbf{A}$ and $\boldsymbol{\pi}$, and concentrate on the contributions from the self-interaction. The sum over modes is converted into an integral in the usual way and after summing over polarisations we obtain the canonical field variables in the form

$$\mathbf{A}(\mathbf{x}) = \frac{1}{(2\pi)^3} \int \left(\mathbf{a}(-\mathbf{k},t) - \mathbf{a}(\mathbf{k},t)^* \right) e^{-i\mathbf{k}\cdot\mathbf{x}} \, d^3k, \tag{C.0.9}$$

$$\boldsymbol{\pi}(\mathbf{x}) = \frac{1}{(2\pi)^3} \int \left(\mathbf{b}(-\mathbf{k},t) + \mathbf{b}(\mathbf{k},t)^* \right) e^{-i\mathbf{k}\cdot\mathbf{x}} \, d^3k, \tag{C.0.10}$$

where the Fourier coefficients are

$$\mathbf{a}(\mathbf{k},t) = \left(\frac{ie}{2c\varepsilon_0} \right) \left(\frac{\chi_a(k)}{k} \right) (\mathbf{1} - \hat{\mathbf{k}}\hat{\mathbf{k}}) : \int_{-\infty}^{+\infty} \mathbf{N}(\mathbf{k},t,t') \, dt', \tag{C.0.11}$$

$$\mathbf{b}(\mathbf{k},t) = \left(\frac{e}{2} \right) \chi_a(k)(\mathbf{1} - \hat{\mathbf{k}}\hat{\mathbf{k}}) : \int_{-\infty}^{+\infty} \mathbf{N}(\mathbf{k},t,t') \, dt', \tag{C.0.12}$$

and

$$\mathbf{N}(\mathbf{k},t,t') = \Theta(t-t') \, \dot{\mathbf{q}}(t')e^{-i\mathbf{k}\cdot\mathbf{q}(t')}e^{-ikc(t-t')}. \tag{C.0.13}$$

In terms of the usual formula for Fourier transformation (B.0.2), the field variables in the wave-vector space are

$$\mathbf{A}(\mathbf{k}) = \mathbf{a}(-\mathbf{k},t) - \mathbf{a}(\mathbf{k},t)^*, \tag{C.0.14}$$

$$\boldsymbol{\pi}(\mathbf{k}) = \mathbf{b}(-\mathbf{k},t) + \mathbf{b}(\mathbf{k},t)^*. \tag{C.0.15}$$

The $\pm \mathbf{k}$ pattern in these relations means that $e^{i\mathbf{k}\cdot\mathbf{q}(t')}$ is a common factor, and so we have

$$\mathbf{A}(\mathbf{k},t) = \left(\frac{e\chi_a(k)}{kc\varepsilon_0} \right) (\mathbf{1} - \hat{\mathbf{k}}\hat{\mathbf{k}}) : \int_{-\infty}^{+\infty} \dot{\mathbf{q}}(t')e^{i\mathbf{k}\cdot\mathbf{q}(t')} \sin[kc(t-t')]\Theta(t-t') \, dt', \tag{C.0.16}$$

$$\boldsymbol{\pi}(\mathbf{k},t) = e\chi_a(k)(\mathbf{1} - \hat{\mathbf{k}}\hat{\mathbf{k}}) : \int_{-\infty}^{+\infty} \dot{\mathbf{q}}(t')e^{i\mathbf{k}\cdot\mathbf{q}(t')} \cos[kc(t-t')]\Theta(t-t') \, dt'. \tag{C.0.17}$$

Evidently

$$\boldsymbol{\pi}(\mathbf{k},t) = \varepsilon_0 \frac{\partial \mathbf{A}(\mathbf{k},t)}{\partial t}, \tag{C.0.18}$$

so that the transverse electric field is

$$\mathbf{E}(\mathbf{k},t) = -\frac{\partial \mathbf{A}(\mathbf{k},t)}{\partial t}, \tag{C.0.19}$$

as expected.

Our method of solution for the field variables using Hamilton's equations is equivalent to the familiar procedure of solving the wave equation

$$\triangle^2 \mathbf{A}(\mathbf{x},t) = -\mu_0 \mathbf{j}^\perp(\mathbf{x},t), \tag{C.0.20}$$

using, for example, the Green's function technique. To see this, we make an easy calculation using (C.0.18) to show that

$$\frac{\partial \boldsymbol{\pi}(\mathbf{k},t)}{\partial t} = -\varepsilon_0 c^2 k^2 \mathbf{A}(\mathbf{k},t) + \mathbf{j}^\perp(\mathbf{k},t), \tag{C.0.21}$$

where

$$\mathbf{j}^\perp(\mathbf{k},t) = e\, \chi_a(k)\, (1-\hat{\mathbf{k}}\hat{\mathbf{k}}) : \dot{\mathbf{q}}(t) e^{i\mathbf{k}\cdot\mathbf{q}(t)}. \tag{C.0.22}$$

Using (C.0.21) with (C.0.18) then gives

$$k^2\, \mathbf{A}(\mathbf{k},t) + \frac{1}{c^2}\frac{\partial^2 \mathbf{A}(\mathbf{k},t)}{\partial t^2} = \mu_0\, \mathbf{j}^\perp(\mathbf{k},t), \tag{C.0.23}$$

which is the Fourier transform of (C.0.20).

In order to complete the Hamiltonian equations of motion for the particle, we must now evaluate (C.0.2) which may be written as a Fourier transform

$$\mathbf{A}_\xi(\mathbf{q}(t)) = \frac{1}{(2\pi)^3}\int \chi_a(k)\mathbf{A}(\mathbf{k})e^{-i\,\mathbf{k}\cdot\mathbf{q}(t)}\,d^3\mathbf{k}, \tag{C.0.24}$$

using (C.0.3). We combine this with (C.0.16) and define a dyadic

$$\mathbf{G}_\xi^\perp(\mathbf{X};t-t') = \left(\frac{c}{(2\pi)^3}\right)\int\left(\frac{\chi_a^2(k)}{k}\right)(1-\hat{\mathbf{k}}\hat{\mathbf{k}})e^{i\mathbf{k}\cdot\mathbf{X}_{t't}}\sin[kc(t-t')]\,d^3\mathbf{k}, \tag{C.0.25}$$

with

$$\mathbf{X}_{t't} = \mathbf{q}(t') - \mathbf{q}(t). \tag{C.0.26}$$

Then the vector potential may be written as

$$\mathbf{A}_\xi(\mathbf{q}(t)) = \left(\frac{e}{\varepsilon_0 c^2}\right)\int_{-\infty}^{+\infty}\mathbf{G}_\xi^\perp(\mathbf{X}_{t't};t-t') : \dot{\mathbf{q}}(t')\Theta(t-t')\,dt'. \tag{C.0.27}$$

The angular integral in (C.0.25) is [3]

$$\frac{1}{4\pi}\int(1-\hat{\mathbf{k}}\hat{\mathbf{k}})_{\sigma\tau}e^{i\mathbf{k}\cdot\mathbf{X}}\,d\Omega = U_{\sigma\tau}\frac{\sin(kX)}{kX} + V_{\sigma\tau}\left(\frac{\cos(kX)}{(kX)^2}-\frac{\sin(kX)}{(kX)^3}\right)$$

$$= \mathcal{F}_{\sigma\tau}(kX), \tag{C.0.28}$$

where the dyadics are

$$\mathbf{U} = 1-\hat{\mathbf{X}}\hat{\mathbf{X}}, \quad \mathbf{V} = 1-3\hat{\mathbf{X}}\hat{\mathbf{X}}, \tag{C.0.29}$$

and we are left with an infinite integral which may be put in the form,

$$\mathbf{G}^\perp(\mathbf{X},\tau) = \frac{c}{2\pi^2 X^2}\int_0^\infty \sin(Tx)\mathbf{H}(x)\,dx, \tag{C.0.30}$$

with $T = c\tau/X$ and

$$\mathbf{H}(x) = \mathbf{U}\sin(x) + \mathbf{V}\left(\frac{\cos(x)}{x} - \frac{\sin(x)}{x^2}\right). \tag{C.0.31}$$

The remaining integrals are standard [4]:

$$\int_0^\infty \sin(Tx)\sin(x)\,dx = \frac{\pi}{2}\left(\delta(1-T) - \delta(1+T)\right), \tag{C.0.32}$$

$$\int_0^\infty \sin(Tx)\left(\frac{\cos(x)}{x} - \frac{\sin(x)}{x^2}\right)dx = \frac{\pi}{2}T\eta(T), \tag{C.0.33}$$

where

$$\eta(T) = \begin{cases} 1 & \text{if } T < 1 \\ \frac{1}{2} & \text{if } T = 1 \\ 0 & \text{if } T > 1 \end{cases}. \tag{C.0.34}$$

Collecting these results together, we get

$$\mathbf{G}^\perp(\mathbf{X},\tau) = \frac{\mathbf{U}}{4\pi X}\left(\delta(X/c - \tau) - \delta(X/c + \tau)\right)$$
$$- \frac{\mathbf{V}}{4\pi X^3}c^2\tau\eta(c\tau/X). \tag{C.0.35}$$

Note that

$$\frac{\delta(X/c \mp \tau)}{X} \equiv \frac{\delta(t' + \frac{|\mathbf{x}-\mathbf{x}'|}{c} - t)}{|\mathbf{x} - \mathbf{x}'|} = G^\pm(\mathbf{x} - \mathbf{x}', t - t'), \tag{C.0.36}$$

where $G^\pm$ are the usual retarded, $(+)$, and advanced, $(-)$, Green's functions for the wave equation. Finally, then the Coulomb gauge space-time Green's function is

$$\mathbf{G}^\perp(\mathbf{x} - \mathbf{x}'; t - t') = \frac{\mathbf{U}}{4\pi}\left(G^+(\mathbf{x} - \mathbf{x}'; t - t') - G^-(\mathbf{x} - \mathbf{x}'; t - t')\right)$$
$$- \mathbf{W}c^2(t - t')\eta\left(\frac{c(t-t')}{|\mathbf{x} - \mathbf{x}'|}\right), \tag{C.0.37}$$

where $\mathbf{W}$ was defined in Appendix B.

Appendix D Normal Ordering of Field Operator Products

According to the results in Chapter 7, the photon operators obey the commutation relation (7.119),

$$[c_{\mathbf{k},\lambda}, c^+_{\mathbf{k}',\lambda'}] = \delta_{\lambda,\lambda'} \delta_{\mathbf{k},\mathbf{k}'}. \tag{D.0.1}$$

In the continuum limit, the Kronecker delta becomes a Dirac delta function,

$$\delta_{\mathbf{k},\mathbf{k}'} \rightarrow \delta^3(\mathbf{k} - \mathbf{k}'). \tag{D.0.2}$$

A field operator variable $\mathsf{X}(\mathbf{x})^\alpha_i$ is a sum of an annihilation operator and a creation operator,

$$\mathsf{X}(\mathbf{x})^\alpha_i = \mathsf{x}(\mathbf{x})^\alpha_i + \mathsf{x}(\mathbf{x})^{*\alpha}_i, \tag{D.0.3}$$

where the subscript i refers to the field variables $\mathbf{A}, \mathbf{B}, \mathbf{E}^\perp$ and the superscript α labels the x, y, z components. The field annihilation operator can be expanded as a linear combination of photon annihilation operators (cf. §7.3):

$$\mathsf{x}(\mathbf{x})^\alpha_i = \sum_{\mathbf{k},\lambda} y_i(\mathbf{x})^\alpha_{\mathbf{k},\lambda} c_{\mathbf{k},\lambda}. \tag{D.0.4}$$

The creation operator $\mathsf{x}(\mathbf{x})^{*\alpha}_i$ is the Hermitian conjugate of $\mathsf{x}(\mathbf{x})^\alpha_i$. If we define

$$L(\mathbf{x})^\tau_{\mathbf{k},\lambda} = \sqrt{\frac{\hbar}{2\varepsilon_0 \Omega}} \hat{e}^\tau_\lambda e^{i\mathbf{k}\cdot\mathbf{x}}, \tag{D.0.5}$$

the coefficients may be factored as

$$y_i(\mathbf{x})^\alpha_{\mathbf{k},\lambda} = F(\mathbf{k})^{\alpha\tau}_i L(\mathbf{x})^\tau_{\mathbf{k},\lambda}, \tag{D.0.6}$$

where ($\omega = kc$)

$$F(\mathbf{k})^{\alpha\tau}_\mathbf{A} = \frac{\delta_{\alpha\tau}}{\sqrt{\omega}}, \quad F(\mathbf{k})^{\alpha\tau}_\mathbf{E} = i\sqrt{\omega}\delta_{\alpha\tau}, \quad F(\mathbf{k})^{\alpha\tau}_\mathbf{B} = i\varepsilon_{\alpha\sigma\tau}\frac{k^\sigma}{\sqrt{\omega}}. \tag{D.0.7}$$

A bilinear product of field operators is usefully related to the corresponding *normal ordered* product. The normal ordering of an operator product requires the transposition of creation and annihilation operators so as to bring all creation operators to the *left* of all annihilation operators. A normal ordered product has zero expectation value in the vacuum state $|0\rangle$, and its expectation value for a coherent state $|\alpha\rangle$ is simply the corresponding classical field expression with α and α^* replacing the photon annihilation and creation operators. Using (D.0.4) we may write

$$\mathsf{X}(\mathbf{x})^\alpha_i \mathsf{X}(\mathbf{x}')^\beta_j = \left(\mathsf{X}(\mathbf{x})^\alpha_i \mathsf{X}(\mathbf{x}')^\beta_j\right)_N + [\mathsf{x}(\mathbf{x})^\alpha_i, \mathsf{x}(\mathbf{x}')^{*\beta}_j]. \tag{D.0.8}$$

The commutator in (D.0.8) can be evaluated using the expansion (D.0.4)

$$[x(\mathbf{x})_i^\alpha, x(\mathbf{x}')_j^{\beta*}] = \sum_{\mathbf{k},\lambda} \sum_{\mathbf{k}',\lambda'} y_i(\mathbf{x})_{\mathbf{k},\lambda}^\alpha y_j(\mathbf{x}')_{\mathbf{k}',\lambda'}^{\beta*} [c_{\mathbf{k},\lambda}, c_{\mathbf{k}',\lambda'}^+]$$

$$= \sum_{\mathbf{k},\lambda} y_i(\mathbf{x})_{\mathbf{k},\lambda}^\alpha y_j(\mathbf{x})_{\mathbf{k},\lambda}^{*\beta}, \tag{D.0.9}$$

after using the commutation relation (D.0.1).

Using (D.0.5) and (D.0.6) the sum over the polarisation states may be made explicit,

$$\frac{\hbar}{2\varepsilon_0\Omega} \sum_{\mathbf{k},\lambda} F(\mathbf{k})_i^{\alpha\tau} F(\mathbf{k})_j^{\beta\sigma*} \hat{e}(\mathbf{k})_\lambda^\sigma \hat{e}(\mathbf{k})_\lambda^\tau e^{i\mathbf{k}\cdot\mathbf{x}-\mathbf{x}'}, \tag{D.0.10}$$

and evaluated using (2.26) from Chapter 4. In the continuum limit, the sum in (D.0.4) is replaced by an integral according to the rule

$$\frac{1}{\sqrt{\Omega}} \sum_{\mathbf{k}} \rightarrow \frac{1}{\sqrt{(2\pi)^3}} \int d^3\mathbf{k}, \tag{D.0.11}$$

and so (D.0.10) reduces to

$$\frac{\hbar}{2\varepsilon_0} \frac{1}{(2\pi)^3} \int_0^\infty k^2\,dk \int F(\mathbf{k})_i^{\alpha\tau} F(\mathbf{k})_j^{\beta\sigma*} (1 - \hat{\mathbf{k}}\hat{\mathbf{k}})_{\sigma\tau} e^{i\mathbf{k}\cdot\mathbf{u}}\,d\Omega(\mathbf{k}), \tag{D.0.12}$$

where $\mathbf{u} = \mathbf{x} - \mathbf{x}'$. Only the coefficient $F(\mathbf{k})_\mathbf{B}^{\alpha\tau}$ depends on the vector $\mathbf{k}$; this can be removed from the integrand by replacing ik^σ with $\nabla_\mathbf{u}^\sigma$ in front of the integral. Then all cases share the same angular integration (Eq. (C.0.28) with $\mathbf{X} = \mathbf{u}$).

The commutator (D.0.9) is therefore

$$\frac{\hbar}{\pi}\left(\frac{1}{4\pi\varepsilon_0}\right) \int_0^\infty k^2 F(k)_i^{\alpha\tau} F(k)_j^{*\beta\sigma} \mathcal{F}(ku)_{\sigma\tau}\,dk, \tag{D.0.13}$$

where now $F(k)_\mathbf{B}^{\alpha\tau} = \varepsilon_{\alpha\nu\tau}\nabla_\mathbf{u}^\nu/\sqrt{\omega}$. The final integration depends on the particular pairing (i, j); for example, with $i = \mathbf{A}, j = \mathbf{E}^\perp$ the result[1] is the transverse delta function, $\delta_{\alpha\beta}^\perp(\mathbf{x} - \mathbf{x}')$. In terms of a dimensionless variable $t = ku$, the integral is

$$\frac{\hbar}{\pi}\left(\frac{1}{4\pi\varepsilon_0}\right) \frac{1}{|\mathbf{x}-\mathbf{x}'|^3} \int_0^\infty t^2 F(t/u)_i^{\alpha\tau} F(t/u)_j^{*\beta\sigma} \mathcal{F}(t)_{\sigma\tau}\,dt, \tag{D.0.14}$$

which shows that for $i = j = \mathbf{A}$, the result varies as $|\mathbf{x} - \mathbf{x}'|^{-2}$, specifically

$$[x(\mathbf{x})_\mathbf{A}^\alpha, x(\mathbf{x}')_\mathbf{A}^{*\beta}] = \left(\frac{\hbar}{2\pi^2\varepsilon_0 c}\right) \frac{u^\alpha u^\beta}{u^2}, \tag{D.0.15}$$

whereas for $i = j = \mathbf{B}$, it varies as $|\mathbf{x} - \mathbf{x}'|^{-4}$.

---

[1] In this case the calculation in all essentials is that required for the evaluation of the equal-time commutation relation $[A(\mathbf{x})^\alpha, E(\mathbf{x}')^{\perp\beta}]$.

Appendix E Calculus for Path-dependent Functionals

First of all, we consider the integration of a 1-form $\boldsymbol{\Phi}(\mathbf{x}) \cdot d\mathbf{x}$, where $\boldsymbol{\Phi}$ is a vector field and $\mathbf{x}$ is a point in $\Re^3$, $(\mathbf{x} = x, y, z)$. The result is a two-point scalar function,

$$\phi(\mathbf{x}, \mathbf{x}_0) = \int_{\mathbf{x}_0}^{\mathbf{x}} \boldsymbol{\Phi}(\mathbf{z}) \cdot d\mathbf{z}, \tag{E.0.1}$$

where $\mathbf{x}_0$ is a fixed point in space. A familiar example is the relationship between work done, W in moving from $\mathbf{x}_0$ to $\mathbf{x}$ under an applied force $\mathbf{F}(\mathbf{x})$,

$$W(\mathbf{x}, \mathbf{x}_0) = \int_{\mathbf{x}_0}^{\mathbf{x}} \mathbf{F}(\mathbf{z}) \cdot d\mathbf{z}. \tag{E.0.2}$$

A *conservative* force satisfies the integrability condition,

$$\frac{\partial F(\mathbf{x})_i}{\partial x_j} - \frac{\partial F(\mathbf{x})_j}{\partial x_i} = 0, \quad i, j = x, y, z, \tag{E.0.3}$$

and then W depends on only the end points, $\mathbf{x}_0, \mathbf{x}$. If (E.0.3) is not satisfied, the relationship (E.0.2) still stands, but W is a multivalued quantity that depends on the path taken from $\mathbf{x}_0$ to $\mathbf{x}$.

Suppose now we evaluate (E.0.1) for one circuit of a closed path

$$\Delta\phi = \oint_C \boldsymbol{\Phi}(\mathbf{z}) \cdot d\mathbf{z}. \tag{E.0.4}$$

According to Stokes's theorem, this is also

$$\Delta\phi = \int_\Gamma (\boldsymbol{\nabla} \wedge \boldsymbol{\Phi}(\mathbf{z})) \cdot d\mathbf{S}, \tag{E.0.5}$$

where Γ is the surface bounded by the curve C; ϕ is single-valued if the Curl of $\boldsymbol{\Phi}$ vanishes on this surface. These considerations tell us that we can always represent the vector field $\boldsymbol{\Phi}$ as the gradient of a scalar field ϕ,

$$\boldsymbol{\Phi}(\mathbf{x}) = \boldsymbol{\nabla}_{\mathbf{x}} \phi(\mathbf{x}), \tag{E.0.6}$$

but we must be aware that $\phi(\mathbf{x})$ will be multivalued if the 1-form $\boldsymbol{\Phi}(\mathbf{x}) \cdot d\mathbf{x}$ is not an exact differential [5], [6]. Line integrals of this type over the Lorentz force, $\mathbf{F}$, the vector potential, $\mathbf{a}$, and the electric polarisation field, $\mathbf{P}$, are important examples in electrodynamics (cf. Chapter 3, §3.3 and Chapter 9).

The evaluation of a derivative of some function involves a comparison of the original quantity with its value after an infinitesimal increment has been made. The derivatives required for (E.0.6) are with respect to $\mathbf{x}$, that is, the *end point* of the path; there are two cases to consider for implementing the increment $d\mathbf{x}$. In the first case, no specific path is specified and we imagine that the end point is augmented to $\mathbf{x} + d\mathbf{x}$, while the remainder of the path is kept fixed; thus, (E.0.6) results from writing [5]–[8]

$$\nabla\phi(\mathbf{x}) = \lim_{\Delta\mathbf{x}\to 0} \frac{\phi(\mathbf{x}+\Delta\mathbf{x},\mathbf{x}_0) - \phi(\mathbf{x},\mathbf{x}_0)}{\Delta\mathbf{x}}. \tag{E.0.7}$$

However, once a definite path is chosen, (E.0.1) can be understood as a single-valued quantity which will satisfy the integrability condition, and a result different from (E.0.6) is obtained. As an example, consider the evaluation of (9.130) required for carrying through the PZW transformation.

The typical line integral in the action integral F is over some path $\mathcal{P}$ ending at a charge located at $\mathbf{x}$. We parameterise the path so that $\sigma = 0$ corresponds to $\mathbf{x}$, and $\sigma = 1$ gives the origin of the path. Then

$$\phi = \int_{\mathcal{P}}^{\mathbf{x}} \mathbf{A}(\mathbf{z}) \cdot d\mathbf{z} = -\int_0^1 A(\mathbf{z})^r \frac{\partial z^r}{\partial\sigma} d\sigma, \tag{E.0.8}$$

using the summation over repeated indices convention. The path is smooth if the tangent vector $(\partial z^r/\partial\sigma)$ is continuous and nowhere zero. The derivative of ϕ is

$$\begin{aligned}
\frac{\partial\phi}{\partial x_s} &= -\frac{\partial}{\partial x_s} \int_0^1 A(\mathbf{z})_r \frac{\partial z^r}{\partial\sigma} d\sigma \\
&= -\int_0^1 \frac{\partial A(\mathbf{z})_r}{\partial z_t} \frac{\partial z^t}{\partial x_s} \frac{\partial z^r}{\partial\sigma} d\sigma - \int_0^1 A(\mathbf{z})_r \frac{\partial}{\partial\sigma}\left(\frac{\partial z^r}{\partial x_s}\right) d\sigma \\
&= \int_0^1 \frac{\partial A(\mathbf{z})_r}{\partial z_m} \frac{\partial z^r}{\partial x_s} \frac{\partial z^m}{\partial\sigma} d\sigma - \int_0^1 \frac{\partial A(\mathbf{z})_m}{\partial z_r} \frac{\partial z^r}{\partial x_s} \frac{\partial z^m}{\partial\sigma} d\sigma - A(\mathbf{z})_r \frac{\partial z^r}{\partial x_s}\bigg|_{\sigma=0}^{\sigma=1},
\end{aligned} \tag{E.0.9}$$

after an integration by parts, and relabelling of the repeated indices in the second integral.

As in Chapter 2, we introduce a field tensor by the definition

$$f(\mathbf{x})_{rs} = \frac{\partial A(\mathbf{x})_s}{\partial x_r} - \frac{\partial A(\mathbf{x})_r}{\partial x_s}, \tag{E.0.10}$$

in terms of which the components of magnetic induction are given by

$$f(\mathbf{x})_{rs} = \varepsilon_{rst} B(\mathbf{x})_t. \tag{E.0.11}$$

Then

$$\frac{\partial\phi}{\partial x_s} = \int_0^1 f(\mathbf{z})_{mr} \frac{\partial z^r}{\partial x_s} \frac{\partial z^m}{\partial\sigma} d\sigma + A(\mathbf{x})_s. \tag{E.0.12}$$

The PZW transformation (Chapter 9) is based on the choice of a straight-line path from some fixed origin $\mathbf{O}$ to a field point $\mathbf{x}$; it may be parameterised as

$$z(\mathbf{x},\sigma)^t = x^t - \sigma q^t, \quad 0 \le \sigma \le 1, \tag{E.0.13}$$

where $\mathbf{q} = \mathbf{x} - \mathbf{O}$. The specification of a straight-line path implies that when we make an infinitesimal displacement of the end point $\mathbf{x}$, the *whole path* moves with the end point while the path origin remains fixed [8], [9]. Now from (E.0.13)

$$\frac{\partial z^r}{\partial x_s} = (1 - \sigma)\delta_{rs}, \tag{E.0.14}$$

so finally

$$\frac{\partial \phi}{\partial x_s} = A(\mathbf{x})_s + \int_0^1 f(\mathbf{z})_{rs}\frac{\partial z^r}{\partial \sigma}(1 - \sigma)\,d\sigma. \tag{E.0.15}$$

We noted in §2.3 that the integral (with a change of sign) can be formally interpreted as a new vector potential; in vector notation it is

$$\int_0^1 (1 - \sigma)\mathbf{q} \wedge \mathbf{B}(\mathbf{z})\,d\sigma = -\mathbf{a}(\mathbf{x}). \tag{E.0.16}$$

It must therefore satisfy Curl $\mathbf{a} = \mathbf{B}$ (where $\mathbf{B}$ is also equal to Curl $\mathbf{A}$). This is verified as follows. We define

$$I(\mathbf{x})_s = \int_0^1 f(\mathbf{z})_{rs}\frac{\partial z^r}{\partial \sigma}(1 - \sigma)\,d\sigma \tag{E.0.17}$$

and calculate its gradient $\partial I(\mathbf{x})_s/\partial x_t = I(\mathbf{x})_{s,t}$. To obtain Curl $\mathbf{a}$, we need the antisymmetric component

$$I(\mathbf{x})_{t,s} - I(\mathbf{x})_{s,t}. \tag{E.0.18}$$

As with the evaluation of the derivative of ϕ, judicious partial integrations are required to extract the required result. The gradient is

$$I(\mathbf{x})_{s,t} = \int_0^1 \frac{\partial f(\mathbf{z})_{rs}}{\partial z_m}\frac{\partial z^m}{\partial x_t}\frac{\partial z^r}{\partial \sigma}(1 - \sigma)\,d\sigma + \int_0^1 f(\mathbf{z})_{rs}(1 - \sigma)\frac{\partial}{\partial \sigma}\frac{\partial z^r}{\partial x_t}\,d\sigma. \tag{E.0.19}$$

Integrating the second term by parts yields

$$\int_0^1 [f(\mathbf{z})_{rs}(1 - \sigma)]\frac{\partial}{\partial \sigma}\frac{\partial z^r}{\partial x_t}\,d\sigma = f(\mathbf{z})_{rs}(1 - \sigma)\frac{\partial z^r}{\partial x_t}\Big|_{\sigma=0}^{\sigma=1}$$
$$- \int_0^1 \frac{\partial}{\partial \sigma}[f(\mathbf{z})_{rs}(1 - \sigma)]\frac{\partial z^r}{\partial x_t}\,d\sigma$$
$$= f(\mathbf{x})_{ts} + \int_0^1 f(\mathbf{z})_{ts}(1 - \sigma)\,d\sigma$$
$$- \int_0^1 \frac{\partial f(\mathbf{z})_{rs}}{\partial z_m}\frac{\partial z^r}{\partial x_t}\frac{\partial z^m}{\partial \sigma}(1 - \sigma)\,d\sigma. \tag{E.0.20}$$

When this is put back into (E.0.19), there are two similar integrals to be subtracted. We use (E.0.14) and the identity

$$f(\mathbf{z})_{rs,t} - f(\mathbf{z})_{ts,r} = f(\mathbf{z})_{rt,s} \tag{E.0.21}$$

to reduce $I_{s,t}$ to

$$I(\mathbf{x})_{s,t} = -f(\mathbf{x})_{ts} + \int_0^1 f(\mathbf{z})_{ts}(1 - \sigma)\,d\sigma + \int_0^1 (1 - \sigma)^2\frac{\partial f(\mathbf{z})_{rt}}{\partial z_s}\frac{\partial z^r}{\partial \sigma}\,d\sigma. \tag{E.0.22}$$

The field tensor is antisymmetric, $f_{rs} = -f_{sr}$, and so from $I_{s,t}$ we obtain the antisymmetric component

$$I(\mathbf{x})_{t,s} - I(\mathbf{x})_{s,t} = 2f(\mathbf{x})_{ts} - 2\int_0^1 f(\mathbf{z})_{ts}(1-\sigma)\,d\sigma$$
$$+ \int_0^1 (1-\sigma)^2 \frac{\partial f(\mathbf{z})_{rs}}{\partial z_t} \frac{\partial z^r}{\partial \sigma}\,d\sigma - \int_0^1 (1-\sigma)^2 \frac{\partial f(\mathbf{z})_{rt}}{\partial z_s} \frac{\partial z^r}{\partial \sigma}\,d\sigma. \quad \text{(E.0.23)}$$

Once again we use the identity (E.0.21) to combine the third and fourth terms into

$$\int_0^1 \left(\frac{\partial f(\mathbf{z})_{ts}}{\partial \sigma} \right) (1-\sigma)^2\,d\sigma, \quad \text{(E.0.24)}$$

and then a final integration by parts yields

$$\int_0^1 \left(\frac{\partial f(\mathbf{z})_{ts}}{\partial \sigma} \right) (1-\sigma)^2\,d\sigma = (1-\sigma)^2 f(\mathbf{z})_{ts}\Big|_{\sigma=0}^{\sigma=1} + 2\int_0^1 f(\mathbf{z})_{ts}(1-\sigma)\,d\sigma$$
$$= -f(\mathbf{x})_{ts} + 2\int_0^1 f(\mathbf{z})_{ts}(1-\sigma)\,d\sigma. \quad \text{(E.0.25)}$$

Combining (E.0.23) and (E.0.25) leaves the desired result:

$$I(\mathbf{x})_{t,s} - I(\mathbf{x})_{s,t} = f(\mathbf{x})_{ts}. \quad \text{(E.0.26)}$$

Because of the $-$ sign in (E.0.16), the Curl of the vector $\nabla\phi$ vanishes.

Now consider the Green's function $\mathbf{g}(\mathbf{x},\mathbf{x}')$, (2.69), for two different paths (C_1, C_2), starting from a fixed origin, $\mathbf{O}$, and ending at the space-point $\mathbf{x}'$,

$$\mathbf{g}(\mathbf{x},\mathbf{x}';\mathbf{O},C) = \int_{\mathbf{O}}^{\mathbf{x}'} \delta^3(\mathbf{z}-\mathbf{x})\,d\mathbf{z}. \quad \text{(E.0.27)}$$

As in Chapter 2, we take C_2 as the straight line between the end points, while C_1 is some other path that need not be further specified. We have

$$\mathbf{g}(\mathbf{x},\mathbf{x}';\mathbf{O},C_1) = \mathbf{g}(\mathbf{x},\mathbf{x}';\mathbf{O},C_2) - \oint \delta^3(\mathbf{y}-\mathbf{x})\,d\mathbf{y}, \quad \text{(E.0.28)}$$

where the integration is over the closed curve $C_2 - C_1$. The *multipole series* for $\mathbf{g}(\mathbf{x},\mathbf{x}')$ is an expansion of $\mathbf{g}(\mathbf{x},\mathbf{x}';\mathbf{O},C_2)$ in powers of $\mathbf{x}' - \mathbf{O}$.

The straight-line path is parameterised as usual as

$$\mathbf{z}(\sigma) = \mathbf{O} + \sigma\mathbf{q}, \quad \mathbf{q} = (\mathbf{x}' - \mathbf{O}) \quad \text{(E.0.29)}$$

with $\mathbf{z}(0) = \mathbf{O}, \mathbf{z}(1) = \mathbf{x}'$, so that

$$\mathbf{g}(\mathbf{x},\mathbf{x}';\mathbf{O},C_2) = \mathbf{q}\int_0^1 \delta^3(\mathbf{O} + \sigma\mathbf{q} - \mathbf{x})\,d\sigma. \quad \text{(E.0.30)}$$

Now

$$\delta^3(\mathbf{O} + \sigma\mathbf{q} - \mathbf{x}) = e^{-\sigma\mathbf{q}\cdot\nabla_\mathbf{x}}\delta^3(\mathbf{O} - \mathbf{x})$$
$$= \sum_{k=0}^{\infty} \frac{(-1)^k \sigma^k (\mathbf{q}\cdot\nabla_\mathbf{x})^k}{k!} \delta^3(\mathbf{O} - \mathbf{x}). \quad \text{(E.0.31)}$$

Putting this expansion into (E.0.30) and integrating term by term yields

$$\mathbf{g}(\mathbf{x},\mathbf{x}';\mathbf{O},C_2) = \mathbf{q} \sum_{k=0}^{\infty} \frac{(-1)^k}{(k+!)!}(\mathbf{q}\cdot\nabla_{\mathbf{x}})^k \delta^3(\mathbf{O}-\mathbf{x}). \tag{E.0.32}$$

From (2.142) the electric polarisation field for this path is

$$\mathbf{P}(\mathbf{x}) = \sum_{n=1}^{N}\sum_{k=0}^{\infty} e_n \mathbf{q}_n \frac{(-1)^k}{(k+!)!}(\mathbf{q}_n\cdot\nabla_{\mathbf{x}})^k \delta^3(\mathbf{O}-\mathbf{x}), \tag{E.0.33}$$

where $\mathbf{q}_n = \mathbf{x}_n - \mathbf{O}$. It should be noted that this is a distribution which is to be used in an integration over $\mathbf{x}$ and we must make use of the property

$$\int \mathrm{d}x f(x) \frac{\mathrm{d}^k}{\mathrm{d}x^k}\delta(x-x') = (-1)^k \left.\frac{\mathrm{d}f^k}{\mathrm{d}x^k}\right|_{x=x'} \tag{E.0.34}$$

which leads to an overall factor of $(-1)^{2k} = 1$ in the result of the integration.

As an example, the interaction energy $\mathcal{E}_{\mathbf{E}}$ for a charge distribution $\rho(\mathbf{x})$ in an electric field can be put in multipolar form,

$$\mathcal{E}_{\mathbf{E}} = -\int \mathbf{P}(\mathbf{x})\cdot\mathbf{E}(\mathbf{x}) = -\sum_{n=1}^{N}\sum_{k=0}^{\infty} e_n \mathbf{q}_n \frac{1}{(k+!)!}(\mathbf{q}_n\cdot\nabla_{\mathbf{x}})^k \mathbf{E}(\mathbf{x})|_{\mathbf{x}=\mathbf{O}}$$

$$\approx -\mathbf{d}\cdot\mathbf{E}(\mathbf{O}) - \tfrac{1}{2}\mathbf{Q}:\nabla\mathbf{E}(\mathbf{O}) - \dots \tag{E.0.35}$$

as in (2.152). The multipole of order k for the charge distribution $\rho(\mathbf{x})$ is the tensor

$$\mathbf{D}^{(k)} = \sum_{n}^{N} e_n \frac{(\mathbf{x}_n - \mathbf{O})^k}{(k+1)!}. \tag{E.0.36}$$

The result (E.0.35) is based on specifying paths of type C_2, that is, straight lines running from $\mathbf{O}$ to the particle positions $\{\mathbf{x}_n\}$. If any of the paths are deformed in any way, that is, become of type C_1, there will be corresponding loop integrals derived from (E.0.28). They vanish only if the field $\mathbf{E}$ is purely longitudinal; if it has a non-zero transverse component (e.g. electromagnetic radiation), such integrals are non-zero. However, since there are infinitely many paths that are *not* of type C_2, we do not discuss the 'evaluation' of loop integrals. Instead one must devise calculations that result in quantities such as cross sections (cf. Chapter 9) that are *path independent*, so that the loop integrals, whatever their value, become irrelevant.

Appendix F Matrix Elements in the Coulomb Gauge

F.1 The Charge–Photon Interaction

Let the particle have a charge e, mass m, and position, $\mathbf{x}$, and momentum, $\mathbf{p}$, operators. In terms of these variables, we can define a charge density,

$$\rho(\mathbf{r}) = e\delta^3(\mathbf{x} - \mathbf{r}), \tag{F.1.1}$$

and a corresponding current density,

$$\mathbf{j}(\mathbf{r}) = \frac{1}{m}\left(\mathbf{p} - \tfrac{1}{2}e\mathbf{A}(\mathbf{x})\right)\rho(\mathbf{r}), \tag{F.1.2}$$

where as usual $\mathbf{A}$ is specifically the Coulomb gauge vector potential. The interaction between the charge and the field is then

$$V = -\int \mathbf{j}(\mathbf{r}) \cdot \mathbf{A}(\mathbf{r})\, d^3\mathbf{r}. \tag{F.1.3}$$

The term linear in e in (F.1.3) can be decomposed into one-photon absorption and emission terms,

$$V^{(1)} = K_a^{(1)} + K_e^{(1)}, \tag{F.1.4}$$

where

$$K_a^{(1)} = -\sum_{\mathbf{q},\sigma} \mathbf{f}_e(\mathbf{q}) \cdot \hat{\boldsymbol{\varepsilon}}(\mathbf{q})_\sigma\, c_{\mathbf{q},\sigma}, \tag{F.1.5}$$

$$K_e^{(1)} = -\sum_{\mathbf{q},\sigma} \mathbf{f}_e(-\mathbf{q}) \cdot \hat{\boldsymbol{\varepsilon}}(\mathbf{q})_\sigma\, c_{\mathbf{q},\sigma}^+, \tag{F.1.6}$$

and

$$\mathbf{f}_e(\mathbf{q}) = \frac{e}{m}\sqrt{\frac{\hbar^2}{2\varepsilon_0 \Omega \mathcal{E}_{\mathbf{q}}}}\,\mathbf{p}e^{i\mathbf{q}\cdot\mathbf{x}}, \quad \mathcal{E}_{\mathbf{q}} = \hbar qc. \tag{F.1.7}$$

Using the reference states for free charges (10.6), and the photon Fock space, the non-zero matrix elements are then

$$\left(K_a^{(1)}\right)_{(n_{\mathbf{q}}-1)n_{\mathbf{q}}}^{\mathbf{p}'\mathbf{p}} = \langle\, \mathbf{p}', \lambda\,[n_{\mathbf{q}} - 1]|K_a^{(1)}|\mathbf{p}, \lambda\,[n_{\mathbf{q}}]\rangle$$

$$= -\frac{e}{m}\sqrt{\frac{\hbar^2 n_{\mathbf{q}}}{2\varepsilon_0 \Omega \mathcal{E}_{\mathbf{q}}}}\ \mathbf{p}\cdot\hat{\boldsymbol{\varepsilon}}(\mathbf{q})_\lambda\ \delta^3(\mathbf{p} + \hbar\mathbf{q} - \mathbf{p}'), \tag{F.1.8}$$

$$\left(\mathsf{K}_e^{(1)}\right)^{\mathbf{p'p}}_{n_{\mathbf{q}}(n_{\mathbf{q}}-1)} = \langle \mathbf{p'}, \lambda\,[n_{\mathbf{q}}]|\mathsf{K}_e^{(1)}|\mathbf{p}, \lambda\,[n_{\mathbf{q}}-1]\rangle$$

$$= -\frac{e}{m}\sqrt{\frac{\hbar^2 n_{\mathbf{q}}}{2\varepsilon_0\Omega\mathcal{E}_{\mathbf{q}}}}\,\mathbf{p}\cdot\hat{\mathbf{e}}(\mathbf{q})_\lambda\,\delta^3(\mathbf{p}-\hbar\mathbf{q}-\mathbf{p'}). \qquad (\text{F.1.9})$$

Thus, the operator $\mathsf{K}_a^{(1)}$ causes the particle to make a transition from the momentum state $\mathbf{p}$ to the momentum state $\mathbf{p'}$ by the absorption of a photon of wave vector $\mathbf{q}$. Similarly, $\mathsf{K}_e^{(1)}$ is responsible for the same transition with the emission of a photon of wave vector $\mathbf{q}$.

We make a similar calculation with the two-photon part of V. We have that

$$\mathbf{A}(\mathbf{x})\cdot\mathbf{A}(\mathbf{x}) = \mathsf{K}_{aa}^{(2)} + \mathsf{K}_{ea}^{(2)} + \mathsf{K}_{ae}^{(2)} + \mathsf{K}_{ee}^{(2)} + \frac{\hbar}{2\varepsilon_0\Omega c}\sum_{\mathbf{q}}\frac{1}{q}, \qquad (\text{F.1.10})$$

where we have arranged the creation operators to stand to the *left* of the absorption operators

$$\mathsf{K}_{aa}^{(2)} = \frac{\hbar}{2\varepsilon_0\Omega c}\sum_{\mathbf{q},\sigma}\sum_{\mathbf{q'},\sigma'}\frac{\hat{\mathbf{e}}(\mathbf{q})_\sigma\cdot\hat{\mathbf{e}}(\mathbf{q'})_{\sigma'}}{\sqrt{qq'}}\,c_{\mathbf{q},\sigma}c_{\mathbf{q'},\sigma'}\,e^{i(\mathbf{q}+\mathbf{q'})\cdot\mathbf{x}}\,,$$

$$\mathsf{K}_{ea}^{(2)} = \frac{\hbar}{2\varepsilon_0\Omega c}\sum_{\mathbf{q},\sigma}\sum_{\mathbf{q'},\sigma'}\frac{\hat{\mathbf{e}}(\mathbf{q})_\sigma\cdot\hat{\mathbf{e}}(\mathbf{q'})_{\sigma'}}{\sqrt{qq'}}\,c_{\mathbf{q},\sigma}^+c_{\mathbf{q'},\sigma'}\,e^{-i(\mathbf{q}-\mathbf{q'})\cdot\mathbf{x}}\,,$$

$$\mathsf{K}_{ae}^{(2)} = \frac{\hbar}{2\varepsilon_0\Omega c}\sum_{\mathbf{q},\sigma}\sum_{\mathbf{q'},\sigma'}\frac{\hat{\mathbf{e}}(\mathbf{q})_\sigma\cdot\hat{\mathbf{e}}(\mathbf{q'})_{\sigma'}}{\sqrt{qq'}}\,c_{\mathbf{q'},\sigma'}^+c_{\mathbf{q},\sigma}\,e^{i(\mathbf{q}-\mathbf{q'})\cdot\mathbf{x}}\,,$$

$$\mathsf{K}_{ee}^{(2)} = \frac{\hbar}{2\varepsilon_0\Omega c}\sum_{\mathbf{q},\sigma}\sum_{\mathbf{q'},\sigma'}\frac{\hat{\mathbf{e}}(\mathbf{q})_\sigma\cdot\hat{\mathbf{e}}(\mathbf{q'})_{\sigma'}}{\sqrt{qq'}}\,c_{\mathbf{q},\sigma}^+c_{\mathbf{q'},\sigma'}^+\,e^{-i(\mathbf{q}+\mathbf{q'})\cdot\mathbf{x}}\,, \qquad (\text{F.1.11})$$

and the last term arises from this 'normal-ordering' of the operator (see Appendix D); it is independent of the particle variables but formally gives an infinite contribution to purely diagonal matrix elements (see Chapter 10). The matrix elements of these operators are then

$$\left(\mathsf{K}_{aa}^{(2)}\right)^{\mathbf{p'p}}_{(n_{\mathbf{q}}-1)(n_{\mathbf{q'}}-1),n_{\mathbf{q}}n_{\mathbf{q'}}'} = \langle \mathbf{p'}, \lambda\,[n_{\mathbf{q}}-1], \lambda'\,[n_{\mathbf{q'}}-1]|\mathsf{K}_{aa}^{(2)}|\mathbf{p}, \lambda\,[n_{\mathbf{q}}], \lambda'\,[n_{\mathbf{q'}}]\rangle$$

$$= y(\mathbf{q},\mathbf{q'},\lambda,\lambda')R_{q,q'}\,\delta^3(\mathbf{p}+\hbar\mathbf{q}+\hbar\mathbf{q'}-\mathbf{p'}), \qquad (\text{F.1.12})$$

$$\left(\mathsf{K}_{ea}^{(2)}\right)^{\mathbf{p'p}}_{n_{\mathbf{q}}(n_{\mathbf{q'}}-1),(n_{\mathbf{q}}-1)n_{\mathbf{q'}}} = \langle \mathbf{p'}, \lambda\,[n_{\mathbf{q}}], \lambda'\,[n_{\mathbf{q'}}-1]|\mathsf{K}_{ea}^{(2)}|\mathbf{p}, \lambda\,[n_{\mathbf{q}}-1], \lambda'\,[n_{\mathbf{q'}}]\rangle$$

$$= y(\mathbf{q},\mathbf{q'},\lambda,\lambda')R_{q,q'}\,\delta^3(\mathbf{p}+\hbar\mathbf{q}-\hbar\mathbf{q'}-\mathbf{p'}), \qquad (\text{F.1.13})$$

$$\left(\mathsf{K}_{ae}^{(2)}\right)^{\mathbf{p'p}}_{(n_{\mathbf{q}}-1)n_{\mathbf{q'}},n_{\mathbf{q}}(n_{\mathbf{q'}}-1)} = \langle \mathbf{p'}, \lambda\,[n_{\mathbf{q}}-1], \lambda'\,[n_{\mathbf{q'}}]|\mathsf{K}_{ae}^{(2)}|\,\mathbf{p}, \lambda\,[n_{\mathbf{q}}], \lambda'\,[n_{\mathbf{q'}}-1]\rangle$$

$$= y(\mathbf{q},\mathbf{q'},\lambda,\lambda')R_{q,q'}\,\delta^3(\mathbf{p}-\hbar\mathbf{q}+\hbar\mathbf{q'}-\mathbf{p'}), \qquad (\text{F.1.14})$$

$$\left(\mathsf{K}_{ee}^{(2)}\right)^{\mathbf{p'p}}_{n_{\mathbf{q}}n_{\mathbf{q'}},(n_{\mathbf{q}}-1)(n_{\mathbf{q'}}-1)} = \langle \mathbf{p'}, \lambda\,[n_{\mathbf{q}}], \lambda'\,[n_{\mathbf{q'}}]|\mathsf{K}_{ee}^{(2)}|\,\mathbf{p}, \lambda\,[n_{\mathbf{q}}-1], \lambda'\,[n_{\mathbf{q'}}-1]\rangle$$

$$= y(\mathbf{q},\mathbf{q'},\lambda,\lambda')R_{q,q'}\,\delta^3(\mathbf{p}-\hbar\mathbf{q}-\hbar\mathbf{q'}-\mathbf{p'}), \qquad (\text{F.1.15})$$

where

$$R_{q,q'} = \sqrt{\frac{n_q n_{q'}}{qq'}}, \quad y(\mathbf{q}, \mathbf{q}', \lambda, \lambda') = \frac{\hbar}{2\varepsilon_0 c} \hat{\boldsymbol{\varepsilon}}(\mathbf{q})_\lambda \cdot \hat{\boldsymbol{\varepsilon}}(\mathbf{q}')_{\lambda'}. \tag{F.1.16}$$

The first, (F.1.12), and fourth, (F.1.15), terms describe two-photon absorption and emission respectively, while the second, (F.1.13), and third, (F.1.14), terms describe scattering of a photon with the charge making a transition $\mathbf{p} \to \mathbf{p}'$.

F.2 The Molecule–Photon Interaction

The matrix elements for the interaction between a molecule and photons can be obtained simply from those given in the preceding section by introducing a sum over the charges in the molecule. The only point that requires comment is the conservation of momentum which involves the centre-of-mass motion of the molecule, but not the internal states. We write the one-photon part of the operator in the form

$$V^{(1)} = M_a^{(1)} + M_e^{(1)}, \tag{F.2.1}$$

where

$$M_a^{(1)} = -\sum_{\mathbf{q},\sigma} \mathbf{f}_{\text{mol}}(\mathbf{q}) \cdot \hat{\boldsymbol{\varepsilon}}(\mathbf{q})_\sigma \, c_{\mathbf{q},\sigma}, \tag{F.2.2}$$

$$M_e^{(1)} = -\sum_{\mathbf{q},\sigma} \mathbf{f}_{\text{mol}}(-\mathbf{q}) \cdot \hat{\boldsymbol{\varepsilon}}(\mathbf{q})_\sigma \, c_{\mathbf{q},\sigma}^+ \tag{F.2.3}$$

and

$$\mathbf{f}_{\text{mol}}(\mathbf{q}) = \sqrt{\frac{\hbar^2}{2\varepsilon_0 \Omega \mathcal{E}_{\mathbf{q}}}} \sum_n \frac{e_n}{m_n} \mathbf{p}_n e^{i\mathbf{q} \cdot \mathbf{x}_n}. \tag{F.2.4}$$

We use (8.49) and (8.52) to introduce the centre-of-mass and internal variables explicitly and then the integral over the centre-of-mass coordinate, $\mathbf{R}$, yields the delta function that expresses the conservation of momentum in the interaction.

$$\langle \mathbf{P}', \phi_{s'} | \sum_n \frac{e_n}{m_n} \mathbf{p}_n e^{i\,\mathbf{q} \cdot \mathbf{x}_n} | \mathbf{P}, \phi_s \rangle = \delta^3(\mathbf{P} + \hbar\mathbf{q} - \mathbf{P}')$$

$$\times \langle \phi_{s'} | \frac{(\mathbf{P} + \hbar\mathbf{q})}{M} \sum_n e_n e^{i\mathbf{q} \cdot \mathbf{r}_n} + \sum_n \frac{e_n}{m_n} \boldsymbol{\sigma}_n e^{i\mathbf{q} \cdot \mathbf{r}_n} | \phi_s \rangle. \tag{F.2.5}$$

The operator part of the first term in the internal state matrix element in (F.2.5) will be recognised as a Fourier component of the molecular charge density in the centre-of-mass frame, while the second term is a current operator involving only the internal variables. In an obvious notation we set

$$\mathbf{J}(\mathbf{P}, \mathbf{k})_{s's} = \frac{\mathbf{P}}{M} \rho_{\text{int}}(\mathbf{k})_{s's} + \mathbf{j}_{\text{int}}(\mathbf{k})_{s's}, \tag{F.2.6}$$

so that

$$\langle \mathbf{P}', \phi_{s'} | \sum_n \frac{e_n}{m_n} \mathbf{p}_n e^{i\,\mathbf{q}\cdot\mathbf{x}_n} | \mathbf{P}, \phi_s \rangle = \mathbf{J}(\mathbf{P} + \hbar\mathbf{q}, \mathbf{q})_{s's} \delta^3(\mathbf{P} + \hbar\mathbf{q} - \mathbf{P}'). \qquad (F.2.7)$$

With this notation, we have

$$\left(\mathsf{M}_a^{(1)} \right)_{(n_\mathbf{q}-1)n_\mathbf{q}}^{\mathbf{P}'\mathbf{P},s's} = \langle \mathbf{P}', s', \lambda[n_\mathbf{q} - 1] | \mathsf{M}_a^{(1)} | \mathbf{P}, s, \lambda[n_\mathbf{q}] \rangle$$

$$= -\sqrt{\frac{\hbar^2 n_\mathbf{q}}{2\varepsilon_0 \Omega \mathcal{E}_\mathbf{q}}} \; \mathbf{J}(\mathbf{P} + \hbar\mathbf{q}, \mathbf{q})_{s's} \cdot \hat{\boldsymbol{\varepsilon}}(\mathbf{q})_\lambda \delta^3(\mathbf{P} + \hbar\mathbf{q} - \mathbf{P}'), \qquad (F.2.8)$$

$$\left(\mathsf{M}_e^{(1)} \right)_{n_\mathbf{q}(n_\mathbf{q}-1)}^{\mathbf{P}'\mathbf{P},s's} = \langle \mathbf{P}', s', \lambda[n_\mathbf{q}] | \mathsf{M}_e^{(1)} | \mathbf{P}, s, \lambda[n_\mathbf{q} - 1] \rangle$$

$$= -\sqrt{\frac{\hbar^2 n_\mathbf{q}}{2\varepsilon_0 \Omega \mathcal{E}_\mathbf{q}}} \mathbf{J}(\mathbf{P} - \hbar\mathbf{q}, -\mathbf{q})_{s's} \cdot \hat{\boldsymbol{\varepsilon}}(\mathbf{q})_\lambda \delta^3(\mathbf{P} - \hbar\mathbf{q} - \mathbf{P}'). \qquad (F.2.9)$$

The reduction of the matrix elements of the two-photon part of V involves similar calculations. It is convenient to write the operator as a product of factors referring to the molecule and to the radiation,

$$\mathsf{V}^{(2)} = \int \gamma(\mathbf{r})\, \mathbf{A}(\mathbf{r}) \cdot \mathbf{A}(\mathbf{r})\, d^3\mathbf{r}, \qquad (F.2.10)$$

where

$$\gamma(\mathbf{r}) = \sum_n \left(\frac{e_n^2}{2m_n} \right) \delta^3(\mathbf{r} - \mathbf{x}_n), \qquad (F.2.11)$$

and $\mathbf{A}(\mathbf{r}) \cdot \mathbf{A}(\mathbf{r})$ is given by (F.1.10) and (F.1.11) with $\mathbf{x}$ replaced by $\mathbf{r}$, and the infinite constant is dropped. Thus, $\mathsf{V}^{(2)}$ can be written as a sum of four terms,

$$\mathsf{V}^{(2)} = \mathsf{M}_{aa}^{(2)} + \mathsf{M}_{ea}^{(2)} + \mathsf{M}_{ae}^{(2)} + \mathsf{M}_{ee}^{(2)}, \qquad (F.2.12)$$

where, for example,

$$\mathsf{M}_{aa}^{(2)} = \int \gamma(\mathbf{r})\, \mathsf{K}_{aa}^{(2)}\, d^3\mathbf{r}, \qquad (F.2.13)$$

and similarly for the other three terms.

The matrix element of γ for the transition $|N\rangle \rightarrow |N'\rangle$ is evaluated using centre-of-mass and internal coordinates and states, and yields

$$\langle \mathbf{P}', s' | \gamma(\mathbf{r}) | s, \mathbf{P} \rangle = e^{i(\mathbf{P}-\mathbf{P}')\cdot\mathbf{r}} \Upsilon(\mathbf{P}' - \mathbf{P})_{s's}, \qquad (F.2.14)$$

where

$$\Upsilon(\mathbf{q})_{s's} = \langle \phi_{s'} | \sum_n \left(\frac{e_n^2}{2m_n} \right) e^{i\,\mathbf{q}\cdot\mathbf{r}_n} | \phi_s \rangle, \qquad (F.2.15)$$

is a matrix element involving only the molecular internal states.

The two-photon operator matrix elements are the same as those following (F.1.11), and so we obtain

$$\left(\mathrm{M}_{aa}^{(2)}\right)_{(n_{\mathbf{q}}-1)(n_{\mathbf{q}'}-1),n_{\mathbf{q}}n_{\mathbf{q}'}}^{\mathbf{P}'\mathbf{P},s's} = \langle \mathbf{P}',s',\lambda[n_{\mathbf{q}}-1],\lambda'[n_{\mathbf{q}'}-1]|\mathrm{M}_{aa}^{(2)}|\mathbf{P},s,\lambda[n_{\mathbf{q}}],\lambda'[n_{\mathbf{q}'}]\rangle$$

$$= y(\mathbf{q},\mathbf{q}',\lambda,\lambda')R_{q,q'}\delta^3(\mathbf{q}+\mathbf{q}'-\mathbf{Q})\Upsilon(\mathbf{Q})_{s's}, \qquad \text{(F.2.16)}$$

$$\left(\mathrm{M}_{ea}^{(2)}\right)_{n_{\mathbf{q}}(n_{\mathbf{q}'}-1),(n_{\mathbf{q}}-1)n_{\mathbf{q}'}}^{\mathbf{P}'\mathbf{P},s's} = \langle \mathbf{P}',s',\lambda[n_{\mathbf{q}}],\lambda'[n_{\mathbf{q}'}-1]|\mathrm{M}_{ea}^{(2)}|\mathbf{P},s,\lambda[n_{\mathbf{q}}-1],\lambda'[n_{\mathbf{q}'}]\rangle$$

$$= y(\mathbf{q},\mathbf{q}',\lambda,\lambda')R_{q,q'}\delta^3(\mathbf{Q}+\mathbf{q}-\mathbf{q}')\Upsilon(\mathbf{Q})_{s's}, \qquad \text{(F.2.17)}$$

$$\left(\mathrm{M}_{ae}^{(2)}\right)_{(n_{\mathbf{q}}-1)n_{\mathbf{q}'},n_{\mathbf{q}}(n_{\mathbf{q}'}-1)}^{\mathbf{P}'\mathbf{P},s's} = \langle \mathbf{P}',s',\lambda[n_{\mathbf{q}}-1],\lambda'[n_{\mathbf{q}'}]|\mathrm{M}_{ae}^{(2)}|\mathbf{P},s,\lambda[n_{\mathbf{q}}],\lambda'[n_{\mathbf{q}'}-1]\rangle$$

$$= y(\mathbf{q},\mathbf{q}',\lambda,\lambda')R_{q,q'}\delta^3(\mathbf{Q}-\mathbf{q}+\mathbf{q}')\Upsilon(\mathbf{Q})_{s's}, \qquad \text{(F.2.18)}$$

$$\left(\mathrm{M}_{ee}^{(2)}\right)_{n_{\mathbf{q}}n_{\mathbf{q}'},(n_{\mathbf{q}}-1)(n_{\mathbf{q}'}-1)}^{\mathbf{P}'\mathbf{P},s's} = \langle \mathbf{P}',s',\lambda[n_{\mathbf{q}}],\lambda'[n_{\mathbf{q}'}]|\mathrm{M}_{ee}^{(2)}|\mathbf{P},s,\lambda[n_{\mathbf{q}}-1],\lambda'[n_{\mathbf{q}'}-1]\rangle$$

$$= y(\mathbf{q},\mathbf{q}',\lambda,\lambda')R_{q,q'}\delta^3(\mathbf{q}+\mathbf{q}'+\mathbf{Q})\Upsilon(\mathbf{Q})_{s's}, \qquad \text{(F.2.19)}$$

with $R_{q,q'}$ and $y(\mathbf{q},\mathbf{q}',\lambda,\lambda')$ as before, and we have introduced the momentum transfer vector for the centre-of-mass motion defined as $\hbar\mathbf{Q} = (\mathbf{P}'-\mathbf{P})$.

Appendix G The Heisenberg Group and Coherent States

In Chapter 5, we introduced some important group theoretical ideas in quantum mechanics with an emphasis on the description of the symmetries of a physical system. There is a different perspective that involves group theory *not* related to physical symmetries which nevertheless goes to the heart of quantum theory since it builds on the fundamental Heisenberg commutation relation (5.19).

Informally, a Lie algebra is a vector space equipped with a multiplication operation of its elements $x, y, \ldots$, called the Lie bracket, $(x, y) \mapsto [x, y]$, that satisfies the Jacobi identity and has $[x, x] = 0$ for all x. The quantum mechanical commutator

$$[\mathsf{x}, \mathsf{y}] = \mathsf{x}\mathsf{y} - \mathsf{y}\mathsf{x} \tag{G.0.1}$$

satisfies this requirement, of course. A Lie algebra is an infinitesimal version of a Lie group; any element of a Lie group g_α which lies in the neighbourhood of the identity element can be written in the form

$$g_\alpha = I + \sum_{r=1}^{p} i\varepsilon_r I_r \tag{G.0.2}$$

with infinitesimal real parameters $\{\varepsilon_1, \varepsilon_2 \ldots \varepsilon_p, |\varepsilon_p| << 1\}$ and generators $\{I_r\}$ that are independent of the parameters [10]. The generators define the algebra through their structure constants,

$$[I_r, I_s] = \sum_{t=1}^{p} C_{rs}^t I_t, \qquad r, s = 1, 2, \ldots p, \tag{G.0.3}$$

with

$$
\begin{aligned}
C_{rs}^t &= -\, C_{sr}^t, \quad \text{antisymmetry} \\
0 &= C_{rs}^t C_{vz}^u + C_{zr}^t C_{vs}^u + C_{sz}^t C_{vr}^u \quad \text{Jacobi identity.}
\end{aligned} \tag{G.0.4}
$$

The Heisenberg algebra $\mathfrak{h}_3$ is the vector space $\mathfrak{R}^3$ with generators $\{I_i, i = 1, 2, 3\}$, and one non-zero independent structure constant, $C_{12}^3 = +1$. It can be realised by three 3×3 upper triangular matrices:

$$I_1 = \begin{bmatrix} 0 & 1 & 0 \\ 0 & 0 & 0 \\ 0 & 0 & 0 \end{bmatrix}, \quad I_2 = \begin{bmatrix} 0 & 0 & 0 \\ 0 & 0 & 1 \\ 0 & 0 & 0 \end{bmatrix}, \quad I_3 = \begin{bmatrix} 0 & 0 & 1 \\ 0 & 0 & 0 \\ 0 & 0 & 0 \end{bmatrix}. \tag{G.0.5}$$

A general element l_α with parameters (x,y,z) is $l_\alpha = xI_1 + yI_2 + zI_3$ which may be represented by the matrix

$$l_\alpha = \begin{bmatrix} 0 & x & z \\ 0 & 0 & y \\ 0 & 0 & 0 \end{bmatrix}. \tag{G.0.6}$$

The commutator of two such elements is then

$$[l_\alpha, l_\beta] = \begin{bmatrix} 0 & 0 & xy' - x'y \\ 0 & 0 & 0 \\ 0 & 0 & 0 \end{bmatrix}. \tag{G.0.7}$$

Equation (G.0.7) points to a clear distinction between (I_1, I_2) and I_3, and suggests that we regard $\Re^3$ as $\Re^2 \times \Re$.

An arbitrary element g_α, of the corresponding Lie group, denoted H_3, can be obtained by successive application of infinitesimal transformations, that is, by evaluation of products of (G.0.2),

$$g_\alpha = \exp\left(i \sum_{r=1}^{p} \alpha_r I_r \right). \tag{G.0.8}$$

Since

$$e^{i2\pi N} = 1, \quad N \in \mathcal{Z}, \tag{G.0.9}$$

the $\{\alpha_r\}$ are periodic and can be chosen so that[1]

$$0 \leq \alpha_r \leq 2\pi, \quad r = 1, 2, \ldots p. \tag{G.0.10}$$

The exponentiation of the matrix in equation (G.0.6) is perfectly well defined, but such representations are not unitary and are not of physical interest.

The appearance of i here is crucial and provides the connection with quantum theory. To see this observe that we can make the identifications

$$I_1 \leftrightarrow \mathsf{x}, \quad I_2 \leftrightarrow \mathsf{p}, \quad I_3 \leftrightarrow i\hbar\mathsf{l} \tag{G.0.11}$$

and realise the algebra $\mathfrak{h}_3$ through the relations

$$[\mathsf{x},\mathsf{p}] = i\hbar\mathsf{l}, \quad [\mathsf{x}, i\hbar\mathsf{l}] = 0. \quad [\mathsf{p}, i\hbar\mathsf{l}] = 0. \tag{G.0.12}$$

The discussion can be readily generalised to $(\Re^{2n} \times \Re)$ to describe the algebra $\mathfrak{h}_{2n+1}$. This mathematical setting for the non-commutation of the basic quantum variables was first recognised by Weyl in a classic text [11]. It leads naturally to consideration of the *infinite*-dimensional representations of H_{2n+1} based on (wave) functions $\psi(x) \in L^2(\Re^{3n})$ – the Schrödinger representation, for which

$$\mathsf{x}\psi(x) = x\psi(x), \quad \mathsf{p}\psi(x) = -i\hbar\frac{\mathrm{d}\psi(x)}{\mathrm{d}x}, \quad i\hbar\mathsf{l}\psi(x) = i\hbar\psi(x). \tag{G.0.13}$$

---

[1] One may think of this as the projection $\Re^2 \times \Re \to \Re^2 \times \mathrm{U}(1)$, where $\mathrm{U}(1)$ is the unit circle in the complex plane, and $\mathcal{Z} : \mathrm{U}(1) = \Re/\mathcal{Z}$.

Thus, a general group element is

$$g_\alpha = \exp(i\theta)\exp(\frac{i}{\hbar}a\mathsf{p})\exp(\frac{i}{\hbar}b\mathsf{x}) \tag{G.0.14}$$

which is valid for any finite number of degrees of freedom.

The exponentiated operators associated with x and p are exactly those described in Chapter 5, using vector notation,

$$\mathsf{U_a} = \exp\left(\frac{i}{\hbar}\mathbf{a}\cdot\mathbf{p}\right), \quad \mathsf{V_b} = \exp\left(\frac{i}{\hbar}\mathbf{b}\cdot\mathbf{x}\right). \tag{G.0.15}$$

With the aid of the Baker–Campbell–Haussdorf formula one has

$$\mathsf{U_a}\mathsf{V_b} = \exp\left[\frac{i}{\hbar}(\mathbf{a}\cdot\mathbf{p}+\mathbf{b}\cdot\mathbf{x})+\frac{1}{2}\frac{i}{\hbar}\mathbf{a}\cdot\mathbf{b}\right]$$

$$\mathsf{V_b}\mathsf{U_a} = \exp\left[\frac{i}{\hbar}(\mathbf{a}\cdot\mathbf{p}+\mathbf{b}\cdot\mathbf{x})-\frac{1}{2}\frac{i}{\hbar}\mathbf{a}\cdot\mathbf{b}\right], \tag{G.0.16}$$

which leads directly to the *Weyl commutation relation*,

$$\mathsf{U_a}\mathsf{V_b} = \exp\left(\frac{i}{\hbar}\mathbf{a}\cdot\mathbf{b}\right)\mathsf{V_b}\mathsf{U_a}. \tag{G.0.17}$$

The annihilation and creation operators introduced in Chapter 7 are defined by a linear transformation of the canonical variables x and p,

$$\sqrt{2\hbar}\begin{bmatrix} \mathsf{c} \\ \mathsf{c}^+ \end{bmatrix} = \begin{bmatrix} 1 & i \\ -i & 1 \end{bmatrix}\begin{bmatrix} \mathsf{x} \\ \mathsf{p} \end{bmatrix}, \tag{G.0.18}$$

and so also can represent $\mathfrak{h}_3$, explicitly

$$[\mathsf{c},\mathsf{c}^+] = \mathsf{I}, \quad [\mathsf{c},\mathsf{I}] = 0, \quad [\mathsf{c}^+,\mathsf{I}] = 0. \tag{G.0.19}$$

In a celebrated paper, Schrödinger [12] described a solution of the wave equation for a harmonic oscillator that had the remarkable property of including the phase space variables describing a solution of the classical equations of motion. The lowest eigenstate of the quantum mechanical harmonic oscillator Hamiltonian is a Gaussian function, $\phi_0(x) \sim \exp(-ax^2/2)$; Schrödinger's solution, which we now call a *coherent state*, is

$$\psi(x) \sim \exp\left(-\frac{(x-x_0)^2}{2} + ip_0 x\right), \tag{G.0.20}$$

where x_0, p_0 are the phase space coordinates of the trajectory of the centroid of the wave packet.

In Chapter 7 coherent states were described in relation to electromagnetic radiation in terms of the eigenstates of the annihilation operator c, (7.47),

$$\mathsf{c}|\alpha\rangle = \alpha|\alpha\rangle, \tag{G.0.21}$$

and it was shown that (G.0.20) can be understood as the action of a Weyl operator on the oscillator ground state wave function. In the notation used there, the ground state of the oscillator is $\langle X|0,0\rangle$ and (G.0.20) is $\langle X|x_0 p_0\rangle$. The identification is completed by putting

$$\alpha = \alpha_R + i\alpha_I = \frac{x_0 + ip_0}{\sqrt{2}}. \tag{G.0.22}$$

The construction of coherent states based on group theoretical ideas is a generalisation of Schrödinger's original idea along two different themes. Firstly, the 'root state' is not restricted to the harmonic oscillator ground state. Secondly, it is not limited to the algebra $\mathfrak{h}_{2n+1}$, so 'generalised coherent states' can be associated with other Lie algebras and their corresponding Lie Groups G. The states of a quantum system are the elements of an Hilbert space $\mathcal{H}$ which carries a unitary irreducible representation Λ of G. The generalised coherent state is formed by the action of an element, g of G acting on some fixed state $|\psi_0\rangle \in \mathcal{H}$,

$$|\psi_g\rangle = \Lambda(g)|\psi_0\rangle, \quad g \in G. \tag{G.0.23}$$

Let T be the maximum subgroup of G that leaves $|\psi_0\rangle$ unchanged up to a phase, with

$$|\psi_t\rangle \equiv \Lambda(t)|\psi_0\rangle = e^{i\theta}|\psi_0\rangle, \quad t \in T \tag{G.0.24}$$

so that $|\psi_t\rangle$ and $|\psi_0\rangle$ describe the *same* state. If T is non-Abelian, the phase θ is a non-trivial gauge parameter.

Every element g of G can be decomposed as a product,

$$g = kt, \tag{G.0.25}$$

where k is an element of the left coset G/T. In general, there is a unitary irreducible representation of k as an exponential operator analogous to the displacement operator $\mathsf{D}(\alpha)$ described in Chapter 7, so one has

$$|\psi_g\rangle = \mathsf{D}_k(\eta)e^{i\theta}|\psi_0\rangle. \tag{G.0.26}$$

The generalised coherent state derived from the group G is then

$$|\Phi(\eta)\rangle = \mathsf{D}_k(\eta)|\psi_0\rangle. \tag{G.0.27}$$

Importantly, the $\{\Phi(\eta)\}$, like Schrödinger's original coherent states, form an overcomplete set, and provide a resolution of the identity [13]

$$\int |\Phi(\eta)\rangle\langle\Phi(\eta)|\,\mathrm{d}\mu(\eta) = \mathsf{I}. \tag{G.0.28}$$

H   Appendix H   An Example of Constrained Dynamics

As an example of constrained Hamiltonian dynamics in action, we consider the simple one-dimensional system described by the following Lagrangian:

$$L = \tfrac{1}{2} m \, \dot{x}^2 + a \, x \, \ddot{x}. \tag{H.0.1}$$

Application of the Euler–Lagrange equation (3.19) for a case involving the particle acceleration yields the equation of motion as

$$(m - 2a) \ddot{x} = 0, \tag{H.0.2}$$

that is, L describes free motion of a particle with a modified mass. L is a highly simplified abstraction of the dynamics of a charged particle interacting with its own electromagnetic field (see §3.8.6). We note in passing that (H.0.1) may be rewritten as

$$L = \tfrac{1}{2}(m - 2a)\dot{x}^2 + a\frac{\mathrm{d}}{\mathrm{d}t}(x\dot{x}), \tag{H.0.3}$$

which leads directly to (H.0.2) in the usual approach since the total time derivative makes no contribution to the variation of the action S and may be dropped.

In order to pass to a Hamiltonian scheme, we introduce two new independent variables

$$q_0 = x, \quad q_1 = \dot{x}, \tag{H.0.4}$$

so that

$$\dot{q}_1 = \ddot{x}, \tag{H.0.5}$$

and a Lagrange multiplier λ which is also regarded as an independent coordinate. The Lagrangian (H.0.1) can then be expressed as [26]

$$L(q_0, q_1, \dot{q}_0, \dot{q}_1, \lambda) = \tfrac{1}{2} m \, q_1^2 + a \, q_0 \, \dot{q}_1 + \lambda (q_1 - \dot{q}_0). \tag{H.0.6}$$

We may understand better the introduction of extra variables by considering the Euler–Lagrange equations for the 'coordinates' $y_i = q_0, q_1, \lambda$ which take the usual form,

$$\frac{\delta L}{\delta y_i} = 0 \; \rightarrow \; \frac{\partial L}{\partial y_i} - \frac{\mathrm{d}}{\mathrm{d}t}\left(\frac{\partial L}{\partial \dot{y}_i}\right), \tag{H.0.7}$$

and lead to

$$y_i = \lambda, \; \rightarrow \quad q_1 - \dot{q}_0 = 0, \tag{H.0.8}$$

$$y_i = q_0, \; \rightarrow \quad a \, \dot{q}_1 + \lambda \; = 0, \tag{H.0.9}$$

$$y_i = q_1, \;\rightarrow\; m\,q_1 + \lambda - a\,\dot{q}_0 = 0. \tag{H.0.10}$$

After elimination of $\dot{\lambda}$ and use of (H.0.8), we find

$$(m - 2a)\,\ddot{x} = 0, \tag{H.0.11}$$

as expected.

It is evident that this is a singular Lagrangian system since the Hessian matrix $\mathbf{W}$ vanishes; in order to pass to the Hamiltonian formalism, we introduce momenta conjugate to the Lagrangian coordinates in the usual way,

$$p_0 = \frac{\partial L}{\partial \dot{q}_0} = -\lambda,$$

$$p_1 = \frac{\partial L}{\partial \dot{q}_1} = a\,q_0,$$

$$\pi = \frac{\partial L}{\partial \dot{\lambda}} = 0, \tag{H.0.12}$$

and define

$$H = \sum_k p_k \dot{q}_k - L(\{q_k, \dot{q}_k\})$$
$$= p_0 \dot{q}_0 + p_1 \dot{q}_1 + \pi \dot{\lambda} - \tfrac{1}{2}m q_1^2 - a\,q_0\,\dot{q}_1 - \lambda\,q_1 + \lambda\,\dot{q}_0. \tag{H.0.13}$$

From the equations (H.0.12), we obtain three primary constraints which may be written as weak equalities:

$$\phi_1 = p_0 + \lambda \approx 0, \tag{H.0.14}$$

$$\phi_2 = p_1 - a\,q_0 \approx 0, \tag{H.0.15}$$

$$\phi_3 = \pi \approx 0. \tag{H.0.16}$$

If we define

$$H_0 = -\tfrac{1}{2}m\,q_1^2 - \lambda\,q_1, \tag{H.0.17}$$

we may write the full Hamiltonian in the standard form,

$$H = H_0 + \sum_m v_m\,\phi_m, \tag{H.0.18}$$

where the coefficients $\{v_m\}$ remain to be determined. The three constraints are second class; however, they are not a minimal set since the matrix of their mutual P.B.s is singular. We have

$$\{\phi_1, \phi_2\} = a, \quad \{\phi_1, \phi_3\} = 1, \quad \{\phi_2, \phi_3\} = 0, \tag{H.0.19}$$

which shows that the 3×3 matrix they form is of rank 2. However, if we define

$$\overline{\phi}_1 = \phi_1, \tag{H.0.20}$$

$$\overline{\phi}_2 = \phi_3, \tag{H.0.21}$$

$$\overline{\phi}_3 = \phi_2 - a\,\phi_3 = p_1 - a\,q_0 - a\,\pi, \tag{H.0.22}$$

we will have two second-class constraints, $(m = 1, 2)$, and one first-class constraint, $(m = 3)$, to be used in (H.0.18), with a matrix of P.B.s of the second-class constraints that is non-singular.

For the consistency of the theory, the constraints must vanish at all times which requires

$$\dot{\overline{\phi}}_m \approx 0, \tag{H.0.23}$$

where we use

$$\dot{\Omega} \approx \{\Omega, H_0\} + \sum_m \overline{v}_m \{\Omega, \overline{\phi}_m\}. \tag{H.0.24}$$

The two second-class constraints yield values $\overline{v}_1 \approx 0$, $\overline{v}_2 = q_1$ for the coefficients $\{\overline{v}\}$, while the first-class constraint yields a new secondary constraint,

$$\dot{\overline{\phi}}_3 \approx \{\overline{\phi}_3, H_0\} \approx 0 \;\rightarrow\; (m - a)\, q_1 + \lambda \approx \overline{\phi}_4 \approx 0. \tag{H.0.25}$$

The new constraint must also vanish for all time, at least weakly, and repeating the above procedure leads to

$$(m - 2a)\, \overline{v}_3 \approx 0, \tag{H.0.26}$$

which requires $\overline{v}_3 \approx 0$; this finishes the checking for consistency.

As a result of these calculations, we have four constraints to consider; they are all second class and so may be turned into strong equalities using the Dirac-bracket construction. The matrix $\mathbf{C}$ in (3.216) is easily constructed with our four constraints taking the place of the $\{\chi_r\}$,

$$\mathbf{C} = \begin{vmatrix} 0 & a & 1 & 0 \\ -a & 0 & 0 & a - m \\ -1 & 0 & 0 & -1 \\ 0 & m - a & 1 & 0 \end{vmatrix} \quad \mathrm{Det}\|\mathbf{C}\| = \frac{1}{(m - 2a)^2}, \tag{H.0.27}$$

and has inverse

$$\mathbf{C}^{-1} = \frac{1}{m - 2a} \begin{vmatrix} 0 & 1 & a - m & 0 \\ -1 & 0 & 0 & 1 \\ m - a & 0 & 0 & -a \\ 0 & -1 & a & 0 \end{vmatrix}. \tag{H.0.28}$$

With the adoption of the Dirac brackets, we may now write the Hamiltonian as

$$H_0 = -\tfrac{1}{2} m\, q_1^2 - \lambda\, q_1, \tag{H.0.29}$$

with the following relations between the canonical variables (all ordinary equalities):

$$p_0 + \lambda \;= 0, \tag{H.0.30}$$

$$\pi \;= 0, \tag{H.0.31}$$

$$p_1 - a\, q_0 - a\, \pi \;= 0, \tag{H.0.32}$$

$$(m - a)\, q_1 + \lambda \;= 0. \tag{H.0.33}$$

The only Dirac bracket of interest is that between q_0 and q_1,

$$[q_0, q_1]^* = \frac{1}{m - 2a} \;\rightarrow\; [q_0, (m - 2a)q_1]^* = 1, \tag{H.0.34}$$

which shows that q_0 and $(m - 2a)q_1$ may be taken as a canonical pair q, p. From the secondary constraint $\overline{\phi}_4 = 0$, we obtain $\lambda = (a - m)q_1$, and so the Hamiltonian (H.0.29) may be expressed entirely in terms of q_1,

$$H = -\tfrac{1}{2}m\, q_1^2 - (a - m)\, q_1^2 = \tfrac{1}{2}(m - 2a)\, q_1^2 \equiv \frac{p^2}{2(m - 2a)}, \tag{H.0.35}$$

with

$$[q, p]^* = 1, \tag{H.0.36}$$

as usual.

References

[1] Feynman, R. P. (1948), Rev. Mod. Phys. **20**, 367.

[2] Klauder, J. R. (2003), arXiv:quant-ph/0303034v1.

[3] Power, E. A. and Zienau, S. (1959), Phil. Trans. Roy. Soc. (London) A**251**, 427.

[4] Polyanin, A. D and Manzhirov, A. V. (2007), *Handbook of Mathematics for Scientists and Engineers*, Chapman Hall/CRC.

[5] Pandres, D. Jr. (1962), J. Math. Phys. **3**, 602.

[6] Khrapko, R. I. (1985), Theor. Math. Phys. (Springer), **65**, 1196.

[7] Mandelstam, S. (1962), Ann. Phys. (N.Y.) **19**, 1.

[8] Schiff, L. I. (1967), Phys. Rev. **160**, 1257.

[9] Woolley, R. G. (1971), Proc. Roy. Soc. (London), A**321**, 557.

[10] Roman, P. (1965), *Advanced Quantum Theory*, Addison-Wesley.

[11] Weyl, H. (1932), *The Theory of Groups and Quantum Mechanics*, translated from the second revised German edition by H. P. Robertson. E. P. Dutton.

[12] Schrödinger, E. (1926), Naturwiss. **14**, 644.

[13] Perelomov, A. M. (1972), Comm. Math. Phys. **26**, 222.

Index

Printed in the United States
by Baker & Taylor Publisher Services